Biological Conversion of Biomass for Fuels and Chemicals
Explorations from Natural Utilization Systems

RSC Energy and Environment Series

Editor-in-Chief:
Professor Laurence Peter, *University of Bath, UK*

Series Editors:
Professor Heinz Frei, *Lawrence Berkeley National Laboratory, USA*
Professor Ferdi Schüth, *Max Planck Institute for Coal Research, Germany*
Professor Tim S. Zhao, *The Hong Kong University of Science and Technology, Hong Kong*

Titles in the Series:
1: Thermochemical Conversion of Biomass to Liquid Fuels and Chemicals
2: Innovations in Fuel Cell Technologies
3: Energy Crops
4: Chemical and Biochemical Catalysis for Next Generation Biofuels
5: Molecular Solar Fuels
6: Catalysts for Alcohol-Fuelled Direct Oxidation Fuel Cells
7: Solid Oxide Fuel Cells: From Materials to System Modeling
8: Solar Energy Conversion: Dynamics of Interfacial Electron and Excitation Transfer
9: Photoelectrochemical Water Splitting: Materials, Processes and Architectures
10: Biological Conversion of Biomass for Fuels and Chemicals: Explorations from Natural Utilization Systems

How to obtain future titles on publication:
A standing order plan is available for this series. A standing order will bring delivery of each new volume immediately on publication.

For further information please contact:
Book Sales Department, Royal Society of Chemistry, Thomas Graham House, Science Park, Milton Road, Cambridge, CB4 0WF, UK
Telephone: +44 (0)1223 420066, Fax: +44 (0)1223 420247
Email: booksales@rsc.org
Visit our website at www.rsc.org/books

Biological Conversion of Biomass for Fuels and Chemicals
Explorations from Natural Utilization Systems

Edited by

Jianzhong Sun
Biofuels Institute, Jiangsu University, Zhenjiang, P.R. China
Email: jzsun1002@hotmail.com

and

Shi-You Ding
Biosciences Center, National Renewable Energy Laboratory, Colorado, U.S.A.
Email: shi.you.ding@nrel.gov

and

Joy Doran-Peterson
Microbiology Department, University of Georgia, U.S.A.
Email: jpeterso@uga.edu

RSCPublishing

RSC Energy and Environment Series No. 10

ISBN: 978-1-84973-424-0
ISSN: 2044-0774

A catalogue record for this book is available from the British Library

Published by The Royal Society of Chemistry,
Thomas Graham House, Science Park, Milton Road,
Cambridge CB4 0WF, UK

Registered Charity Number 207890

For further information see our web site at www.rsc.org

Preface

Decomposition of plant biomass plays a unique and vital role in carbon recycling on earth, but it is also indispensable for many life systems from microorganisms to animals, as a fundamental carbohydrate supply or an important intermediate in food chains. Biomass degradation in nature is generally considered to be a decay processes carried out by microbial communities, mainly consisting of bacteria and fungi. Although most animals lack the capability to use biomass directly, some cellulose-feeding animals subsist on biomass as their main or only food and these animals possess an incredible conversion efficiency. The ability of these animals, from arthropods to mammals, such as termites and cows, to feed on woody or herbaceous plants and detritus, has stimulated extensive investigations into the mechanisms of how these animals efficiently digest the structural and recalcitrant lignocellulose in their foods. Among these animals, termites are the most efficient degrader of cellulose and consume the greatest quantities of plant biomass each year. With these explorations, scientists, long fascinated by the humble termite's ability to turn wood into energy for life, could possibly advance biochemical processes to convert biomass to biofuels and chemicals in industry. Discovering novel lignocellulolytic enzymes and their associated novel genes, and understanding novel lignocellulolytic systems/mechanisms that may apply to a nature-inspired processing via biomimetics in the modern biorefinery of biomass for fuels and chemicals, are especially promising. The extent of biomimetic benefits from cellulose-feeding animals or other unique cellulolytic systems is only just beginning to gain recognition and has scarcely been technically and economically evaluated.

To meet some intractable challenges facing the world in biological conversion of biomass for fuels and chemicals, with a total of 19 unique chapters, this book reviews recent advances in fundamental understanding of

RSC Energy and Environment Series No. 10
Biological Conversion of Biomass for Fuels and Chemicals: Explorations from Natural Utilization Systems
Edited by Jianzhong Sun, Shi-You Ding and Joy Doran-Peterson
© The Royal Society of Chemistry 2014
Published by the Royal Society of Chemistry, www.rsc.org

the chemical and physical properties of biomass and its decomposition systems in nature, current development of state-of-the-art technologies to improve biomass conversion efficiency, and perspectives on the development of integrated biorefinery to produce cost competitive biofuels from lignocellulosic biomass. Chapter 1 serves as a comprehensive introduction that includes an illustration of the main theme of the book as well as the particular contents of each chapter. Following chapters discussing biomass structure, chemistry and modification (Chapters 2 to 6), natural biomass utilization systems and the applications of these systems/mechanisms to overcome current bottlenecks in industrial biocatalyst processing to generate a product are further presented (Chapters 7 to 19). As a result, this book will meet the needs of academic communities and a variety of industrial groups focused on rapid acceleration of progress in lignocellulosic biofuels and bio-chemicals industries. This book is intended to provide researchers and students with a comprehensive introduction/review to this emerging and a multi- disciplinary field, while also functioning as an important reference for those already active in the areas of biofuels and bio-chemical-related industries.

Jiangzhong Sun
Shi-You Ding
Joy Doran-Peterson

Contents

RSC Energy and Environment Series No. 10
Biological Conversion of Biomass for Fuels and Chemicals: Explorations from Natural Utilization Systems
Edited by Jianzhong Sun, Shi-You Ding and Joy Doran-Peterson
© The Royal Society of Chemistry 2014
Published by the Royal Society of Chemistry, www.rsc.org

**Chapter 5 Advances in the Genetic Manipulation of Cellulosic
Bioenergy Crops for Bioethanol Production 53**
Chuansheng Mei and Callista Rakhmatov

CHAPTER 1

Biomass and its Biorefinery: Novel Approaches from Nature-Inspired Strategies and Technology

JIANZHONG SUN,*[a] SHI-YOU DING[b] AND
JOY DORAN-PETERSON[c]

[a] Biofuels Institute, Jiangsu University, Zhenjiang 212013, P. R. China;
[b] Biosciences Center, National Renewable Energy Laboratory, Golden,
CO 80401, USA; [c] University of Georgia, Department of Microbiology,
Athens, GA 30602, USA
*Email: jzsun1002@hotmail.com

1.1 Current Technologies in the Biological Conversion of Biomass to Biofuels

Biofuels have been around since modern cars were invented. Early engines were designed to be able to run on ethanol or vegetable oil. However, the discovery of huge deposits of petroleum led to the development of refinery-based fossil fuels, and have provided a cheap supply of gasoline and diesel for decades. Although the concept of biofuels has never been forgotten, industrial-scale biofuel production has yet to be developed. Recently, increased gasoline prices, environmental concerns, and the security of energy supplies have drawn much

RSC Energy and Environment Series No. 10
Biological Conversion of Biomass for Fuels and Chemicals: Explorations from Natural Utilization Systems
Edited by Jianzhong Sun, Shi-You Ding and Joy Doran-Peterson
© The Royal Society of Chemistry 2014
Published by the Royal Society of Chemistry, www.rsc.org

public attention. Many countries have therefore promoted the research and development (R&D) of alternative fuels in response to these concerns. One of the promising routes towards alternative energy is biofuels produced from lignocellulosic biomass through biological conversion processes, in which non-food resources, such as agricultural and forestry waste, and energy crops, are utilized as feedstock. Processes that include thermochemical pretreatment, enzymatic saccharification, and microbial fermentation are used to convert polysaccharides in biomass into alcohol fuels, such as ethanol for transportation.[1] Although carbohydrates are an energy source for many organisms in nature, the natural decay process by microorganisms is slow, albeit efficient.[2] In order to establish biorefinery and achieve economically competitive biofuel production from biomass, technical challenges and barriers must be overcome to accelerate the development of the growing biofuel industry. The key components involved are sustainable feedstock supply, optimized bioconversion technologies, and integrated biorefinery on an industrial scale.

The R&D of feedstock supply and logistics has been focused on developing techniques to provide sustainable biomass supply (quantity) and quality in production, harvest, storage, handling, and transportation. Preprocessing treatments, such as reduction of particle size, drying, blending, and densification, have also been tested to standardize the feedstock supply chain, reduce transportation costs, and to upgrade biomass reactivity in down-stream operations.

Current biomass conversion technology normally consists of three steps: thermochemical pretreatment, enzymatic saccharification, and microbial fermentation. Pretreatment is a process that conditions biomass feedstock to be amenable to cellulolytic enzymes to break down structural polysaccharides to fermentable sugars. Physical pretreatment may include mechanical milling to reduce biomass particle size, steam explosion, and hydrothermolysis. Thermochemical pretreatment normally involves a dilute acid (*e.g.*, H_2SO_4) or base (*e.g.*, NaOH) to promote hydrolysis by removing hemicellulose and/or lignin. Many other solvents have also been explored and shown to disrupt biomass and promote hydrolysis, such as alkaline H_2O_2, ozone, organosolv, glycerol, dioxane, phenol, ethylene glycol, concentrated mineral acid, ammonia, metal complexes, and others. For any given feedstock, some methods may be more effective than others. Many pretreatment technologies have been developed aiming to achieve a high yield of sugars and limit the degradation products that may inhibit microbial fermentation. Some of these methods, while effective, are not economically feasible.[3] Due to the limited understanding of the complex structure of biomass, pretreatment is still considered to be the most costly step when the overall cost of the process is taken into account.[4] A recent study seemed to shed light on the rational design of the pretreatment process, suggesting a focus on eliminating the lignin while leaving the structural polysaccharides (*i.e.*, hemicellulose as well as cellulose) intact within the plant cell walls. Such pretreatment would leave an open structure that allows easy access by the enzymes and rapid digestion of polysaccharides.[5]

Enzymatic saccharification is a process which uses an enzyme cocktail that contains a mixture of cellulases, hemicellulases, and accessory enzymes to depolymerize polysaccharides into simple sugars. Currently, commercial enzyme mixtures are developed based exclusively on the fungal cellulases, *i.e.*, *Trichoderma reesei* enzymes. After decades of strain development and study of *T. reesei* genetics, the enzyme performance and production has been significantly improved, and the cost is acceptable.

Biomass hydrolysates from pretreatment and enzymatic hydrolysis are usually a mixture of different sugar monomers, oligomers, and degraded intermediates. Conventional fermentation microorganisms can utilize hexose sugars (glucose, mannose, and galactose) but not pentose sugars (xylose and arabinose), and their growth is normally inhibited by the degraded compounds. Research has been focused on the development of a strain that can co-ferment both C6 and C5 sugars, and on improvement of tolerance to the inhibitor. Today, biomass saccharification and fermentation are usually performed as an integrated process, such as simultaneous saccharification and fermentation (SSF) which consolidates enzyme hydrolysis and microbial fermentation of cellulose products into one process step using cellulases produced elsewhere. Simultaneous saccharification and co-fermentation (SSCF) that integrate hemicellulose hydrolysis products into SSF is a second such integrated process. Another concept that has been proposed is known as "consolidated bioprocessing" (CBP), and combines enzyme production, biomass saccharification, and fermentation as a single step. This may theoretically further reduce the process cost, however, challenges remain in the development of strain(s) that produce highly active enzymes (cellulases, hemicellulases and accessory enzymes), and are capable of fermenting these mixed hydrolysis products into biofuels with high yield.[6,7]

It is worth noting that the proposed biorefinery of biomass to biofuels is a man-made process and does not occur in nature. In other words, natural organisms may effectively mediate one of the steps in the course of converting biomass to fuels, but there is no single microorganism that can utilize biomass as carbon and as an energy source to produce a form of biofuels, such ethanol, as the main end-product. Significant application of mechanical, chemical, and biological engineering is, therefore, required for biorefinery, the environmental impact of which has yet to be assessed. Leaning from natural systems is the key to advancing the aforementioned biological processes and ensure the economic success of biorefinery.

1.2 Natural Biomass Utilization Systems and their Potential for Biorefinery

Degradation of lignocellulosic biomass in nature is generally considered to be a microbial deconstruction process carried out by a variety of microorganisms or microbial communities, including bacteria and fungi. The three major components of lignocellulose, cellulose, hemicellulose, and lignin, all require separate classes of enzymes to cleave their polymeric forms into shorter chains or monomers for further conversion processes. Explorations of several biomass utilization systems (from a single microorganism to comprehensive digestion systems) are presented in Chapters 7–14. Individual microorganisms capable of degrading plant-cell-wall polymers (usually cellulose and hemicellulose) followed by conversion of those polymers into a single product are desirable for industrial processes. However, single microorganisms with the ability to degrade multiple polymers usually produce multiple products *via* fermentation of the resultant sugars. In addition to the fermentable carbohydrates, lignin can constitute a significant percentage of plant biomass on a weight basis and is a complex polymer of phenylpropane units cross-linked to each other with different chemical bonds. Some individual organisms, predominantly the white-rot fungi, produce enzymes to deconstruct the lignin fraction, mediated by extra-cellular lignin and manganese peroxidases.[8] Actinomycetes can also deconstruct lignin, but they are much less efficient and usually degrade less than 20% of the total lignin present.[9–11] Representative lignocellulose-degrading organisms and their average percentage of plant-cell-wall deconstruction is depicted for cellulose, hemicellulose and lignin in Figure 1.1.

Organisms surviving on lignocellulosic biomass possess many adaptations, including specialized gut systems, which can be considered to be natural bioreactors. Examination of natural biomass-utilization systems, in order to identify mechanisms, enzymes, and/or organisms for further improving managed industrial processes for biomass conversion is the main focus of the book.

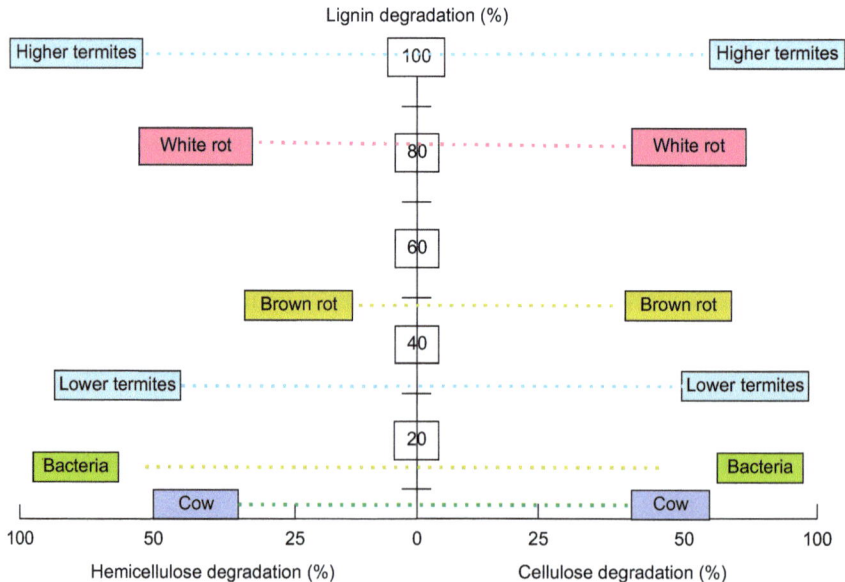

Figure 1.1 Representative lignocellulolytic systems and their conversion efficiency for the three main polymers of biomass.

Lignocellulose-degrading bacteria have been isolated from natural environments and if they also ferment the hydrolytic products of their enzyme activities, they may be referred to as "consolidated bioprocessing bacteria" (CBP bacteria). *Clostridium thermocellum*[12,13] and *Clostridium phytofermentans*[14] are the most thoroughly studied of the CBP bacteria and Chapter 7 summarizes the technologies used to study plant biomass fermentation using the single model bacterium *C. phytofermentans*. This mesophilic organism isolated from forest soil and grows on both the soluble and insoluble parts of plant biomass by first enzymatically digesting the plant polysaccharides and then fermenting the resulting sugars to mainly ethanol and acetate.[14,15] Industrial processes would be relatively simple if a robust CBP organism could be isolated or genetically engineered for the conversion of biomass to biofuel. However, a CBP organism with very fast conversion rates and highly efficient fermentation to produce a single product does not exist, yet. This chapter provides a detailed analysis of molecular techniques used to study the genome, transcriptome, and proteome of a single microorganism and provides insight into improving fermentation yields for *C. phytofermentans* and other CBP bacteria.

In addition to characterizing individual microorganisms capable of degrading plant cell walls and isolating individual enzymes from lignocellulose-degrading bacteria or insects, there is a great deal of interest in studying whole systems where a consortium of microorganisms with multiple capabilities interact, oftentimes with the host, to degrade lignocellulose. There are multiple examples of lignocellulose-conversion processes in nature including herbivorous mammals and lignocellulose-degrading insects, and these are used as examples in Chapters 8–14.

Chapter 8 details lignocellulose degradation in termite symbiotic systems and notes that termites, notorious for the tremendous damage they cause, are now considered an informative bioreactor because they can dissimilate 74–99% of the cellulose and 65–87% of the hemicellulose that they ingest (Figure 1.1).[16] To achieve these high conversion percentages in a short timeframe, termites have evolved alimentary tracts and symbiotic microbial systems to constitute a small-scale, yet highly efficient bioreactor, consisting of a grinding machine (mandible and proventriculus), a reaction chamber (digestive tract), and the microbial community and their own enzymes. The authors then describe how, within this micro-scale bioreactor, mechanical and enzymatic action, *via* the microbial community or protist symbionts, work together with the insect to achieve extensive lignocellulose degradation.

A survey of functional genes from cellulose-feeding insects is presented in Chapter 9, highlighting the many plant-cell-wall-deconstructing enzymes with high specific activities that have been isolated from herbivorous and xylophagous insect species. One recurring theme in these chapters and in the literature in general is that the wealth of sequence information that is now available has now caused a bottleneck in our ability to clone, identify, characterize, evaluate, and over-produce each enzyme. A major limitation is the capacity to evaluate the biochemical activities of gene products,[17] and substantial yield increases and cost reductions will be required to make large-scale fermentation of lignocellulosic biomass possible.[18] The authors of Chapter 9 also caution against using only homologues defined by amino acid sequence homology so that we do not miss the opportunity to define novel enzymes with potential industrial relevancy.

Chapter 10 describes the biological pretreatment of biomass by wood-feeding termites and examines the contributions of the insect itself to biomass deconstruction. The authors suggest that the wood-degrading process that has evolved in termite gut systems supports a new concept that complete degradation or removal of lignin may not be required for enhanced cellulose hydrolysis. Instead, the authors discuss the potential of selective modification of lignin functional groups and ether linkages, which play critical roles in lignin recalcitrance to enzymatic cellulose hydrolysis, to result in efficient pretreatment. This nature-inspired pretreatment scheme can be implemented under ambient conditions with no heat, no pressure, needing few chemicals, and results in mainly lignin modification. If such a method could be developed for industrial processes, this would represent a breakthrough in pretreatment technology.

Chapter 11 introduces the depth and breadth of wood-feeding insects with potential for biofuel production by providing an inventory of wood-feeding insects and evidence for their ability to digest cellulose. Cellulose digestion has been demonstrated in insect species from many diverse taxonomic groups including: various wood-feeding insects (woodroaches – Dictyoptera, lower and higher termites – Isoptera, various beetles – Coleoptera, and wood wasps – Hymenoptera), detritus/litter-feeding insects (*e.g.*, immature stages of leaf-shedding aquatic insects – Trichoptera, Diptera, and Plecoptera, silverfish – Thysanura, crickets – Orthoptera), and forage-feeding insects (beetles – Coleoptera). More than 20 families representing 10 distinct insect orders, *e.g.*, Thysanura, Plecoptera, Dictyoptera, Orthoptera, Isoptera, Coleoptera, Trichoptera, Hymenoptera, Phasmida, and Diptera have representative members capable of digesting cellulose.[19]

Chapter 12 presents the characteristics of the wood-feeding cockroach that make it an efficient mini bioreactor that involves a suite of specialized enzymes which synergistically break down the matrix of cellulose, hemicelluloses, and lignin in plant cell walls. Both termites and wood cockroaches convert over 90% of lignocellulose into fermentable sugars in their hindguts, making them an ideal habitat from which to identify novel enzymes specifically adapted to hydrolyze biomass.

A survey of selected natural lignocellulose-degrading systems is presented in Chapter 13, with the focus on -omics technologies and systems biology approaches for the reverse design of biocatalysts for biofuel production. The authors indicate that only a few fungi have high lignin-degradation selectivity that could greatly increase cellulose hydrolysis and saccharification efficiency, and that these processes are aerobic. Processes such as anaerobic degradation use mainly mesophilic, rumen-derived bacteria.[20,21] Compared to anaerobic systems, aerobic bacterial pretreatment has a higher efficiency in degrading high lignin biomass.[22,23] Despite long processing times and low saccharification efficiency *via* fungal pretreatment methods, recent biological pretreatment by termite systems indicate that it only takes 4–5 hours to move from modification of lignin components in the foregut/midgut to the hindgut for biomass hydrolysis and processing.[24] Exploring the best performance parameters for multiple systems may enable scientists to reverse design a more efficient industrial process using a combination of different natural biocatalytic systems.

Chapter 14 provides an in-depth study of the ruminant animal as a natural biomass conversion platform, noting that over 60 years ago, Hungate developed the analogy of the ruminant as a self-feeding, self-replicating, mobile, cellulose-degrading bioreactor.[25] Also discussed are the many different micro-organisms with potential applications in bioconversion processes that have been isolated and characterized from ruminants. The author also points out that the ease with which ruminal fermentations are run in bioreactors provides support for integrating these innovations into an industrial volatile fatty acids platform for fuel and chemical production.

1.3 Perspectives on Breakthroughs in Biorefinery for Fuels and Chemicals

It has been generally accepted that alternative and renewable fuels and chemicals derived from lignocellulosic biomass will offer the potential to reduce our dependence on petroleum and other traditional non-renewable energy sources. However, breakthrough technologies in biomass conversion are still very limited, from both theoretical and application aspects, in realizing large-scale success due to the intractable barriers in efficiency and processing economy. We could potentially benefit from a review of our ongoing strategies and technology development and exploring other lignocellulolytic systems in nature, such as wood-feeding termites, cows, or other sound biological systems that could reveal new insights.[19] Such animal lignocellulolytic systems can process cellulosic biomass efficiently with their highly specialized digestive systems, which can truly be considered as highly efficient natural bioreactors.[26] The biggest challenge in biological conversion of biomass is to acquire a deeper understanding of biomass recalcitrance as well as the conversion processing mechanisms from the advanced lignocellulolytic systems evolved by some animals.[27,28] With some of the most intractable issues facing the world regarding efficient and economic conversion of cellulosic biomass, this book intends to demonstrate the opportunities, breakthroughs, remaining challenges and perspectives in employing nature-inspired lignocellulolytic systems for the development of robust, effective, and environmentally friendly biorefineries.

1.3.1 Understanding Biomass Recalcitrance to Deconstruction

The collective resistance that lignocellulosic materials pose to deconstruction by microbes and enzymes is generally referred to as "biomass recalcitrance".[27] This term is used in several contexts that may, or may not reflect the same underlying structural elements. For example, it can be used to highlight the substantial differences in severity required for the dilute acid hydrolysis of lignocellulose and starch and to explain why pretreated lignocellulose requires 100 times more enzyme for complete saccharification than pretreated starch does. In addition, "recalcitrance" is also used when describing the kinetic phenomena in which the rate of cellulose digestion slows during extended reaction times. It has been reported that a variety of natural factors, including both chemical and structural elements, is believed to contribute to the recalcitrance of lignocellulosic feedstock conversion to chemicals or enzymes. In the context of biorefinery, these chemical and structural features of biomass would affect liquid penetration and/or enzyme accessibility and activity and, thus, conversion costs.[1]

Plant biomass types, particularly terrestrial higher plants, have evolved superb mechanisms for resisting assault on their structural sugars from microorganisms and most animals. It has been speculated that the recalcitrance trait developed in terrestrial higher plants during evolution, in part, as a consequence of their moving from the protection of the aquatic environment.[27] As a comparison, lower plants, such as green algae growing in aquatic environments as well as other non-flowering plants, virtually lack the well-developed plant cell walls providing protection against attack by microbes and enzymes. In addition, their cell wall structures typically lack lignin elements to protect structural sugars [Figure 1.2(a)]. Although little is known from current references about the evolutionary steps or the intermediate forms presented during evolution, higher plants have indeed developed many systems to cope with the challenges of physical, chemical, and biological factors in nature to extract the structural sugars buried in the plant-cell-wall matrix. These defense systems against biological attacks include, but are not limited to, the epidermis, or outer layer of plant anatomy, as well as the structure and organization of vascular

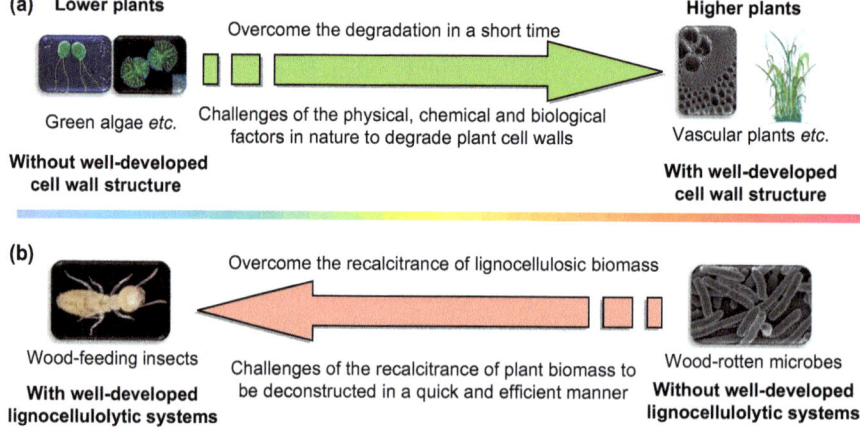

Figure 1.2   Evolution of plant biomass recalcitrance and the co-evolution of various lignocellulolytic systems in nature. (a) Evolution of plant biomass recalcitrance. (b) Co-evolution of various lignocellulolytic systems in nature.

tissues and even of the cell walls. Further information regarding this subject is available from the recent reviews by Himmel *et al.*[1,27] and Foston and Ragauskas,[29] and need not be repeated in detail here.

Research efforts to redirect the evolutionarily imposed protection of plants' cell-wall polysaccharides are now underway, mainly using genetic approaches to modify cell-wall characteristics and to also express the foreign cellulolytic enzymes in plant cells, thus permitting more efficient and economic biological/chemical hydrolysis processes, as well as improved agronomic productivity.[28] With these research innovations, we may potentially optimize both plant-cell-wall production and its down-stream deconstruction in ways not normally achievable in nature.

To acquire a deeper understanding of biomass recalcitrance and access to an efficient and economic conversion, five chapters (Chapters 2–6) of this book describe the state-of-the-art, as well as promising novel approaches to overcoming fundamental science and engineering barriers to reducing the recalcitrance of biomass as well as the enzymatic cost involved in the processing. Chapter 2 presents a short overview of the lignocellulose structure and chemistry so as to offer a better understanding of the deconstruction mechanism of the biomass to biofuels and other bioproduct processes. Chapter 3 reviews various microscopy and spectroscopy tools used to characterize plant-cell-wall structure, including optical microscopy, atomic force microscopy, and electron microscopy, which will provide insight into the fundamental mechanisms that contribute the native recalcitrance of lignocelluloses. Chapter 4 primarily deals with the challenges of better understanding how the structure of lignin changes during pretreatment processing and also demonstrates recent advances in the fundamental characteristics of ethanol organosolv lignin *via* NMR spectroscopy techniques and molecular weight analysis. Chapter 5 assesses advances in the genetic manipulation of cellulosic bioenergy crops for ethanol production that utilize dedicated energy plants as the biofactories to produce foreign lignocellulosic enzymes, which can reduce exogenous enzyme loading for industry, thereby reducing the cost of biorefinery processing. Chapter 6 is concerned primarily with the diverse distribution of lignocellulosic resources, their composition, current application as feedstock and the state-of-the-art in modification of plant-cell-wall structure to increase biomass digestibility and its conversion efficiency through genetic engineering of energy crops. These five chapters provide current advances in the understanding of plant-cell-wall structure and future strategies for modifying cell walls to improve the current conversion efficiency and economy.

1.3.2 Exploring and Integrating Nature-Inspired Technology

Different glycolytic activities exist in various biological systems, such as from bacteria, fungi, marine isopods, cellulose-feeding arthropods, and cellulose-feeding mammals, and these may have co-evolved with the recalcitrance development of plant cell walls as a chemical or structural defense. In general, as we have proposed in Figures 1.2(a) and (b), each party in a co-evolutionary relationship exerts selective pressure on the other, thereby affecting each other's evolution. Clearly, different biological systems may present a different strategy to cope with the recalcitrance of plant cell walls under the chronically selective pressure in a particular environment, which eventually leads to a unique cellulolytic system that may function with different efficiencies in time and different competences in the degradation of cellulose and hemicellulose, as well as lignin (Figure 1.1).

For example, wood-feeding termites efficiently deconstruct various major polymer components of biomass within 24 hours with their well-developed lignocellulolytic systems. The evolutionary direction for various biological systems in the utilization of lignocellulosic feedstocks appears to be directly opposed to the evolution path taken by the plant kingdom (Figure 1.2). Although lignocellulolytic activities were originally thought to be restricted to plants, bacteria, and fungi, there is rapidly accumulating evidence for the existence of animal lignocellulases (cellulases, hemicellulases, and lignases), especially in wood-feeding termites.[19,26] In particular, these cellulose-feeding animals have had to develop a strategy and superb enzyme system to overcome the recalcitrance of lignocellulosic biomass and get the carbohydrate energy embedded in lignocellulose in a quick and efficient manner. These diverse and highly adapted digestive systems can efficiently deconstruct various lignocellulosic feedstocks, will inform us of unique lignocellulolytic systems, and will help us develop a novel and nature-inspired system for modern biorefinery.

Recent exploration of these unique lignocellulolytic systems using advanced molecular biotechnologies, including meta-genomics, proteomics, transcriptomics, and synthetic biology, have brought new insight into the mechanisms of biomass deconstruction within these efficient, but complicated digestive gut systems. However, the integration of nature-inspired technologies into the modern biorefinery system is another important challenge and task in our research agenda. To meet this demand, this book has made an effort to address this obligation in Chapters 15–19.

Chapter 15 describes a unique insect gut system, *Tipula abdominalis* (in the larval stages), as a natural bioreactor, where these larvae host a diverse hindgut microbiota presenting some new microorganisms with novel enzymatic activities. The inventory of gut microorganisms, followed by characterization of selected isolates, then identification of novel enzymes from one of the isolates is presented. Using this novel enzyme in fermentations of pectin-rich biomass for ethanol production is described as an example of applying insect-associated microbial enzymes to improve an existing bioconversion process. Exploring various insect lignocellulolytic systems could lead to the discovery of a variety of novel biocatalysts and the genes that encode them, as well as associated unique mechanisms for efficient biomass conversion.

With thousands of new genes identified from a variety of lignocellulolytic systems in nature, including those from cellulose-feeding animal origins and their gut microbiota (most of them are culture-independent), gene expression from these origins has become a critical bottleneck for identification, characterization, evaluation, and over-expression of each single enzyme. Chapter 16 reviews the most up to date technologies, as well as promising novel approaches, such as the Hsh expression system invented by the authors for *E. coli*, to overcome fundamental science and engineering barriers to enabling heat–shock induction repression in industrial fermenters, soluble expression of aggregation-prone protein, and *in situ* error-prone PCR. This chapter also introduces the tools and technologies developed in recent years for improvements in recombinant enzyme production and the properties desired for cost-efficient industrial processes.

Lignin-unlocking processes by wood-feeding termites, an important biological pretreatment step, operate in a specific gut physiochemical environment with a variety of lignolytic oxidases and some non-biological co-factors in their gut system, such as redox potential, pH, metal elements and H_2O_2 which need to be eventually integrated and converted into a novel nature-inspired technology system. Chapter 17 provides a summary and research update for cellulose-dissolving systems that would inevitably affect the subsequent enzymatic hydrolysis of the regenerated cellulose. In contrast to traditional pretreatment methods, cellulose solvents that mimic the activity of lignin-degrading enzymes from a biological system have some unique advantages, including milder operation conditions, which can function well with some important co-factors, fewer degradation by-products, and higher cellulose accessibility to cellulase.

It is instructive to investigate the solutions and strategies resulting from nature-inspired evolution for overcoming biomass recalcitrance, and then applying promising strategies to help solve those intractable issues in developing an efficient and economic biomass conversion process for fuels and chemicals. Chapter 18 aims to discuss what we can learn from natural biomass-utilization systems for developing novel bioreactors; the authors review the current state-of-the-art biomimetics and its potential for nature-inspired technology and innovation.

Overall process economics and sustainability perspectives are critical for the development of a robust, effective, and environmentally friendly biorefinery system and for potentially realizing industrial-level applications. Chapter 19 conducts a techno-economic analysis and life-cycle assessment (LCA) of lignocellulosic biomass to sugars using various pretreatment technologies; the authors report the process economic analysis and LCA of the six leading pretreatment processes using hybrid poplar as the biomass feedstock. They also further introduce an integrated and commercial-scale lignocellulosic sugar process module that has been proposed as the foundation of the cost analysis for a specific biomass pretreatment process.

These five chapters of this book mainly focus on integrating the advanced nature-inspired strategies and technologies into a modern biorefinery system for the viable biological conversion of biomass.

1.3.3 Perspectives for the Advanced Biological Conversion of Biomass

The biological conversion of lignocellulosic biomass has long been recognized as a promising approach to access the fermentable sugars for biofuels and other bioproducts. However, in practice, the current state of technology with respect to biomass conversion is still far away from being suitable for large-scale application due to its efficiency and processing economics.[26] It is the property of biomass recalcitrance that is largely responsible for the high cost and low efficiency of lignocellulose conversion.[1] Thus, our inevitable challenge is to acquire a deeper understanding of biomass recalcitrance and the optimized conversion mechanisms presented by some unique lignocellulolytic systems in nature, which may provide real solutions to these problems. We have much to learn from the sound cellulolytic systems in nature that have successfully cracked the code for lignocellulose deconstruction using amazing strategies and mechanism development throughout evolutionary history.

With the science and engineering issues facing the world regarding the efficient and economic conversion of lignocellulosic biomass, and the cutting edge technologies that have emerged in the past decade, this book comes at a critical and timely moment. The ultimate goal of this book is to acquire a deeper understanding of biomass recalcitrance to deconstruction and to develop a novel approach for modern biorefinery processing with nature-inspired strategies and technologies. Using biomimetic strategies from a systems biology approach combined with other advanced technologies may pave the way for future breakthroughs and innovations in associated areas of industrial biotechnology. It is further hoped that this book can promote productive dialog and collaboration between scientists working in the various disciplines needed to address this global challenge.

References

1. M. E. Himmel, S.-Y. Ding, D. K. Johnson, W. S. Adney, M. R. Nimlos, J. W. Brady and T. D. Foust, *Science*, 2007, **315**, 804.
2. H. Wei, Q. Xu, L. E. Taylor, J. O. Baker, M. P. Tucker and S.-Y. Ding, *Curr. Opin. Biotechnol.*, 2009, **20**, 330.
3. N. Mosier, C. Wyman, B. Dale, R. Elander, Y. Y. Lee, M. Holtzapple and M. Ladisch, *Bioresour. Technol.*, 2005, **96**, 673.
4. L. R. Lynd, M. S. Laser, D. Brandsby, B. E. Dale, B. Davison, R. Hamilton, M. Himmel, M. Keller, J. D. McMillan, J. Sheehan and C. E. Wyman, *Nat. Biotechnol.*, 2008, **26**, 169.
5. S.-Y. Ding, Y. S. Liu, Y. N. Zeng, M. E. Himmel, J. O. Baker and E. A. Bayer, *Science*, 2012, **338**, 1055.
6. L. R. Lynd, W. H. Van Zyl, J. E. McBride and M. Laser, *Curr. Opin. Biotechnol.*, 2005, **16**, 577.
7. L. R. Lynd, P. J. Weimer, W. H. Van Zyl and I. S. Pretorius, *Mol. Biol. Rev.*, 2002, **66**, 506.
8. T. K. Kirk and R. L. Farrell, *Annu. Rev. Microbiol.*, 1987, **41**, 465.
9. D. L. Crawford, *FEMS Symp.*, 1986, **34**, 715.
10. T. Richard, The effect of lignin on biodegradability, http://compost.css.cornell.edu/calc/lignin.html.

11. M. Basaglia, G. Concheri, S. Cardinali, M. B. Pasti-Grigsby and M. P. Nuti, *Can. J. Microbiol.*, 1992, **38**, 1022.
12. L. R. Lynd, W. H. Van Zyl, J. E. McBride and M. Laser, *Curr. Opin. Biotechnol.*, 2005, **16**, 577.
13. X. Shao, M. Jin, A Guseva, C. Liu, V. Balan, D. Hogsett, B. E. Dale and L. R. Lynd, *Bioresour. Technol.*, 2011, **102**, 8040.
14. T. A. Warnick, B. A. Menthe and S. B. Leschine, *Int. J. Syst. Evol. Microbiol.*, 2002, **52**, 1155.
15. A. C. Tolonen, A. C. Chilaka and G. M. Church, *Mol. Microbiol.*, 2009, **74**, 1300.
16. C. Rouland-Lefèvre, in *Termites: Evolution, Sociality, Symbioses, Ecology*, T. Abe, D. E. Bignell and M. Higashi, Kluwer Academic Publishers, Dordrecht, 2000, p. 289.
17. G. Banerjee, J. S. Scott-Craig and J. D. Walton, *Bioenerg. Res.*, 2010, **3**, 82.
18. H. Wu, J. Pei, Y. Jiang, X. Song and W. Shao, *Biotechnol. Lett.*, 2010, **32**, 795.
19. J. Z. Sun and X. G. Zhou, in *Recent Advances of Entomological Research: From Molecular Biology to Pest Management*, ed. T. X. Liu and L. Kang, Springer, Berlin, 2011, p. 434.
20. Z.-H. Hu and H.-Q. Yu, *Process Biochem.*, 2005, **40**, 2371.
21. Z.-H. Hu, S.-Y. Liu, Z.-B. Yue, L.-F. Yan, M.-T. Yang and H.-Q. Yu, *Environ. Sci. Technol.*, 2007, **42**, 276.
22. A. Arora, L. Nain and J. K. Gupta, *World J. Microbiol. Biotechnol.*, 2005, **21**, 303.
23. A. Mshandete, L. Björnsson, A. K. Kivaisi, S. T. Rubindamayugi and B. Mattiasson, *Water Res.*, 2005, **39**, 1569.
24. J. Ke, D. Laskar, D. Gao and S. Chen, *Biotechnol. Biofuels*, 2012, **5**, 11.
25. A. T. Hendricks and W. M. Zeeman, *Bioresour. Technol.*, 2009, **100**, 10.
26. J. Z. Sun and M. E. Scharf, *Insect Sci.*, 2010, **17**, 163.
27. M. E. Himmel and S. K. Picataggio, in *Biomass Recalcitrance: Deconstructing the Plant Cell Wall for Bioenergy*, ed. M. E. Himmel, Blackwell, New Delhi, 2008, p. 1.
28. A. Brune, *Trends Biotechnol.*, 1998, **16**, 16.
29. M. Foston and A. Ragauskas, *Ind. Biotechnol.*, 2012, **8**, 191.

CHAPTER 2

Overview of Lignocellulose: Structure and Chemistry

SHI-YOU DING

Biosciences Center, National Bioenergy Center, National Renewable Energy Laboratory, Golden, CO 80401, USA
Email: shi.you.ding@nrel.gov

2.1 Introduction

In the Earth's ecosystem, autotrophs are organisms that use energy and simple chemicals in the environment to synthesize complex organic compounds (biomass) that can be used by heterotrophs as an energy and carbon source. On land, higher plants are the major producers of biomass through the process of photosynthesis, in which plants capture sunlight, split water atoms, release oxygen, and use hydrogen to reduce carbon dioxide. At this point in the process, energy and carbon are stored as sugars that are often further polymerized in the plant to form polysaccharides, such as starch and cellulose. The plant cell walls contain the major material that plants produce. Within that material, many components—including polysaccharides, lignin, proteins, minerals, and others— inter-connect to form complex matrix structures. Lignocelluloses are dead plant matter which are primarily composed of cell wall material.

Lignocellulose materials have been used as an energy source since humans first learned to burn wood and dead plants for cooking and heating. Currently, lignocelluloses in agricultural and forestry residues, dedicated energy plants, and industrial wastes such as bagasse are considered to be a renewable source of transportation fuel; however, instead of simply burning them, advanced

RSC Energy and Environment Series No. 10
Biological Conversion of Biomass for Fuels and Chemicals: Explorations from Natural Utilization Systems
Edited by Jianzhong Sun, Shi-You Ding and Joy Doran-Peterson
© The Royal Society of Chemistry 2014
Published by the Royal Society of Chemistry, www.rsc.org

technology must be developed that can efficiently process these highly heterogeneous materials into fuels. The concept of biorefinery has been proposed to use sustainable technologies that can produce transportation fuels from lignocelluloses, benefiting the environment by reducing greenhouse gas emissions.[1,2] These technologies can be categorized into two pathways. The first is the biochemical pathway, which includes sequential processes, such as thermochemical pretreatment to precondition biomass feedstock, enzymatic saccharification to hydrolyze polysaccharides into simple sugars, and microbial fermentation to metabolize sugars to ethanol or other alcoholic fuels. The second pathway is the thermochemical pathway, in which heat and chemical catalysts break down the polymeric materials in the plant cell walls and produce synthetic gas or bio-oil. Scientists and engineers are now engaged in optimizing these processes, aiming to reduce operational costs, increase the yields of desirable fuels and chemicals, and achieve 'zero-waste' utilization of biomass. This chapter is intended to summarize the current research into plant-cell-wall structure and chemistry with respect to a greater understanding of the deconstruction mechanism of the biomass-to-biofuels process.

2.2 Cell Wall Biogenesis

Research into the model plant Arabidopsis (*Arabidopsis thaliana*), has suggested that an estimated 10% of the plant genome, approximately 2500 genes, has a putative function that involves cell-wall biosynthesis, modification, and metabolism. Annotation of these genes is largely based on the homologous comparison of prokaryotic sequences in other organisms. Although most of these gene functions are yet to be elucidated experimentally, the biochemical activities include the generation of substrates, polymerization, trafficking control, and cell-wall modification and rearrangement. It can be concluded that cell-wall biosynthesis is regulated in a spatially and temporally complex manner. Further studies with a systems-based approach are needed to fully understand how these chemically and structurally different polymeric materials assemble to form the dynamic structure of cell walls. In this regard, comprehensive reviews are available in the literature.[3,4] Here, I intend to give an update about how cellulose is synthesized and incorporated with other major matrix polymers, *e.g.*, hemicelluloses, pectins, and lignins. The models presented here are somewhat hypothetical and are based primarily on the observation of nano-scale imaging and analysis of the enzymatic digestibility of the plant cell walls.

Figure 2.1 illustrates the major steps of the biosynthesis of cell-wall components and their assembly. In general, cell-wall biogenesis can be described as four major processes: (1) cellulose synthesis that forming macrofibrils; (2) hemicellulose and pectin synthesis and modification; (3) macrofibril splitting when cells elongate and expand; and (4) monolignol synthesis and lignin polymerization. Cellulose (Figure 2.2) is synthesized by multi-enzyme complexes called "rosettes", which are proposed to contain 36 cellulose synthases (CesA) in each rosette that is embedded in the plasma membrane (PM). The

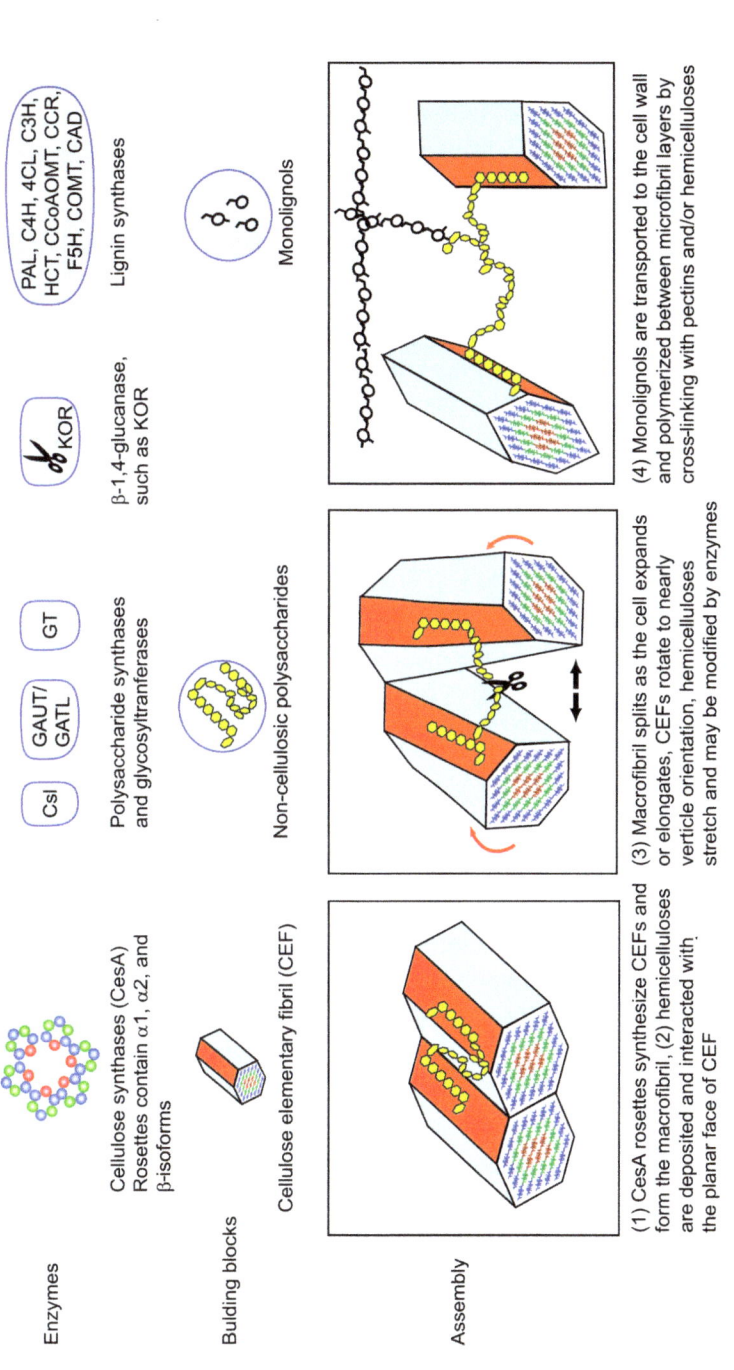

Figure 2.1 The major steps of synthesis and assembly of cell-wall polymers (*i.e.*, cellulose, hemicelluloses/pectins, and lignins). In addition to these synthases, many other proteins may also be involved in cell-wall biogenesis, such as cellulose synthase interactive protein (CSI), tracheary element differentiation-related (TED) proteins, COBRA-like proteins, chitinase-like proteins, and probably many other proteins yet to be discovered. In addition, the cytoskeleton is associated with the cellulose-synthesis process by playing a role in controlled microfibril orientation.[4]

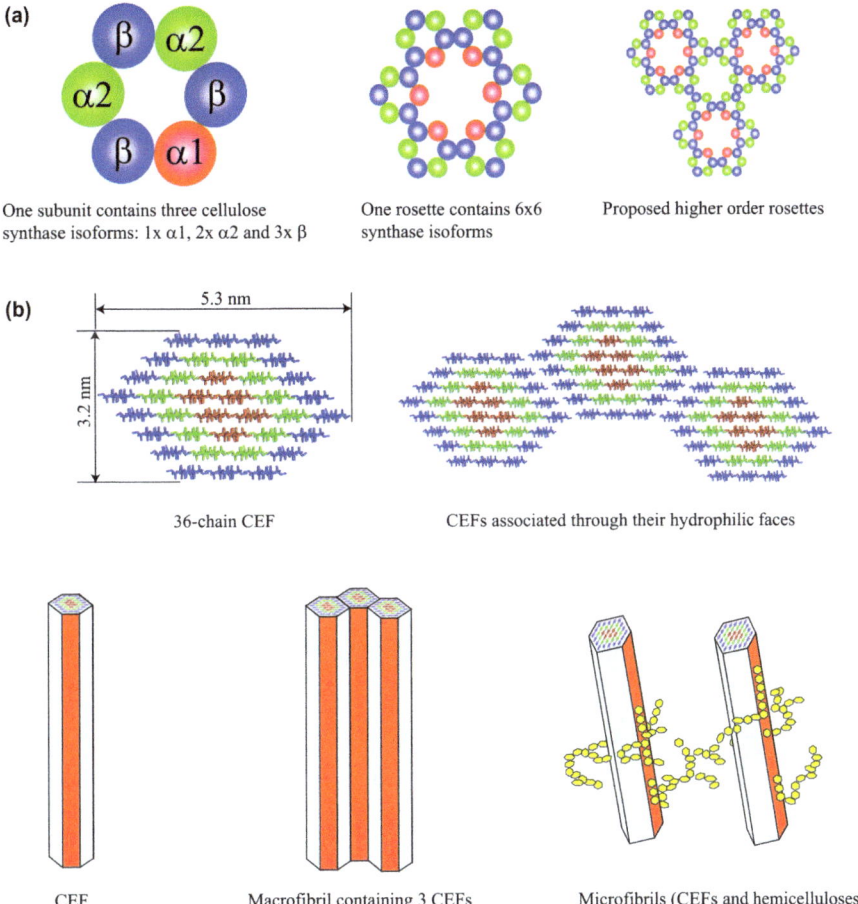

One subunit contains three cellulose synthase isoforms: 1x α1, 2x α2 and 3x β

One rosette contains 6x6 synthase isoforms

Proposed higher order rosettes

36-chain CEF

CEFs associated through their hydrophilic faces

CEF

Macrofibril containing 3 CEFs

Microfibrils (CEFs and hemicelluloses)

Figure 2.2 Cellulose synthase (CesA) and cellulose fibrils. (a) A 36-mer rosette contains (1) α1, (2) α2, and (3) β-CesA isoforms and a proposed higher order of rosette array. (b) Cellulose elementary fibril (CEF), macrofibril, and microfibril. A 36-chain CEF contains 18 chains (blue) on the surface, 12 chains (green) in the transition, and 6 chains (red) in the core. The chain arrangement is based on known native cellulose I.[9,10] The dimensions of hexagonal CEF are 3.2×5.3 nm.

most-accepted model of the rosette is composed of six identical subunits, in which each subunit contains (1) α1, (2) α2, and (3) β-isoforms. Each CesA enzyme synthesizes 1 linear β-1,4-linked glucan chain and 36 glucan chains form the rigid cellulose elementary fibril (CEF). Cellulose synthesis is believed to be a rapid process, and many rosettes spin CEFs simultaneously in the PM. These CEFs may initially coalesce through the hydrophilic faces to form ribbon-like bundles, called macrofibrils, which have been observed by atomic force microscopy (AFM) in fresh growing cells.[5] During cellulose synthesis, hemicelluloses are synthesized by cellulose-synthase-like (Csl) enzymes and glycosyltransferases (GTs), and extruded from Golgi vesicles to the wall. The similarity of the backbone structure of hemicelluloses, such as xyloglucan and mixed-linkage β-1,3 or β-1,4-glucan, facilitate the initial binding between hemicelluloses and the planar face of the CEF. As the cell expands or elongates, the macrofibril splits and hemicelluloses stretch, causing the CEF to rotate to nearly vertical orientation.[6] A β-1,4-glucanase, such as KORRIGAN (KOR), which was found to play a critical role in cell elongation, may function to cleave hemicelluloses to allow further splitting of CEFs. Pectins are also synthesized in Golgi by galacturonosyltransferase (GAUT), GAUT-like (GATL), and other GTs. Hemicelluloses and pectins are modified by GTs to form cross-linked matrices that surround the CEFs.

Cells that have secondarily thickened walls are lignified. Many enzymes are involved in the synthesis of the monolignols,[7] such as the following:

- phenylalanine ammonialyase (PAL)
- cinnamate-4-hydroxylase (C4H)
- 4-coumarate : CoA ligase (4CL)
- *p*-coumarate-3-hydroxylase (C3H)
- *p*-hydroxycinnamoyl–CoA : quinate/shikimate
- *p*-hydroxycinnamoyltransferase (HCT)
- caffeoyl-CoA-*O*-methyltransferase (CCoAOMT)
- cinnamoyl-CoA-reductase (CCR)
- ferulate-5-hydroxylase (F5H)
- caffeic acid *O*-methyltransferase (COMT)
- cinnamyl alcohol dehydrogenase (CAD)

The monolignols are incorporated into highly complicated lignin polymers containing three units: guaiacyl (G), syringyl (S), and *p*-hydroxyphenyl (H). A fourth unit is the recently reported caffeyl alcohol (C).[8] Lignification occurs at the later stage of secondary wall thickening and most likely takes place between layers of CEF hemicelluloses.

2.3   The Cell Wall Structure in Biomass

Lignocellulosic biomass is primarily composed of dead plant cell walls. The wall structures are dramatically modified after maturation, senescence, and harvest. Unlike in living plants, most of the cytoplasmic components and cell membrane are degraded during the process of plant senescence. The cell walls in biomass can be generally classified as three types (Table 2.1). The primary cell wall (PW) is the cell that never undergoes secondary thickening in the wall. This type of cell can be easily identified in plant tissues; such cells are relatively small in size (*ca.* 50–100 μm in diameter) and polyhedron shaped with thin walls (~100 nm). The PWs are composed primarily of polysaccharides (~90%, mostly cellulose) and proteins (~10%). The CEFs under high-resolution AFM appear to be macrofibrils that contain a number of CEFs arranged in parallel. These CEFs have been found aligned together through their hydrophilic faces,

Table 2.1 Summary of the structural features of the primary (PW), parenchyma-type secondary (pSW), and sclerenchyma-type secondary (sSW) walls.

Wall types	PW	pSW	sSW
Cell size and shape	*ca.* 50–100 μm	*ca.* 200–300 μm	Varies, fibers can be as long as several mm
Cell shape	Polyhedron	Polyhedron	Fibers or tubes
Wall thickness	~100 nm	*ca.* 2–5 μm	*ca.* 5–10 μm or larger
Cell types	Non-thickened parenchyma, guard cells in stomata, companion cells in the phloem	Thickened parenchyma, collenchyma, sieve elements	Fibers, sclereids, tracheids, vessels
Surface structure	Disordered ribbon-like macrofibrils	Mixed small macrofibrils and microfibrils	Coated by a warty layer
Lignification	Non-lignified	Partially lignified	Fully lignified
Accessibility to cellulases	Fully accessible	Accessible on surface	Not accessible
Digestibility by cellulases	Fully digestible	Partially digestible	Not digestible

so that the macrofibrils appear to be structured like ribbons (Figure 2.3). Hemicelluloses are not major components in the PW, therefore the surface of the macrofibril appears fairly clean and highly accessible to the carbohydrate-binding module (CBM) that specifically recognizes the planar faces of crystalline cellulose.[6,11–15]

The second type of cell wall in biomass is the parenchyma-type secondary wall (pSW). After vegetative growth, especially in grass, the cell walls in parenchyma tissue usually are secondarily thickened to enforce the strength of the plant body. This type of cell is large (*ca.* 200–300 μm in diameter) and polyhedron shaped with thick walls (*ca.* 2–5 μm). Lignifications may also occur during cell expansion and elongation. However, the pSW may not be completely lignified after plant maturation. The surface of the pSW contains mixed small macrofibrils and mostly individual microfibrils that are composed of only one CEF with hemicelluloses associated on the surface. These microfibrils are arranged in parallel and embedded in the matrix polymer networks (*i.e.*, hemicelluloses and pectins). The CEFs appear vertically oriented as planar face-to-planar face with matrix polymer bridges between them (Figure 2.3). The surface of the pSW is partially accessible to the cellulase binding, in which the macrofibrils are fully accessible and the microfibrils are partially available because the matrix polymers block the planar face of the CEF.

The third type of cell wall is the sclerenchyma-type secondary wall (sSW), which includes fibers in the vascular bundle (VB) sheath, inter-fascicular fibers, xylem vessels and tracheids, and hypodermis sclerenchyma. This type of cell provides mechanical support and transports water and minerals in plants.

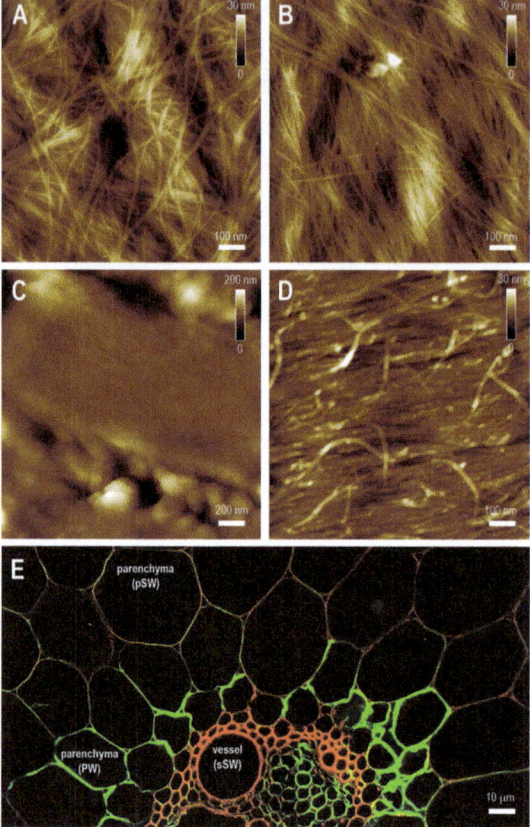

Figure 2.3   Three types of cell walls in mature maize stem. Atomic force micrographs of (a) the primary cell wall (PW) showing macrofibrils, (b) a parenchyma-type secondary cell wall (pSW) showing mixed macrofibrils and micro-fibrils, (c) a sclerenchyma-type secondary wall (sSW) surface showing no fibril structure, (d) a longitudinal section of the fiber wall showing microfibrils heavily coated by matrix polymers, and (e) green-fluores-cence-protein-tagged CBMs showing the accessibility to different wall types, strong binding to PW, weak binding to pSW, and no binding to sSW.

The sSWs are thick (*ca.* 5–10 μm or larger) and extremely elongated—as long as several mm. The dimensions of the sSWs vary between cell types, for instance, the fiber cells are narrow (~ 10 μm) and the vessel cells are large (*ca.* 50–300 μm). The sSWs are completely lignified after primary growth of plants, and account for the majority of biomass, especially in woody plants. The inner surface of the sSW is further covered by a warty layer, so that the microfibril structure cannot be observed in untreated biomass. The sections of the sSW show a similar microfibril structure and arrangement as the surface of the pSW, except the diameters of the microfibril appear slightly larger, possibly because of the association of more matrix polymers.

2.4 The Cell Wall Chemistry in Biomass

Lignocellulosic biomass is composed of polysaccharides ($\sim 70\%$ dry weight) and lignins ($\sim 25\%$ dry weight) and a small amount of proteins and minerals. The polysaccharides are cellulose, hemicelluloses, and pectins. Lignins are a group of highly branched phenylpropanoid polymers found in lignified cell walls.

Cellulose is the most abundant biopolymer on Earth, accounting for approximately 50% of the dry weight of lignocellulosic biomass. Cellulose is one of the most important natural polymers with wide applications in our daily life, such in as pulp and paper, textiles, food, and biomedical and many other industrial materials. Recently, cellulose has also been considered to be the renewable source of glucose for the production of biofuels and other by-products.[16] Cellulose is composed of simple, linear chains of glucose residues that are linked by β-1,4-glycosidic bonds. However, the physical properties of cellulose are complex, depending on the degree of polymerization, the number of chains in a single cellulose fibril, and the inter- and intra-chain hydrogen bonding and van der Waals interactions. Based on these physicochemical properties, cellulose can form different allomorphs, such as I_α, I_β, II, III, and IV. Furthermore, the cellulose fibers in nature are associated with hemicelluloses, and are often bundled to form larger macrofibrils or aggregates.

The basic structure of native cellulose in higher plant cell walls is 36-chain CEFs. Although the crystalline structure of the CEF has not been solved, two major allomorphs of native cellulose, *i.e.*, I_α from freshwater alga *Glaucocystis nostochinearum*, and I_β from the tunicate *Halocynthia*, structures have been determined by using synchrotron and neutron diffraction techniques. Many features of fundamental structure are common between cellulose I_α and I_β. The cellulose chains lie parallel and are inter-connected by hydrogen-bond networks. The intra-chain H-bonds between a hydroxyl group and the next ring oxygen (O3–H$\cdots$O5) and between hydroxyl groups (O2–H$\cdots$O6) lock each glucose residue to form the flat ribbon structure of the cellulose chain. The inter-chain H-bonds between hydroxyl groups of the neighboring chain (O6–H$\cdots$O3) facilitate the interaction between chains forming the cellulose sheets. The cellulose crystal is formed by stacking planar cellulose sheets through the van der Waals interaction and weak H-bonds of C–H$\cdots$O between neighboring sheets. The difference between cellulose I_α and I_β is the relative position of each chain or glucose residue. In I_α, all chains are identical but alternate glucose units in each chain, whereas in I_β, two distinct kinds of chains are arranged in alternating sheets. A mostly accepted hypothesis has been proposed that these two crystalline forms of cellulose co-exist in native cellulose materials. However, other researchers have also observed disorder in plant-cell-wall CEF surface chains, which blurs the distinction between the I_α and I_β allomorphs.[17] In addition, the reported number of cellulose chains in a single CEF is inconsistent, based on the data collected using different techniques, with numbers ranging from 18 to 36 chains.[18] The CEF is normally presented as a hexagonal shape in its cross-section, containing 36 chains. The cellulose chains arrange as 3 layers, 18 in the surface, 12 in the transition, and 3 in the core. In a mature cell, many other polysaccharides, such as hemicelluloses and pectins, are deposited in the wall and are further rearranged and modified; therefore, it is expected that the 18 surface chains are disordered to some extent.

Hemicelluloses and pectins are commonly called "matrix polysaccharides" because of their highly branched chemistry and amorphous structure. Unlike cellulose, matrix polysaccharides do not form a fibrillar structure because of the mixed linkages of glucose, such as the mixed-linkage β-1,3- and −1,4-glucans, or because of the branched structure with different glycosyl residues. Major hemicelluloses in biomass include xylan, xyloglucan, glucuronoxylan, arabinoxylan, and glucomannan. Approximately 20 different glycosyl residues are involved in hemicellulose structure. Pectins are α-1,4-linked-galacturonic acid polysaccharides, including homogalacturonans, rhamnogalacturonan I (RG-I), rhamnogalacturonan II (RG-II), and substituted galacturonans. Although hundreds of genes have been identified to be putatively involved in the biosynthesis of these matrix polysaccharides, resulting in complex chemical structures in the cell walls, the biodegradation of these polymers is relatively efficient compared with cellulose. Extensive reviews about the complex chemistry and the structure of hemicelluloses and pectins are available in the literature.[19]

Lignins are a group of hydrophobic polymers that contain three major units, called guaiacyl (G), syringyl (S), and *p*-hydroxyphenyl (H). The biological functions of lignin are to provide mechanical support for the cell walls and to restrict water transport through vascular tissue to ensure efficient water conduction in the plant. Lignins may also play a role in fighting off pathogen attack,[20] which is generally considered to be one of the most important limiting factors in the enzymatic cell-wall saccharification process.[6] The primary structure of lignins remains obscure; the three monolignol units (*i.e.*, G, S, and H) are polymerized *via* radical oxidation and coupling during the cell-wall lignification process.[21] The dimerization of the two subunits occurs primarily at their β-positions forming β–β, β–O4, and β–O5 covalent bonds, and the dimer is dehydrogenated and further coupled with another radicalized monomer. The hydrophobic chemistry and heterogeneity of these phenolic compounds are in fact the barriers that hinder the penetration of biocatalysts (enzymes) from accessing their polysaccharide substrates.

2.5   Conclusion and Future Directions

The cell walls in a living plant are dynamic and highly heterogeneous in chemistry, structure, and function. The complexity of such chemical and structural heterogeneity in cell walls is somewhat overstated when translated to the field of lignocellulose research. In fact, compared with the cell-wall biosynthesis that may involve at least hundreds of enzymes and proteins, the enzymes that are required to effectively degrade (solubilize) the cell walls could be as few as three. Such an example would be an endo-glucanase, an exo-glucanase, a β-glycosidase, and a limited number of hemicellulases. The synthases are highly specific in their catalytic activities on glycosyl bonds, whereas the hydrolases are much less specific. For instance, a β-1,4-glucanase could effectively hydrolyze a β-1,4-linked glycosidic bond in any polysaccharide.

The digestibility of the cell walls in biomass by cellulolytic enzymes is strongly correlated with the cell-wall structure. In untreated biomass, the PWs are completely degradable, and there is partial and no digestibility in pSWs and sSWs, respectively. The most important difference between these three types of cell walls is lignin content; the PWs are not lignified, and the pSWs and the sSWs are partially and completely lignified, respectively. A chemical treatment that removes lignins without changing polysaccharides (*e.g.*, acid chlorite treatment at room temperature) enables complete digestion of all types of cell walls, suggesting that lignins are the important factor that hinders the enzyme accessibility to structural polysaccharides.[6]

It is now believed that the architecture of the cell-wall network plays a much more important role than the primary chemistry of these polymers with respect to biomass digestibility. A deeper understanding of primary polymer chemistry in the cell walls is important to adding industrial value to end-products produced by biomass. However, such understanding is limited in providing the knowledge needed to improve the enzymatic process of saccharification. The nano-scale microfibril networks formed by CEFs and matrix polysaccharides are highly accessible to cellulolytic enzymes, especially to the fungal-free enzymes;[6] the small physical size of these enzymes (5–10 nm) matches the porosity of the CEF network, which permits enzyme penetration inside the microfibril network and efficient digestibility. However, the enzyme accessibility to polysaccharides is hindered by the lignin layers formed between the CEF-hemicellulose lamellae in untreated biomass. Current state-of-the-art technologies for bioprocessing biomass to biofuels require a thermochemical pretreatment step. Many pretreatment approaches have been developed that aim to enhance biomass digestibility by enzymes; these approaches may affect cell-wall accessibility by directly modifying (*i.e.*, oxidizing) or delocalizing lignin. That is, the dilute acid pretreatment hydrolyzes hemicelluloses, resulting in lignin migration and aggregation at elevated temperature. Nevertheless, an optimal pretreatment method should: (1) maximize polysaccharide accessibility to enzymes; and (2) minimize modification of cell-wall architecture or sugars. Therefore, efforts to improve pretreatment should be focused on the development of chemical specificity to lignin modification at low temperature (sugars may be degraded at elevated temperature, which produces compounds that inhibit saccharification and fermentation in later steps). Effectively delocalizing lignin with chemical catalysts remains a big challenge because of the structural barriers involved in plant tissues and cell-wall architecture. One promising approach is the genetic modification of lignin biosynthesis; engineering energy plants with desirable lignin content and composition that do not affect plant growth and can be degraded with a designed chemical pretreatment upon harvest.[22–24] Finally, a systematic effort must be considered that coordinates improvement of the cocktail of cellulolytic enzymes, fermentation strain development, and product marketing.

Acknowledgements

This work was supported by the US Department of Energy (contract no. DE-AC36-08-GO28308 with the National Renewable Energy Laboratory). The authors acknowledge research support from the BioEnergy Science Center, a US Department of Energy (DOE) Bioenergy Research Center, supported by the Office of Biological and Environmental Research in the DOE Office of Science (grant no. ER65258 for instrumentation and data analysis).

References

1. M. E. Himmel, S. Y. Ding, D. K. Johnson, W. S. Adney, M. R. Nimlos, J. W. Brady and T. D. Foust, *Science*, 2007, **315**, 804.
2. A. J. Ragauskas, C. K. Williams, B. H. Davison, G. Britovsek, J. Cairney, C. A. Eckert, W. J. Frederick, J. P. Hallett, D. J. Leak, C. L. Liotta, J. R. Mielenz, R. Murphy, R. Templer and T. Tschaplinski, *Science*, 2006, **311**, 484.
3. A. Endler and S. Persson, *Mol. Plant*, 2011, **4**, 199.
4. M. S. Doblin, F. Pettolino and A. Bacic, *Funct. Plant Biol.*, 2010, **37**, 357.
5. S. Y. Ding and M. E. Himmel, *J. Agric. Food Chem.*, 2006, **54**, 597.
6. S. Y. Ding, Y. S. Liu, Y. N. Zeng, M. E. Himmel, J. O. Baker and E. A. Bayer, *Science*, 2012, **338**, 1055.
7. R. Vanholme, B. Demedts, K. Morreel, J. Ralph and W. Boerjan, *Plant Physiol.*, 2010, **153**, 895.
8. F. Chen, Y. Tobimatsu, D. Havkin-Frenkel, R. A. Dixon and J. Ralph, *Proc. Natl. Acad. Sci. U. S. A.*, 2012, **109**, 1772.
9. Y. Nishiyama, P. Langan and H. Chanzy, *J. Am. Chem. Soc.*, 2002, **124**, 9074.
10. Y. Nishiyama, J. Sugiyama, H. Chanzy and P. Langan, *J. Am. Chem. Soc.*, 2003, **125**, 14300.
11. Y. S. Liu, J. O. Baker, Y. N. Zeng, M. E. Himmel, T. Haas and S. Y. Ding, *J. Biol. Chem.*, 2011, **286**, 11195.
12. D. J. Dagel, Y. S. Liu, L. L. Zhong, Y. H. Luo, M. E. Himmel, Q. Xu, Y. N. Zeng, S. Y. Ding and S. Smith, *J. Phys. Chem. B*, 2011, **115**, 635.
13. Y. S. Liu, Y. N. Zeng, Y. H. Luo, Q. Xu, M. E. Himmel, S. J. Smith and S. Y. Ding, *Cellulose*, 2009, **16**, 587.
14. Q. Xu, M. P. Tucker, P. Arenkiel, X. Xi, G. Rumbles, J. Sugiyama, M. E. Himmel and S. Y. Ding, *Cellulose*, 2009, **16**, 19.
15. J. Lehtio, J. Sugiyama, M. Gustavsson, L. Fransson, M. Linder and T. T. Teeri, *Proc. Natl. Acad. Sci. U. S. A.*, 2003, **100**, 484.
16. C. You, H. Chen, S. Myung, N. Sathitsuksanoh, H. Ma, X. Z. Zhang, J. Li and Y. H. Zhang, *Proc. Natl. Acad. Sci. U. S. A.*, 2013, **110**(18), 7182–7187.
17. R. H. Atalla, J. W. Brady, J. F. Mattews, S.-Y. Ding and M. E. Himmel, in *Biomass Recalcitrance, Deconstructing the Plant Cell Wall for Bioenergy*, M. E. Himmel, Blackwell Publishing, Oxford, 2008, p. 188.

18. A. N. Fernandes, L. H. Thomas, C. M. Altaner, P. Callow, V. T. Forsyth, D. C. Apperley, C. J. Kennedy and M. C. Jarvis, *Proc. Natl. Acad. Sci. U. S. A.*, 2011, **108**, E1195.

19. D. Mohnen, M. Bar-Peled and C. Somerville, in *Biomass Recalcitrance, Deconstructing the Plant Cell Wall for Bioenergy*, M. E. Himmel, Blackwell Publishing, Oxford, 2008, p. 94.

20. T. Kawasaki, H. Koita, T. Nakatsubo, K. Hasegawa, K. Wakabayashi, H. Takahashi, K. Urnemura, T. Urnezawa and K. Shimamoto, *Proc. Natl. Acad. Sci. U. S. A.*, 2006, **103**, 230.

21. N. Terashima, M. Yoshida, J. Hafren, K. Fukushima and U. Westermark, *Holzforschung*, 2012, **66**, 907.

22. C. X. Fu, J. R. Mielenz, X. R. Xiao, Y. X. Ge, C. Y. Hamilton, M. Rodriguez, F. Chen, M. Foston, A. Ragauskas, J. Bouton, R. A. Dixon and Z. Y. Wang, *Proc. Natl. Acad. Sci. U. S. A.*, 2011, **108**, 3803.

23. F. Chen and R. A. Dixon, *Nat. Biotechnol.*, 2007, **25**, 759.

24. F. Chen, Y. Tobimatsu, D. Havkin-Frenkel, R. A. Dixon and J. Ralph, *Proc. Natl. Acad. Sci. U. S. A.*, 2012, **109**, 1772.

CHAPTER 3

Advances in the Measurement and Characterization of Biomass Structure and Processing

YU-SAN LIU[a,b] AND SHI-YOU DING*[a,b]

[a] Biosciences Center, National Renewable Energy Laboratory, Golden, CO 80401, USA; [b] BioEnergy Science Center, Oak Ridge National Laboratory, Oak Ridge, TN 37831, USA
*Email: shi.you.ding@nrel.gov

3.1 Introduction

The recalcitrance of lignocellulosic plant biomass makes the cost of biofuel higher than that of fossil fuel. Plant cell wall has a highly complex macro-molecular structure mainly consisting of crystalline cellulose that is embedded in an amorphous matrix of cross-linked lignin and hemicelluloses. To break down the lignin–hemicellulose barrier, thermochemical and/or biological pre-treatment is usually applied. Therefore, to reduce the cost of the lignocellulose-to-ethanol process, we need to control the cost of feedstock, thermochemical pretreatment, enzymes, and fermentation. In contrast to cornstarch for human food, non-food plant biomass feedstock includes forestry waste and grasses, as well as agricultural residues (corn stover). Deconstructing the plant cell wall alters the cell-wall structure making it more permeable to enzymes, and therefore understanding the ultra-structure of the plant cell wall and how it changes during the pretreatment process, provides insight into fundamental

RSC Energy and Environment Series No. 10
Biological Conversion of Biomass for Fuels and Chemicals: Explorations from Natural Utilization Systems
Edited by Jianzhong Sun, Shi-You Ding and Joy Doran-Peterson
© The Royal Society of Chemistry 2014
Published by the Royal Society of Chemistry, www.rsc.org

mechanisms that contribute to the native recalcitrance of cell walls. In this chapter, we review the various microscopy and spectroscopy tools used to characterize plant-cell-wall structure, including optical microscopy, atomic force microscopy, and electron microscopy. We start with the principle of the instruments followed by their application in biomass conversion. The instruments can also be combined with one other to make them more powerful and able to characterize more aspects of biomass.

3.2 Atomic Force Microscopy

Atomic Force Microscopy (AFM) is a type of scanning probe microscopy with unprecedented atomic-level resolution. It can image a sample's surface providing a three-dimensional (3D) profile by measuring the force between the tip and sample, which follows Hooke's Law. The sharp tip with radius of a few nm is supported on the end of a flexible cantilever. The tip is used to scan across the sample surface by 'gently' touching it, and thus providing sample morphology information. AFM images not only represent the topography of the sample surface at very high magnification, but also measure the attractive and repulsive forces between the scanning probe tip and the sample surface, thus providing both height and phase images. For biological applications, AFM has the capability of operating in an aqueous environment and enables the observation of dynamic molecular events in real time and under real physiological conditions without the need for complex sample preparation procedures (*e.g.*, staining, freezing, chemical extraction, dehydration or embedding), which is essential for conserving native biomass structure. To probe plant tissues, "tapping mode" (or "intermittent contact mode") is usually used to prevent damage to these soft materials. Unlike contact mode, the probe does not contact the sample surface, but oscillates up and down, near its resonance frequency, above the adsorbed fluid layer on the surface during scanning.

In biomass conversion, AFM provides not only plant-cell-wall architecture (primary plant cell wall, secondary plant cell wall, and fibrils) but can also trace the interaction dynamics of biocatalysts and substrates. Several different cellulose and plant-cell-wall microfibrils have been characterized, such as *Valonia* (primarily I_α crystals ~ 20 nm in diameter), tunicate (primarily in the I_β form), bacterial microcrystalline cellulose (ribbon-like cellulose microfibrils), maize parenchyma cell wall (cellulose in the form of small microfibrils 3–5 nm in diameter), Avicel (microfibrils shown as small crystals or bundles ranging from 100 to 200 nm in length, and 5 to 10 nm in diameter) and phosphoric acid treated cellulose (amorphous).[1–3] The detailed structure of the primary cell wall of maize parenchyma and the cellulose synthesis process have also been studied by AFM: well-shaped pits, 2 μm wide and 150–200 nm deep, are found in the cell wall; parallel-microfibrils are embedded in the wall matrices forming cell wall lamella; and macrofibrils were found to exist only in the uppermost layer of the native primary cell wall and appeared to be bundles of elementary fibrils.[4]

Real-time AFM images were acquired to study *T. reesei* cellobiohydrolase I (CBH I) hydrolyze crystalline cellulose (*Valonia*) over the course of 11 hours. CBH I is bound to, and moves on the cellulose hydrophobic faces. The authors observed cellulose structural changes over time: a decrease in width; an increase in surface roughness; and a change in the shape of the fiber ends.[5] High-speed AFM (1–4 frames per second) enabled the study of the sliding movement of *Tr*Cel7A on crystalline cellulose (*Cladophora* sp.), and an average velocity of 3.5 nm s^{-1} was measured.[6]

3.3 Fluorescence Microscopy

Fluorescence microscopy is a type of optical microscopy. The sample is illuminated with light of a short wavelength (high energy) which causes fluorescence. The light emitted by fluorescence is of a longer wavelength (lower energy) than the illumination, and is then detected through a microscope objective and an image is recorded using a CCD camera. The basic system components include a light source (lamp or laser), filter sets (excitation, dichroic, emission), and a detection device (camera). The filter sets are chosen to match the excitation and emission spectra of the labeled fluorophore. The sample can be fluoresced in its natural form (primary fluorescence or autofluorescence) or labeled with fluorophores (dyes, fluorescence proteins, quantum dots *etc.*). In plant biology, various types of fluorescence microscopy like epifluorescence microscopy, confocal microscopy, and total internal reflection fluorescence microscopy (TIRF) are used widely.

Auto-fluorescence is an intrinsic optical property of biological materials, where the materials fluoresce in a similar manner to the fluorophore dyes or markers used to label biological objects of interest. The intrinsic auto-fluorescence of biomass is mainly from lignin and phenolic compounds. Lignin is a complex compound with strong auto-fluorescence in the visible as well as far-infrared regions.[7] The broad emission spectrum of lignin contains contributions from its three monolignols: *p*-hydroxyphenyl (H), guaiacyl (G), and syringal (S). Auto-fluorescence offers a direct, nearly native method to visualize the distribution of lignin in the cell wall without the need of labor-intensive and time-consuming chemical and immunological labeling. Thus, auto-fluorescence imaging provides insight into lignin location and its relative concentration (intensity). Greater lignin accumulation in the middle lamella and cell corners has been observed, and lignin-rich sclerenchyma cells are near the epidermis.[8] In addition, dynamic tracking of lignin solubilization during an ionic liquid pretreatment of switchgrass has been reported.[8] The ionic liquid, EmimAc (1-*n*-ethyl-3-methylimidazolium acetate), is an effective biomass pretreatment method to dissolve lignocellulose and later help hydrolyze the resulting liquor into sugars. After 2 h of EmimAc ionic liquid treatment, the organized switchgrass plant-cell-wall structure has been completely broken down. A rapid swelling of the secondary plant cell walls within 10 min at 120 °C, disrupting the inter- and intra-molecular hydrogen bonding between cellulose fibrils and lignin was observed. The swelling was followed by complete dissolution of the biomass over 3 h.[8]

In addition to the presence of auto-fluorescence, labeling cell-type-specific cell-wall polymers can elucidate the sophisticated polymer configurations. Due to the complex polymers located in the plant cell wall, understanding the spatial distribution of cell-wall components is challenging. The molecular probes known as "carbohydrate-binding modules" (CBMs) and monoclonal antibodies can be used to recognize cell-wall macromolecules.

3.3.1 Carbohydrate-Binding Modules

CBMs are non-catalytic proteins found in many glycoside hydrolases. They serve as attachment devices for glycoside hydrolases and their target substrates and therefore are usually used as polysaccharide-specific probes. CBMs are classified into three types according to the functions. Type A CBMs, or "surface-binding" CBMs, comprise a planar hydrophobic surface, which binds specifically to insoluble crystalline cellulose. Type B CBMs, or "glycan-chain-binding" CBMs, bind to single-chain polysaccharides. Type C CBMs, or "small-sugar-binding" CBMs specifically bind to the ends of polysaccharides or oligosaccharides.[9] Researchers have identified CBMs that specifically bind to crystalline cellulose,[10–14] xylan,[15] and amorphous cellulose.[9,16]

To make CBMs visible under fluorescence microscopy, fluorescent tags need to be incorporated into them. Using fluorescent quantum dots (QDs) and fluorescent proteins (FPs) to label CBMs has been reported.[13,14,17] NAC/histidine-capped CdSe/ZnS QDs have been bound to a CBM (*Ac*CBM2 or *Ct*CBM3) fusion protein with dual hexahistidine-tags at its N- and C- termini. The CBM–His probes bound to crystalline cellulose, *Valonia*, were examined by TEM and fluorescence microscopy.[13,14] Three different types CBM (*Ct*CBM3, *Ct*CBM6, *Sl*CBM20) each exhibiting different carbohydrate specificities were each fused with either green fluorescent protein (GFP) or red fluorescent protein (RFP) and employed for dual-labeling fluorescence microscopy studies of primary cell walls and various carbohydrate target molecules. *Ct*CBM3 and *Sl*CBM20 were specifically bound to crystalline cellulose and starch, respectively. *Ct*CBM6 had a strong interaction with xylan and xylodextrins and a weak interaction with cellodextrins, non-crystalline cellulose, and xyloglucan.[17] The fluorescent CBM probes were used for characterizing both native complex carbohydrates and pretreated biomass.

3.3.2 Monoclonal Antibodies

Immunofluorescence monoclonal antibodies (Abs) against polysaccharides have been applied to investigate polymer distribution in plants by confocal microscopy.[18–20] Two rat monoclonal antibodies, LM10 (rat immunoglobulin class IgG2c) and LM11 (class IgM), bind to transverse sections of *Silene* and tobacco stems, flax hypocotyls, and wheat grain. A study showed that LM10 is specific to non- or low-substituted xylans, whereas LM11 binds to wheat arabinoxylan and unsubstituted xylans indicating the presence of both epitopes in the secondary cell walls of xylem, but differences in occurrence in other cell types.[19] In another study, the rat monoclonal Abs JIM7, JIM5, LM5 and LM6 and the mouse monoclonal Ab 2F4 were used to map pectic motif distribution in sugarbeet root cell walls. Control sections were treated in parallel but without the primary Abs. All cell walls remained unlabeled when the primary Abs were omitted from control sections. The results showed that pectic polysaccharides are spatially regulated in sugarbeet root cell walls and the spatial patterns vary between cell types implying that structural variants of pectic polymers are involved in the modulation of cell-wall properties.[18] In addition to fluorescent dyes, immunolabeling techniques with Abs can also be applied for conjugation with gold (immunogold labeling) for TEM characterization.[18,20]

3.4   Fourier Transform Infrared Microspectroscopy

Infrared spectroscopy is a technology that is used to detect molecular vibrational energy (or frequency) which lies in the infrared (IR) region. Infrared light is passed through molecules and the intensity of the transmitted light is measured at each frequency. When molecules absorb IR radiation, transitions occur from a ground vibrational state to an excited vibrational state. In order for a molecule to be IR active there must be a change in dipole moment as a result of the vibration that occurs when IR radiation is absorbed, and hence asymmetric vibrations cause the most intense IR absorption. Fourier transform is a data-processing technique that turns raw data (interferogram) into the desired result (the sample's spectrum): light output as a function of IR wavelength (or equivalently, wavenumber). FTIR spectra of pure compounds are generally so unique that they are like a molecular 'fingerprint'. The most useful fingerprint region lies from 4000 to 400 cm^{-1}. FTIR spectroscopy can be upgraded to FTIR microspectroscopy when equipped with a microscope. This enables the visualization of the sample, the choice of a specific region for analysis, and two-dimensional (2D) data acquisition by raster scanning the sample simultaneously. IR spectra are acquired at each pixel of 1D or 2D maps, and chemical maps can thus be derived. FTIR microspectroscopy can be used to directly characterize polysaccharide and aromatic molecules in the cell wall with minimal sample preparation.[21,22] It can also map the chemical distribution in plant cells and thus reveal intrinsic heterogeneities in a single cell wall or in many different cell-wall types within a tissue.[15,21] FTIR microspectroscopy can be more powerful when integrated with the attenuated total reflectance (ATR) technique, synchrotron sources, a focal plane array (FPA) IR detector, or chemometric analysis.

3.4.1   Attenuated Total Reflectance Fourier Transform Infrared Microspectroscopy

ATR uses the property of total internal reflection resulting in an evanescent wave. An IR beam is passed through an ATR crystal with a high refractive index, such that at a certain angle the light undergoes multiple internal reflections in contact with the sample. This reflection creates the evanescent wave which extends into the sample. The penetration depth into the sample is typically between 0.5 and 2 µm, and therefore, there must be direct contact between the sample and the ATR crystal surface. The ATR crystals must have refractive index values greater than that of the sample, typically, between 2.38 and 4.01 at 2000 cm^{-1}. Crystals made of zinc selenide (ZnSe), germanium (Ge), and diamond are often used in ATR-FTIR to study plant materials due to faster sampling, improved reproducibility and reliability, and less impedance by water absorption.

In biomass-conversion studies, ATR-FTIR has been applied to study switchgrass pretreated by an ionic liquid and dilute acid.[23] The ratio of two peaks 1510 : 900 cm^{-1} representing the lignin-to-cellulose ratio was compared between the ionic liquid and dilute acid pretreatment methods. The ratio decreased from 0.49 (untreated switchgrass) to 0.13 (ionic liquid pretreatment), but increased to 0.68 for dilute acid pretreatment indicating that ionic liquid pretreatment is a promising alternative to dilute acid pretreatment for switchgrass.

ATR-FTIR has also been used to measure crystallinity changes in cellulose I polymers.[24] After normalization the obtained spectra of the absorbance of the O – H in-plane deformation band at 1336 cm^{-1}, the lateral order index (LOI, α 1429/893) and hydrogen bond intensity (HBI, α 3336/1336) can be used to interpret qualitative changes in cellulose crystallinity. Kljun *et al.*[24] used ATR-FTIR and X-ray diffraction (XRD) to observe crystallinity changes in cotton fibers after sodium hydroxide (NaOH) treatment, and compared the results with CBM probes. They found close agreement with changes in crystallinity observed with ATR-FTIR techniques compared to CBM probes when NaOH concentrations greater than 2.0 mol dm^{-3} were used.

3.4.2 Synchrotron-Radiation-Based Fourier Transform Infrared Microspectroscopy

Synchrotron-radiation-based FTIR (SR-FTIR) microspectroscopy uses synchrotron emission as its IR source, which is 100–1000 times brighter than a standard globar (conventional thermal IR source). Due to the greater brightness of synchrotron IR light, it enables the beam size to be reduced below 10 µm without a significant loss of photons, and thus allows a very small area to be explored providing higher accuracy and precision. The technique allows faster data collection, reaches the diffraction limit, and provides a very good signal-to-noise (S/N) ratio with high ultra-spatial resolution.[25–27] The main drawbacks are the requirement for a bright synchrotron beam and special instrumentation, and the difficulties in sample preparation for some feed materials. Using the SR-FTIR technique, the intensities and distributions of the biological components (such as lignin, proteins, lipids, structural and non-structural carbohydrates and their ratios) in the microstructure of plant tissue within cellular dimensions could be imaged.[28–33]

3.5 Raman Spectroscopy

Raman spectroscopy is based on inelastic scattering (Raman scattering) of a photon that interacts with molecular vibrations, resulting in an energy shift (Raman shift) of the photon. Raman shift provides information about vibrational, rotational and other low-frequency transitions in molecules. To be Raman active, the polarizability of the molecule must change with the vibrational motion. Thus, Raman spectroscopy complements IR spectroscopy. A Raman microspectroscopy system typically consists of an excitation source (laser), a spectrophotometer, a CCD detector, and a confocal microscope. It is widely used because it offers label-free chemical contrast with high resolution and chemical specificity,[34–37] and can be used in solid, liquid, and gas samples. Several potential bioenergy crop candidates have been characterized by Raman microspectroscopy including black spruce wood,[34] poplar wood,[36] flax stem,[38] *Arabidopsis thaliana* and *Poplus trichocarpa* stem wood,[37] and corn stover and *Eucalyptus globulus*.[39] Raman signals are normally weak, and needs long acquisition time (0.1–1 s per pixel). To overcome these drawbacks, many different ways of sample preparation, sample illumination and scattered light detection have been invented.

3.5.1 Coherent Anti-Stokes Raman Scattering Microscopy

Coherent anti-Stokes Raman scattering (CARS) microscopy produces a signal in which the emitted waves are coherent with one another. As a result, a CARS signal is orders of magnitude greater than that from spontaneous Raman scattering. Unlike Raman spectroscopy using a single continuous wave (CW) laser, CARS generally requires two pulsed laser sources to focus on the sample, a pump beam with frequency ω_p, and a Stokes beam with frequency ω_S. Two laser beams are temporally and spatially overlapped to generate anti-Stokes signals at frequency $\omega_{aS} = 2\omega_p - \omega_S$ such that the frequency difference $(\omega_p - \omega_S)$ is tuned to match a particular Raman active molecular vibration frequency.[40,41] CARS microscopy provides a contrast mechanism based on molecular vibrations, which are intrinsic to the samples, as well as high spatiotemporal resolution, and is free of background noise from one-photon-excited fluorescence.[42] It also offers much greater sensitivity and faster acquisition times than traditional Raman spectroscopy.[43]

CARS imaging of lignin distribution in the cell wall has been demonstrated in the lignocellulosic biomass alfalfa[44] and corn stover.[45] The $1600\,\text{cm}^{-1}$ Raman mode was used to represent the lignin signal in wild-type (WT) and two lignin-down-regulated alfalfa lines: shikimate hydroxycinnamoyl transferase (HCT) and coumaroyl shikimate 3-hydroxylase (C3H) for CARS imaging. A reduction of the CARS signal in the cell corners and the compound middle lamella (the middle lamella + primary walls) areas was observed.[44]

3.5.2 Stimulated Raman Scattering Microscopy

In stimulated Raman scattering (SRS) microscopy, the pump beam with frequency ω_p and the Stokes beam with frequency ω_S are incident on the sample. If the frequency difference ($\omega_p - \omega_S$) matches a molecular vibration, stimulated excitation of vibration transitions occurs. The intensity of the pump beam is annihilated (stimulated Raman loss: SRL) and the Stokes beam is created (stimulated Raman gain: SRG). SRS cannot occur when the frequency difference ($\omega_p - \omega_S$) does not match any vibrational resonance that absorbs the difference in energy from the beams. Thus, SRS does not have a non-resonant background signal. The major advantages of SRS are identical spectral response to spontaneous Raman scattering but with orders of magnitude larger signal which is linearly dependent on the analyte concentration, fast image acquisition (50 μs per pixel) and a non-resonant background.[46] SRS thus is an ideal technique for investigating biomass-conversion processes since it has the capability of real-time monitoring, it is label-free and non-invasive, and has high spatial resolution.

The real-time image delignification process by acid chlorite in corn stover has been demonstrated.[47] The Raman peaks at $1600\,\text{cm}^{-1}$ (aromatic stretching of lignin) and $1100\,\text{cm}^{-1}$ (C–C and C–O stretching) were used to represent lignin and cellulose respectively for SRS imaging. After applying acid chlorite (0.1 M HCl (aq) and 10% $NaClO_2$) for 53 min, the lignin signal reduced significantly (more than 8×) whereas the cellulose signals remained at the same level.[47] This real-time SRS imaging technique demonstrates the possibility of studying enzymatic breakdown cellulose processes or the *in vivo* imaging of lignification and cellulose biosynthesis in plant cell wall development.

3.6   Transmission Electron Microscopy

In transmission electron microscopy (TEM) a high-energy electron beam emitted by an electron gun is transmitted through a very thin sample in a vacuum to image and determine the internal structure of materials. The electrons are focused with electromagnetic lenses and the image is observed on a fluorescent screen, or recorded on photographic film or by digital camera. Because the wavelength of electrons is much smaller than that of light, the resolution for TEM images is a 1000× better than those obtained with a light microscope. Thus, TEM can reveal the finest details of internal structure at atomic resolution. In order to allow electrons to pass through the sample, the sample needs to be ultra-thin (ideally 100–300 nm), and hence requires sophisticated preparation techniques, such as chemical fixation, shadowing or staining with high-contrast compounds, resin embedding, or microtoming. However, these preparation procedures are usually considered to alter the native sample structure. Compositional analysis of a material can be obtained when TEM is equipped with energy-dispersive spectroscopy (EDS). Images obtained from TEM are 2D projections of a 3D volume of the material under study, but for structural biology, cryo- or cold-stage transmission electron microscopy (cryo-TEM) can resolve of multi-subunit proteins and clusters and tomography provides reconstruction of complex images in a 3D profile, and in their native state.

A recent study visualized corn stover cell-wall pretreated with ammonia fiber expansion (AFEX) by 3D electron tomography.[48] Nanoporous tunnel-like networks that enhance enzyme accessibility are embedded in cellulosic microfibrils. The estimated exposed pore surface area was between 0.005 and $0.05\,nm^2$ per nm^3 which resulted in AFEX-pretreated cell wall enhancing the enzymatic hydrolysis yield by 4–5 fold over untreated cell walls.

3.7   Scanning Electron Microscopy

In scanning electron microscopy (SEM) electrons are focused into a small beam by a series of electromagnetic lenses in the SEM column under vacuum, and scan the sample in a raster scan pattern. When the electrons hit the sample, secondary electrons are produced by inelastic scattering and backscattered electrons are produced by elastic scattering. These electrons are collected by a secondary detector or a backscatter detector, converted to a voltage, and amplified. A 2D distribution of the voltage signal is displayed on the monitor, which is a representation of the topography of the sample. SEM has a larger depth of field than light microscopy, which allows a large amount of the sample to be in focus at one time, and thus can produce the characteristic 3D appearance useful for understanding the surface structure of a sample. SEM also produces images of high resolution (less than 1 nm), which means that closely spaced features can be examined at a high magnification. Sample preparation is easier than for TEM since most SEMs only require the sample to be electrically conductive. Similar to TEM, qualitative and quantitative chemical analysis information can be obtained by using EDS.

SEM has been widely used to study the structure of biomass which usually coated with a thin layer of electron-dense material, such as carbon or atomized gold. A cross-section of a native maize stem tissue showed vascular bundles and pith tissues, as well as diverse cell sizes, shapes, and cell walls.[1] After thermochemical pretreatment of corn stover stalks, lignin coalescence and a range of lignin droplets are formed in the xylem and sclerenchyma cells.[49] A recent SEM study showed significant structural differences between corn leaves, stalk pith, and stalk rind before and after liquid hot water pretreatment and correlated the changes to sugar yields from enzymatic hydrolysis of the corn fractions.[50]

References

1. M. E. Himmel, S. Y. Ding, D. K. Johnson, W. S. Adney, M. R. Nimlos, J. W. Brady and T. D. Foust, *Science*, 2007, **315**, 804.
2. J. M. Yarbrough, M. E. Himmel and S. Y. Ding, *Biotechnol. Biofuels*, 2009, **2**, 11.
3. Y. H. P. Zhang, S. Y. Ding, J. R. Mielenz, J. B. Cui, R. T. Elander, M. Laser, M. E. Himmel, J. R. McMillan and L. R. Lynd, *Biotechnol. Bioeng.*, 2007, **97**, 214.
4. S. Y. Ding and M. E. Himmel, *J. Agric. Food Chem.*, 2006, **54**, 597.

5. Y. S. Liu, J. O. Baker, Y. N. Zeng, M. E. Himmel, T. Haas and S. Y. Ding, *J. Biol. Chem.*, 2011, **286**, 11195.
6. K. Igarashi, A. Koivula, M. Wada, S. Kimura, M. Penttila and M. Samejima, *J. Biol. Chem.*, 2009, **284**, 36186.
7. V. De Micco and G. Aronne, *Biotech. Histochem.*, 2007, **82**, 209.
8. S. Singh, B. A. Simmons and K. P. Vogel, *Biotechnol. Bioeng.*, 2009, **104**, 68.
9. A. B. Boraston, P. Chiu, R. A. J. Warren and D. G. Kilburn, *Biochemistry*, 2000, **39**, 11129.
10. J. Lehtio, J. Sugiyama, M. Gustavsson, L. Fransson, M. Linder and T. T. Teeri, *Proc. Natl. Acad. Sci. U. S. A.*, 2003, **100**, 484.
11. D. N. Bolam, A. Ciruela, S. McQueen-Mason, P. Simpson, M. P. Williamson, J. E. Rixon, A. Boraston, G. P. Hazlewood and H. J. Gilbert, *Biochem. J.*, 1998, **331**, 775.
12. J. Tormo, R. Lamed, A. J. Chirino, E. Morag, E. A. Bayer, Y. Shoham and T. A. Steitz, *EMBO J.*, 1996, **15**, 5739.
13. Y. S. Liu, Y. N. Zeng, Y. H. Luo, Q. Xu, M. E. Himmel, S. J. Smith and S. Y. Ding, *Cellulose*, 2009, **16**, 587.
14. Q. Xu, M. P. Tucker, P. Arenkiel, X. Ai, G. Rumbles, J. Sugiyama, M. E. Himmel and S. Y. Ding, *Cellulose*, 2009, **16**, 19.
15. L. McCartney, A. W. Blake, J. Flint, D. N. Bolam, A. B. Boraston, H. J. Gilbert and J. P. Knox, *Proc. Natl. Acad. Sci. U. S. A.*, 2006, **103**, 4765.
16. P. Tomme, A. L. Creagh, D. G. Kilburn and C. A. Haynes, *Biochemistry*, 1996, **35**, 13885.
17. S. Y. Ding, Q. Xu, M. K. Ali, J. O. Baker, E. A. Bayer, Y. Barak, R. Lamed and J. Sugiyama, G. Rumbles and M. E. Himmel, *Biotechniques*, 2006, **41**, 435.
18. F. Guillemin, F. Guillon, E. Bonnin, M. F. Devaux, T. Chevalier, J. P. Knox, F. Liners and J. F. Thibault, *Planta*, 2005, **222**, 355.
19. L. McCartney, S. E. Marcus and J. P. Knox, *J. Histochem. Cytochem.*, 2005, **53**, 543.
20. M. Eder, R. Tenhaken, A. Driouich and U. Lutz-Meindl, *J. Phycol.*, 2008, **44**, 1221.
21. N. C. Carpita, M. Defernez, K. Findlay, B. Wells, D. A. Shoue, G. Catchpole, R. H. Wilson and M. C. McCann, *Plant Physiol.*, 2001, **127**, 551.
22. M. C. McCann, M. Bush, D. Milioni, P. Sado, N. J. Stacey, G. Catchpole, M. Defernez, N. C. Carpita, H. Hofte, P. Ulvskov, R. H. Wilson and K. Roberts, *Phytochemistry*, 2001, **57**, 811.
23. C. L. Li, B. Knierim, C. Manisseri, R. Arora, H. V. Scheller, M. Auer, K. P. Vogel, B. A. Simmons and S. Singh, *Bioresour. Technol.*, 2010, **101**, 4900.
24. A. Kljun, T. A. S. Benians, F. Goubet, F. Meulewaeter, J. P. Knox and R. S. Blackburn, *Biomacromolecules*, 2011, **12**, 4121.
25. H. Y. N. Holman, K. A. Bjornstad, M. P. McNamara, M. C. Martin, W. R. McKinney and E. A. Blakely, *J. Biomed. Opt.*, 2002, **7**, 417.

26. N. S. Marinkovic, R. Huang, P. Bromberg, M. Sullivan, J. Toomey, L. M. Miller, E. Sperber, S. Moshe, K. W. Jones, E. Chouparova, S. Lappi, S. Franzen and M. R. Chance, *J. Synchrotron Radiat.*, 2002, **9**, 189.
27. L. M. Miller, P. Dumas, N. Jamin, J. L. Teillaud, J. Miklossy and L. Forro, *Rev. Sci. Instrum.*, 2002, **73**, 1357.
28. C. I. Lacayo, A. J. Malkin, H. Y. N. Holman, L. A. Chen, S. Y. Ding, M. S. Hwang and M. P. Thelen, *Plant Physiol.*, 2010, **154**, 121.
29. J. M. Estevez, P. V. Fernandez, L. Kasulin, P. Dupree and M. Ciancia, *Glycobiology*, 2009, **19**, 212.
30. P. Yu, *Br. J. Nutr.*, 2004, **92**, 869.
31. P. Yu, D. A. Christensen, C. R. Christensen, M. D. Drew, B. G. Rossnagel and J. J. McKinnon, *Can. J. Anim. Sci.*, 2004, **84**, 523.
32. P. Q. Yu, J. J. McKinnon, C. R. Christensen, D. A. Christensen, N. S. Marinkovic and L. M. Miller, *J. Agric. Food Chem.*, 2003, **51**, 6062.
33. K. M. Dokken and L. C. Davis, *J. Agric. Food Chem.*, 2007, **55**, 10517.
34. U. P. Agarwal, *Planta*, 2006, **224**, 1141.
35. N. Gierlinger, L. Sapei and O. Paris, *Planta*, 2008, **227**, 969.
36. N. Gierlinger and M. Schwanninger, *Plant Physiol.*, 2006, **140**, 1246.
37. M. Schmidt, A. M. Schwartzberg, P. N. Perera, A. Weber-Bargioni, A. Carroll, P. Sarkar, E. Bosneaga, J. J. Urban, J. Song, M. Y. Balakshin, E. A. Capanema, M. Auer, P. D. Adams, V. L. Chiang and P. J. Schuck, *Planta*, 2009, **230**, 589.
38. D. S. Himmelsbach, S. Khahili and D. E. Akin, *Vib. Spectrosc.*, 1999, **19**, 361.
39. L. Sun, B. A. Simmons and S. Singh, *Biotechnol. Bioeng.*, 2011, **108**, 286.
40. W. M. Tolles, J. W. Nibler, J. R. McDonald and A. B. Harvey, *Appl. Spectrosc.*, 1977, **31**, 253.
41. A. M. Zheltikov, *J. Raman Spectrosc.*, 2000, **31**, 653.
42. J. X. Cheng and X. S. Xie, *J. Phys. Chem. B*, 2004, **108**, 827.
43. J. X. Cheng, A. Volkmer, L. D. Book and X. S. Xie, *J. Phys. Chem. B*, 2001, **105**, 1277.
44. Y. N. Zeng, B. G. Saar, M. G. Friedrich, F. Chen, Y. S. Liu, R. A. Dixon, M. E. Himmel, X. S. Xie and S. Y. Ding, *BioEnergy Res.*, 2010, **3**, 272.
45. C. L. Evans and X. S. Xie, *Annu. Rev. Anal. Chem.*, 2008, **1**, 883.
46. C. W. Freudiger, W. Min, B. G. Saar, S. Lu, G. R. Holtom, C. W. He, J. C. Tsai, J. X. Kang and X. S. Xie, *Science*, 2008, **322**, 1857.
47. B. G. Saar, Y. N. Zeng, C. W. Freudiger, Y. S. Liu, M. E. Himmel, X. S. Xie and S. Y. Ding, *Angew. Chem., Int. Ed.*, 2010, **49**, 5476.
48. S. P. S. Chundawat, B. S. Donohoe, L. D. Sousa, T. Elder, U. P. Agarwal, F. C. Lu, J. Ralph, M. E. Himmel, V. Balan and B. E. Dale, *Energy Environ. Sci.*, 2011, **4**, 973.
49. B. S. Donohoe, S. R. Decker, M. P. Tucker, M. E. Himmel and T. B. Vinzant, *Biotechnol. Bioeng.*, 2008, **101**, 913.
50. M. Zeng, E. Ximenes, M. R. Ladisch, N. S. Mosier, W. Vermerris, C.-P. Huang and D. M. Sherman, *Biotechnol. Bioeng.*, 2012, **109**, 398.

CHAPTER 4

Lignin Modification to Reduce the Recalcitrance of Biomass Processing

BASSEM B. HALLAC AND ARTHUR J. RAGAUSKAS*

Institute of Paper Science and Technology, School of Chemistry and Biochemistry, Georgia Institute of Technology, Atlanta, GA 30332, USA
*Email: arthur.ragauskas@chemistry.gatech.edu

4.1 Introduction

Finding an alternative source of energy to fossil fuels is undoubtedly one of the most important necessities and challenges that our society is currently addressing. The urgency to develop new energy-production technologies emerges from the growing global energy demand and concerns about the negative effects of increasing greenhouse gas emissions from fossil fuels.[1–4] Therefore, any new form of energy must be sustainable, renewable, environmentally friendly, and economically/commercially feasible. Lignocellulosic bioethanol is currently being explored as a substitution to fossil fuels, because such materials are readily available, avoid issues surrounding 'food or fuel', and have the potential to have a relatively small environmental impact.[5]

Producing lignocellulosic ethanol is challenging because lignocellulosic biomass is resistant to chemical and biological degradation. Lignocellulosic biomass is composed of three biopolymers: cellulose, hemicellulose, and lignin, which together form a complex and rigid structure.[1,6] To reduce biomass

RSC Energy and Environment Series No. 10
Biological Conversion of Biomass for Fuels and Chemicals: Explorations from Natural Utilization Systems
Edited by Jianzhong Sun, Shi-You Ding and Joy Doran-Peterson
© The Royal Society of Chemistry 2014
Published by the Royal Society of Chemistry, www.rsc.org

recalcitrance, a pretreatment stage is required, so that the biomass can become more accessible to enzymes. Pretreatment is considered to be the most intensive operating/operating cost component of cellulosic ethanol production, and has a significant impact on the efficiency of enzymatic hydrolysis and subsequent fermentation.[7-9] Therefore, research is heavily focused on understanding the effect of pretreatment technologies on the fundamental characteristics of lignocellulosic biomass. Improving our fundamental knowledge of pretreatment technologies will lead to significant advances in the field of sustainable low-cost cellulosic biofuel production.[7]

One way to achieve favorable overall process economics is to adapt the concept of "biorefinery", in which all components of the biomass are fully used to make a range of fuels, chemicals, materials, heat, and power.[1-3] High-value-added chemicals can be extracted from lignocellulosic materials, cellulose and hemicellulose are converted to fermentable sugars, and lignin can be utilized not only as an in-house fuel source but also for co-product applications. Lignin has been used as a polymer substitution for bio-based materials, a precursor for biofuels, and a source for making carbon fiber. From this perspective, understanding how the structure of lignin changes during pretreatment is essential in order to identify future applications for this valuable biopolymer. The focus of this chapter is to discuss the fundamental characteristics of ethanol organosolv lignin (EOL) as studied by two NMR spectroscopy techniques, and molecular weight analysis.

4.2   Lignin

Lignin is an amorphous, cross-linked, and three-dimensional phenolic polymer consisting of methoxylated phenylpropane structures.[2] A schematic representation of a proposed softwood lignin structure is depicted in Figure 4.1.[10] Lignin is considered to be the most recalcitrant biopolymer in the plant cell wall and performs three main functions. First, lignin decreases the permeability of water across the cell walls, which is an important role in the transport of water and nutrients.[11] Second, it provides rigidity and structural support to the cell wall to resist compression and bending[11] Third, lignin can protect the cell wall from microorganisms by providing resistance against the penetration of destructive enzymes through the cell wall.[11]

Lignin can be biosynthesized from three monolignols: coniferyl alcohol, sinapyl alcohol, and *p*-coumaryl alcohol.[2,13] The polymerization process is initiated by the oxidation of the monolignol phenolic hydroxyl groups. The oxidation itself has been shown to be catalyzed *via* an enzymatic route.[2,13] It is believed that both peroxidases and laccases are involved in lignin synthesis, where laccase is primarily responsible for the initial polymerization of monolignols to oligolignols, while peroxidases, on the other hand, catalyze the reactions of oligolignols leading to the extended lignin macropolymer.[10] The enzymatic dehydrogenation is initiated by an electron transfer that yields reactive monolignol species with free radicals. A monolignol with a free radical can then couple with another monolignol with a free radical to generate a dilignol.

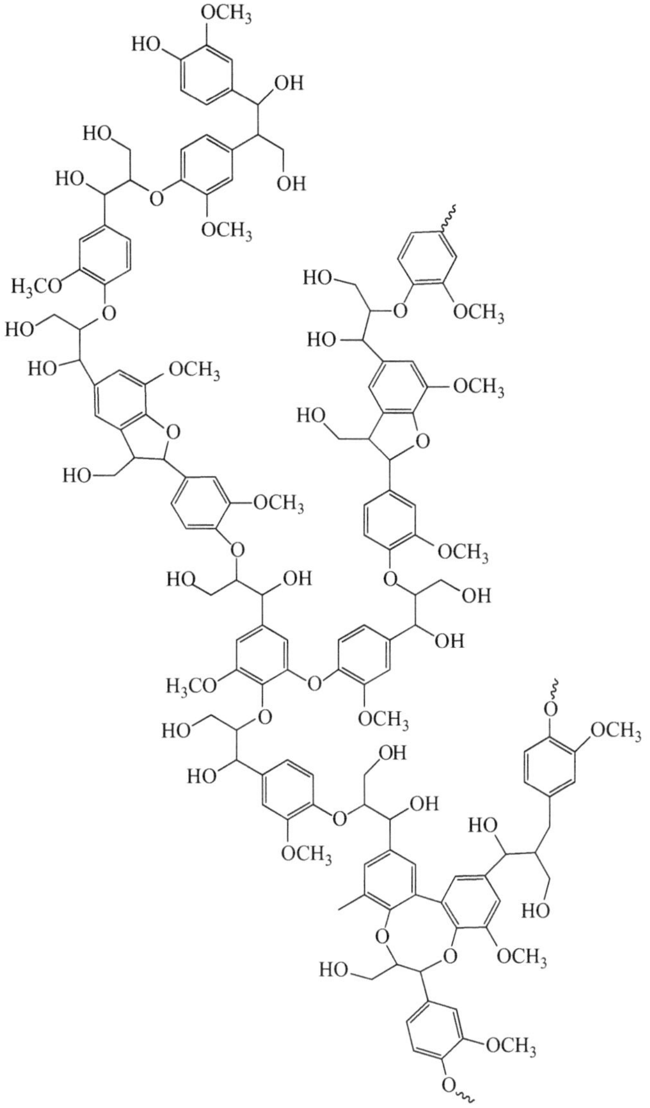

Figure 4.1 Example of the structure of native softwood lignin.[12]

Subsequent nucleophilic attack by water, alcohols, or phenolic hydroxyl groups on the benzyl carbon of the quinone methide intermediate will restore the aromaticity of the benzene ring.[2] The generated dilignols will then undergo further polymerization. Weight-averaged molecular weights ($\bar{M}_W$) of milled wood lignin (MWL) from several species are summarized in Table 4.1.[14,15]

Inter-unit linkages include: β–O–aryl ether (β–O4'), resinol (β–β'), phenylcoumaran (β–5'), biphenyl (5–5'), and 1,2-diaryl propane (β–1').[2,16] Typical proportions of lignin linkages in softwood and hardwood species are

Table 4.1 Weight-averaged molecular weights $(\bar{M}_W)$ of milled wood lignin (MWL) from several species.

Species	$\bar{M}_W$ of milled wood lignin/g mol^{-1}
Buddleja. davidii	16800
Douglas fir	7400
White fir	8300
Redwood	5900
Southern pine	14900
Eucalyptus globulus	6700
Switchgrass	5000

Table 4.2 Proportions of different types of lignin inter-unit linkages per 100 aromatic units in softwood and hardwood lignin.

Linkage type	Spruce lignin[17]	Eucalyptus grandis lignin[18]
β–O–4′	45	61
α–O–4′	16	—
β–β′	2	3
β–5′	9	3
5–5′	24–27	3
β–1′	1	2
4–O–5′	—	9
Dibenzodioxocin	7	—

Table 4.3 Molar percentages of guaiacyl (*g*), syringyl (*s*), and *p*-hydroxyphenyl (*h*) units in several biomass lignins.

Origin	g	s	h
Wheat straw[19]	45	46	9
Loblolly pine[15]	87	0	13
Spruce[17]	98	2	0
Beech[20]	56	40	4
Eucalyptus globulus[21]	14	84	2
Alamo switchgrass[14]	51	41	8

summarized in Table 4.2. Softwood lignin is composed mainly of guaiacyl units and trace numbers of *p*-hydroxyphenyl units, while hardwood lignin comprises guaiacyl (G) and syringyl (S) units with minor numbers of *p*-hydroxyphenyl (H) units. Grass lignin typically contains significant amounts of all three lignin units. Table 4.3 summarizes typical H : G : S ratios for lignin from biomass.

4.3 Ethanol Organosolv Pretreatment

Organosolv pretreatment is of interest because it has several advantages: (1) organic solvents are easy to recover by distillation and recycle for pretreatment; (2) the recovered lignin after pretreatment has desirable characteristics that can be used for several co-products; and (3) the pretreatment utilizes of all the biomass components, which makes it feasible for biorefining lignocellulosic biomass.[22,23]

In an organosolv pretreatment, the internal bonds between lignin and hemicellulose are broken, hemicelluloses and lignin are hydrolyzed by treating the biomass in an organic or aqueous–organic solvent mixture with the addition of an inorganic acid catalyst, such as H_2SO_4 or HCl.[24] Typically, solvents such as methanol, ethanol, acetone, ethylene glycol, triethylene glycol, and phenol are used in the organosolv process.[24] After pretreatment, the hydrolyzed lignin dissolves in the organophilic phase and is recovered as the filtrate (EOL), the cellulose is recovered as the solid residue, and the hemicellulose is recovered in the water-soluble fraction as monomeric and oligomeric sugars.[24,25] It has been shown that organosolv lignin has a very low carbohydrate content, as indicated by the absence of signals between δ 102–90 ppm in its ^{13}C NMR spectrum.[26,27]

4.4 Analytical Techniques used for Ethanol Organosolv Lignin Characterization

Two common NMR techniques are used to study the structure of lignin: quantitative ^{13}C and ^{31}P NMR spectroscopies; as well as gel-permeation chromatography (GPC).

4.4.1 Quantitative ^{13}C NMR Spectroscopy

^{13}C NMR spectroscopy is a powerful technique that is frequently used to determine the number of lignin inter-unit linkages and moieties, providing a comprehensive overview of the structure of the lignin macromolecule. Quantitative ^{13}C NMR analysis requires a number of conditions to be fulfilled. First, the lignin sample must be free of contaminants, such as carbohydrates and extractives. Also, the lignin–solvent solution must be made as concentrated as possible to maximize the signal-to-noise ratio and minimize baseline and phasing distortions. Typically, the solvent used is DMSO-d_6 and the sample is prepared as 80 mg of lignin dissolved in 500 μL of DMSO-d_6. Generally, the spectra are acquired at 50 °C to reduce the viscosity. A 12 s pulse delay has been used which is 5× the longest lignin–carbon T_1 relaxation time.[28] Finally, the inverse-gated decoupling sequence is applied, which involves turning off the proton decoupler during the recovery between pulses so that the nuclear Overhauser effect (NOE) is avoided.[28] An example of a quantitative ^{13}C NMR spectrum of lignin is presented in Figure 4.2.

Information from quantitative ^{13}C NMR can be gathered *via* the number of oxygenated aromatic carbons (δ 160.0–140.0), aromatic carbon–carbon (δ 140.0–123.0), aromatic methine carbons (δ 123.0–103.0), and aliphatic carbons structures (δ 90.0–58.0), as well as methoxyl content (δ 58.0–54.0).[29,30] The

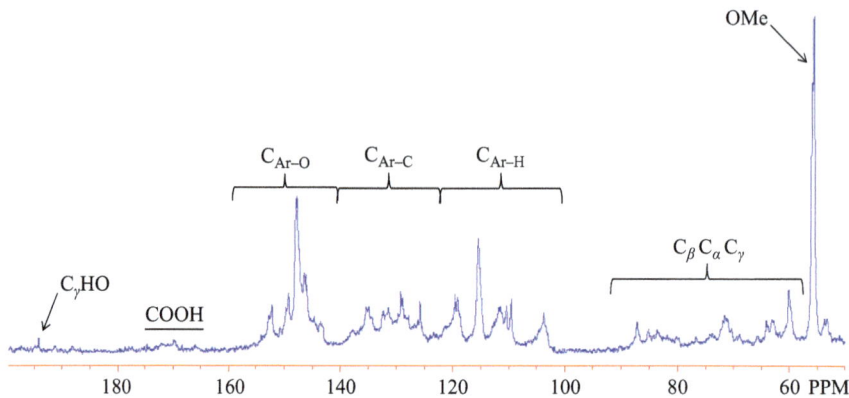

Figure 4.2   Quantitative ^{13}C NMR spectrum of an ethanol organosolv lignin.

region between δ 160 and 103 corresponds mainly to the six aromatic carbons, and is set as the reference for quantifying the lignin structures as well as the functional groups.[31] So the integral values for the structural moieties are reported per aryl group.[29] For instance, the number of β–O–aryl ether (β–O–4′) units can be estimated from the integral value between δ 61.5–58.0. The degree of condensation (DC) in lignin is determined in two ways depending on whether it is a hardwood or softwood. For softwood, the DC value is calculated as 3 minus the total area of the protonated aromatic region.[30] For hardwood, DC can be determined by subtracting the experimental value of C_{Ar-H} from the theoretical value of C_{Ar-H}.[18] The experimental value is the integral of the tertiary aromatic carbon region (δ 123.0–103.0), and the theoretical value is calculated from the following equation:[18]

$$\text{Theoretical } C_{Ar-H} = 2s + 3g + 2h$$

4.4.2   Quantitative ^{31}P NMR Spectroscopy

^{31}P NMR Spectroscopy is a technique that has been exploited to determine the number and location of hydroxyl functional groups in lignin. The method involves derivatization of lignin with the phosphorylating agent 2-chloro-4,4,5,5-tetramethyl-1,3,2-dioxaphospholane (TMDP). The reaction of TMDP with hydroxyl functional groups is illustrated in Figure 4.3. TMDP reacts with hydroxyl functional groups to give phosphate products which are resolvable by ^{31}P NMR into separate regions: aliphatic, phenolic, guaiacyl, and carboxylic acid hydroxyl groups.[32] Figure 4.4 illustrates a quantitative ^{31}P NMR spectrum of an EOL. Typically, quantitative ^{31}P NMR spectra are acquired on dry lignin ($\sim$25 mg) dissolved in a solvent mixture consisting of 1.6 : 1 (v/v) anhydrous pyridine/deuterated chloroform with chromium(III) acetylacetonate ($\sim$3.6 mg mL^{-1}) as a relaxation agent and cyclohexanol ($\sim$4.0 mg mL^{-1}) as an

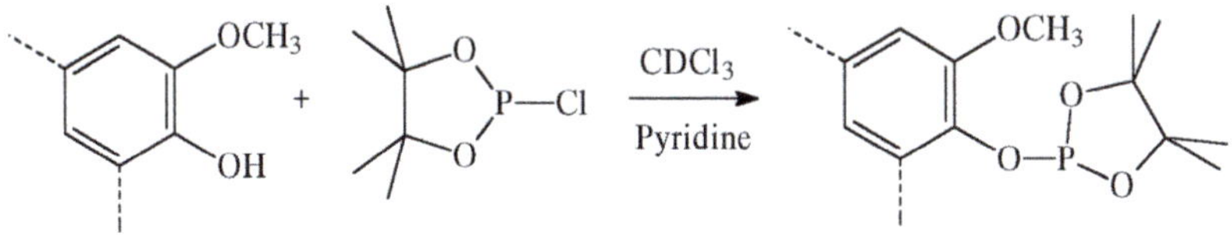

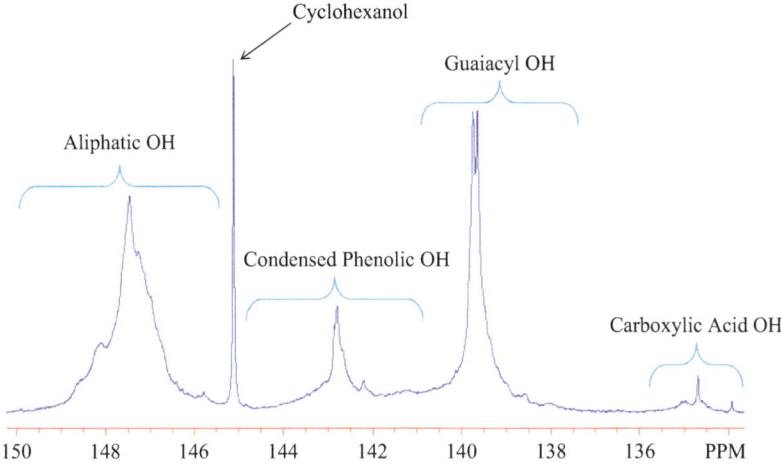

Figure 4.3 Derivatization of phenolic structures with 2-chloro-4,4,5,5-tetramethyl-1,3,2-dioxaphospholane (TMDP).

Figure 4.4 Quantitative ^{31}P NMR spectrum of an ethanol organosolv lignin.

internal standard.[32,33] The dissolved lignin is then derivatized with 100 μL of TMDP. The acquisition conditions use an inverse-gated decoupling pulse sequence and a 25 s pulse delay.

4.4.3 Gel-Permeation Chromatography

A common way to determine the molecular weight of lignin is using the gel-permeation chromatography (GPC) technique. GPC provides information on the weight-averaged molecular weight $(\bar{M}_w)$ the number-averaged molecular weight $(\bar{M}_n)$, and the polydispersity index $D = (\bar{M}_w/\bar{M}_n)$. GPC does not give the absolute molar mass because the calculations are based on the molecular weights of a set of standards which is frequently a set of well-defined polystyrene standards with varying molecular weights. A certain range of the standards are used, depending on the nature of the analysis, in order to construct a calibration curve, which is then used to obtain the molecular weight of lignin. Lignin is usually acetylated prior to analysis to allow dissolution in tetrahydrofuran (THF).[34] Typically, the lignin samples ($\sim$ 20 mg) are dissolved in a 1 : 1 acetic anhydride/pyridine mixture (1.00 mL) and kept for 24 h at room temperature. The solvent mixture is then concentrated under reduced pressure at 50 °C, dissolved in chloroform (50 mL), washed with water (3×20 mL), and dried over anhydrous MgSO$_4$. Afterwards, the mixture is filtered, and the chloroform is removed with a rotary evaporator. Acetylated lignin is dissolved in THF (1.0 mg mL^{-1}) and detected using a UV detector set at 270 nm.

4.5 The Chemistry of Ethanol Organosolv Delignification

Delignification is the chemical breakdown of the lignin macromolecule prior to dissolution[35] It has been shown that the organosolv process delignifies lignocellulosic biomass *via* cleavage of ether linkages, mainly α- and β-ether bonds.[35] McDonough has nicely summarized the different mechanism through which these bonds are cleaved during organosolv pretreatment.[35] Solvolytic cleavage *via* a quinone methide intermediate, solvolytic cleavage by a nucleophilic substitution (S$_N$2) mechanism, and cleavage through the formation of a

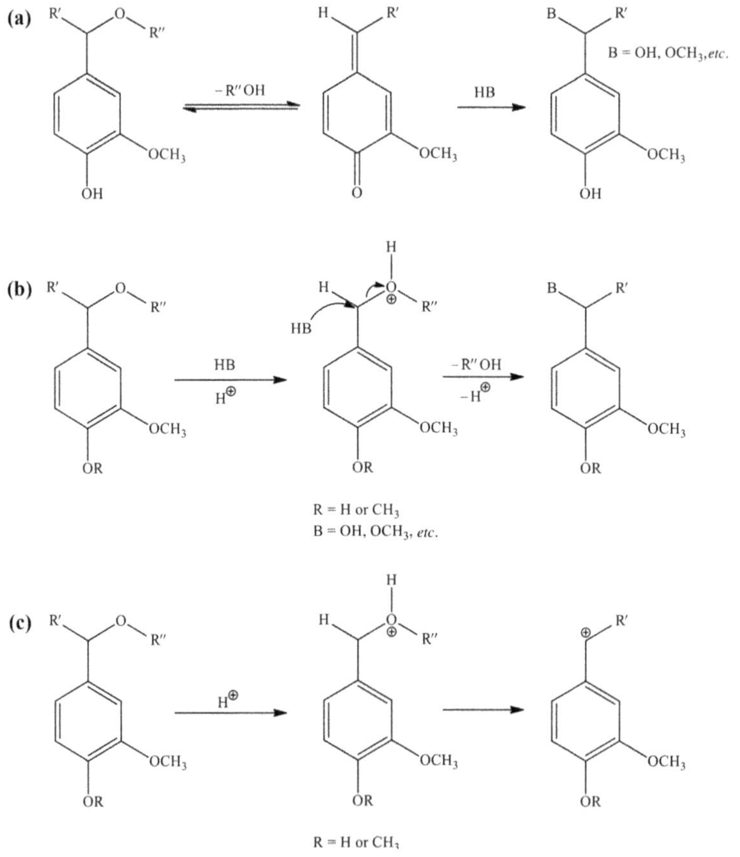

Figure 4.5 Reaction mechanisms for the cleavage of α-ether bonds in lignin. (a) Solvolytic cleavage *via* a quinone methide intermediate. (b) Solvolytic cleavage by a nucleophilic substitution (S$_N$2) mechanism. (c) Cleavage through the formation of a benzyl carbocation.

benzyl carbocation are all possible pathways for the hydrolysis of α-ether bonds (Figure 4.5).[35] As for β-ether bonds hydrolysis, the possible pathways are: cleavage through the formation of a benzyl carbocation, solvolytic cleavage to form ω-hydroxyguaiacylacetone or Hibbert's ketones (Figure 4.6), solvolytic cleavage with the elimination of formaldehyde (Figure 4.7), and homolytic cleavage (Figure 4.8).[26,35]

Different studies have measured the content of β–O–4′ inter-unit linkages before and after ethanol organosolv pretreatment, and the data showed significant scission of this bond (Table 4.4).[26,36,37] About 50 to 57% of the β–O–4′ units were cleaved during EOP (Ethanol Organosolv Pretreatment). Depending on the EOP conditions employed, the hydrolysis of β–O–4′ bonds will increase with increasing pretreatment severity (Figure 4.9).[38] Studies have differed in their proposed mechanism for β–O–4′ bond cleavage. A homolytic β–O–4′ bond cleavage was proposed for *Buddleja davidii* lignin,[26] β–O–4′ bond cleavage through the formation of a benzyl carbocation was proposed for loblolly pine

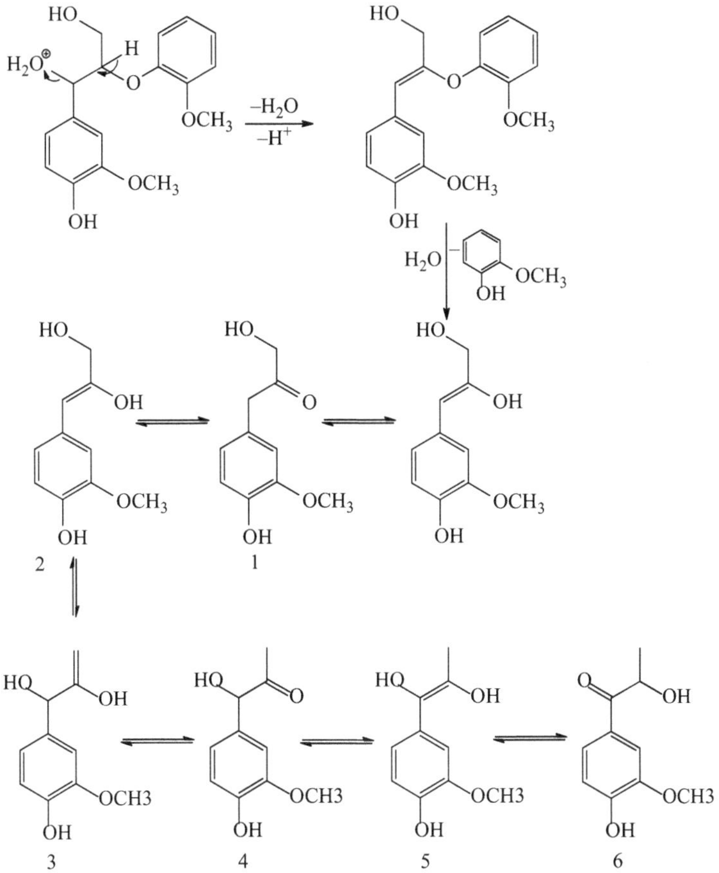

Figure 4.6　Solvolytic cleavage of the β-ether bond to form ω-hydroxyguaiacylacetone or Hibbert's ketones (compounds 1–6).

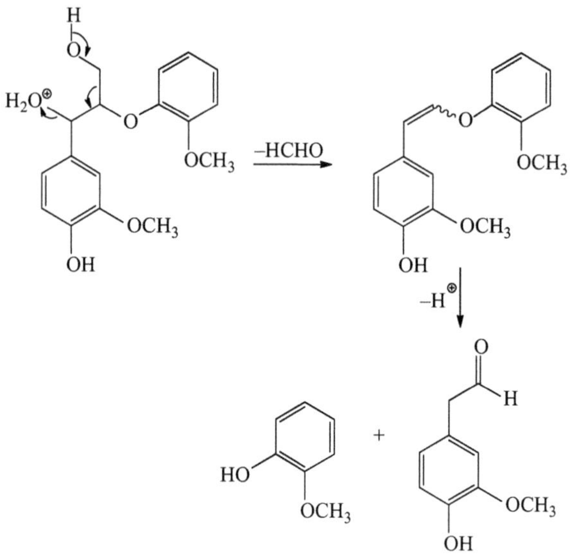

Figure 4.7 Solvolytic cleavage of the β-ether bond with the elimination of formaldehyde.

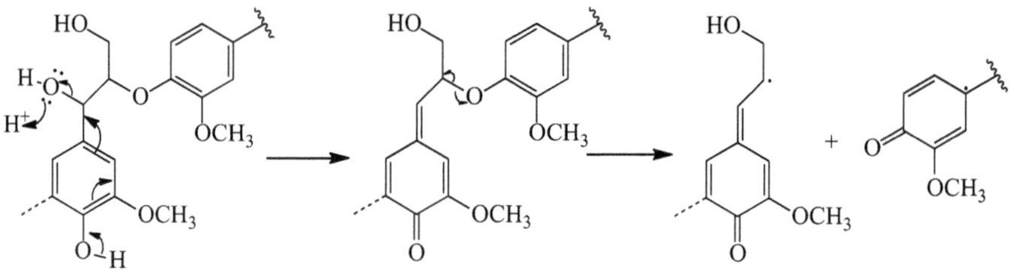

Figure 4.8 Homolytic cleavage of the β-ether bond.

lignin,[36] and solvolytic cleavage of the β–O–4′ bond to form Hibbert's ketones as well as cleavage through the formation of a benzyl carbocation were proposed for *miscanthus* lignin.[38] It appears that the degradation pathway for lignin is species-dependent and pretreatment conditions dependent.

Lignin condensation is another reaction that occurs during organosolv pretreatment. Lignin condensation is considered to be a counterproductive process because it leads to the formation of inter-molecular bonds between lignin fragments, at a time when lignin must be depolymerized to be removed from the cell wall. Condensation reactions result from the formation of the carbocation, normally located at the C_α of the side-chain, which can then possibly bind with an electron-rich carbon atom in the aromatic ring of another lignin unit (*i.e.*, through the free C5 or C6 positions).[26,35,39] Under acidic pretreatment conditions, lignin can undergo either depolymerization or repolymerization with a carbocation as the common intermediate (Figure 4.10).[39]

Table 4.4 The extent of β–O–4′ inter-unit linkage cleavage during EOP for several lignocellulosic biomass lignins.

	β–O–4′ Content (number/aryl group)	
Origin	*Untreated lignin*	*EOL*
Miscanthus[a]	0.45	0.19
Buddleja davidii[b]	0.56	0.24
Loblolly pine[c]	0.60	0.30

[a]EOP conditions: 1.2% w/w H_2SO_4 and 65% v/v ethanol at 190 °C for 60 min.
[b]EOP conditions: 1.5% w/w H_2SO_4 and 65% v/v ethanol at 195 °C for 60 min.
[c]EOP conditions: 1.1% w/w H_2SO_4 and 6% v/v ethanol at 170 °C for 60 min.

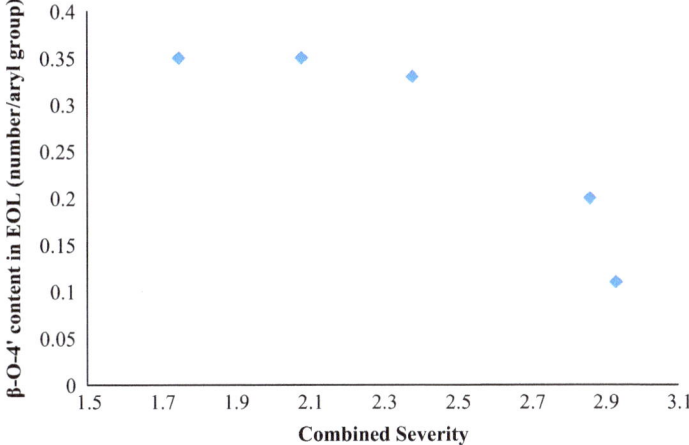

Figure 4.9 The effect of ethanol organosolv pretreatment severity on the β–O–4′ content in *Miscanthus x giganteus* EOL.

The fundamental reactions of depolymerization and repolymerization of lignin during organosolv pretreatment have been demonstrated on model compounds as well as on native wood.[26,35,39]

The degree of lignin condensation (DC) in fact increases during EOP. For instance, after EOP, the DC value for *Buddleja davidii* increased from 45% in untreated lignin to 82% in EOL,[26] and for Loblolly pine, it increased from 0.4% in untreated lignin to 1.1% in EOL.[36]

Upon cleavage of the β–O–4′ inter-unit linkage, the molecular weight of lignin decreases, forming lignin fragments that are soluble in ethanol. This is the main characteristic of EOP that facilitates delignification. Table 4.5 illustrates the effect of EOP on the molecular weight of several lignins.[26,36,37] The results clearly indicate that EOP significantly alters the molecular weight of lignin. On the other hand, the distribution of the molecular weights appears to be species-dependent and pretreatment conditions dependent. After EOP, The polydispersity indices of *Buddleja davidii* lignin increased, of *Miscanthus x giganteus* decreased, and of Loblolly pine remained constant.[26,36,37] The change

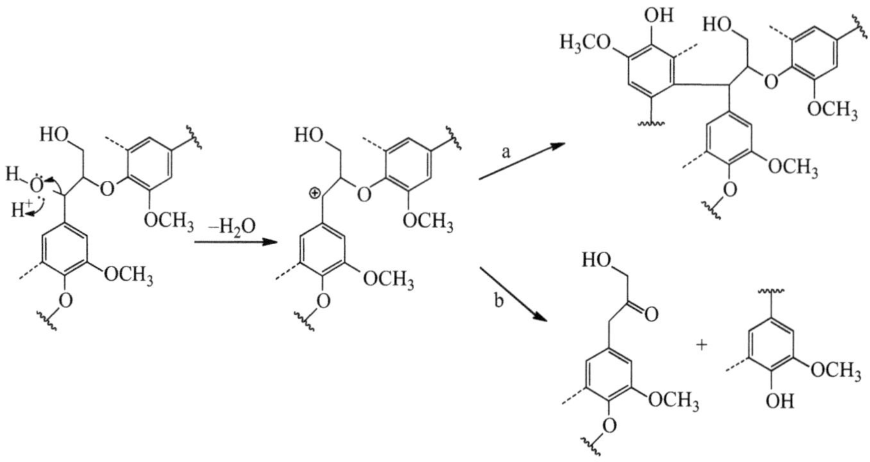

Figure 4.10 Possible reactions pathways for lignin during ethanol organosolv pretreatment. (a) Repolymerization/condensation, and (b) depolymerization.

Table 4.5 The effect of ethanol organosolv pretreatment on the weight-averaged molecular weight ($\bar{M}_{\mathrm{w}}$), the number-averaged molecular weight ($\bar{M}_{\mathrm{n}}$), and the polydispersity index $D = (\bar{M}_{\mathrm{w}}/\bar{M}_{\mathrm{n}})$ of several lignins.

Origin		$\bar{M}_{\mathrm{W}}/g\ mol^{-1}$	$\bar{M}_{\mathrm{n}}/g\ mol^{-1}$	D
Buddleja davidii	Untreated	16 800	7260	2.31
	EOL[a]	2490	645	3.86
Miscanthus	Untreated	13 700	8300	1.65
	EOL[b]	7060	4690	1.51
Loblolly pine	Untreated	13 500	7590	1.77
	EOL[c]	5410	3070	1.77

[a]EOP conditions: 1.5% w/w H_2SO_4 and 65% v/v ethanol at 195 °C for 60 min.
[b]EOP conditions: 1.2% w/w H_2SO_4 and 65% v/v ethanol at 190 °C for 60 min.
[c]EOP conditions: 1.1% w/w H_2SO_4 and 65% v/v ethanol at 170 °C for 60 min.

in polydispersity index is due to the simultaneous competition between degradation (depolymerization) and condensation (repolymerization) of lignin, which can increase, decrease or not affect the polydispersity of lignin.[39] The favored pathway would probably be dependent on the structure of lignin as well as the pretreatment conditions employed. However, this issue of depolymerization and repolymerization has not yet been resolved in the literature.

During EOP, the amount of aliphatic OH decreases, while the amount of condensed, guaiacyl, and carboxylic OH increases (Table 4.6). Due to condensation, the content of condensed phenolic OH increases. The guaiacyl OH content increases and the aliphatic OH decreases due to the cleavage of the lignin inter-unit linkages, mainly β–O–4′ units, as shown in Figure 4.11.[26] The homolytic cleavage of β–O–4′ bonds *via* the quinone methide intermediate

Table 4.6 Hydroxyl content of untreated lignin and EOL from various biomass species.

| Origin | | Hydroxyl content/mmol g^{-1} lignin | | | |
		Aliphatic	Condensed phenolic	Guaiacyl	Carboxylic
Buddleja davidii	Untreated	4.51	0.27	0.43	0.03
	EOL[a]	1.86	1.07	1.66	0.15
Miscanthus	Untreated	4.00	1.88	0.67	0.13
	EOL[b]	1.19	3.20	1.33	0.22
Loblolly pine	Untreated	7.30	1.40	0.50	0.00
	EOL[c]	4.20	5.30	1.40	0.30

[a]EOP conditions: 1.5% w/w H_2SO_4 and 65% v/v ethanol at 195 °C for 60 min.
[b]EOP conditions: 1.2% w/w H_2SO_4 and 65% v/v ethanol at 190 °C for 60 min.
[c]EOP conditions: 1.1% w/w H_2SO_4 and 65% v/v ethanol at 170 °C for 60 min.

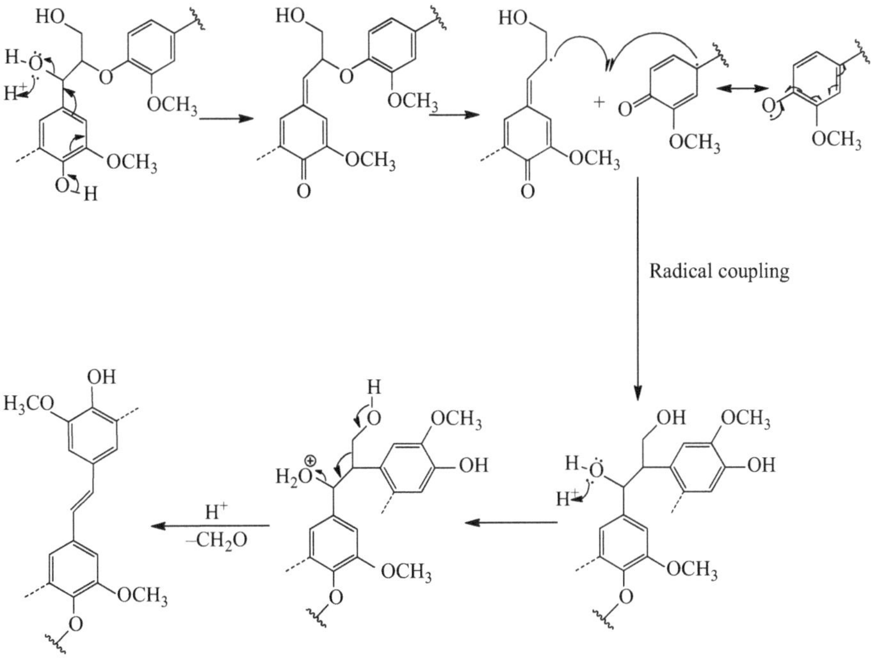

Figure 4.11 Degradation pathway of β–O–4′ units during EOP.

causes an increase in guaiacyl OH, and the degradation of β–1′ inter-unit linkages to give stilbene structures through the loss of the γ-methylol group as formaldehyde decreases the amount of aliphatic OH.[26]

4.6 Structural Features of Ethanol Organosolv Lignin

Organosolv lignins are environmentally advantageous because they contain insignificant levels of sulfur and sodium, indicating very low ash content (<1.0%).[22,40] Both hardwood and softwood organosolv lignins have low

Table 4.7 The weight-averaged molecular weight $(\bar{M}_w)$, the number-averaged molecular weight $(\bar{M}_n)$, the polydispersity index $D = (\bar{M}_w / \bar{M}_n)$, and the functional groups of pine and poplar EOL.[41,42]

EOL^a	Molecular weight and polydispersity index^b			Functional group^c/mmol g^{-1} lignin	
	$\bar{M}_w$/g mol^{-1}	$\bar{M}_n$/g mol^{-1}	D	ArOH	AlkOH
Pine	1280	3010	2.35	3.41	4.43
Poplar	1093	2105	1.93	3.48	3.85

^{a}Ethanol organosolv lignin produced under 1.10% w/w H_2SO_4, 65% v/v ethanol, 170 °C, and 60 min for pine, and 1.25% w/w H_2SO_4, 50% v/v ethanol, 180 °C, and 60 min for poplar.
^{b}M_w = weight average molecular weight; M_n = number average molecular weight; M_w/M_n = polydispersity;
^{c}ArOH = phenolic hydroxyl group; AlkOH = aliphatic hydroxyl group.

molecular weights and narrow polydispersities; low polydispersity is important in controlling the variability in co-product applications.[22] These lignins show a low glass-transition temperature and exhibit flow when heated. They have high solubilities in organic solvents and are very hydrophobic and practically in-soluble in water. Also, softwood and hardwood organosolv lignins have a high level of phenolic hydroxyl groups, relative to lignins prepared using other processes.[22] Table 4.7 summarizes the characteristics of pine and poplar EOL.

An unanswered challenge that still remains in the open literature is the extent of recovery and yield of low-molecular-weight oligomeric lignin during the ethanol organosolv pretreatment process. These are two critical parameters for accomplishing the economic viability of down-stream processing and the conversion of lignin to a high-value product.

4.7   Applications of Ethanol Organosolv Lignin

With the high phenolic content of EOL, this type of lignin can be used for the production of phenolic, epoxy, and isocyanate resins.[22] In addition, high phenolic content is desirable for anti-oxidant activity, which makes EOL a possible source of products with high radical-scavenging potential.[42] It is known that polyphenols inhibit the oxidation of low-density proteins, which in turn reduces the risk of heart disease, and that they also have anti-inflammatory and anti-carcinogenic properties.[36] Furthermore, lower molecular-weight and polydispersity values have also been correlated with higher anti-oxidant activity.[42] The conversion of EOL to a potential fuel precursor for green gasoline/diesel *via* catalytic hydrogenolysis has also been demonstrated.[34] EOL has a number of potential applications and the focus should be on researching and developing these applications to an industrial level.

4.8   Conclusions

The characteristics of ethanol organosolv lignin have been well studied using quantitative ^{13}C and ^{31}P NMR spectroscopy, as well as gel-permeation chromatography. The results indicate that the delignification mechanism and the resulting EOL are dependent on the biomass species and the pretreatment conditions used. EOL has a low molecular weight, high phenolic content, and a high degree of condensation.

If one can envision the concept of biorefinery, in which all components of biomass are fully used to make a range of fuels, chemicals, materials, heat, and power, then lignin must be utilized not only as an in-house fuel source but also for co-product applications. With its desired characteristics, EOL has the potential to be used as a polymer substitute, a precursor for biofuels, and a source for making carbon fiber. These characteristics include high purity, low molecular weight distribution, and high phenolic content.

References

1. A. J. Ragauskas, C. K. Williams, B. H. Davison, G. Britovsek, J. Cairney, C. A. Eckert, W. J. Frederick Jr, J. P. Hallett, D. J. Leak, C. L. Liotta, J. R. Mielenz, R. Murphy, R. Templer and T. Tschaplinski, *Science*, 2006, **311**, 484.
2. Y. Pu, D. Zhang, P. M. Singh and A. J. Ragauskas, *Biofuels, Bioprod. Biorefin.*, 2007, **2**, 58.
3. A. J. Ragauskas, M. Nagy, D. H. Kim, C. A. Eckert, J. P. Hallett and C. L. Liotta, *Ind. Biotechnol.*, 2006, **2**, 55.
4. P. Sannigrahi, Y. Pu and A. Ragauskas, *Curr. Opin. Environ. Sustainability*, 2010, **2**, 383.
5. K. David and A. J. Ragauskas, *Energy Environ. Sci.*, 2010, **3**, 1182.
6. Y. P. Zhang, *J. Ind. Microbiol. Biotechnol.*, 2008, **35**, 367.
7. B. Yang and C. E. Wyman, *Biofuels, Bioprod. Biorefin.*, 2008, **2**, 26.
8. T. W. Jeffries and Y. Jin, *Appl. Microbiol. Biotechnol.*, 2004, **63**, 495.
9. R. P. Chandra, R. Bura, W. E. Mabee, A. Berlin, X. Pan and J. N. Saddler, *Adv. Biochem. Eng. Biotechnol.*, 2007, **108**, 67.
10. F. S. Chakar, PhD thesis, Institute of Paper Science and Technology, 2000.
11. K. V. Sarkanen and C. H. Ludwig, *Lignin: Occurrence, Formation, Structure and Reactions*, Wiley, New York, 1971.
12. M. Nagy, PhD thesis, Georgia Institute of Technology, 2009.
13. F. S. Chakar and A. J. Ragauskas, *Ind. Crops Prod.*, 2004, **20**, 131.
14. R. Samuel, Y. Pu, B. Raman and A. J. Ragauskas, *Appl. Biochem. Biotechnol.*, 2010, **162**, 62.
15. A. Guerra, I. Filpponen, L. A. Lucia and D. S. Argyropoulos, *J. Agric. Food Chem.*, 2006, **54**, 9696.
16. W. Boerjan, J. Ralph and M. Baucher, *Annu. Rev. Plant Biol.*, 2003, **54**, 519.
17. E. A. Capanema, M. Y. Balakshin and J. F. Kadla, *J. Agric. Food Chem.*, 2004, **52**, 1850.
18. E. A. Capanema, M. Y. Balakshin and J. F. Kadla, *J. Agric. Food Chem.*, 2005, **53**, 9639.

19. L. V. Kanitskaya, A. V. Rokhin, D. F. Kushnarev and G. A. Kalabin, *Vysokomol. Soedin.*, 1998, **40**, 800.
20. J. W. Choi, O. Faix and D. Meier, *Holzforschung*, 2001, **55**, 185.
21. P. C. Pinto, D. V. Evtuguin and C. P. Neto, *Ind. Eng. Chem. Res.*, 2005, **44**, 9777.
22. X. Pan, C. Arato, N. Gilkes, D. Gregg, W. Mabee, K. Pye, Z. Xiao, X. Zhang and J. Saddler, *Biotechnol. Bioeng.*, 2005, **90**, 473.
23. X. Zhao, K. Cheng and D. Liu, *Appl. Microbiol. Biotechnol.*, 2009, **82**, 815.
24. M. Galbe and G. Zacchi, *Adv. Biochem. Eng. Biotechnol.*, 2007, **108**, 41.
25. X. Pan, N. Gilkes, K. Kadla, K. Pye, S. Saka, D. Gregg, K. Ehara, D. Xie, D. Lam and J. Saddler, *Biotechnol. Bioeng.*, 2006, **94**, 851.
26. B. B. Hallac, Y. Pu and A. J. Ragauskas, *Energy Fuels*, 2010, **24**, 2723.
27. J. J. Villaverde, J. Li, M. Ek, P. Ligero and A. de Vega, *J. Agric. Food Chem.*, 2009, **57**, 6262.
28. D. Roberted. *Carbon-13 Nuclear Magnetic Resonance Spectroscopy*, Springer, New York, 1992.
29. K. M. Holtman, H. Chang and J. F. Kadla, *J. Agric. Food Chem.*, 2004, **52**, 720.
30. K. M. Holtman, H. Chang, H. Jameel and J. Kadla, *J. Wood Chem. Technol.*, 2006, **26**, 21.
31. B. B. Hallac, P. Sannigrahi, Y. Pu, M. Ray, R. J. Murphy and A. J. Ragauskas, *J. Agric. Food Chem.*, 2009, **57**, 1275.
32. A. Granata and D. S. Argyropoulos, *J. Agric. Food Chem.*, 1995, **43**, 1538.
33. Y. Pu and A. J. Ragauskas, *Can. J. Chem.*, 2005, **83**, 2132.
34. M. Nagy, K. David, G. J. P. Britovsek and A. J. Ragauskas, *Holzforschung*, 2009, **63**, 513.
35. T. J. McDonough, *Tappi J*, 1993, **76**, 186.
36. P. Sannigrahi, A. J. Ragauskas and S. J. Miller, *Energy Fuels*, 2010, **24**, 683.
37. R. El Hage, N. Brosse, L. Chrusciel, C. Sanchez, P. Sannigrahi and A. Ragauskas, *Polym. Degrad. Stab.*, 2009, **94**, 1632.
38. R. L. Hage, N. Brosse, P. Sannigrahi and A. Ragauskas, *Polym. Degrad. Stab.*, 2010, **95**, 997.
39. J. Li, G. Henriksson and G. Gellerstedt, *Bioresour. Technol.*, 2007, **98**, 3061.
40. A. Lindner and G. Wegener, *J. Wood Chem. Technol.*, 1988, **8**, 323.
41. X. Pan, D. Xie, R. W. Yu and J. N. Saddler, *Biotechnol. Bioeng.*, 2008, **101**, 39.
42. X. Pan, J. F. Kadla, K. Ehara, N. Gilkes and J. N. Saddler, *J. Agric. Food Chem.*, 2006, **54**, 5806.

CHAPTER 5

Advances in the Genetic Manipulation of Cellulosic Bioenergy Crops for Bioethanol Production

CHUANSHENG MEI*[a] AND CALLISTA RAKHMATOV*[b]

[a] Institute for Sustainable and Renewable Resources, Institute for Advanced Learning and Research, Danville, VA 24540, USA and Departments of horticulture and Forest Resources and Environmental Conservation, Virginia Tech, Blacksburg, VA 24061, USA; [b] Biosystems and Agricultural Engineering, Michigan State University, East Lansing, MI 48824, USA
*Email: chuansheng.mei@ialr.org; ransomca@msu.edu

5.1 Introduction

5.1.1 The Need/Demand for Ethanol Fuel in the USA

Energy security is very important to countries that import most of their fossil fuels, such as the USA and China, the two countries in the World with the largest energy consumption. The limited supply of fossil fuels in the World, rapidly increasing demands as economic development improves in developing countries such as China, and political turmoil in oil-producing countries are all driving this concern. Also, burning petroleum-based oils causes global warming and environmental pollution by releasing CO_2. Alternative energy sources,

RSC Energy and Environment Series No. 10
Biological Conversion of Biomass for Fuels and Chemicals: Explorations from Natural Utilization Systems
Edited by Jianzhong Sun, Shi-You Ding and Joy Doran-Peterson
© The Royal Society of Chemistry 2014
Published by the Royal Society of Chemistry, www.rsc.org

especially sustainable and renewable energy, must be developed. Biofuels, such as bioethanol from sugarcane and corn grain, have been widely used in Brazil and the USA, respectively.

Ethanol fuel is generally used in low-level blends of up to 10%, and manufacturers have begun producing vehicles that can utilize blends with higher percentages. In addition, the increase in flexible-fuel vehicles has provided another market for ethanol in the form of blends up to 85% (E85).[1–4]

The steep growth in ethanol consumption has also been driven by federal legislation aimed at reducing fossil oil consumption, enhancing energy security, and reducing greenhouse gas emissions. Currently, a 10% blend of ethanol in gasoline is mandated in the USA. In the Energy Independence and Security Act of 2007, 36 billion gallons of ethanol will be needed per year by 2022. Accordingly, the USA became the World's largest producer of ethanol in 2005 and produced over 13 billion gallons in 2010.[5,6] As of 2012, the USA has 139 biomass plants that are operational, 15 under construction, and 39 proposed (both for construction and conversion).[7]

5.1.2   The Potential for Corn Ethanol Fuel Production

Most ethanol in the USA is produced using corn grain (starch) as the feedstock.[2] The current capacity (14.6 billion gallons) is nearing its cap of 15 billion gallons by 2015 derived from corn (starch) set by the Energy Independence and Security Act of 2007.[8,9] The use of corn grain as a fuel has caused many problems, including increasing food and feed prices and a low energy return. Furthermore, growing corn is highly intensive and requires fertilizers, pesticides, irrigation and fertile land. The N_2O released from nitrogen fertilizer applications is also a potential greenhouse gas causing global warming.[10] Therefore, the Act stipulates that by 2022, the most ethanol (21 billion gallons) must be derived from non-cornstarch (*e.g.*, lignocellulose) feedstocks.[9,11,12]

5.1.3   Cellulosic Ethanol Potential

In the USA it was estimated that approximately 1 billion tons per year of crop and forest residues, energy crops and other lignocellulosic residues could become available for production of ethanol, yielding about 108.5 billion gallons.[13,14] Corn (maize) stover, rice straw, fast-growing perennial grasses (such as switchgrass, *Miscanthus* and giant reed), and other energy crops (such as poplar) have high amounts of lignocellulosic biomass and have been recommended for use as lignocellulosic feedstocks.[15]

Some authors estimate that over 10–50 billion tons of lignocellulosic biomass could be available globally per year.[16] If we assume that 85 gallons can be produced per ton,[17] then 850–4250 billion gallons of ethanol could be produced, mostly from agricultural waste (waste crops, crop residues and other crop waste), replacing about 32% of the current use of gasoline if E85 is used.[14] The potential production could be much higher if supplemented by dedicated energy crops and if the systems, infrastructure and technology were in place.[14,17] However, current ethanol production from lignocellulosic biomass is not cost-effective and could not compete with fossil oils.

5.1.4 The Ethanol Production Process

Ethanol is produced by yeast through the fermentation of sugars (usually sucrose or glucose). The carbon (sugar) source, or "feedstock", is usually a plant material. Sugarcane and corn (maize) grains are the most widely used feedstocks today. Production of ethanol from sugary feedstocks such as sugarcane is relatively straightforward: the sugar can be easily extracted and directly used for fermentation. Starchy feedstocks such as corn (maize) grains must first be milled, and the starch hydrolyzed to glucose (saccharification) by α-amylase before it can be fermented. These technologies are mature and widely used in commercialization in the USA for cornstarch feedstock and in Brazil for sugarcane feedstock (Figure 5.1).

However, the process of producing ethanol from lignocellulosic feedstocks is more complicated. Before hydrolysis, the feedstocks must first undergo a pretreatment process (Figure 5.1). It is difficult to extract the simple sugars necessary for fermentation from cellulose because cellulose is a structural component of the plant cell wall and forms microfibrils, which are wrapped in hemicellulose (composed of other sugars) and embedded in a matrix of lignin and other substances, such as pectins. The location of the cellulose makes the cellulase enzymes inaccessible to their substrates. Therefore, lignocellulosic plant material must be first pretreated to disrupt the lignin and modify the structure of the cellulose (decrease its crystallinity and increase its accessibility by exposing the chain ends). Then, large numbers of hydrolysis enzymes (due to their low specific activity) *i.e.*, hemicellulases and cellulases are required to completely hydrolyze cellulose into its constituent unit, glucose, compared to the hydrolysis of starchy feedstocks.

5.1.5 The Plant Cell Wall

Understanding the structure and components of the plant cell wall is the key to knowing how the deconstruction process works. Lignocellulosic biomass is mostly composed of the plant cell wall, which is a complex, highly organized

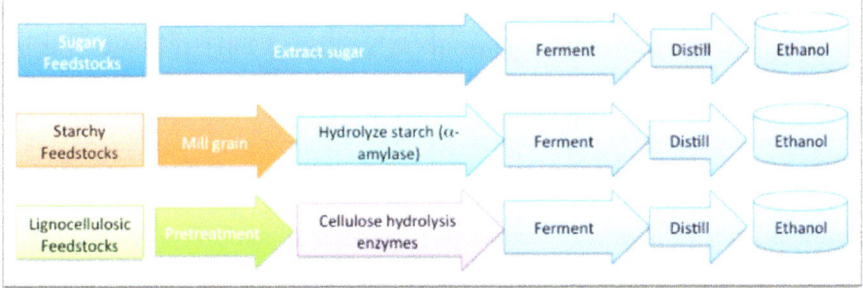

Figure 5.1 Procedures for ethanol production from different feedstocks.

structure made of various components such as polysaccharides, proteins, aromatic substances and other compounds. Polysaccharides are the main component and the structural framework of the cell wall. These polysaccharides are composed of long chains of monosaccharides that are covalently linked. They can also have long side-chains. In addition to glucose, there are other 10 common monosaccharide sugars that are found in plant cell walls, all derived from glucose, that make up these polysaccharide chains. These are rhamnose, galactose, galacturonic acid, glucuronic acid, apiose, xylose, arabinose, mannose, mannuronic acid, and fucose.[18] The major cell-wall components are cellulose, hemicellulose and lignin. Pectins, structural proteins and other substances are also present in the cell wall,[15] but they are present in much smaller amounts and will not be discussed here.

5.1.5.1   Cellulose

Cellulose is a long, linear polysaccharide chain of several hundred, and up to 15 000, $\beta(1 \rightarrow 4)$-linked anhydrous D-glucose (dextrose) units.[15] Microfibrils, the structural component of the primary and secondary cell wall of plants, are made up of cellulose chains.[20] The $\beta(1 \rightarrow 4)$ linkage means that each unit is orientated 180° relative to the unit it is attached to. Cellulose does not branch or coil, making it rigid and rod-like, while starch, on the other hand, has $\alpha(1 \rightarrow 4)$ linkages, and its glucose polymers are coiled and branched, resulting in a more amorphous structure. Microfibrils consist of crystalline and amorphous regions, which are staggered. In crystalline regions of microfibrils, cellulose chains are lined up parallel to each other, and the glucose molecules form hydrogen bonds with other glucose molecules, either on the same or on a nearby chain, binding them together and causing them to be tightly packed. In the amorphous regions of microfibrils, the structure is more loose. Cellulose is the most common organic compound on Earth. It accounts for 15–30% of the dry mass of primary cell walls, up to 40% of secondary cell walls, and about 33% of overall plant matter.[18]

5.1.5.2   Hemicellulose

Hemicellulose is also a polysaccharide, but unlike cellulose, which is crystalline, long and unbranched, hemicellulose is more random and amorphous, consisting of shorter chains (500–3000 sugar units), and is branched. It coats the cellulose microfibrils and cross-links them together. Unlike cellulose, which is made only of anhydrous glucose, hemicellulose can be made of any of several heteropolymers. Based on the components of the backbones, hemicellulose can be classified into three major types, namely xylan (xylose backbone), xyloglucan (glucose backbone), and galactomannan (mannose backbone).[21] Most of the sugars in hemicellulose (xylose and arabinose) are pentoses, whereas cellulose is made of glucose, which is a hexose. Hemicellulose represents 20–40% of the total dry weight of plant matter,[18] and since hemicellulose mostly consists of xylans, xylose (the backbone of xylans) is the second most abundant sugar after glucose on Earth.[22]

5.1.5.3   Lignin

Lignin is not made of sugar molecules, but aromatic compounds (phenylpropanoids). It is a very complex chemical compound and consists of three different monolignol units (*p*-hydroxylphenyl [H], guaiacyl [G], and syringyl [S]), cross-linked together. These units differ in the degree of methylation of their phenylpropane unit,[23] and their ratios determine recalcitrance of lignin degradation. Lignin is sticky and aids in binding hemicelluloses to microfibrils. It also provides structure and strength, protects the cell against invading pathogens, and waterproofs the cell wall.[24] It is mostly present in secondary cell walls, and accounts for 10–25% of the total dry weight.[18]

5.2   Cell Wall Degradation

Cell-wall degradation, also known as "deconstruction" or "hydrolysis", is a natural process carried out by various organisms in which the various cell-wall components are broken down into smaller constituent units. This process is useful in carbon recycling in nature. Several steps are necessary for cell-wall degradation:

(1) removal or breakdown of lignin and other substances that might inhibit enzymatic reactions and block access of cellulase enzymes to the cellulose;
(2) removal or hydrolysis of hemicelluloses to expose the cellulose to cellulase enzymes; and
(3) hydrolysis of cellulose to glucose by cellulase enzymes.

5.2.1   Natural Processes

Many microorganisms (bacteria and fungi) are able to break down plant cell walls, including anaerobes (for example, those found in the rumen) and aerobes (for example, those able to decompose dead plant matter). These organisms usually produce a series of enzymes able to hydrolyze different polysaccharides, and these enzymes act synergistically to decrystallize cellulose and completely degrade the cell wall into its constituent units.[25] We do not know exactly how many enzymes are involved in the complete degradation of the plant cell wall. In general, three major types of enzymes are involved in, namely, ligninases, hemicellulases, and cellulases.

5.2.1.1   Ligninases (Lignin-Modifying Enzymes)

A "ligninase" is the generic name for any enzyme that breaks down lignin compounds. Because ligninase enzymes are not hydrolytic but oxidative, the term "lignin-modifying enzymes" has been suggested, but the term "ligninase" is still commonly used. White-rot fungi are thought to be the most effective lignin-degrading microbes in nature,[26] especially *Phanerochaete chrysosporium* and *Trametes versicolor*. These fungi produce three major families of lignin-modifying enzymes: laccases (EC1.10.3.2), manganese-dependent per-oxidases (MnP, EC1.11.1.7), and lignin peroxidases (LiP, EC1.11.1.14).[27,28]

5.2.1.2   Hemicellulases

Since hemicelluloses mainly consist of xylan, we will focus more on the details of xylanases, which degrade xylan hemicellulose. Generally, xylanases include endo-1,4-β-xylanase (EC3.2.1.8), β-xylosidase (EC3.2.1.37), β-arabinofur-anosidase (EC3.2.1.39), β-glucuronidase (EC3.2.1.131), and acetylxylan ester-ase (EC3.1.1.72).[29] Among them, endo-1,4-β-xylanase is the most important because it cleaves the backbone of xylan into xylooligosaccharides, which is a key step in depolymerizing xylan. Xylanases can also be classified into two groups, family 10 and family 11 glycosylhydrolases (GH). Xylanases in the GH 10 family can cleave xylan backbones with branches and have lower substrate specificity, while xylanases in the GH 11 family cleave unsubstituted xylan chains and are more specific towards substrates.[30,31]

5.2.1.3   Cellulases

Cellulases are responsible for the hydrolysis of cellulose to fermentable sugar (glucose) and can be divided into endo-glucanases, exo-glucanases, and β-glu-cosidases. Endo-glucanases, also called "endocellulases" or "endo-1,4-β-gluca-nases" (EC3.2.1.4), randomly cleave the internal bonds of polysaccharides to release short oligosaccharides. Then, exo-glucanases, also called "exo-cellulases" or "cellobiohydrolases" (CBH) (EC3.2.1.91), hydrolyze oligosaccharides to cel-lobiose from either the reducing end or non-reducing end. CBH1 progressively cleaves oligosaccharides from the reducing end, and CBH2, from the non-reducing end. Finally, β-glucosidases, also called "cellobiases" (EC3.2.1.21) are responsible for degrading cellobiose into the simple sugar, glucose.

5.2.2   Biorefinery

Ethanol can be produced from lignocellulosic biomass through biological (cellulolysis) or thermochemical (gasification) processes. The thermochemical approach involves transforming the biomass into carbon monoxide and hydrogen gases, followed by fermentation with the *Clostridium ljungdahlii* bacterium to ethanol. This method will not be discussed further in this chapter. The biological approach consists of the following three major steps:

(1) pretreatment – prepares the biomass for hydrolysis;
(2) hydrolysis by enzymes – breaks down the cellulose to hexose sugar monomers; and
(3) fermentation by yeast.

5.2.2.1   Pretreatment

Both physical and chemical pretreatments are currently required to recover the maximum cellulose for conversion. Physical pretreatment (also called "size reduction") involves chopping up the biomass into smaller pieces (for example, milling or using sawdust). Chemical pretreatment involves freeing the cellulose from the cell-wall components for cellulase digestion.[32,33]

As mentioned above, cellulose is strong and crystalline and resists hydrolysis. Pretreatment is therefore necessary to break up its crystalline structure, and also remove or disrupt the lignin and hemicellulose.[24] Crystalline cellulose can be converted to amorphous cellulose in water at a temperature of 320 °C and a pressure of 25 MPa.[34] It can also be broken down into glucose units chemically by treating it with concentrated acids at high temperature. So far, the major pretreatments that have been developed include: dilute acid, acid flow-through, ammonia fiber expansion (AFEX), ammonia recycle percolation, steam water explosion, lime, organosolv pulping, sulfite pretreatment to overcome the recalcitrance of lignocellulose (SPORL), alkaline wet oxidation, and ozone pretreatment.[24,32,35–38] On the other hand, hemicellulose is much less structured and strong, and can be easily hydrolyzed by dilute acids or bases.

5.2.2.2 Hydrolysis

Cellulose hydrolysis (cellulolysis) is the process used to break down cellulose molecules into sugars. There are two methods, *i.e.*, chemical and enzymatic methods. The chemical process involves the use of dilute acid and will not be discussed further. The enzymatic reaction involves applying a mixture of cellulases (described in detail above) to break down the cellulose into fermentable hexose sugars (glucose monomers). Currently, hemicellulases are not used to hydrolyze hemicellulose to pentose sugars in industrial ethanol production. However, microorganisms able to ferment pentose sugars to ethanol have been recently discovered.[39]

5.2.2.3 Fermentation

Once the fermentable sugars are obtained, they are fermented by yeast and/or other microorganisms to ethanol. Baker's yeast (*Saccharomyces cerevisiae*) has been used for centuries, but it uses hexose (six-carbon) sugars such as glucose. A considerable amount (over 20% in grasses)[40] of pentose (five-carbon) sugars (such as xylose and arabinose) are present in the sugar solution from the hemicellulose portion of the lignocellulosic biomass. Currently these sugars are not utilized but several microorganisms have been found to have the ability to ferment pentose sugars to ethanol.[39]

5.3 The Current Status of Ethanol Production from Lignocellulosic Biomass

As of 2008, several commercial-scale lignocellulosic biomass ethanol plants were under construction or operational in the USA. The operational facilities had an annual capacity of 3.17 million gallons (12 million liters). The 26 new plants would add 21.13 million gallons more (80 million liters).[41] Recently,

POET LLC, a US biofuel company headquartered in Sioux Falls SD, received a $105 million loan guarantee from the USA Department of Energy and is building a cellulosic ethanol plant that will use corn cobs, leaves, husks and corn stover as feedstocks with a capacity of up to 25 million gallons per year.[42]

Iogen in Canada is the World's first demonstration pilot plant for the production of ethanol from agricultural residues and employs advanced microorganisms and fermentation systems, which could ferment both hexose and pentose sugars into ethanol; it has a capacity of 1.6 million gallons (6 million liters) per year.[43]

Plants with a total annual capacity of nearly 500 million liters (about 131 million gallons) are operating or under construction in Denmark, Italy, Spain, and the UK.[41,44–50] Pilot plants are also under construction or operating in Germany, Norway and Sweden.[49,51]

5.3.1 Challenges to Cellulosic Ethanol Production

The use of plant biomass to produce fermentable sugars for ethanol fuel production is a promising and appealing alternative technology, and advances have been made in improving the ethanol yield and decreasing the costs associated with its production.[52,53] However, major barriers still hinder commercial-scale production using current cellulosic ethanol technology, including the high cost of pretreatment and the large amount of microbial cellulase enzymes required.[24,54]

Pretreatment of lignocellulosic biomass is necessary prior to hydrolysis to disrupt or remove the lignin and break up the crystalline structure of the cellulose. New pretreatment technologies have been developed. However, current pretreatment processes use heat and chemical means to remove lignins and hemicelluloses,[55] which produce excessive waste.[17] Replacing the current technology with the use of enzymes such as hemicellulases and ligninases would be friendlier to environment. Therefore, research is being conducted on fungal ligninases and hemicellulases to potentially decrease the necessity (and thus the cost) of pretreatment.

Hydrolysis enzymes (cellulases) account for about 30–50% of the total cost of the bioconversion of lignocellulose to ethanol.[56,57] In comparison, they account for only 8% of the cost of producing ethanol from starch grain.[15,57,58] Production of these enzymes on a large scale using biomass crops as biofactories may be possible due to recent advancements in plant biotechnologies. In this chapter, we will explore the problems, challenges, and solutions to ethanol production from lignocellulosic materials, with a focus on utilizing plants as biofactories for hydrolysis enzyme production.

5.4 Plants as Biofactories for Hydrolysis Enzyme Production

The huge amount of enzyme needed to hydrolyze cellulose to fermentable sugars is still considered the biggest challenge in bioethanol production from lignocellulosic biomass. One scenario described by Howard *et al.*[17] is that about 3.6 million tons of cellulase enzymes would be needed to degrade 425 million tons of lignocellulosic biomass to meet a goal of 36 billion gallons of bioethanol production, which would replace 30% of transportation fuel in 2022. They estimated that $30 billion of capital investment would be required to build numerous microbial fermenters ($10 million each) with a projected goal of $0.10–0.20 per gallon of ethanol production.[17] Alternative methods of producing enzymes must be explored. The least expensive way to produce foreign proteins is to use plants as biofactories, so the production of hydrolysis enzymes within plants, especially bioenergy crops, is a potential alternative to the current enzyme production method by microbes.[17] In addition, enzyme production in the same biomass crops to be used as feedstocks could reduce exogenous enzyme loading, thereby reduce the cost of bioethanol production.

5.4.1 Plants as Biofactories

Plants have been used as biofactories to produce many valuable products, such as pharmaceutical vaccines, due to the great progress made in biotechnology during recent decades.[58a] Therefore, it may be possible to use plants as biofactories to produce the hydrolysis enzymes used in biothanol production. The heterologous expression of cellulases and hemicellulases in plants has been summarized in previous reviews.[59,60] As plant biotechnology has improved in recent years, more and more promising results have been achieved, and these will be described in more detail in Section 5.5.

5.4.2 Enzyme Resources: Bacteria, Fungi, and Insects (Termites)

In nature, bacteria, fungi, and insects recycle carbon and nitrogen by degrading lignocellulosic biomass. From these organisms, many genes encoding enzymes for lignocellulose degradation have been isolated and functionally studied. In general, three major categories of enzymes are considered necessary to completely degrade lignocellulose biomass, *i.e.*, cellulases, hemicellulases, and ligninases.

5.4.2.1 Cellulases

The most studied cellulase is endo-1,4-β-glucanase E1 from bacterial *Acidothermus cellulolyticus* E1. Other endo-glucanase genes have been isolated from other microorganisms, such as an endo-glucanase gene from *Ruminococcus albus*,[61,62] the cel5A gene from *Thermotogs meritima*,[63] and a membrane-bound endo-glucanase gene from hybrid poplar KOR (*Populus alba* x *P. grandidentata*).[64]

The exo-glucanase genes encoding CBH1 and CBH2 were isolated from the fungi *Penicillium* sp. and *Trichoderma* sp. respectively.[65] In addition, the CBH1 gene was isolated from *Trichoderma reesei* and used to transform maize plants.[66,67]

The fungus *Aspergillus niger* has abundant β-glucosidase genes and is widely used in industry because it can co-culture with *Saccharomyces cerevisiae* for ethanol production in a simultaneous saccharification and fermentation technique.[68] A β-glucosidase gene from *Butyrivibrio fibrisolvens* H17c and a hyperthermostable β-glucosidase gene from *Thermotoga maritima* have also been reported.[69,70] With a proteomics strategy using two-dimensional PSGE in-gel activity assay and tandem mass spectrometry, Kim *et al.*[71] discovered that a novel β-glucosidase from *Aspergillus fumigatus* had a much higher heat tolerance than the previous β-glucosidase from *A. niger* and *A. oryzae*, keeping 90% of its original activity at 65 °C for 6 h and 65% for 19 h.[71]

The best-studied microbial cellulase producers are *Trichoderma* spp., particularly *Trichoderma reesei*, which have commonly been employed in industry to produce cellulases and hemicellulases.[68] Insects, particularly wood-feeding termites, are also good resources of cellulose enzymes, and numerous potential cellulolytic enzymes from insects have been reviewed.[72] Some enzymes have been characterized with heterologous expression in *E. coli*. The genes are potential candidates to be used in the bioconversion of lignocellulose biomass to fermentable sugars. However, these genes have not been used in the heterologous expression of plants yet.

Endo-glucanases, CBHs and β-glucosidases act synergistically to completely hydrolyze cellulose to single sugars. β-Glucosidase acts synergistically with endo-glucanase Cel5A on cellulose degradation, and exhibited a 37% increase of glucose production over Cel5A alone.[70] In addition, CBH and β-glucosidase are synergistic in cellulose degradation, as β-glucosidase removes cellobiose, which is a product and inhibitor of CBH enzymes. Most cellulase enzymes have both a catalytic domain and a carbohydrate-binding domain (CBD) or module (CBM). Although the catalytic domain alone has a high activity against cellulose, both the catalytic domain and the CBM together have a much higher activity for degrading crystalline (also known as "insoluble") cellulose substrates.

Cellulases usually have a lower activity, 1–2 orders of magnitude lower compared with other carbohydrate hydrolases, such as amylases.[40] A useful website about the Carbohydrate-Active enzymes (CAZy) Database (www.cazy.org) has extensive information on the structural features and carbohydrate-binding domains of glycoside hydrolases. Two specific methods, direct evolution and rational design, for how to improve cellulose-specific activity were described in more detail by Howard *et al.*[17]

5.4.2.2   Hemicellulases

Hemicellulases, like their complex substrates described in Section 5.1.5.2, are enzymes that are responsible for completely degrading complex substrates. Many hemicellulase genes have been identified in microorganisms. In a mini review by Polizeli *et al.*,[73] some xylanases produced by different microorganisms and their characterization were described. Recently, more xylanase genes have been discovered and their functionalities studied.[73] Some belong to the GH 10 family, such as xyl10B from *T. maritima*,[74] xynA from the thermophilic bacterium *Dictyoglomus thermophilum*, xyl6E7 from fungus-growing termites, close to *Macrotermes annandalei*,[29] bsx from *Bacillus* sp. NG-27, and xynB from *Clostridium stercorarium*.[75,76] Other xylanases belonging to the GH 11 family, such as xynB from *Dictyoglomus thermophilum*, xynA from *Clostridium thermocellum* F1,[77] and another two from *Pseudomonas cellulos*[78] have also been discovered. Some xylanase enzymes can tolerate very high temperatures, such as xyl10B, which has an optimal temperature of 90 °C, is still active at 105 °C, and has a low activity at 20–30 °C,[79,80] This temperature range for activity is a good characteristic for heterologous expression in plants that are to be used for lignocellulosic hydrolysis. On the other hand, the xyl6E7 enzyme has a high activity, is substrate-specific with a wide range of pH stabilities, but has extremely low thermostability.[29]

Thermophilic microorganisms and wood-feeding termites are good resources for seeking out potential hydrolysis enzymes. Van Fossen *et al.*[81] found two key GH-based genomic loci, the *xynB-xynF* GH locus and the *celA-manB* GH locus, from the genome of the thermophilic bacterium *Caldicellulosiruptor saccharolyticus*, that were able to effectively deconstruct plant biomass.[81] Lignocellulose-feeding insects, particularly wood-feeding termites, show unbelievable wood degradation abilities and have evolved a powerful system to break down lignocellulosic biomass.[82] Through genomic and meta-genomic research, many cellulose degrading genes from termites have been identified and their functionalities characterized. For instance, a β-1,4-endo-glucanase gene and a β-glucosidase gene from termites were studied with heterologous expression in *E. coli*.[83] The thermostability of the enzymes was improved for potential industry use.[84,85]

5.4.2.3 Ligninases

The white-rot fungus *Phanerochaete chrysosporium* is a lignin-digester and produces LiP, MnP and laccase enzymes. One laccase gene from *P. chrysosporium* was used to transform maize plants.[86] Two laccase genes from the ascomycete *Melanocarpus albomyces* and the basidiomycete *Pycnoporus cinnabarinus* were transformed to rice plants.[87] The MnP gene from *P. chrysosporium* was also transformed to maize plants.[88]

5.4.3 Thermostable/Thermophilic Enzymes

Thermostable enzymes are able to tolerate moderate heat (usually around 55 °C) without loss of activity, and thermophilic enzymes require high temperatures for activity, usually with optimal temperatures between 60 and 80 °C.[73] Extremophilic/hyperthermophile organisms grow well at 80–122 °C, and enzymes isolated from these organisms are able to tolerate higher temperatures, with an optimal temperature of 90 °C or higher.

Heterologous expression of hydrolase genes could have adverse effects on host plants, such as sterility or abnormal growth due to degradation of the plant cell walls. One strategy to overcome these negative effects is to produce thermostable/thermophilic/extremophilic enzymes, which remain inactive during plant growth and development and become active at temperatures up to more than 50 °C during the hydrolysis phase of bioethanol production.

Several genes encoding thermostable/thermophilic/extremophilic cellulases and hemicellulases have been identified, such as endo-glucanase E1 from *A. cellulolyticus*, which has an optimal temperature of 80 °C, and xylanase xyl10B from *Thermotoga maritime*, which has an optimal temperature of 90 °C.[79,80] XynA and xynB enzymes from transgenic *Arabidopsis* have an optimal temperature of 85 °C.[89]

5.4.4   Subcellular Targeting

Subcellular localization is not only important for foreign protein accumulation, but also critical to host plants because it separates the enzymes from their substrates. Subcellular localization is also imperative because the right sub-cellular localization usually aids in correct folding and glycosylation, results in reduced degradation, and could increase the stability of hydrolysis enzymes.[90,91] Different signal sequences have been used to successfully target hydrolase enzymes to the cytoplasm, vacuoles, apoplast (inter-cellular areas), endoplasmic reticulum (ER), chloroplast, mitochondria, and peroxisome. This strategy could prevent hydrolase enzymes from accessing their substrates and so minimize negative effects on host plants. For example, endo-glucanase has successfully been targeted to the apoplast in tobacco and maize,[92,93] chloroplasts in tobacco and potato,[63,90,92,94,95] mitochondria in maize,[96] ER in tobacco and maize,[95,96] vacuoles in potato,[94] and peroxisomes in tobacco.[90] Hemicellulases, such as xylanases, have effectively been localized into the apoplast in *Arabidopsis*,[89] chloroplast in tobacco,[90] and so on. β-Glucosidase has been targeted into vacuoles in maize.[69] Ligninases, such as MnP and laccase have been targeted to the apoplast of maize plants.[88]

It is not clear why and where hydrolysis enzymes should be targeted. Targeting results mostly depend on empirical experiments. Therefore, different targeting strategies should be tested for each hydrolysis enzyme in different species. For example, Hood *et al.*[97] found that endo-glucanase E1expressed in maize had a high activity when targeted to vacuoles and ER with an accumulation of more than 16% total soluble proteins (TSPs), but no activity when targeted to the apoplast. However, a high accumulation of CBH1 was found in apoplast and ER (up to 16% TSPs), while no activity was observed when it was targeted to vacuoles.[97] Another example is that higher levels of CBH1 and CBH2 accumulated in transgenic sugarcane plants when targeted to the vacuole compared to the ER, while a higher level of endo-glucanase enzyme was produced with chloroplast targeting, compared with ER- and vacuole-targeting.[65]

Multi-targeting could be better for some hydrolysis enzymes. Bae *et al.*[90] suggested that hydrolase enzymes be dual targeted to maximize foreign protein accumulation in plants. Xylanase enzyme accumulation increased by 60 and 140% of TSPs when targeted to both the chloroplast and the peroxisome, compared with the enzyme targeted to the chloroplast only and the peroxisome only, respectively.[90]

In addition to chloroplast targeting, direct chloroplast genome transformation was used to achieve much higher levels of expression and accumulation of hydrolysis enzymes in tobacco plants.[98,99] This topic will be discussed in Section 5.6.5.

5.5 Successful Genetic Engineering of Cell Wall Degradation Enzymes in Plants

Enzyme production costs have dropped significantly due to the tremendous improvement of hydrolysis enzyme activity and genetic improvement of microorganism strains. It was reported that the Novozymes and Genencor companies could produce enzymes at 20 cents per gallon of ethanol production in 2005.[100] However, microbial fermentation for producing commercial amounts of hydrolysis enzymes requires very high capital investment for building fermenters and for operation. In the long-term, it would be a potential alternative to produce hydrolysis enzymes in plants because it is comparatively inexpensive, there is no high capital investment requirement for production, and it is easy to scale up or down.[17] In this section, we will discuss recent developments in this area in more detail, focusing on three major kinds of enzymes used in lignocellulose degradation, namely cellulases, hemicellulases, and ligninases.

5.5.1 Transgenic Plants with Cellulase Genes

Three different cellulase enzymes (endo-glucanases, cellobiohydrolases/exoglucanases (CBHs), and β-glucosidases) are responsible for cellulose degradation to fermentable sugars. In early experiments, scientists tried to transform tobacco plants with cellulase or hemicellulase genes for enhanced digestibility by ruminants or monogastric animals.[61,62,101,102] Genetic engineering of plants with cellulase genes has mainly focused on the endo-glucanase E1 gene from *A. cellulolyticus*. The full-length and/or catalytic domain (cd) of the E1 gene has been successfully expressed in tobacco and potato,[94,103] maize,[93,96,97] duckweed,[104] and *Arabidopsis*.[105] The maize E1cd transgenic plants produced up to 2% of TSPs with apoplast and mitochondria localization.[93,96] *Arabidopsis* transgenic plants transformed with the E1cd targeted to the apoplast, produced about 26% E1 of TSPs without adverse effects on growth and development.[105] An endo-glucanase cel5A gene was expressed in tobacco plants with the enzyme targeted to chloroplasts using the Rubisco small subunit transit peptide, and the transgenic lines showed high levels of the hyperthermostable Cel5A enzymes, up to 5.2% of TSPs in leaf extracts.[63] Moreover, a membrane-bound endo-β-1,4-glucanase was expressed in hybrid poplar, a potential bioenergy crop.[64] An endo-glucanase gene was also transformed to another bioenergy crop, sugarcane, with a constitutive promoter and the green-tissue-specific promoter Zm-PepC with no abnormal phenotypes observed.[65] Additionally, endo-glucanase enzymes not only degrade cellulose but also play a role in cellulose synthesis. For example, tobacco and maize plants over-expressing endo-glucanase E1, driven by a constitutive promoter and targeted to the cell wall, showed a reduction of cell-wall degradation recalcitrance, and were more amenable to pretreatment and enzyme hydrolysis. This may be because endo-glucanase enzymes alter the cellulose structure during cell-wall synthesis.[106]

Most cellulase enzymes consist of a cd and CBM/CBD linked by a flexible linker region.[107] CBMs are present in both the cellulases and xylanases of microorganisms.[108] CBMs or CBDs can be fused at either the N- or the C-termini of the cd of cellulase and may play several roles, including regulation of the gene expression level, RNA and protein stability, and activity against insoluble substrates.[107] Although a cd alone has high activity, the combination of both catalytic and cellulose-binding domains demonstrates much higher cellulolytic activity, especially on insoluble substrates. For example, when expressed in *E. coli*, Cel5A fused with CBM6 (a CBM of *C. stercorarium* xylanase A) or CBM1 from *T. reesei*, exhibited a 14–18-fold increase in activity against the insoluble substrate avicel,[109] compared to *T. maritime* Cel5A alone. The results clearly indicate that Cel5A protein without CBMs was much less effective for the degradation of crystalline cellulose substrates, but no differences were found for the hydrolysis of the soluble cellulose substrate, carboxymethyl cellulose. Recently, transgenic tobacco plants expressing the *T. maritima* cel5A and CBM6-cel5A genes in different subcellular localizations were generated. The highest levels of enzymes were observed when targeted to the chloroplast, with 4.5% Cel5A of TSPs and 5.2% Cel5A-CBM6 of TSPs. In addition, the enzymes from CBM6-cel5A transgenic plants produced 33% more sugars against insoluble substrates compared to enzymes from transgenic plants expressing only the cel5A gene.[110] In another example, when an endo-glucanase gene of *Ruminococcus albus* was fused with CBM6, the enzyme CBM-Cel5A produced 4× more reducing sugar against an insoluble substrate, compared with only Cel5A enzyme.[111] The CBM has pivotal roles both in cellulose degradation and in modifying crystalline cellulose.[112] Over-expression of microbial CBM in plants was able to disrupt crystalline cellulose and increased cellulose hydrolysis.[113] Transgenic potato plants over-expressing CBMs from *Clostridium cellulovorans* also showed improved plant growth.[114]

Expression of cellobiohydrolases was reported in tobacco,[98,115] potato,[115] and maize[97] with apoplast localizations and no abnormal morphology. Harrison *et al.*[65] reported that recombinant cellobiohydrolase (CBH1 and CBH2) and endo-glucanase genes were individually transformed into sugarcane, and the transgenic plants accumulated high levels of enzymes in the mature leaves with the green-tissue-specific promoter, maize phosphoenol pyruvate carboxylase, Zm-PepC.[65] All these transgenic plants showed normal growth and development. The Zm-PepC promoter was able to direct active CBH1 enzyme to the green and senescent leaves of mature sugarcane. Furthermore, the CBH1 gene from *T. reesei* was transformed to maize with a seed-specific promoter and accumulated up to 5.2% of TSPs in seeds without an abnormal phenotype.[97]

β-Glucosidase is responsible for digesting disaccharides into simple sugars, which can be fermented into ethanol. Maize was transformed with the β-glucosidase gene from *Butyrivibrio fibrisolvens* H17c, and the enzyme accumulated up to 3.1% of TSPs.[69] The bacterial β-glucosidase gene from *T. maritime* was expressed in tobacco plants and the enzyme accumulated up to 4.5% of TSPs in the cytosol and 5.8% of TSPs in the chloroplast. The thermostable enzyme that was produced was affected by drying at room temperature, but retained activity for up to three days after freezing and lyophilization.[70]

5.5.2 Transgenic Plants with Xylanase Genes

Xylanases are an important group of hemicellulase enzymes. Xylanase genes have been expressed in wheat seeds[116] and rice[75] but caused abnormal plant growth, such as shriveled wheat seed and stunted and infertile rice plants. Similar results were obtained by Gray *et al.*[76] when expressing two xylanase genes, *bsx* from *Bacillus* sp. NG-27 and *xynB* from *Clostridium stercorarium*, in maize under the constitutive rice ubiquitin 3 promoter and grain-specific rice glutelin 4 promoter. The enzymes BSX and xynB accumulated up to 4.0% and 16.4% of TSPs, respectively, but the plants were stunted and had sterile grains. In contrast, two codon-optimized endoxylanases, xynA and xynB, from the thermophilic bacterium *Dictyoglomus thermophilum*, were expressed in *Arabidopsis* under the control of the constitutive CaMV 35S promoter and targeted to the apoplast. The transgenic plants showed normal growth and development, and both enzymes extracted from dried stems had high activity,[75,77] which means that drying does not affect enzyme activity. Furthermore, Zhang *et al.*[117] evaluated different transgenic maize plants over-expressing either the endo-glucanase gene, the endoxylanase gene, or both of them, and found that there were no differences in glucan and xylan amounts between transgenic and control plants. However, under mild pretreatment below 75 °C with in-house enzyme cocktails, hydrolysis of pretreated transgenic materials produced up to 141% more glucose and 172% more xylose than hydrolysis of pretreated control materials. The same hydrolysis rate was achieved with a 25% reduction of exogenous enzyme loading with transgenic materials compared with control materials. Ethanol production also increased by 55% from transgenic materials over control materials.[117] Their results suggest that production of xylanase *in planta* should be considered as an enzyme pretreatment to improve biomass hemicellulose and cellulose hydrolysis.

Xylanase expression in monocot plants often results in abnormal morphology.[75,116] However, dicot plants (such as *Arabidopsis*, described above) show normal growth and development with xylanase expression.[75,101] One explanation for this difference is that xylan is the major hemicellulose in the cell walls of monocot plants while dicot plants contain xyloglucans as the major hemicellulose.[118] Another explanation might be differences in the signal processing of microbial genes between the monocot and dicot plants. In one study, the

Clostridium stercorarium xynB gene (GH 10 family), was expressed in both tobacco and rice, both with and without a signal sequence. The transgenic tobacco and rice plants expressing the xynB gene alone had a normal phenotype. When the xynB gene with the signal sequence was expressed in plants, the tobacco plants had a normal phenotype, while the rice plants showed abnormal growth.[75] Similar results were obtained in experiments with rice and wheat. The gene encoding the cd of xynA1 (without a signal sequence) from thermostable *Clostridium thermocellum* F1 was transformed to rice with a constitutive promoter, and the gene expression was detected in rice straw and grains with normal plant growth, and the enzyme activity was retained in the desiccated grains.[77]

5.5.3 Transgenic Plants with Ligninase Genes

Lignin is a major barrier in lignocellulosic biomass conversion, and usually pretreatment is required to remove it. Intensive research has been devoted to the improvement of biomass quality, *i.e.*, reducing the lignin content or changing the lignin composition for easy degradation by genetic engineering technology. A few exciting results have been achieved in switchgrass, a promising bioenergy crop, by repression of the lignin biosynthesis genes, 4CL or CAD, *via* RNAi knockout.[119–121] The details of this process will not be discussed in this chapter, instead we will focus on the genetic engineering of ligninase genes into plants to allow them to benefit from the pretreatment process, so less severe pretreatment is needed, which is more environmentally friendly than current pretreatments. Compared with the genetic engineering of cellulase and hemicellulase genes in plants, there are fewer reports on heterologous expression of ligninase genes in plants. Laccases, which have been reported to improve lignocellulose degradation, have been the subject of a few studies. Since the activity of laccases generates free radicals, active oxygen localization will be critical to host plants. Hood *et al.*[86] found a fungal laccase gene had a high expression level (up to 0.8% of TSPs) with the maize embryo-preferred promoter, globulin 1 and cell-wall targeting, but seeds underwent browning and had limited germination. Two fungal laccases from *Melanocarpus albomyces* and *Pycnoporus cinnabarinus* were transformed to rice under a constitutive promoter and with targeting to the endosperm of rice seeds, and up to 1% of TSPs was obtained.[87] MnP plays an important role in lignin degradation. An MnP gene from the white-rot fungus *Phanerochaete chrysosporium* was transformed to maize plants and accumulated up to 14% of TSPs in seeds, with minimal negative effects on plant growth.[88] However, the percentage of the enzyme MnP is relatively low on the basis of dry biomass.

5.6 Challenges and Perspectives

In current bioethanol production from lignocellulosic biomass, the need for pretreatment and enzymes for hydrolysis are two major hindrances. The huge amount of enzymes required for degrading lignocellulosic biomass is still considered a key roadblock to commercial lignocellulosic ethanol production.[122] Because cellulases have a lower activity for cellulose degradation than amylase for starch hydrolysis, many more cellulase enzymes (40–100×) are needed to digest the same amount of feedstock.[123] It was estimated that about 8.5–15 kg of enzymes are required to degrade 1 ton of lignocellulosic biomass.[17,60]

As plant biotechnology continually improves, it may become possible for plants to be used as biofactories for producing hydrolysis enzymes, and great progress has been made towards this, as described in the above sections. In addition, the genetic engineering technology of bioenergy crops such as switchgrass and poplar has been greatly improved,[124,125] but the research on large-scale production of hydrolase enzymes in plants is still in its infancy. In order to use plants as biofactories for the large-scale production of hydrolysis enzymes required for industrial lignocellulosic bioethanol production, many problems still need to be solved, including abnormal growth and development of transgenic plants, low levels of expression of foreign hydrolases, and the stability of hydrolase enzyme activity post-harvest or during pretreatment.

5.6.1 Low Levels of Foreign Protein Accumulation

In general, transgenic plants show low levels of exogenous protein accumulation, often only 0.1–2.0% of TSPs. In order to increase hydrolase enzyme accumulation, several strategies have been developed including: (1) gene codon optimization and structure modification for expression in plants; (2) subcellular localization of foreign proteins in plants (as described above); (3) the use of tissue-specific and/or induced promoters; (4) selection of plants with high transgene expression by conventional breeding; and (5) the use of chloroplast transformation instead of nuclear transformation, which will be discussed in Section 5.6.5

5.6.1.1 Gene Codon Optimization

There is preferred codon usage among organisms, particularly between different kingdoms. From analysis of the available genome sequences, it is obvious that codon usage preference in higher plants is very different to codon usage preference in microorganisms. For example, when the wild type Bt gene from the bacterium *Bacillus thuringiensis* was expressed in plants, transgene expression was very low because there is a great difference in codon usage preference between the microbial wild-type Bt gene and plant genes.[126] After modifications based on codon usage preference in plants, the modified gene expression dramatically increased, and the Bt protein was able to accumulate to a high level. So far, Bt cotton and Bt corn have been widely planted in the field for insect resistance. Many hydrolysis genes from microorganisms have been modified based on codon usage preference and have been introduced to plants for research. Each gene should be optimized for gene expression in the particular host plant.

5.6.1.2   Subcellular Localizations

Subcellular targeting for foreign proteins is critical for achieving high levels of accumulation and for ensuring normal growth and development of host plants. Different proteins have different preferred subcellular localizations; for example, endo-glucanase prefers the vacuole, while CBH1 favors the apoplast.[97] Dual targeting or multi-targeting may be a better way to obtain high expression levels of foreign proteins. As mentioned above, targeting xylanase to both chloroplasts and peroxisomes simultaneously, resulted in much higher protein accumulation than with either alone.[90] Much work needs to be done to achieve the subcellular localization of individual genes.

5.6.1.3   Conventional Breeding

Hood *et al.*[67] developed a successful conventional breeding method to enhance exogenous protein (hydrolase enzyme) accumulation in maize. Plants expressing high levels of transgenes were back-crossed to elite and high-oil maize germplasm, self-pollinated, and the progeny with the highest exogenous protein accumulation were selected. A 14-fold increase for E1 and more than a 40-fold increase for CBH1 were obtained in the T6 generation compared with the T1, and a 20-fold increase in laccase enzyme was achieved in 5 generations.[67] In addition, the germination rate was improved in laccase transgenic seeds.[86] Using this method, great improvements have been made in the accumulation of several foreign proteins in transgenic plants, but the mechanisms are still unknown.

5.6.2   Plant Growth and Development

Biotechnology is a powerful tool for the improvement of various agricultural traits, such as herbicide resistance, insect resistance, drought tolerance, and yield increase, but there are frequently problems to overcome. For instance, non-target effects can cause unintended worsening of traits other than the targeted trait; or whole plant growth can be slowed, stunted or dwarfed; or plants can be less fertile or even sterile; and so on. In order to overcome abnormal plant morphologies caused by foreign protein expression in plants, a few strategies have been developed, such as the use of induced promoters or tissue-specific promoters, which will be described in Section 5.6.3, and the use of thermostable/thermophilic/extremophilic enzymes, which were described in Section 5.4.3. A new technology developed by the Agrivida Company involves transforming bioenergy crops with hydrolase genes but keeping the enzymes inactive until harvest. When lignocellulosic biomass is processed in bioethanol production, the enzymes in the feedstock are activated to degrade cellulose and hemicellulose into fermentable sugars. The technology inserts "inteins" into the frame with the host protein and the intein translates with the host protein together as a precursor. Inteins are protein introns and selfish DNA elements. These precursors carry out an auto-catalytic protein splicing reaction forming two products: the host protein and the intein.[127] With this proprietary intein-modified enzyme technology, production costs could be reduced by over 30%, and exogenous enzyme loadings could be decreased by over 75%, relative to current industry protocols.[128] Since foreign enzymes remain dormant, the transgenic plants show normal growth and development.

5.6.3 Induced Promoters/Tissue-Specific Promoters

The use of constitutive promoters to drive heterologous gene expression in plants could result in high transgenic expression, since these promoters are highly active in all plant growth and development stages. These promoters include the CaMV 35S promoter, the ubiquitin promoter, and the actin promoter. However, the high expression level of exogenous genes in plants can influence normal plant growth and development, so transgenic plants often exhibit abnormal morphologies (as discussed above). One alternative is to use induced promoters to avoid continuous expression of transgenes, which will lessen the burden on plant metabolism and save energy for plant growth.

There are several induced promoters, including chemical- and growth-induced promoters. Chemical-induced promoters are promoters regulated by certain chemicals. One chemical, ecdysone, was used to turn on the ecdysone receptor for regulating transgene expression. Yang *et al.*[129] used a chemical-induced promoter to drive a gene involved in the lignin biosynthesis pathway in *Arabidopsis*, which improved biomass quality due to lignin composition changes without negative effects on plant growth and development.

Tissue-specific promoters are only active in specific tissues. A seed-specific promoter was successfully used to express hydrolysis enzymes in maize seeds, and a high expression level was achieved with normal plant growth.[86,97] In addition to tissue-specific promoters, growth-regulating promoters, which only express during certain growth stages, are also important. For example, senescence-induced promoters are usually highly active during the end of the growing season. The promoter of SAG12, a senescence-associated gene, was identified from *Arabidopsis*[130] and was tightly controlled during leaf senescence.[131] Recently, a promoter from monocot maize, SEE1, a senescence-enhanced protease belonging to the same cysteine endopeptidase family as SAG12, was found to be functional when tested in transgenic maize plants.[132] These promoters are useful because they function only during the end of the growing season, so they will be less likely to cause any negative effects on plants. For bioenergy crops, which are harvested at the end of the growing season, these promoters could be used to drive gene expression during the end of the growing season, and the enzymes in bioenergy feedstocks could be used for feedstock degradation (auto-hydrolysis) after pretreatment, which would make enzymes more easily able to access their substrates.

5.6.4 Combinations of Different Enzymes in Plants

To completely degrade lignocellulosic biomass, three major groups of enzymes are required: cellulases, hemicellulases, and ligninases. Each group also has various enzymes. For example, cellulases include endo-glucanases, exo-glucanases, and glucosidases. Therefore, expression of only one gene in plants, even with a high expression level, is not enough to accomplish the deconstruction of lignocellulosic biomass to fermentable sugars. In addition, hydrolase enzymes have synergistic actions in the degradation of lignocellulosic biomass. In the ideal case, an individual plant would produce two or more different kinds of hydrolase genes.

One way to express multiple genes in single plants is through traditional breeding: generating individual transgenic plants expressing different genes, then making crosses between different transgenic plants and selecting progeny having both transgenes. It is possible to get 2–3 transgenes into one plant, but very difficult to get more or most of the hydrolase genes into one plant in this way. Another way is to place a few of the genes in one construct and then transform the plants with the construct. Using this method, one could dramatically reduce the number of transformations.

A more promising method is to engineer multi-functional enzymes consisting of several cds for transformation. For example, a multi-functional hemicellulase gene chimera with a flexible peptide linker was engineered and expressed in *E. coli*. The chimera was constructed by linking the xylanase domain (which also has endo-glucanase activity) from the *Clostridium thermocellum* xynZ and gene encoding an enzyme with dual-functional arabinofuranosidase/ xylosidase activity from a compost starter mixture. The chimeric enzymes had four different functional activities and showed better hydrolysis on natural xylans and corn stover compared with the individual enzyme mixture.[133] The linker is vital for maintaining enzyme activity. A good linker should have a flexible secondary structure, have no protease-recognition sites, and provide sufficient space for keeping each cd right-folding.[134] Also, another two chimeric enzymes were constructed by linking the cd of xylanase to either arabinofuranosidase or a xylosidase by a flexible peptide linker. The genes were expressed in *E. coli*,[135] and the enzymes showed similarity to the mixture of individual parental enzymes. Recently, a multi-functional chimeric hydrolase gene cassette was used to transform tobacco plants, which produced 1.9% of TSPs.[134] It may be possible to engineer a binary vector that contains three to six gene cassettes.[136] If a binary vector has four trimeric gene-expression cassettes, it could possibly produce up to 12 enzymes in one transformation.

When different enzymes are combined, the ratio should also be considered. For example, TSPs containing endo-glucanases and CBH1 enzymes from transgenic plants in a ratio of 1 : 4 gave the best result in hydrolysis of the CMC (soluble cellulose) substrate in the laboratory.[66] A similar result was reported by Baker *et al.*[137] Regarding the ratios of enzymes, one could consider expressing different enzymes using diverse promoters (strong, medium, or weak), or using various subcellular targeting strategies to produce varying amounts of different enzymes.

5.6.5   Chloroplast Transformation

In addition to nuclear transformation, which is commonly known, another strategy is chloroplast transformation, which transfers genes into the chloroplast genome. Chloroplast transformation has many advantages over nuclear transformation, such as high expression levels of transgenes, transgene containment because of maternal inheritance in most crops, multiple gene engineering *via* multi-gene operons, and lack of gene silencing due to site-specific integration.[138]

Chloroplast transformation could enable the production and accumulation of foreign proteins as high as 70% of TSPs.[139] Several papers have been published on the transformation of the chloroplast genome with biomass degradation enzyme genes. For example, tobacco transformed with E1cd accumulated about 12% of TSPs in leaves.[140] Homotransplastomic (meaning that every chloroplast genome has the transgene) tobacco lines with endo-glucanase showed normal growth and reproduction.[99] Compared with enzymes produced by *E. coli*, the enzymes expressed in transplastomic tobacco plants showed higher activity, higher temperature stability and wider pH optima. The plant crude enzyme cocktails released more than 36-fold more glucose from filter paper, pine wood or citrus peel than commercial cocktails. Producing these highly active enzymes in plants could greatly reduce the cost of lignocellulosic bioethanol production. The tobacco chloroplast genome has been transformed with several genes: xyl10B from *T. maritime*, and homotransplastomic plants accumulated up to 13% of TSPs with no adverse morphology;[74] the cel6A gene, encoding an endo-glucanase enzyme from *Thermobifida fusca*, which accumulated up to 10.7% of TSPs;[141] and a β-glucosidase gene from the thermophilic bacterium *T. fusca*, which obtained up to 12% of TSPs.[76] In an experiment with hydrolysis enzymes, four genes encoding Bgl1C (β-glucosidase), Cel6B (exoglucanase), Cel9A (endo-glucanase), and Xeg74 (xyloglucanase) enzymes were individually transformed into the tobacco chloroplast genome. The enzyme cocktail from the extracts of four different transgenic plants was able to efficiently degrade pretreated wheat straw, and an almost five-fold increase of glucose release was observed compared with extracts from wild-type plants.

An important regulator element controls transgene expression in the chloroplast genome, the down-stream box (DB) region, located immediately down-stream of the start codon, consisting of 10–15 codons.[76,141] Experimental results showed that the DB from the NPTII gene was best for β-glucosidase enzyme accumulation, compared with those from TetC and GFP genes[76] while the DB from the TetC gene exhibited up to 10.7% *T. fusca* Cel6A of TSPs,[141] a 10-fold increase compared with the DB from the rbcLgene.[98] Only 0.1% of TSPs was obtained in transgenic plants with nuclear transformation.[115] So far, the mechanism of the down-stream box is not clear; it could affect translation efficiency and mRNA stability.[143,144] In addition, different 3′ untranslated regions are important for mRNA accumulation and stability in chloroplast transgenes.[145]

Chloroplast transgenes often have a high expression level, which may cause abnormal growth and development, or severe growth reduction.[140] For example, the over-expression of hydrolase enzymes in the tobacco chloroplast genome resulted in pigment-deficient mutant phenotypes,[99,146] which greatly affected photosynthesis ability. Transgene expression in the chloroplast genome is usually constitutive. Recently, Verhounig *et al.*[142] designed a synthetic riboswitch (part of an mRNA molecule directly binding to a small target molecule and affecting the gene expression),[147] which was able to induce chloroplast transgene expression.

Chloroplast genome transformation has shown promise for the production of hydrolase enzymes used in the bioconversion of lignocellulosic biomass to fermentable sugars. Tobacco transplastomic plants have the potential to produce large amounts of thermostable cell-wall-degrading enzymes as described above. Although several species, including cotton, soybean, carrot, lettuce, cabbage, eggplant, sugarbeet, tomato, and poplar have been reported to be successful with chloroplast transformation,[138] some techniques are limited to a few laboratories, and the results are not easily repeated by others. Currently only tobacco chloroplast transformation is relatively successful. Therefore, chloroplast transformation technology needs to be developed in other plant species, particularly in bioenergy crops.

5.6.6   Enzyme Activity

Although some enzymes are thermostable, thermophilic, or extremophilic, some of them lose their activity during pretreatment of lignocellulosic biomass. For example, when transgenic maize plants containing endo-glucanase E1 enzyme were pretreated using the AFEX method, the E1 enzyme lost more than 70% of its activity.[148] Drying the plant material also causes the loss of enzymatic activity. For instance, the *Thermotoga maritima* hyperthermostable β-glucosidase has an optimal temperature of 80 °C but lost more than 70% of its activity after drying at room temperature.[70] Therefore, the enzymatic activity of the heterologous proteins expressed in plants needs to be tested after plant material drying and/or the pretreatment process. In the case of activity loss due to pretreatment or material drying, the enzymes should be extracted and then used in hydrolysis.[149] Some enzymes do remain high activity in dry materials, such as xynA and xynB, which retained their high activity in dry stems of *Arabidopsis*.[89]

5.7   Future Directions

Current technology for bioethanol production from lignocellulosic biomass is neither cost-effective nor competitive to fossil fuels. In order to make bioethanol production competitive, future research should be focused on friendlier pretreatment technology and effective production of hydrolysis enzymes in feedstocks.

5.7.1   Pretreatment Technology

For pretreatment, less severe methods should be emphasized with the addition of ligninase and hemicellulase enzymes. As we discussed above, feedstocks containing hydrolysis enzymes can be pretreated at temperatures below 75 °C (mild pretreatment),[117] which will save energy and reduce pretreatment costs. As chemical pretreatment pollutes the environment, replacing it with ligninase and hemicellulase enzymes will reduce environmental pollution.[17] In addition, hemicellulases release pentose sugars, which could be fermented to ethanol by different yeast strains.[150] Therefore, the use of all sugars (hexoses and pentoses) will definitely increase the competitiveness of lignocellulosic bioethanol production.[151]

5.7.2 *In Planta* Hydrolysis Enzyme Production

Although microbial production of hydrolysis enzymes is efficient in industrial use, the cost of building fermenters and operating them is huge. *In planta* production of the hydrolysis enzymes used in lignocellulosic bioconversion is the least expensive method in the long term. However, large-scale enzyme production from plants is still far off, and many technologies need to be further improved. The following directions could be considered when techniques are developed.

5.7.2.1 *Potential Genes*

As more genome sequences become available, genes encoding various enzymes involved in lignocellulose degradation will be more easily mined out and be more efficiently modified for increasing thermostability and specific activity. Thermophilic/extremophilic microorganisms and wood-feeding termites are good resources for discovering genes with potential use. The criteria for choosing potential gene candidates are those encoding thermostable/thermophilic/ extremophilic enzymes with little or low activity during plant growth stages, but high activity post-harvest or during drying.

5.7.2.2 *Heterologous Expression in E. coli*

Once potential candidate genes have been identified, a good way to quickly characterize an enzyme is by heterologously expressing the gene in *E. coli* and then modifying the enzyme. The proteins produced by *E. coli* can be used in various enzymatic kinetics experiments.

5.7.2.3 *Heterologous Expression in Model Plants*

After the potential hydrolysis enzymes are characterized chemically and physically, heterologous expression should be conducted in model plants, such as tobacco and *Arabidopsis*, since these plants are readily transformed. Many parameters could be optimized in relatively short periods of time, such as codon optimization, removal of signal sequences, subcellular localizations, the use of induced promoters, and multi-functional chimeric genes, *etc*. After the transgenic plants are generated, the plant growth and morphology, protein accumulation levels, enzymatic activity, and activity remaining after harvest or during drying should be evaluated.

5.7.2.4   Heterologous Expression in Bioenergy Crops

Although most bioenergy crops are not easy to genetically engineer compared with model plants, some bioenergy crops, such as switchgrass and hybrid poplar, are amenable. Production of hydrolysis enzymes in bioenergy crops will greatly reduce costs of bioethanol production from these bioenergy crops.

References

1. J. Goettemoeller and A. Goettemoeller, in *Biofuels, Biorefineries, Cellulosic Biomass, Flex-Fuel Vehicles, and Sustainable Farming for Energy Independence*, Prairie Oak Publishing, *Maryville*, 2007, p. 56.
2. *Ethanol Market Penetration*, http://www.afdc.energy.gov/afdc/ethanol/market.html.
3. *Ethanol Industry Outlook*, http://ethanolrfa.org/pages/annual-industry-outlook.
4. http://www.afdc.energy.gov/afdc/vehicles/index.html.
5. *Historic U. S. Fuel Ethanol Production*, http://www.ethanolrfa.org/pages/statistics#A.
6. *World Fuel Ethanol Production*, http://ethanolrfa.org/pages/World-Fuel-Ethanol-Production.
7. http://biomassmagazine.com/blog/article/2011/10/biomass-power-map-sneak-peek.
8. Ethanol Production Mandates, http://www.e85prices.com/.
9. *Energy Information Administration*, http://www.eia.doe.gov/ask/renewables_faqs. asp#ethanol_affect_fuel_economy.
10. P. J. Crutzen, A. R. Mosier, K. A. Smith and W. Winiwarter, *Atmos. Chem. Phys. Discuss.*, 2007, **7**, 11191.
11. Using Biofuel Tax Credits to Achieve Energy and Environmental Policy Goals, http://www.cbo.gov/ftpdocs/114xx/doc11477/07-14-Biofuels.pdf.
12. Federal Biomass Policy, http://www1.eere.energy.gov/biomass/federal_biomass.html.
13. R. D. Perlack, L. L. Wright, A. F. Turhollow, R. L. Graham, B. J. Stokes, and D. C. Erbach, *Biomass as a Feedstock for a Bioenergy and Bioproducts Industry: The Technical Feasibility of a Billion-Ton Annual Supply*. US Department of Energy and US Department of Agriculture, 2005.
14. S. Kim and B. E. Dale, *Biomass Bioenergy*, 2004, **26**, 361.
15. M. Knauf and M. Moniruzzaman, *Int. Sugar J.*, 2004, **106**, 147.
16. N. Greene, F. E. Celik, B. Dale, M. Jackson, K. Jayawardhana, H. Jin, E. D. Larson, M. Laser, L. Lynd, D. MacKenzie, J. Mark, J. McBride, S. McLaughlin and D. Saccardi, *Growing Energy: How Biofuels Can Help End America's Oil Dependence*, 2004.

17. J. A. Howard, Z. Nikolov and E. E. Hood, in: *Plant Biomass Conversion*, ed. E. E. Hood, P. Nelson and R. Powell, John Wiley & Sons Inc., 2011, p. 227.
18. N. Carpita and M. McCann, in: *Biochemistry & Molecular Biology of Plants*, ed. B. Buchanan, W. Gruissem and R. L. Jones, John Wiley & Sons, 2002, p. 52.
19. R. L. Crawford, *Lignin Biodegradation and Transformation*. New York, John Wiley and Sons, 1981, ISBN 0-471-05743-6.
20. D. M. Updegraff, *Anal. Biochem.*, 1969, **32**, 420.
21. van den J. Brink and de R. P. Vries, *Appl. Microbiol. Biotechnol.*, 2011, **91**, 1477.
22. A. T. Hendriks and G. Zeeman, *Bioresour. Technol.*, 2009, **100**, 10.
23. N. A. Eckardt, *Plant Cell*, 2002, **14**, 1185.
24. N. Mosier, C. Wyman, B. E. Dale, R. Elander, Y. Y. Lee, M. Holtzapple and M. Ladisch, *Bioresour. Technol.*, 2005, **96**, 673.
25. R. A. J. Warren, *Annu. Rev. Microbiol.*, 1996, **50**, 183.
26. T. M. D'Souza, C. S. Merritt and C. A. Reddy, *Appl. Environ. Microbiol.*, 1999, **65**, 5307.
27. K. Boominathan and C. A. Reddy, *Proc. Natl. Acad. Sci. USA*, 1992, **89**, 5586.
28. C. F. Thurston, *Microbiol.*, 1994, **140**, 19.
29. N. Liu, Y. Yan, M. Zhang, L. Xie, Q. Wang, Y. Huang, X. Zhou, S. Wang and Z. Zhou, *Appl. Environ. Microbiol.*, 2011, **77**, 48.
30. A. Pollet, J. A. Delcour and C. M. Courtin, *Crit. Rev. Biotechnol.*, 2010, **30**, 176.
31. P. Biely, M. Vrsanska, M. Tenkanen and D. Kluepfel, *J. Biotechnol.*, 1997, **57**, 151.
32. H. B. Klinke, A. B. Thomsen and B. K. Ahring, *Appl. Microbiol. Biotechnol.*, 2004, **66**, 10.
33. L. Olsson and B. Hahn-Hägerdal, *Enzyme Microbiol. Technol.*, 1996, **18**, 312.
34. S. Deguchi, K. Tsujii and K. Horikoshi, *Chem. Commun.*, 2006, **31**, 3293.
35. T. Eggeman and R. T. Elander, *Bioresour. Technol.*, 2005, **96**, 2019.
36. C. E. Wyman, B. E. Dale, R. T. Elander, M. Holtzapple, M. R. Ladisch and Y. Y. Lee, *Bioresour. Technol.*, 2005, **96**, 2026.
37. C. E. Wyman, B. E. Dale, R. T. Elander, M. Holtzapple, M. R. Ladisch and Y. Y. Lee, *Bioresour. Technol.*, 2005, **96**, 1959.
38. J. Y. Zhu, X. J. Pan, G. S. Wang and R. Gleisner, *Bioresour. Technol.*, 2009, **100**, 2411.
39. A. K. Chandel, G. Chandrasekhar, K. Radhika, R. Ravinder and P. Ravindra, *Biotechnol. Mol. Biol. Rev.*, 2011, **6**, 8.
40. H. Jørgensen, J. B. Kristensen and C. Felby, Biofuels, *Bioprod. Biorefin.*, 2007, **1**, 119.
41. *Renewables Global Status Report*, http://www.ren21.net/REN21Activities/Publications/GlobalStatusReport/GSR2010/tabid/5824/Default.aspx.
42. http://www.ens-newswire.com/ens/jul2011/2011-07-12-091.html.

43. http://www.iogen.ca/company/about/index.html.
44. http://www.ethanolfuelresource.com/index.php.
45. http://www.inbicon.com/About_inbicon/News/Data/Pages/ All_systems_go_ at_world%E2%80%99s_largest_ cellulosic_ethanol_plant.aspx.
46. http://www.biogasol.dk/ About-Us-60.aspx.
47. http://dinby.dk/bolderslev/ensted-vaerket-kan-producere-bio-ethanol?minby= bolderslev#minby:bolderslev.
48. http://www.maabjergenergyconcept.dk/delprojekter.aspx.
49. http://www.ethanolproducer.com/articles/7659/worldundefineds-largest-cellulosic-ethanol-plant-breaks-ground-in-italy.
50. http://www.biofuelreview.com/index.php?option=com_content&task = view&id = 1062.
51. http://www.newsdesk.se/pressroom/policom_consulting_ab/pressrelease/ view/american-interest-in-sekabs-cellulosic-ethanol-technology-279045.
52. W. M. Ingledew, in *The Alcohol Textbook*, ed. T. P. Lyons, D. Kelsall and J. Murtagh, Nottingham University Press, Nottingham, 1995, p. 55.
53. L. R. Lynd, W. H. van Zyl, J. E. McBride and M. Laser, *Curr. Opin. Biotechnol.*, 2005, **16**, 577.
54. M. A. Kabel, M. J. E. C. van der Maarel, G. Klip, A. G. J. Voragen and H. A. Schols, *Biotechnol. Bioeng.*, 2006, **93**, 56.
55. M. Taherzadeh and K. Karimi, *BioResources*, 2007, **2**, 707.
56. http://sustainable-energy.ksu.edu/files/cse//Wang.pdf.
57. H. Chen and W. Qiu, *Biotechnol. Adv.*, 2010, **28**, 556.
58. (a) J. K. Ma, R. Chikwamba, P. Sparrow, R. Fischer, R. Mahoney and R. M. Twyman, *Trends Plant Sci.*, 2005, **10**, 580; (b) R. L. Howard, E. Abotsi, E. L. J. van Rensburg and S. Howard, *Afr. J. Biotechnol.*, 2003, **2**, 602.
59. M. B. Sainz, *In Vitro Cell. Dev. Biol.: Plant*, 2009, **45**, 314.
60. L. E. Taylor, A. Dai, S. R. Decker, R. Brunecky, W. S. Adney, S.-Y. Ding and M. E. Himmel, *Trends Biotechnol.*, 2008, **26**, 413.
61. T. Kawazu, T. Ohta, K. Ito, M. Shibata, T. Kimura, K. Sakka and K. Ohmiya, *J. Ferment. Bioeng.*, 1996, **82**, 205.
62. T. Kawazu, J. L. Sun, M. Shibata, T. Kimura, K. Sakka and K. Ohmiya, *J. Biosci. Bioeng.*, 1999, **88**, 421.
63. S. Kim, D.-S. Lee, I. S. Choi, S.-J. Ahn, Y.-H. Kim and H.-J. Bae, *Transgenic Res.*, 2010, **19**, 489.
64. V. J. Maloney and S. D. Mansfield, *Plant Biotechnol. J.*, 2010, **8**, 294.
65. M. D. Harrison, J. Geijskes, H. D. Coleman, K. Shand, M. Kinkema, A. Palupe, R. Hassall, M. Sainz, R. Lloyd, S. Miles and J. L. Dale, *Plant Biotechnol. J.*, 2011, **9**, 884.
66. S.-H. Park, C. Ransom, C. Mei, R. Sabzikar, C. Qi, S. Chundawat, B. Dale and M. Sticklen, *J. Chem. Technol. Biotechnol.*, 2011, **86**, 633.
67. E. E. Hood, S. P. Devaiah, G. Fake, E. Egelkrout, K. Teoh, D. V. Requesens, C. Hayden, K. R. Hood, K. M. Pappu, J. Carroll and J. A. Howard, *Plant Biotechnol. J.*, 2012, **10**, 20.

68. R. P. Rao, N. Dufour and J. Swana, *In Vitro Cell. Dev. Biol.: Plant*, 2011, **47**, 637.

69. C. B. Ransom, *Production and Analysis of Biologically-Active Cellulases for Ethanol Fuel in Maize Biomass*, PhD thesis, Michigan State University, 2007.

70. S. R. Jung, S. Y. Kim, H. H. Bae, H. S. Lim and H. J. Bae, *Bioresour. Technol.*, 2010, **101**, 7144.

71. K.-H. Kim, K. M. Brown, P. V. Harris, J. A. Langston and J. R. Cherry, *J. Proteome Res.*, 2007, **6**, 4749.

72. J. D. Willis, C. Oppert and J. L. Jurat-Fuentes, *Insect Sci.*, 2010, **17**, 184.

73. M. L. T. M. Polizeli, A. C. S. Rizzatti, R. Monti, H. F. Terenzi, J. A. Jorge and D. S. Amorim, *Appl. Microbiol. Biotechnol.*, 2005, **67**, 577.

74. J. Y. Kim, M. Kavas, W. M. Fouad, G. Nong, J. F. Preston and F. Altpeter, *Plant Mol. Biol.*, 2011, **76**, 357.

75. T. Kimura, T. Mizutani, J.-L. Sun, T. Kawazu, S. Karita, M. Sakka, Y. Kobayashi, K. Ohmiya and K. Sakka, *Biosci. Biotechnol. Biochem.*, 2010, **75**, 954.

76. B. N. Gray, H. Yang, B. A. Ahner and M. R. Hanson, *Plant Mol. Biol.*, 2011, **76**, 345.

77. T. Kimura, T. Mizutani, T. Tanaka, T. Koyama, K. Sakka and K. Ohmiya, *Appl. Microbiol. Biotechnol.*, 2003, **62**, 374.

78. K. Emami, T. Nagy, C. M. G. A. Fontes, L. M. A. Ferreira and H. J. Gilbert, *J. Bacteriol.*, 2002, **184**, 4124.

79. T. K. Ihsanawati, T. Kaneko, C. Morokuma, R. Yatsunami, T. Sato, S. Nakamura and N. Tanaka, *Proteins*, 2005, **61**, 999.

80. C. Winterhalter and W. Liebl, *Appl. Environ. Microbiol.*, 1995, **61**, 1810.

81. A. L. Van Fossen, I. Ozdemir, S. L. Zelin and R. M. Kelly, *Biotechnol. Bioeng.*, 2011, **108**, 1559.

82. J. Z. Sun and X. G. Zhou, in *Recent Advances of Entomological Research: from Molecular Biology to Pest Management*, ed. T. X. Liu and L. Kang, Springer, Berlin, 2010, p. 434.

83. J. Ni, M. Takehara and H. Watanabe, *Biosci. Biotechnol. Biochem.*, 2005, **69**, 1711.

84. J. Ni, M. Takehara, M. Miyazawa and H. Watanabe, *Protein Eng., Des. Sel.*, 2007, **20**, 535.

85. J. Ni, G. Taokuda, M. Takehara and H. Watanabe, *Appl. Entomol. Zool.*, 2007, **42**, 457.

86. E. E. Hood, M. R. Bailey, K. Beifuss, M. Horn, M. Magallanes-Lundback, C. Drees, D. E. Delaney, R. Clough and J. A. Howard, *Plant Biotechnol. J.*, 2003, **1**, 129.

87. C. de Wilde, E. Uzan, Z. Zhou, K. Kruus, M. Andberg, J. Buchert, E. Record, M. Asther and A. Lomascolo, *Transgenic Res.*, 2008, **17**, 515.

88. R. C. Clough, K. Beifuss, J. Lane, K. Pappu, K. Thompson, M. R. Bailey, D. E. Delaney, R. Harkey, C. Drees, J. A. Howard and E. E. Hood, *Plant Biotechnol. J.*, 2006, **4**, 53.

89. B. Borkhardt, J. Harholt, P. Ulvskov, B. K. Ahring, B. Jørgensen and H. Brinch-Pedersen, *Plant Biotechnol. J.*, 2010, **8**, 363.
90. H. Bae, D.-S. Lee and I. Hwang, *J. Exp. Bot.*, 2006, **57**, 161.
91. M. Sticklen, *Crop Sci.*, 2007, **47**, 2238.
92. Z. Dai, B. S. Hooker, D. B. Anderson and S. R. Thomas, *Transgenic Res.*, 2000, **9**, 43.
93. G. Biswas, C. Ransom and M. Sticklen, *Plant Sci.*, 2006, **171**, 617.
94. Z. Dai, B. S. Hooker, D. B. Anderson and S. R. Thomas, *Mol. Breeding*, 2000, **6**, 277.
95. Z. Dai, B. S. Hooker, R. D. Quesenberry and S. R. Thomas, *Transgenic Res.*, 2005, **14**, 627.
96. C. Mei, C.-H. Park, R. Sabzikar, C. Qi, C. Ransom and M. B. Sticklen, *J. Chem. Technol. Biotechnol.*, 2009, **84**, 689.
97. E. E. Hood, R. Love, J. Lane, J. Bray, R. Clough, K. Pappu, C. Drees, K. R. Hood, S. Yoon, A. Ahmad and J. A. Howard, *Plant Biotechnol. J.*, 2007, **5**, 709.
98. L.-X. Yu, B. N. Gray, C. J. Rutzke, L. P. Walker, D. B. Wilson and M. R. Hanson, *J. Biotechnol.*, 2007, **131**, 362.
99. D. Verma, K. Anderson, S. Jin, N. D. Singh, P. E. Kolattukudy and H. Daniell, *Plant Biotechnol. J.*, 2010, **8**, 332.
100. N. Moreira, *Sci. News*, 2005, **168**, 218.
101. K. Herbers, I. Wilke and U. Sonnewald, *Biotechnol.*, 1995, **13**, 63.
102. K. Herbers, H. J. Flint and U. Sonnewald, *Mol. Breeding*, 1996, **2**, 81.
103. T. Ziegelhoffer, J. A. Raasch and S. Austin-Phillips, *Mol. Breeding*, 2001, **8**, 147.
104. Y. Sun, J. J. Cheng, M. E. Himmel, C. D. Skory, W. S. Adney, S. R. Thomas, B. Tisserat, Y. Nishimura and Y. T. Yamamoto, *Bioresour. Technol.*, 2007, **98**, 2866.
105. M. T. Zeigler, S. R. Thomas and K. J. Danna, *Mol. Breeding*, 2000, **6**, 37.
106. R. Brunecky, M. J. Selig, T. B. Vinzant, M. E. Himmel, D. Lee, M. J. Blaylock and S. R. Decker, *Biotechnol. Biofuels*, 2011, **4**, 1.
107. I. Levy and O. Shoseyov, *Biotechnol. Adv.*, 2002, **20**, 191.
108. I. Levy, Z. Shani and O. Shoseyov, *Biomol. Eng.*, 2002, **19**, 17.
109. S. A. Mahadevan, S. G. Wi, D. S. Lee and H. J. Bae, *FEMS Microbiol. Lett.*, 2008, **287**, 205.
110. S. A. Mahadevan, S. G. Wi, Y. O. Kim, K. H. Lee and H. J. Bae, *Transgenic Res.*, 2011, **20**, 877.
111. H. J. Bae, G. Turcotte, H. Chamberland, S. Karita and L.-P. Vezina, *FEMS Microbiol. Lett.*, 2003, **227**, 175.
112. G. Lopez-Casado, B. R. Urbanowicz, C. M. B. Damasceno and J. K. C. Rose, *Curr. Opin. Plant Biol.*, 2008, **11**, 329.
113. M. Abramson, O. Shoseyov and Z. Shani, *Plant Sci.*, 2010, **178**, 61.
114. L. Safra-Dassa, Z. Shani, A. Danin, L. Roiz, O. Shoseyov and S. Wolf, *Mol. Breeding*, 2006, **17**, 355.
115. T. Ziegelhoffer, J. Will and S. Austin-Phillips, *Mol. Breeding*, 1999, **5**, 309.

116. J. Harholt, C. Inga, I. C. Bach, S. Lind-Bouquin, K. J. Nunan, S. M. Madrid, H. Brinch-Pedersen, B. Preben, P. B. Holm and H. V. Scheller, *Plant Biotechnol. J.*, 2010, **8**, 351.

117. D. Zhang, A. L. Van Fossen, R. M. Pagano, J. S. Johnson, M. H. Parker, S. Pan, B. N. Gray, E. Hancock, D. J. Hagen, H. A. Lucero, B. Shen, P. A. Lessard, C. Ely, M. Moriarty, N. A. Ekborg, O. Bougri, V. Samoylov, G. Lazar and R. M. Raab, *Bioenergy Res.*, 2011, **4**, 276.

118. N. C. Carpita and D. M. Gibeaut, *Plant J.*, 1993, **3**, 1.

119. C. Fu, J. R. Mielenz, X. Xiao, Y. Ge, C. Hamilton, M. Rodriguez, F. Chen, M. Foston, A. Ragauskas, J. Bouton, R. A. Dixon and Z.-Y. Wang, *Proc. Natl. Acad. Sci. U. S. A.*, 2011, **108**, 3803.

120. C. Fu, X. Xiao, Y. Xi, Y. Ge, F. Chen, J. Bouton, R. A. Dixon and Z.-Y. Wang, *Bioenergy Res.*, 2011, **4**, 153.

121. B. Xu, L. L. Escamilla-Treviño, N. Sathitsuksanoh, Z. Shen, H. Shen, Y. H. P. Zhang, R. A. Dixon and B. Zhao, *New Phytol.*, 2011, **192**, 611.

122. A. Margeot, B. Hahn-Hagerdal, M. Edlund, R. Slade and F. Monot, *Curr. Opin. Biotechnol.*, 2009, **20**, 1.

123. S. T. Merino and J. Cherry, *Adv. Biochem. Eng. Biotechnol.*, 2007, **108**, 95.

124. R. Li and R. Qu, *Biomass Bioenergy*, 2011, **35**, 1046.

125. J. Song, S. Lu, Z. Z. Chen, R. Lourenco and V. L. Chiang, *Plant Cell Physiol.*, 2006, **47**, 1582.

126. E. E. Murray, J. Lotzer and M. Eberle, *Nucleic Acids Res.*, 1989, **17**, 477.

127. http://bioinfo.weizmann.ac.il/~pietro/inteins/.

128. http://www.agrivida.com/news/releases/2010december13.html.

129. J. Yang, F. Chen, O. Yu and R. N. Beachy, *Plant Physiol. Biochem.*, 2011, **49**, 103.

130. C. M. Griffiths, S. E. Hosken, D. Oliver, J. Chojecki and H. Thomas, *Plant. Mol. Biol.*, 1997, **34**, 815.

131. S. Gan and R. M. Amasino, *Science*, 1995, **270**, 1966.

132. P. R. H. Robson1, I. S. Donnison1, K. Wang, B. Frame, S. E. Pegg, A. Thomas and H. Thomas, *Plant Biotechnol. J.*, 2004, **2**, 101.

133. Z. Fan, J. R. Werkman and L. Yuan, *Biotechnol. Lett.*, 2009, **31**, 751.

134. Z. Fan and L. Yuan, *Plant Biotechnol. J.*, 2010, **8**, 308.

135. Z. Fan, K. Wagschal, C. C. Lee, Q. Kong, K. A. Shen, I. B. Maitiand and L. Yuan, *Biotechnol. Bioeng.*, 2009, **102**, 684.

136. S. M. Chung, E. L. Frankman and T. Tzfira, *Trends Plant Sci.*, 2005, **10**, 357.

137. J. O. Baker, C. I. Ehrman, W. S. Adney, S. R. Thomas and M. E. Himmel, *Appl. Biochem. Biotechnol.*, 1998, **70–72**, 395.

138. H. Daniell and K. J. Edwards, *Plant Biotechnol. J.*, 2011, **9**, 526.

139. M. Oey, M. Lohse, B. Kreikemeyer and R. Bock, *Plant J.*, 2009, **57**, 436.

140. T. Ziegelhoffer, J. A. Raasch and S. Austin-Phillips, *Plant Biotechnol. J.*, 2009, **7**, 527.

141. B. N. Gray, B. A. Ahner and M. R. Hanson, *Biotechnol. Bioeng.*, 2009, **102**, 1045.

142. A. Verhounig, D. Karcher and R. Bock, *Proc. Natl. Acad. Sci. U. S. A.*, 2010, **107**, 6204.
143. H. Kuroda and P. Maliga, *Nucleic Acids Res.*, 2001, **29**, 970.
144. H. Kuroda and P. Maliga, *Plant Physiol.*, 2001, **125**, 430.
145. S. Tangphatsornruang, I. Birch-Machin, C. A. Newell and J. C. Gray, *Plant Mol. Biol.*, 2011, **76**, 385.
146. K. Petersen and R. Bock, *Plant Mol. Biol.*, 2011, **76**, 311.
147. http://en.wikipedia.org/wiki/Riboswitch.
148. F. Teymouri, H. Alizadeh, L. Laureano-Preze, B. Dale and M. Sticklen, *Appl. Biochem. Biotechnol.*, 2004, **116**, 1183.
149. C. Ransom, V. Balan, G. Biswas, B. Dale, E. Crockett and M. Sticklen, *Appl. Biochem. Biotechnol.*, 2007, **137–140**, 207.
150. A. Matsushika, H. Inoue, T. Kodaki and S. Sawayama, *Appl. Microbiol. Biotechnol.*, 2009, **84**, 37.
151. M. E. Himmel, S. Y. Ding, D. K. Johnson, W. S. Adney, M. R. Nimlos, J. W. Brady and T. D. Foust, *Science*, 2007, **315**, 804.

CHAPTER 6

The Diversity of Lignocellulosic Biomass Resources and their Evaluation for Use as Biofuels and Chemicals

PENG CHEN[a] AND LIANGCAI PENG*[a,b]

[a] National Key Laboratory of Crop Genetic Improvement, Biomass and Bioenergy Research Centre, and College of Life Science and Technology, Huazhong Agricultural University, Wuhan 430070, P. R. China;
[b] College of Plant Science and Technology, Huazhong Agricultural University, Wuhan 430070, P. R. China
*Email: lpeng@mail.hzau.edu.cn

6.1 The Diversity of Lignocellulosic Biomass Resources

According to an International Energy Agency report, a 50% reduction in CO_2 emissions by 2050 requires at least a $4\times$ increase in the use of bioenergy, reaching 1.5×10^{20} J per year, equivalent to 20% of the World's primary energy. Most of this increase will likely come from lignocellulosic biomass, and 1.5×10^{10} tons per year of biomass with a conversion efficiency of 60% and an energy content of 17 MJ kg^{-1} dry matter will be required. Lignocellulose is the most abundant biomaterial on Earth, and a conservative estimation of global biomass production could be around 10 Mg per hectare per year.[1,2] According

RSC Energy and Environment Series No. 10
Biological Conversion of Biomass for Fuels and Chemicals: Explorations from Natural Utilization Systems
Edited by Jianzhong Sun, Shi-You Ding and Joy Doran-Peterson
© The Royal Society of Chemistry 2014
Published by the Royal Society of Chemistry, www.rsc.org

to predictions by the Department of Energy (DOE) and the Department of Agriculture in the USA, biomass production will reach 10×10^{12} tons per year by 2050, and lignocellulosic ethanol will constitute 30% of the total liquid fuel. Therefore, lignocellulosic ethanol has the potential to meet most global transportation fuel needs with lower agricultural input and lower net CO_2 emissions than fossil fuels, and its replacement of first-generation bioenergy will resolve the conflict between energy demand and food supply. However, the main challenges associated with the development of lignocellulosic biofuels include maximization of biomass yield per hectare per year, improvement of biomass quality and maintenance of sustainability while minimizing agricultural input, and prevention of competition with food production.

Sources of lignocellulosic biomass include forestry products, crop residues, energy plants such as switchgrass, *Miscanthus*, sugarcane and arundo donax, and other waste products. Among all of the energy plants, the one with highest annual dry-matter production is Napier Grass (*Pennisetum purpureum*) in Salvador (88 MT per hectare per year) and *E. polystachya* in the Amazon floodplains (100 MT per hectare per year).[3] In the temperate zone, *M. giganteus* with a peak biomass of 30 MT per hectare per year is of considerable interest. Marker-assisted breeding and transgenic techniques are being used to select energy crops with high *biomass e.g.*, 150–250 tons per hectare. Higher yields are likely from sugarcane–*Miscanthus* hybrids, energy canes and Napier grass under proper conditions.[4] The leading lignocellulosic feedstocks today are corn stover and rice straw, as well as perennial grasses such as switchgrass and *Miscanthus*, and woody lignocelluloses such as poplar and eucalyptus. In addition, species with a high tolerance to drought or cold conditions are also being developed (Table 6.1).

Table 6.1   Bioethanol production from various lignocellulosic energy plants.[16]

Lignocellulosic energy plants	Biomass dry yield/ MT per hectare per year	Bioethanol production/ L per hectare	Caloric value/ $MJ\ kg^{-1}$	Water requirements/ mm per year	Nitrogen requirements/ kg per hectare per year	Tolerance to drought
Rice straw	25	500	17.6	700–1200	45–340	Low
Corn stover	30	900	17.7	500–800	90–120	Low
Wheat straw	30	500	18.5	300–400	200–500	Low
Sweet sorghum	60–105	2700	15.0	400–800	18–27	High
Switchgrass	10–25	5000	17.0	100–150	0–210	High
Miscanthus	15–40	4600–12400	9–17	750–1200	0–15	Moderate
Reed canarygrass	15–35	n.a.	5–19	1000	0–140	Moderate
Agave spp.	10–34	3000–10500	n.a.	300–800	0–12	High
Poplar	5–11	1500–3400	17.9	700–1050	0–50	Moderate

n.a. = not available.

6.1.1 Residues of Food Crops

6.1.1.1 Wheat (Triticum aestivum), Rice (Oryza sativa) and Maize (Zea mays)

Wheat, rice and maize are major food crops that make up about 75% of the total worldwide agricultural residues. In China, three crop residues constitute about 70% of the total biomass at 7.5×10^8 tons per year, equivalent to 3.5×10^8 tons of standard coal.[5] About 23% of the crop residues are used for forage, 4% for industry materials, and 0.5% for biogas, but the majority is burned or wasted (37% is directly combusted by farmers, 15% is lost during collection and the remaining 20.5% is discarded or directly burnt in the field).[6] In fact, significant amounts of crop residue need to be left on the ground for soil conservation and sustainable yield production, and only about 15–20% of total residue could be available for bioenergy purposes.[7]

Bread wheat (*Triticum aestivum*) is a small grain hexaploid ($2n = 6x = 42$) which originates from ancient crescents that existed both as diploids ($2n = 2x = 14$) and tetraploids ($2n = 4x = 28$). Wheat grain has evolved and been selected for more than a thousand years, with an improved lodging resistance and high yield. Wheat straw can be combusted for the heat power in industry, but it can also be used for bioethanol production. About 731 million tons of rice straw is produced annually,[8] which could be potentially converted into about 205 billion liters of bioethanol each year, which is the largest amount from a single biomass feedstock.[9] Maize (*Zea mays*) is a large-grain diploid cereal originating from Central America, and its grain was used in the USA during the early development of the bioethanol industry, comparable to sugarcane in Brazil. As a C4 plant, maize provides a greater biomass residue than C3 plants such as wheat and rice. In recent years, a greater number of traits has been modified in maize than in wheat, including pest and disease resistance, lodging, and dry mass partitioning. In contrast to wheat and rice, maize shows less of a response to nitrogen fertilizer, therefore mutual shading and abiotic stress tolerance have been the targets of genetic breeding to improve its yield.[10]

6.1.1.2 Sweet Sorghum (Sorghum bicolor L. Moench)

Sweet sorghum is a breeding line from the ordinary grain sorghum species. It grows fast and has a high tolerance to drought, submergence and salt stresses. Its stalk contains 17–21% sugar content, and it weighs 60–80 tons per hectare.[11] By estimate, sweet sorghum could potentially produce 20 million tons of bioethanol in the alkaline soils in the north of China.[12] With a high yield of biomass (60–75 tons per hectare) and fermentable sugars and a relatively low input, sweet sorghum has received considerable attention as a desirable feedstock for bioethanol production in China. Currently, various germplasm collection and genetic breeding strategies are being used to screen out the specific sweet sorghum materials/varieties either with high sugar content in their stalk/grain or with high biomass digestibility. For instance, the brown midrib (*bm*) mutants of sweet sorghum have showed a reduced lignin content and high lignocellulosic digestibility.[13,14] Recently, the Chinese Academy of Sciences (CAS) organized a worldwide collection of potential energy plants,[15] and Dr Jing and his colleagues at the Institute of Botany, have collected hundreds of sweet sorghum germplasm materials. Dr Peng's research team at the Biomass and Bioenergy Center of Huazhong Agricultural University have collaborated to screen out several promising sweet sorghum materials that have high sugar content and cost-effective biomass digestibility.

6.1.2 Non-Food Crop Plants

Due to concerns regarding food security, food crops are questionable sources of sustainable energy plants, therefore non-food or dedicated energy plants are of most interest worldwide. The use of non-food energy crops has several advantages, including less occupation of land, better cost-competitive prices, and a wide range of feedstock resources.[7]

6.1.2.1 Herbaceous Lignocellulosic Plants

Switchgrass is a widely adapted endemic species of the North American native ecosystem. It is a cross-pollinated and short day plant with two ecotypes, the upland ecotypes are mostly octaploids ($2n = 8x = 72$). Switchgrass is a C4 plant with high productivity across a wide geographical range on diverse agricultural sites in the USA, especially on relatively poor-quality sites where water and nutrients are limiting factors. Its powerful underground rhizome also recycles nutrients back to the soil which in the long run provides soil stability and reduces energy input, with a relatively low management cost. Therefore, switchgrass is top of the list among the 18 perennial grass species investigated by the USA herbaceous energy crop research project,[16] and its considerable within-species variability suggests space to increase yield through further breeding. The first registered switchgrass species "Shawnee" has an annual yield of 14–20 tons per hectare, which could produce 5000 L bioethanol at a conversion rate of 75%. A study of four switchgrass genotypes revealed a similar cellulose and hemicellulose composition but significant differences in lignin content, especially the ratio of lignin monomers and amounts of *p*-coumaric and ferulic acids,[17] which favors the use of alkaline-based pretreatment on switchgrass biomass because the base-catalyzed saponification would readily cleave *p*-coumaric acid and ferulic esters to facilitate removal of lignin in favor of subsequent enzymatic hydrolysis.[18–20] In addition, the full genome of switchgrass has been sequenced and marker-assisted breeding is being exploited to improve the biomass yield.

Miscanthus is a C4 perennial plant originating from East Asia, and has the highest biomass yield among all of the herbaceous energy plants. The height of *M. giganteus* can reach 7–10 m and theoretical biomass production could even reach 22 tons per hm^2. Because of its high calorific value ($18.2\,MJ\,kg^{-1}$), it is widely used in Europe in the electricity and gas production industries. The estimated energy ratio (energy output of a system *vs.* the energy input needed to operate it) of *Miscanthus* grown under UK conditions is 37.5 : 1, which is greater than that of switchgrass, willow and reed canarygrass whose energy ratios were estimated to be 35 : 1, 22 : 1 and 18 : 1, respectively.[21] *Miscanthus* also exhibits better combustion qualities due to its low moisture and ash content, especially for winter harvest materials, based on the reports from The European Miscanthus Improvement Project.[22] Data from this project based on 15 *Miscanthus* genotypes show that *M. giganteus* and *M. sacchariflorus* have higher lignin contents and significantly lower hemicellulose contents than the *M. sinensis* genotypes. The lignin and cellulose contents of all genotypes are higher in the winter harvest, based on proportion of dry weight, and *M. giganteus* contains the highest concentration of cellulose throughout all European countries. Little variation was observed between *M. giganteus* genotypes, which suggests that the *M. giganteus* hybrids currently cultivated in Europe are probably derived from the same parental germplasm.[22]

C4 plants have a significantly higher photosynthetic efficiency yield ($50 \, \text{gm}^{-2} \, \text{day}^{-1}$) than C3 plants ($30–40 \, \text{gm}^{-2} \, \text{day}^{-1}$), and intrinsically greater water use efficiency ($220–350 \, \text{g}$ of dry matter per liter of H_2O) than C3 species ($400–1000 \, \text{g}$ of dry matter per liter of H_2O).[23] Hence, *Miscanthus* can store around 30% of its total dry matter in its root and rhizome. High water and nutrient usage efficiency reduce the demand for fertilizers during the long-term growth of *Miscanthus* in the field. For instance, *Miscanthus* cultivation in Rothamsted, UK, during a 14-year period in which no nitrogen was added to the field, showed a stable output with all above-ground biomass removed each year.[24] The biomass yields from *Miscanthus* are generally higher compared to short rotation coppice from woody plants, which has a low energy input (around 9 MJ per hectare compared to 21 MJ per hectare from wheat production).[25] However, to achieve the expected yields, *Miscanthus* planting densities need to reach around 20 000 plants per hectare, which requires a massive propagation of seedlings either by rhizome production, micro-propagation techniques or stem cutting production systems because the species presently used, *M. giganteus*, is a triploid ($2n = 3x = 57$) and is self-sterile. As a drawback, *Miscanthus* must be propagated vegetatively *via* rhizome cuttings due to its self-incompatibility. The aim of a breeding program for *Miscanthus* should include the seed sterility of hybrids and better over-winter survival. The present species, *M. giganteus*, is most likely a hybrid of *M. sacchariflorus* and *M. sinensis*, as it has the rhizome morphology and flowering period in between these two parents. In southern and central Europe, *M. giganteus* has the highest yield but in northern Europe *M. sinensis* genotypes or even some C3 plants outperform *M. giganteus* when the soil temperature drops below $-3.5 \, °C$.[26]

Miscanthus mainly originates from East Asia and nearby Pacific islands, and more than 17 species have been identified worldwide.[27,28] Because the genetic origin of *Miscanthus* is in East Asia, Hunan Agricultural University and Wuhan Botany Garden CAS have collected more than 1400 natural *Miscanthus* accessions including four major species (*Miscanthus sacchariflorus*, *Miscanthus lutarioriparius*, *Miscanthus sinensis*, and *Miscanthus floridulus*) across mainland China. Each species covers different ecological and regional types, which can be attributed to a diverse germplasm resource. In order to select-out desirable *Miscanthus* as energy plants, Dr Peng's laboratory from Huazhong Agricultural University has investigated 300 representative accessions including biomass yield, cell-wall composition/structure, biomass digestibility and biofuel productivity (unpublished data). Several *Miscanthus* materials have been selected and are under development as desirable energy plants, and a genetic model has been proposed for the genetic modification of *Miscanthus* to produce plants with high biomass digestibility and improved ecological adaptation.

Alfalfa is one of the world's oldest forage crops, and it is considered as a dual-use plant from which leaves are separated for high-protein feed and the lignified stems can be used for energy production. As a lignocellulosic energy plant, alfalfa has special characteristics: soluble carbohydrates, especially pectin, constitutes 40% of its non-cellulosic sugar, therefore no pretreatment is needed and the sugar–ethanol conversion rate is much higher. Another key factor making alfalfa a promising energy crop is a high-value by-product which can be used as an adhesive to replace the highly toxic phenol formaldehyde. Genetic selection has focused on creating lines with stiff stems of increased inter-node length and the generation of biomass-type germplasm with higher cell-wall polysaccharides, which could contribute to a greater stem dry matter yield and theoretical ethanol yields compared with hay-type alfalfas.

According to the USA Herbaceous Energy Crop Research Project, giant reed and reed canarygrass have also been identified as good candidate herbaceous energy plants among the 17 species tested.[70] Reed canarygrass is a tall, coarse and erect C3 grass, but it is a highly self-sterile allotetraploid plant ($2n = 4x = 28$). Reed canarygrass has adapted to grow fairly well in cool temperate climates; and is quite resistant to drought conditions. It has been used as a forage crop, and the current breeding program for reed canarygrass includes reducing the nitrogen and potassium requirements and increasing the lignin content to produced thicker stalks.[29] Reed canarygrass breeding for industrial use has been carried out and the European Giant Reed Network was established in 1997. The results of the survey suggest reed canarygrass as a promising energy plant for northern Europe based on the following advantages: it has already adapted to short vegetation periods and low temperatures; overwintering is safe; seed establishment is possible; and its biomass is of high quality. Giant reed is recommended for Mediterranean regions for the following reasons: it has already adapted to the site conditions; it has a high biomass yield; it requires low irrigation and nitrogen inputs; it has a high resistance to drought; moderate genetic variability and the possibility of delayed harvest once a year.[16]

6.1.2.2 Woody Lignocellulosic Plants

Trees provide potentially higher calorific values for biofuel production than agricultural crops, because trees can achieve an energy conversion factor of 16 which is 8× that of corn and 2× that of sugarcane. Trees can be grown on marginal agricultural land, and can also play roles in retaining the biodiversity of the local ecosystem. In particular, short-rotation woody plants have been suggested as ideal energy plants due to their fast growth during the junior period, low occupation to vast areas, high disease resistance and low management costs. With a coppice production approach, poplar (*Populus* spp.), willow (*Salix* spp.), eucalyptus, pine (*Pinus salius*), silver maple (*Acer saccharinum*), sweetgum (*Liquidambar styraciflua*), sycamore (*Platanus occidentalis*), black lotus (*Robinia pseudoacacia*), alder (*Alnus* spp.), Chinese tallow (*Sapium sebiferum*) and mesquite (*Prosopis* spp.) have been suggested as potential energy plants.

Poplars and willows constitute the family *Saliceae*. Being woody species, they have high CO_2 exchange rates and photosynthetic capacities, with short-rotation coppice (SRC) in which trees are cut-back/coppicing at 3–5 year intervals, a theoretical maximum growth could be achieved due to the removal of apical dominance. Among the 330–500 species, the shrub willow (*S. viminalis* in Europe and *S. eriocepala* in North America) is the most suitable as a bioenergy resource. Coppiced willows allow the growth of auxiliary buds, and in the second and third years more than 50 sprouting shoots develop, reaching a maximum of leaf area index (LAI) near midsummer. During coppicing, the plants can be cut near ground level, minimizing the loss of mineral nutrients, soil erosion, organic carbon emissions and investment into regrowing the roots. Willow (*Salix* spp.) and eucalyptus with nitrogen-fixing symbionts and mycorhizzal associations help to optimize the nutrient recycling and minimize inputs. The stem characteristics and coppicing response together with disease resistance have been the selection criteria for idea willow lines. For high yield purposes, two different strategies have been proven to be effective: one with a large number of thin stems; the other with fewer but thicker stems. For a willow breeding program, more aspects need to be addressed such as vegetative growth regulation, nutrient partitioning, water usage and over-winter performance in northern zones. For SRC use of willow, planting densities of 15 000–18 000 stools per hectare and a rotation cycle of three years have been widely adopted to gain optimal productivity.[10]

Poplar as a hardwood is a model plant for research into wood formation with the whole genome sequences available for several species. Transgenic techniques for poplar are mature, which makes it possible to select plants of interest *via* molecular breeding. Similar to willow, high biomass was achieved with contrasting growth strategies in poplar. However, there are a number of different traits: (1) poplars have a similar response to coppicing as willows, but competition for light results in high shoot mortality, therefore poplars are usually grown at lower densities (10 000 plants per hectare) and sprouting shoots need to be cut away in the second and third years; (2) the rotation cycle is longer, an average of 6–7 years compared to 3–5 years for willow; (3) poplar and willow roots share similarities, but their carbon and nitrogen partitioning systems are different; and (4) water is especially important for poplar growth, although clones and species with different water stress tolerances have been identified.[10] Poplar and willow are both hardwoods. Hardwoods have higher hemicellulose content (35%) than softwood (28%), which makes them more susceptible to alkaline-based pretreatment during the biochemical conversion process. Xylose is released as the major pentose sugar upon hemicellulose degradation, which is further converted to furfural in dilute acid. The lignin content in hardwoods ranges from 20 to 40% on a dry biomass basis, and its residue is often recycled to provide heat during the conversion process.[30]

Pine has a higher combined sugar content than eucalyptus, black locust, hybrid poplar and switchgrass,[8] implying a higher potential for bioethanol production. Being a softwood, it has a lower hemicellulose content and higher lignin content than hardwood, which makes it more suitable for electricity production *via* combustion.[31]

Eucalyptus is an important resource as a raw material for the pulp industry. During the recycling of plant fibers, hemicellulose and lignin are removed from the material which, in turn, results in irreversible structural changes that affect the digestibility of the reused plant fibers. In a study of eucalyptus fibers with different hemicellulose contents but similar lignin contents, the cellulose crystallinity remained relatively constant, but the area of accessible fibril surface decreased with decreasing hemicellulose content. Fibril size was not affected, but fibril aggregate size increased by 24%.[32] AFM imaging has shown that more grooves appear on the surface of the fiber wall with the least hemicellulose content and cellulose crystallinity, and the mean cross-sectional area increases gradually with decreasing hemicellulose content.[32]

6.2 Conversion of Biomass from Lignocellulosic Materials to Biofuels and Chemicals

Biomass is principally made up of three components of plant cell walls: cellulose (30–45%), a β-1,4-glucan polymer that is crystalline, hemicelluloses (20–30%), branched polymers that are composed of mainly xylose and other five-carbon sugars; and lignins (25–35%), non-carbohydrates that inter-link other polymers into a robust cell-wall structure and architecture (Table 6.2).[33] The purpose of the lignocellulosic biomass conversion process is to utilize the carbohydrate and aromatic polymers in the plant secondary cell wall for the production of biofuels or chemicals. For bioenergy production, reduction of the higher oxygen contents of the cell-wall components (*e.g.*, through fermentation of hexose and pentose sugars to ethanol; production of higher alcohols and hydrocarbons) and consolidation to higher energy density molecules are key challenges. Due to inherent plant-cell-wall recalcitrance (the barrier resulting from the complex structure of cellulose, hemicellulose and lignin in the plant secondary cell wall), the current conversion efficiency is only about 40% from cellulosic feedstock into ethanol, far below what it was from first-generation feedstocks such as corn and sugarcane, for which simple fermentation converts about 90% of the feedstock's simple sugars to ethanol. The multiple and near-term routes for overcoming recalcitrance can be broadly separated into biochemical and thermochemical conversion methods (Table 6.3). In biochemical conversion, various pretreatments (physical or chemical) are applied to the lignocellulosic raw material to break down the cell

Table 6.2 Sugar content and cell-wall composition of various lignocellulosic plants.[1]

Energy Plant	Cell-wall composition			Sugar content					Ash	Other extractives
	Cellulose	Hemicellulose	Lignin	Glucose	Xylose	Galactose	Arabinose	Mannose		
Corn stover	43	28	4	36.4	18.0	1.0	3.0	0.6	14	2
Wheat straw	39	30	14	38.2	21.2	0.7	2.5	0.3	12	5
Rice straw	34	24	11	34.2	24.5	6.5	11.7	n.a.	14	17
Sweet sorghum	43	35	17	49.7	23.3	0.5	2.2	n.a.	0.3	n.a.
Switchgrass	32	36	23.5	37.5	29.0	1.4	3.8	0.8	5–10	10
Miscanthus	34	22	18	31.0	20.4	0.9	2.8	5.8	2–4	20
Alfalfa	39	9	2.9	32.8	10.0	1.9	2.1	2.0	6.8	8
Reed	29	17	2	26.0	15.3	1.6	3.1	0.6	5–7	15
Hardwood (poplar)	45	30	20	49.9	17.4	1.2	1.8	0.5	0.6	5
Softwood (pine)	42	27	28	33.9	10.5	3.4	3.0	8.9	0.5	3

n.a. = not available.

Table 6.3 Overview of various physiochemical pretreatments.[43]

Pretreatment category	Temperature/ °C	Reaction time/min	Chemical reagent	% Residual cellulose	% Residual hemicellulose	% Residue lignin	% Glucan conversion	% Xylan conversion
Dilute acid	160–220	1–30	H₂SO₄	85–95	5–25	80–90	92	93
Steam explosion	180–290	1–15	None or SO₂	95–99	5–60	50–60	n.a.	n.a.
Hot water	160–230	10–30	None	90–99	45–60	n.a.	91	81
Lime	25–160	120– weeks	CaO	97–99	65–97	40–50	94	76
ARP	160–180	10–30	NH₄OH	90–99	40–70	15–60	90	88
AFEX	40–180	5–45	NH₃ or NH₄OH	100	100	100	96	91

n.a. = not available.

wall to create relatively more accessible cellulose and hemicellulose for enzymatic saccharification steps. Thermochemical conversion is further delineated into two regimes based on the operating temperature of pyrolysis and gasification.

6.2.1 Biochemical Conversion

Biochemical conversion uses low-severity thermochemical treatment (pretreatment) at temperatures between 100 and 200 °C to partially break down the cell wall for enzymatic accessibility, followed by enzymatic degradation of sugar chains, and final fermentation which converts a mixture of hexose and pentose into ethanol. The combined steps of pretreatment and enzymatic hydrolysis are responsible for overcoming biomass recalcitrance and releasing sugar from polysaccharides during biochemical conversion. Reduction of the carbohydrate streams to fuels, which is the last step, is an area of active research spanning the co-fermentation of pentose and hexose sugars to ethanol,[34] metabolic pathway engineering for production of higher alcohols and hydrocarbons,[35,36] and the application of catalytic routes to fuels.[37,38] Pretreatment and enzymatic saccharification were first conducted in individual compartments, with the fast development of combined processing technologies and consolidated bioprocessing (CBP) methods, the whole procedure was now conducted within the same container. The heat/acid-based pretreatment is the most researched, promising and close-to-commercial process, however many problems still exist, such as the toxic byproducts generated during pretreatment which subsequently inhibit enzymatic saccharification, and the co-fermentation of pentose and hexose derived from the celluloses and hemicelluloses of lignocellulosic biomass by microorganisms.

Pretreatment accounts for about 18% of the total cost, current techniques include ammonia recycle percolation (ARP); AFEX-ammonia fiber expansion; the use of ammonia water, dilute acid, AFEX (ammonia fiber explosion), pH controlling and lime pretreatment. Some of these techniques such as dilute acid and pH control, release polysaccharides, whereas AFEX for example releases basically glucose and xylose. Adding xylanase during pretreatment could improve conversion of xylan by up to 30%. The main factors influencing pretreatment are the DP (degree of polymerization), the efficiency of the reducing ends, the accessibility of the β-glycan bond and the CRI (crystallinity index). Taking poplar as an example, lime and SO_2 work better than AFEX in pretreatment, as indicated by the lower demand for exogenous enzymes and the higher conversion efficiency.

Enzymes used in lignocellulosic biochemical conversion include the CBH1 and CBH2 enzymes from *Trichoderma reesei*. Due to the extremely complex ultra-structure of plant cell walls, cellulose embedded with hemicelluloses and lignin is difficult to access which is the main reason for low conversion efficiencies and high pretreatment costs. The use of crude hemicellulases improves the saccharification efficiency, suggesting a synergistic effect between cellulase and hemicellulase. The current xylose yield is 50–80%, and no gene product has been shown *in vitro* to contain xylanase activity. Therefore research into hemicellulose-degradation enzymes is an important area for the biomass conversion industry.

High processing temperatures require a higher energy input and heat-resistant/anti-corrosive pressure equipment. However, excessively high temperatures may also aggravate the degradation of useful components and the formation of inhibitors. The currently available pretreatment techniques cannot yet meet the requirements for efficient lignocellulose-to-ethanol conversion. Low-temperature alkali (LTA) pretreatment has been used on sweet sorghum bagasse and showed efficient removal of the majority of the lignin, therefore greatly enhancing the enzymatic digestibility of the cellulose. The *bm* mutant is more susceptible to LTA pretreatment, leading to higher enzymatic hydrolysis efficiency.[39]

6.2.2 Thermochemical Conversion

Thermochemical conversion includes combustion, gasification, pyrolysis, liquefaction, hydrothermal upgrading, *etc.*[7] Direct combustion can employ a wide variety of systems including pile burners (the most common), stocker (or grate-fired) combustors and fluidized-bed combustors. Gasification involves the partial oxidation of the biomass in order to convert it into a gaseous fuel. Biomass is more reactive than coal and is usually gasified at temperatures between 800 and 1000 °C at 20–30 bar.[2] Pyrolysis is a thermal destructive distillation of biomass in the near-absence of oxygen at a temperature of around 5000 °C, and yields solid fuel-charcoal. Liquefaction is a low-temperature, high-pressure process using a catalyst. Hydrothermal upgrading converts biomass to a crude mixture in water at a high pressure and moderate temperature. A related concept is syngas, which refers to the mixture of CO/H_2 from gasified biomass that can be converted into liquid fuel using special microbes such as *Clostridium ljungdahlii*.[40]

The advantages of thermochemical conversion are low residence time and the ability to handle varied feedstocks in a continuous manner. However, the conversion process is not selective. The advantage of the gasification process is that not only the polysaccharides but also the carbon present in lignin is available for fuel production. For these processes, the composition of cellulose, hemicellulose and lignin and their cross-linking status does not play a role, but low water and ash content are desirable. Thermochemical processing options appear to be more promising for the conversion of the lignin fraction of cellulosic biomass than biological options. However, finding a cost-effective all-thermochemical process has been difficult.[41,42]

6.2.3 New Methods for Biomass–Energy Conversion

In an ionic liquid treatment, both the anions and cations of the ionic liquid are thought to participate in cell-wall and cellulose solubilization, with the anions playing a more dominant role.[6] The ionic liquid (IL) is applied at low temperatures (normally at 100 °C) and a low steam pressure is used to dissolve and distill the cellulose and lignin, but currently research on the recovery of ILs and isolation of the dissolved lignin–hemicellulose after pretreatment is lacking. Supercritical fluid techniques use CO_2, ethanol, acetone or water under supercritical conditions as a solvent for lignocellulosic biomass processing. Most chemical pretreatments result in significant xylan de-acetylation, which yields improved xylan hydrolysis, but the impact of de-acetylation on the cellulose–hemicellulose association is unknown. Studies of the use of both ILs and AFEX have revealed that the middle lamella and outer secondary grass cell walls are the most prone to disruption during pretreatment. Most acidic and oxidative pretreatments result in an increase of cellulose crystallinity and a reduction in its degree of polymerization—the formation of crystalline cellulose III$_I$ caused a four- to five-fold higher saccharification rate compared to cellulose I$_\beta$. As for lignin composition and redistribution, AFEX has been shown to subtly alter the distribution of lignin and hemicellulose without altering the core lignin chemistry.[43]

6.3   Usage of Lignocellulosic Biomass

Lignocellulosic biomass can be converted into solid fuels such as pellets/briquettes for heat/electricity by direct burning, into liquid fuels such as ethanol (or butanol), biogases such as methane, and various chemicals (DME, fatty acid/bio-oil, *etc.*). With the appropriate technology, biomass could also be converted into valuable products such as chemical precursors, cheap energy sources for fermentation, as well as improved animal feeds and human nutrients.[44]

6.3.1   Use in Bioenergy Production

6.3.1.1   *Liquid Fuel*

In a sugar–ethanol conversion process with the current technology, about 300 L of ethanol would be expected from a metric ton of switchgrass or poplar, with the potential to reach more than 380 L using improved techniques. Similar amounts of fuels such as butanol, alkanes, terpenes or other prospective biofuels are expected on an energy basis.[4] Compared to ethanol, butanol has the advantages of; having a higher energy density; being less volatile therefore easier to use together with gasoline; being less corrosive which makes it more suitable for existing pipelines; separating less easily in the presence of water, so it is better adapted to the present distribution system. The industry facilities required to produce butanol are rather similar to those currently being used for bioethanol production, therefore the major problem butanol production is the toxicity of butanol to most bacteria and yeast strains when the butanol concentration is above 2%.[35] However, during recent years, superior butanol-producing microbial strains have been developed through metabolic engineering, and companies (DuPont and BP) are also working together to develop an advanced butanol-based biofuel from sugarbeet feedstock.

Both bioethanol and biobutanol are alcohols made by the fermentation of carbohydrates. Biodiesel consists of hydrocarbons (generally 16–20 carbons in length) made by the transesterification of vegetable oils and animal fats, and is the most common jet fuel worldwide. At present, only bioethanol and biodiesel are produced on an industrial scale, making up more than 90% of the biofuel market.[6] The majority of the bioethanol supply comes from the sugarcane bioethanol industry which is most successful in Brazil. In Europe, wheat, sugarbeet and waste from the wine industry are predominately used. The world's largest biodiesel producer is the European Union, accounting for 53% of all biodiesel production in 2010. According to the International Energy Agency, biofuels have the potential to meet more than a quarter of world's demand for transportation fuels by 2050.

6.3.1.2 Solid Fuel

Solid fuel or the burning of biomass for electricity is quite well developed in some European countries. The steam that is produced from biomass combustion powers steam turbines and generates electricity. The main issue to be considered is the transportation cost of sparsely distributed biomass resources. Biomass transportation has to be within a 50 km radius around the plant site to be cost-effective. Currently the cost of biomass-generated electricity is around $0.1 kWh^{-1} which is much higher than conventional methods, therefore the economic profit needs to be improved for the development of the utilization of biomass as a solid fuel.

6.3.1.3 Gaseous Fuel

Hydrogen can be produced from biomass by pyrolysis, gasification, steam gasification, steam reforming of bio-oils and enzymatic decomposition of sugars, but the yield of H_2 from biomass is relatively low (16–18% dry biomass based). Two main biological processes to produce biohydrogen exist:[42,45] one is dark fermentation that leads to the production of H_2, CO_2 and some simple organic compounds. The other process produces H_2 by light-driven processes (*e.g.*, direct biophotolysis, indirect biophotolysis, and photofermentation). Fermentative H_2 production systems usually consist of one acidogenic reactor and one methanogenic reactor, and the latter one is needed to release the product of previous step which is the main limiting factor. In mixed fermentation systems, organic waste and crop residues can be used for the production of methane. The technique for the utilization of rice and wheat straw is well established in China, and one cubic meter of methane can be converted from about 2.5 kg of biomass of crop residues. Hydrocarbons can be made from sugars of lignocellulosic biomass through microbial fermentation or liquid-phase catalysis, or directly from woody biomass through pyrolysis or gasification. Traditional pyrolysis simply heats the biomass in the absence of air which produces a very acidic intermediate product that needs to be stabilized and upgraded in subsequent catalytic step, and advanced pyrolysis now uses catalysts to convert biomass into high-octane gasoline-range aromatics in a single and inexpensive step. Gasification also uses whole biomass but converts it spontaneously at very high temperatures into a mixture of CO and H_2, or syngas, as a starting material for subsequent Fischer–Tropsch synthesis. Hydrocarbons contain a higher energy density and give much higher mileage than ethanol or E85 ethanol (85% ethanol, 15% gasoline), and will work with existing infrastructure. Because the present production cost of lignocellulosic bioethanol is still about twice as that of cornstarch, biogas companies claim that hydrocarbon commercial production is 5–7 years away at a capacity of 100 million gallons per year at a competitive price with petroleum.[46]

6.3.2 Use in Bio-Based Chemical Production

Various compounds can be obtained during lignocellulosic biomass processing.[47] Biomass carbohydrate-converted products such as alcohols, carboxylic acids, and esters are also stereo- and regiochemically pure. Their down-stream applications include solvents, plastics, lubricants, fragrances, *etc.* In 2004, the DOE published a report about 12 chemicals that are derived from biomass conversion, which could be used as building-block chemicals: Four Carbon 1,4-Diacids (Succinic, Fumaric, and Malic); 2,5-furan dicarboxylic acid; 3-hydroxypropionic acid; aspartic acid; glucaric acid; glutamic acid; itaconic acid; levulinic acid; 3-hydroxybutyrlactone; glycerol; sorbitol and xylitol/arabinitol.[48] Since the value of the chemical industry is comparable to the fuel industry, the coproduction of value-added biochemicals is beneficial for the whole economy of the biomass industry. Biorefinery, a new concept arising for the comprehensive utilization of biomass resources, has developed very quickly during the last 10 years. Through a combination of technologies, biorefinery should integrate the biomass-conversion process to produce a range of fuels, power, materials and chemicals.[49–51] An integrated biorefinery could optimize the use of biomass for the production of biofuel, bioenergy, and biomaterials for both short- and long-term sustainability.[2]

The complexity of biomass allows its main components to be processed into a wide range of products, for example, lactate can be produced during hexose and pentose fermentation by *B. coagulans*, and xylitol or furfural ($C_5H_4O_2$) can be obtained from the fermentation of xylose or xylans. Crude glycerol, a co-product during biodiesel synthesis, was once considered a waste material, but it is now used as a low-cost building block for conversion to the higher value propylene glycol,[52] and in the production of epichlorohydrin in the manufacture of epoxy resins and elastomers.[53] Due to its versatility, glycerol is considered to be a good substitute for many commonly used petrochemicals,[54] such as acting as a substrate for microbial production of 1,3-propanediol.[55] Acetylpropionic acid, generated as a by-product in lignocellulosic processing, is a basic substrate in the biotech industry, and it can be used in hexose conversion and the production of resins, plasticizers and fabrics. Lignin residue can be converted to benzene, xylene, toluene and other aromatic compounds.[56]

However, in order to make biochemicals from these key components, the first step is the fractionation of cellulose, hemicellulose and lignin. It has been shown that if hemicellulose could be separated from lignin, more efficient production of fuels (*e.g.*, ethanol) could be achieved, and higher value chemicals (*e.g.*, polyesters) could be produced.[57] A new approach to pretreatment which aims not only to improve the subsequent enzymatic hydrolysis, but also the key constituents of lignocellulosic biomass is required. Concentrated phosphoric acid can completely dissolve cellulose fibers and disrupt hydrogen bonds between crystalline cellulose, and a cellulose organic solvent-based fractionation technique utilizing acetone has also been presented.[58] Since there is almost a limitless number of combinations of anions and cations for ILs, and both cellulose and lignin can be dissolved in a variety of ILs, there is the possibility of finding an IL that works best with a particular source of material. Other extraction methods include the use of supercritical CO_2, near-critical water, and gas-expanded liquids.[59,60] The current costs of many carbohydrates and their derivatives are already competitive with petrochemicals, plus there are many more special products that are only available from the biorefinery of biomass, such as novel gasoline blends dependent on C5 to C10 hydrocarbons. Controlled elimination of water from sugars generates hydroxymethylfurfural (HMF), levulinic acid and other organic acids, and the aldol-crossed condensation between HMF and acetone leads to the production of alkanes from C9 to C15 in the presence of a catalyst.[37] Levulinic acid is produced *via* the conversion of hexose sugars in acid; and it can be used as a building block for other specialty chemicals or in resins, plasticizers and textiles.[61,62]

Applications in synthetic biology open up new options for engineering non-native metabolic pathways of certain bacteria for the production of small molecules from sugars or other intermediates from biomass, *e.g.*, by aqueous phase reforming (APR), which generates H_2 and hydrocarbon fuels from water-soluble oxygenated compounds derived from sugars, sugar alcohols, and glycerol from various feedstocks. The synthetic engineering of bacteria from *Clostridium*, *Desulfovibrio*, *Escherichia*, *Trdiumbutyricum*, *Azobobacter*, *Citrobacter*, *Klebsiella*, *Enterobacter*, *Anabaena*, *Aquifex* and *Acetomicrobium* has been reported for H_2 production *via* fermentation. Several photosynthetic bacteria have also been identified including *Rhodobacter sphaeroides*, *Rhodobacter capsulatus*, *Rhodopseudomonas palustris* and *Rhodospirillum rubrum*. The primary products of APR are H_2 and hydrocarbon fuels, while the byproducts include CO_2. APR occurs at temperatures and pressures where the water–gas shift reaction is favorable, making it possible to generate H_2 with low amounts of CO in a single chemical reactor. Unlike steam-reformation processes, APR produces H_2 from liquid-phase solutions, resulting in considerable energy savings. The process also minimizes the undesirable decomposition reactions typically encountered when carbohydrates are heated to elevated temperatures. Furthermore, the reactor and catalysts can be altered to generate high-energy hydrocarbons from biomass-derived compounds, for example HMF in the aqueous phase can be separated *via* the immiscible organic phase.[63,64] HMF is a top value-added platform chemical, and can be further used for the production of alkanes. Aimilar conversion and separation routes are currently being exploited by researchers for other monomer sugars or small molecule intermediates.

Succinate is one of the most in-demand building-block chemicals according to studies from the DOE,[48] and it is derived *via* the fermentation of glucose.[47] 1,3-Propanediol is a key building block for polypropylene terephtahlate which is not available from petrochemicals, and it is produced *via* the fermentation of glycerol. Itaconic acid is produced *via* the fermentation of carbohydrates by fungi and is considered to be a biofriendly substitute for acrylic and methacrylic acid in polymers and in styrene–butadiene systems.[48,62]

Biorefinery is "the sustainable processing of biomass into a spectrum of marketable products", and various small molecules could be generated from lignocellulosic biomass such as isoprene, acrylic acid, 3-hydrocrylic acid, lactic acid, acetic acid, 1,4-butanediol, and dihydrobutyric acid which can be used for the production of biopolymers such as polylactic acid (PLA), polybutelene succinate (PBS), polytrimethylene terephthalate (PTT), polyhydroxyalkanoate (PHA) *etc.* Ethylene, which can be polymerized into the widely used polyethylene, has been shown to be synthesized directly from biomass hydrolysates using bacteria that occur naturally in soil and on the surface of fruits,[65] although the amount of ethylene currently produced during biorefinery is only a negligible contribution because most of it still comes from petrochemical industry. Lactic acid is easily produced by several kinds of bacteria through fermentation, and it can be converted to several important chemicals including methyl lactate, lactic acid, and polyacic acid which is a biodegradable substituent for polyethylene terephthalates (PETs). Currently, all lactic acid manufacture is based on carbohydrate fermentation, and *Lactobacillus lactis* and *Aspergilus niger* are most commonly used bacteria for the commercial production of lactic acid.[66] Lactic acid could be converted to acrylic acid, which is a major commodity monomer with an international demand of over 1.6 billion kg per year in 2007.[53] Acrylic acid and its derivatives are key components in the manufacture of polymers including coating and absorbent materials, detergents and textiles. Some microorganisms express acrylic acid pathways, suggesting the possibility of direct fermentation pathway from biomass to acrylic acid.[67]

One of the best illustrations of polymer production *via* biological conversion from biomass is the family of polyhydroxyalkanoates (PHAs). PHAs are a family of natural polymers with over 150 different hydroxyalkanoates produced by a range of bacteria (*e.g.*, *Burkholderia cepacia*) from different genera. PHAs are formed as intra-cellular granules which perform carbon and energy storage for up to 90% of the dry cell weight.[68–70] and their wide range of properties enables them to be used in large quantities in the plastics industry. Since PHAs are synthesized directly as polymers, research into making short- and medium-length PHAs from various carbon sources, either by constructing a 'super' bacterium or *via* the use of microbial consortia can been conducted.[70,71]

Polymerization can be achieved either chemically *via* catalysts or biologically with the addition of enzyme(s). Maize biorefinery in the USA generates more than 30 different products, and all parts of plants are used including oil, proteins, starches and fibers. The DOE plans to reach the following usages of biomass by the year 2030: electricity 5%; liquid fuel 20%; and biochemical

25%. The increase in the proportion of biomass-derived electricity from 2010–2030 is the smallest compared to the increase in the use of biomass-derived biochemicals and liquid fuels. Commodity chemicals can also be directly extracted from biomass, and one of the most notable examples is ferulic acid, which can be extracted in high percentages from corn fiber, and is used in the production of vanillin and guaiacol.

6.4   The Systematic Evaluation of Bioenergy Resources

Although the net profit for the production of lignocellulosic bioethanol at present is not great when compared to first-generation bioethanol from corn and other crops, other issues need to be included in the evaluation of energy plants (Table 6.4). An important factor is the carbon-sequestering ability of energy plants and its long-term impact on the global environment. In general, carbon neutrality is difficult to achieve, even for a biomass crop, if all the energy input including that associated with harvesting and transport is considered. The total carbon mitigation calculated over 15 years for a *Miscanthus* crop averages around 6 tons of carbon per year.[72] Switchgrass and *Miscanthus* have powerful underground rhizomes, and their carbon-sequestering ability is 20–30× that of corn, which would reduce the greenhouse effect more effectively in the long-term. Another issue is the conservation of biodiversity and environmental factors. Approximately 18% of the world's terrestrial surface is semi-arid, and the arable land in most countries is limited to large-scale cultivation of energy plants. However, sweet sorghum and *Miscanthus* have been recommended for growth as a first priority on the marginal land of China because of their diverse natural germplasm resources, rich biomass, efficient lignocellulose degradation and effective adaptation to various environmental conditions. In particular, sweet sorghum is suitable for growing in the north of China, whereas *Miscanthus* is suitable for the south. The use of species with a high drought resistance such as certain *Agave* species would be a solution for deserts and regions with minimum rainfall.[4] *A. sisalana* in Tanzania produced 58 wet MT per hectare per year and *A. salmiana* grown in Mexico yielded 10 dry MT per hectare per year. Considering more of the land that has fallen out of agricultural production worldwide, the land available for cultivation of *Agave* is vast.

The criteria for good biofuel resources are: (1) they do not compete with food crops; (2) they do not lead to land clearing, (3) they offer real greenhouse gas reductions; and (4) they maximize social benefits.[73] The leading lignocellulosic feedstock resources today are from corn stover and rice straw, as well as perennial grasses such as switchgrass and *Miscanthus*, and woody trees such as poplar and eucalyptus.[74] Both perennial grasses and woody trees can be sustainable, but cause some mineral nutrient removal during harvesting. Hence, it is essential to recycle nutrients from biomass-processing facilities back to the land, and it is also desirable to use genetically diverse or species-diverse plantations to support ecosystem health and sustainable biomass production. By-products or co-products that arise during the generation of biofuel feedstocks

Table 6.4 Evaluation of lignocellulosic energy plants.

Plant species	Establishment of labor demand	Regeneration/propagation	Disease resistance	Harvest frequency	Genome sequence available	Transgenic techniques
Maize	Expensive	Seed	Low	Annual	Yes	Yes
Wheat	Expensive	Seed	Low	Annual	Yes	Yes
Rice	Expensive	Seed	Low	Annual	Yes	Yes
Sweet sorghum	Expensive	Seed	High	Annual	Yes	Yes
Switchgrass	Easy	Rhizome, seed, tissue culture	High	Annual	In progress	Yes
Miscanthus	Easy	Rhizome, seed, tissue culture	High	Annual	In progress	Yes
Reed canarygrass	Easy	Cuttage or plant division	Moderate	Annual	No	No
Alfalfa	Easy	Seed	Low	Annual	In progress	Yes
Poplar	Expensive	Cuttage, tissue culture	Moderate	6–7 years	Yes	Yes
Eucalyptus	Easy	Seed, cuttage, grafting, tissue culture	Moderate	20–24 months	In progress	Yes
Willow	Easy	Cuttage, tissue culture?	Moderate		In progress	No

need be assessed together with the impact of biofuel production on food supply, greenhouse gas emission and environmental consequences of land clearing and potential biodiversity reduction. Production of biofuels from waste materials may release chemicals such as dioxins and heavy metals that could result in unintended public health consequences. Retention of crop residues in soils, including the biomass produced from crops, is essential to the ecosystem and sustainable agronomic productivity. Soils are a source of greenhouse gases (CO_2, CH_4 and N_2O) when erosion is accelerated and when under-management creates negative carbon and nutrient budgets.[75] Therefore recent research has suggested leaving substantial quantifies of crop residues on the land, although conservative removal rates can still provide as much sustainable biomass as dedicated perennial crops grown on degraded land. Alternatively, double crops (crops grown between the summer growing seasons of conventional row crops and harvested for biofuels before row crops are planted in the spring) and mixed cropping systems in which food and energy crops are grown simultaneously have been presented as options to avoid soil erosion and land clearing.[73] Trees offer promise where they grow well on degraded lands, and are also positive for a great variety of ecosystem services such as biodiversity promotion, nutrient retention and flood protection.[76]

For a better and more realistic evaluation of a certain type of lignocellulosic resource, the full life cycle of biofuel production, transformation and combustion needs to be considered. The best solutions should be continually updated to evaluate the extent to which various objectives are being achieved, such as energy gains, greenhouse gas reduction, and preservation of biodiversity and maintenance of food security. Policies should then be presented with a combination of consideration from scientists, economists and technologists. Bioenergy systems need to be established locally with a complete infrastructure for a harvesting, transporting, storing and refining systems, and it would make no sense to produce biomass in the USA, Brazil or South Asia and then to export it to the rest of the world, which would result in subsequent CO_2 release, when the use of biomass is has been implemented to reduce CO_2 release. A lot of other issues need to be addressed such as the availability of resources and competing use, and social, economic and environmental factors.

6.5 Using Genetic Breeding to Improve Lignocellulosic Biomass Characteristics

In general, an increase in plant biomass production can be achieved through traditional breeding or transgenic approaches in the following ways: (1) increasing the conversion rate from solar energy by manipulating photosynthetic pathways; (2) enhancing plant disease and pest resistance; (3) raising adaptability to various abiotic stress conditions such as drought and cold; (4) reducing the plant demand for fertilizer; (5) delaying plant flowering for long vegetative growth; (6) increasing cellulose production; and (7) improving plant cell-wall structures for cost-effective biomass digestibility and biofuel production. Lignocellulose–ethanol production, because of biomass recalcitrance, is under development.[74,37] Although a great effort has been made to increase the lignocellulose–ethanol conversion rate, three crucial factors are still causing difficulties: biomass pretreatment, lignocellulosic enzymatic hydrolysis and sugar fermentation. The recalcitrance is determined by the cellulose crystallinity, the hemicellulose character and the lignin linking styles of the plant cell walls.[78,79] Extreme pretreatment conditions such as strong acids/bases, or extreme temperatures/pressures, lead to economic loss for biofuel production and also cause secondary environmental pollution. Therefore the discovery of energy crops is a bottleneck-breaking solution. Without a doubt, the characterization of germplasm resources is both initial and essential work. Not only can it be used to indirectly discover valuable genetic materials for energy crop breeding, but it can also directly select-out energy plants.

Plant secondary cell walls are present only in certain tissues, and consist of cellulose (20–50% of dry weight), hemicellulose (15–35%), and lignin (10–30%).[1,33] The chemical composition of cell walls differs significantly between monocots (*e.g.*, corn stover and switchgrass) and dicots (*e.g.*, *Arabidopsis* and poplar), and directly influences the degradation efficiency due to the different susceptibilities of lignocellulosic biomass.[25] However, all cellulose is either crystalline or amorphous, and the most abundant crystalline polymorph in higher plants is cellulose I_β. Hemicelluloses are much more heterogeneous, and can be classified structurally into four distinct classes: xylans, mannans, β-glucans with mixed linkages (β-1-, 3-β-1- or 4-glucosyl backbones) and xyloglucans. The most abundant hemicellulose found in monocots is glucuronoarabinoxylan, whereas in dicots it is galactoglucomannan.[43] Both cellulose and hemicellulose are used as substrates for enzymatic saccharification, and a mixture of hexose and pentose released from this procedure is further subjected to fermentation for the production of ethanol in the lignocellulose–ethanol industry. Lignin is released as residue product, or recycled for heating during the procedure.

6.5.1 Improvement of Cellulose Accessibility

Cellulose in plant secondary cell walls is the most abundant biomass resource on Earth. However, due to the complex structure of the plant cell wall, the main problem with utilization is the difficulty of enzyme access to the cellulose molecules, which are surrounded by a network of hemicellulose, lignin and pectin. Therefore improvement of cellulose accessibility by changing the cell-wall components and structure has been the focus of the genetic manipulation of energy plants. Two strategies have been applied:

(1) Reducing the cellulose crystallinity character by ectopic expression of CBMs (carbohydrate-binding motifs); CBMs between cellulose molecules could induce the depolymerization of crystalline cellulose without hydrolysis reactions,[80] therefore increasing the efficiency of subsequent endo-glucanase hydrolysis. CBM expression in transgenic plants has been demonstrated in poplar, *Arabidopsis* and rice,[81] over-expression led to a 30% increase in rumen digestibility of the rice straw. Expansins and

expansin-like proteins (*e.g.*, swollenins) found in plants serve as endogenous CBMs and both proteins are potential candidates for enhancing sugar release upon saccharification. Another protein that can degrade and modify cellulose, hemicellulose and lignin is the non-hydrolytic hemoflavoenzyme cellobiose dehydrogenase (CDH), yet reduced crystallinity may result in increased disease susceptibility and reduced fitness of woody plants and grasses, but might also increase the expansion and growth rates due to the flexible orientation of cellulose microfibrils.

(2) Modification of hemicellulose and pectin; altering hemicellulose levels and their side-chains are among the main approaches to increasing cellulose accessibility. Transgenic ryegrass expressing ferulic acid lipase showed better degradation by xylanase by influencing the cross-linking between polysaccharides and ferulic acids.[82] XTH (xyloglucan endotransglucosylase/hydrolase) could be used to loosen the cell-wall structure, and degrade xylan, but it also participates in the reassembly of hemicellulose, therefore modifying the hemicellulose–cellulose network.[1] Silencing of the *PoGT47C* gene in poplar, a glycosyltransferase homologous to *Arabidopsis FRA8* involved in hemicellulose biosynthesis, caused an increase in glucose yield following enzymatic hydrolysis, indicating that reducing xylan content leads to improved saccharification.[68] PME (pectin methylesterase) influences cell-wall rigidity through pectin modification, and it has been shown that down-regulation of PME could drastically increase biomass production and improve saccharification in transgenic *Arabidopsis*.[83]

6.5.2 Increasing Cellulose Content and Total Biomass

Since the first cellulose biosynthetic gene *CesA* was identified in cotton, the *CesA* and *Csl* superfamily have been considered for use in biomass enhancement. However, attempts to increase cellulose content by *CesA* gene overexpression have not been successful so far, possibly due to the fact that several *CesA* genes are required for the assembly and function of the cellulose–synthase complex. Several non-*CesA* genes involved in cell-wall synthesis could also be used for elevating cellulose production, such as Korrigan, Cobra and Kobito.[84] Importantly, as the major transcription factors are identified for regulating secondary cell-wall synthesis in *Arabidopsis*, we may directly improve the quantity and quality of biomass by altering those gene expression times and levels in energy plants. Recently the *wrky* transcriptional factor in *Arabidopsis* was identified as a negative regulator for the genes responsible for secondary cell-wall formation, and the corresponding loss-of-function *wrky* mutant showed a 50% increase of biomass density in the stem tissue of *Arabidopsis*.[85]

Unbranched hemicellulose forms hydrogen bonds with the surface of cellulose fibrils, whereas the side-chains of branched hemicellulose are covalently bonded to hemicellulose or lignin, forming an enzyme-impenetrable cross-linked network, also known as the "lignin carbohydrate complex" (LCC). There are two major hemicelluloses in grasses: MLG (β-1, 3-β-1, 4-glucan) and GAX (β-1,

4-linked xylose backbone with single arabinose and glucuronic acid side-- chains.[86,87] Because GAX links tightly to lignins, we can use MLG to replace GAX by expressing the *CslF* and *CslH* genes that have been characterized to catalyze MLG biosynthesis. Recent findings about three glycosyltransferase (TaGT) proteins that participate cooperatively in GAX polymer synthesis in wheat, will extend the effort towards cell-wall remodeling.[88] One major goal of genetic manipulation of lignocellulosic plants is to reduce the ester and phenolic linkages within this network, making it possible for the saccharification enzyme to gain access to the cellulose and hemicellulose, and improve the overall yield of biofuel. Strategies to improve the overall biomass yield include expression of transcriptional factors, use of glycan synthatase and over-expression of different glycoside hydrolases (*e.g.*, endo-cellulase and glucan hydrolases belonging to the GH9 family) and also the CBMs mentioned above, to modify cellulose de- position and cross-linking with hemicelluloses. The combined over-expression of *SuSy*, which targets increased transport of UDP-glucose as a substrate for cel- lulose synthesis, with UDP-glucose pyrophophorylase (UGPase) together sig- nificantly increases plant height and overall biomass. Fructose is known as an inhibitor of *SuSy*, therefore decreasing the fructose level by over-expression of sucrose phosphate synthase which recycles fructose and generates sucrose would in turn promote *SuSy* activity.

6.5.3 Manipulation of Lignin Content and Composition

Examination of natural variation in alfalfa, switchgrass, canarygrass, and sor- ghum has shown that decreased lignin levels improve *in vitro* enzyme hy- drolysis.[89,90] Lignin pathway modification in alfalfa generates transgenic lines with increased enzymatic sugar release, essentially proportional to the extent of lignin down-regulation.[91] Transgenic switchgrass with a down-regulated caffeic acid *O*-methyltransferase (*COMT*) gene in the lignin pathway revealed a normal growth phenotype, reduced the lignin content and altered the lignin composition, improved the saccharification efficiency and increased ethanol production.[92]

Approaches to the modulation of lignin content by down-regulation of lignin biosynthetic genes are under debate due to the impact on the mechanical strength of the plants and environmental stresses. Transgenic alfalfa plants with down-regulation of six genes along the lignin biosynthetic pathway yielded nearly twice as much cell-wall sugar, but increased saccharification came at the expense of low biomass yield and severe growth defects.[91] Numerous studies in other plant species also demonstrated dwarfing and collapsed xylem pheno- types upon down-regulation of lignin biosynthetic genes.[93] Techniques have been proposed that temporarily loosen the lignin without the undesirable physiological effects, for example, the introduction of tyrosine-rich peptides into poplar trees demonstrated higher susceptibility to protease digestion but no change in the total lignin content or overall plant morphology.[94]

Lignins are primarily composed of guaiacyl (G, 35–49%) and syringyl (S, 40– 61%) units and hydroxycinnamates (4–15%). Ferulic and coumaric acids are also present in plant cell walls.[95] Although lignocellulose biodegradation is restricted by both lignins and phenolic acid esters, the ratio of coniferyl lignin to syringyl lignin is a crucial factor in determining the degree of biomass recalcitrance.[91] In addition, esterified phenolic acids including ferulic and *p*-coumaric acids, constitute a major chemical limitation for non-lignified cell-wall biodegradation in grasses.[77] Several key enzymes in lignin metabolism, such as C3H, C4H, 4-CL and CCoAOMT, have been characterized in dicot plants,[96,97] but only the *brown midrib* mutant with known lesions in lignin biosynthesis has been extensively studied in corn, sorghum, and millet crops.[28] It is not very clear yet which are the key genes in regulating lignin biosynthesis and esterified phenolic acid formation in grasses; the down-regulation of the switchgrass *COMT* gene modestly reduced the lignin content (up to ∼15%) and the S : G ratios to a similar degree, improved cell-wall digestibility was also reported in transgenic maize with low S : G ratios.[92] When the *CCR* gene was down-regulated in poplar, this resulted in a more easily digested plant cell wall and the release of twice as much sugar.[98]

Recently, we have found several rice and maize mutants that show differences in lignin composition and straw/stalk biomass degradation, distinct from the previously identified cell-wall mutants that showed abnormal phenotypes such as dwarfism, irregular xylem and even lethality.[48] These mutants displayed normal agronomic traits and grain yield but remarkable alteration of cell-wall composition. We have also identified natural *Miscanthus* varieties with large variations in biochemical conversion efficiencies through a survey of over 1400 natural germplasms of *Miscanthus* in China (Figure 6.1 and Table 6.5).

6.5.4   Exogenous Expression of Hydrolytic Enzymes in Transgenic Plants

In addition to plant cell-wall manipulation, attempts to over-express cell-wall-degrading enzymes in transgenic plants have also been successful. Bacterial cellulases or xylanases have been demonstrated to be thermostably expressed in various cellular compartments without adverse effects on plant growth.[99] In recent years, several laboratories have attempted to express microbe cellulase genes, and determine the hydrolysis activity in transgenic plants. They did not observe any visible side-effects in plant growth and biomass yield.[100] The catalytic domain of E1 1,4-β-endo-glucanase from *Acidothermus* has been expressed at high concentrations in *Arabidopsis*, tobacco, potato, corn and rice;[101,102] exogenous α-glucosidase and β-glycosidase have also been tested for cell-wall hydrolysis in the pulp and paper industries. Our laboratory has started to ectopically express fungi-specific cellulase genes in selected rice and maize mutants using inducible gene promoters. In addition, we are on the way to transforming the lignin-hydrolysis genes of white-rot fungi into energy crops.

Most studies have reported heterologous protein levels of up to 2% of the total soluble protein, and these enzymes are expressed in the vacuole rather than in the cell wall. It has been estimated that 20–40% of the cost of lignocellulosic ethanol production come from the usage of hydrolytic enzymes, yet

Table 6.5 Cell-wall components by different pretreatments from mutants of rice, wheat and *Miscanthus*. [103]

		Cell-wall composition			Heat pretreatment[a]		Alkaline pretreatment[a]		Acid pretreatment[a]	
		Cellulose	Hemicellulose	Lignin	C6-Sugar	C5-Sugar	C6-Sugar	C5-Sugar	C6-Sugar	C5-Sugar
Rice	RC12	49.98	26.70	23.32	32.52	4.58	48.95	21.14	43.00	25.17
	RC15	38.56	34.74	26.70	19.47	5.39	29.91	27.81	42.24	32.40
	RC46	47.35	29.41	23.23	25.69	6.35	45.66	25.29	38.43	26.46
	RG65	51.27	26.36	22.36	38.07	5.91	34.37	19.49	41.83	22.70
	WT	49.90	24.68	25.42	18.55	4.31	35.32	18.79	36.10	19.74
Wheat	CM4	47.12	25.78	27.10	n.a.	n.a.	25.52	25.03	35.66	23.29
	CM31	45.32	22.92	31.76	n.a.	n.a.	41.05	18.72	19.39	21.00
	CM43	44.01	24.80	31.19	n.a.	n.a.	34.59	28.26	29.45	25.45
	WT	45.18	28.05	26.77	n.a.	n.a.	31.68	28.08	19.71	28.41
Miscanthus	MI10	28.25	38.74	33.01	19.07	3.87	47.07	24.54	31.98	31.31
	MI108	33.33	38.11	28.56	8.59	2.60	45.86	26.06	30.61	33.96
	MI1	44.50	27.95	27.56	3.29	1.24	21.78	14.08	14.56	24.63

n.a. = not available.
[a]Degradation efficiency is calculated as the total soluble hexose or pentose sugar (% of total cell wall) released from pretreatment followed by 0.4% cellulase digestion.

Figure 6.1   Natural *Miscanthus* germlines with high or low degradation efficiency. (a)
and (b) Plant morphology of *Miscanthus* (M1 and M108) growing in the
field, (c,d) scanning electron micrograph of M108 which has a smooth stem
transverse section and low biomass degradation efficiency, (e,f) scanning
electron micrograph of M1 which has a rough stem transverse section and
relatively high biomass degradation efficiency (scale bar = 50 µm).

only a few studies have examined auto-hydrolysis of the modified biomass from
transgenic plants, therefore, at present, it is hard to justify the use of this kind
of technique on the basis of real production cost.

6.5.5   Integration of Soluble Polymers

Introducing new polysaccharides into plant cell walls improves the solubility of
biomass resources without impacting plant growth. The new polysaccharide is
soluble, forming channels during biomass processing and extraction, therefore
creating easy access for solvents and enzymes. Examples of such exogenous
polysaccharides include EPS (exo-polysaccharide) from bacteria, algae cell
walls, seaweed, and even plant viruses. Hyaluronan synthase from *Chlorella*
viruses introduced into tobacco plants demonstrated enhanced cellulose hydrolysis upon heat and acid pretreatments,[74] but few of the genes and biosynthetic pathways have been fully elucidated.

6.6   Concluding Remarks

Lignocellulosic plants with high biomass yields and suitable agronomic traits
have been identified in various climatic regions, and they exist at various levels
of domestication and cultivar selection. While draft genome sequences and
transformation protocols for some of the species are being developed, the
breeding programs for some undomesticated species have just begun. Lignocellulosic biomass has been identified as a feedstock for "second-generation"
bioenergy, however there will be no simple clear-cut transition from the first to
the second generation. Some first-generation feedstocks, such as sugarcane for
bioethanol production, will remain competitive for many years. Owing to the

abundance of cell wall material generated by plants, lignocellulosic biomass is considered a valuable resource for the supply of biofuels and other chemicals. Nature has already created plant cells with diverse walls, and because the traits that are desirable for biofuel use or biochemical use vary depending on the processing method, and these traits differ from those needed for traditional uses such as food and wood production, much discussion has focused on modifying cell-wall properties that will allow a balance to be achieved between the various applications.

With the abundance of biomass resources available, the development of conversion technologies for biofuels and valuable biochemicals represents an important issue. Advances have been made towards understanding the factors that contribute to lignocellulose recalcitrance, and more genes that can be manipulated for cell-wall modification are being investigated. However, the effect of the manipulation of cell-wall traits in dedicated biofuel crops needs to be evaluated with respect to the overall fitness of the genetically engineered crop. Major breakthroughs regarding our understanding of plant cell-wall biosynthesis and structure will help us to overcome the recalcitrance of lignocellulose and enable its efficient and cost-competitive conversion to biofuels and biochemicals, as well as achieve a sustainable biomass yield. It is important to bear in mind that although biotechnology will be essential for the successful generation of energy plants, it should go hand-in-hand with traditional breeding efforts aimed at maintaining or enhancing the important agronomic traits that initially made these plants good candidate for biofuel production.

References

1. M. Pauly and K. Keegstra, *Plant J.*, 2008, **54**, 559.
2. A. J. Ragauskas, C. Williams, B. Davison, G. Britovsek, J. Cairney, C. Eckert, W. Frederick, J. P. Hallett, D. J. Leak, C. L. Liotta, J. R. Mielenz, R. Murphy, R. Templer and T. Tschaplinski, *Science*, 2006, **311**, 484.
3. S. P. Long, M. B. Jones and M. J. Roberts, *Primary Productivity of Grass Ecosystems of the Tropics and Sub-tropics*, Chapman Hall, London, 1992.
4. C. Somerville, H. Youngs, C. Taylor, S. C. Davis and S. P. Long, *Science*, 2010, **329**, 790.
5. X. Cheng, W. B. Zhu and G. H. Xie, *J. Nat. Resour*, 2009, **24**, 842.
6. H. Liu, K. L. Sale, B. M. Holmes, B. A. Simmons and S. Singh, *J. Phys. Chem. B*, 2010, **114**, 4293.
7. B. Antizar-Ladislaol and J. Turrion-Gomez, *Biofuels, Bioprod. Biorefin.*, 2008, **2**, 455.
8. M. Balat, H. Balat and C. Oz, *Science*, 2008, **34**, 551.
9. K. Karimi, G. Emtiazi and M. J. Taherzadeh, *Enzyme Microb. Technol.*, 2006, **40**, 138.
10. A. Karp and I. Shield, *New Phytol.*, 2008, **179**, 15.
11. N. C. Carpita and M. C. McCann, *Plant Physiol.*, 2010, **153**, 1362.

12. Y. C. Shi, *Serials of Renewable Development Strategies in China: Biomass Volume*, China Electricity Publishing House, Beijing, 2008.
13. H. G. Jung and M. S. Allen, *J. Anim. Sci.*, 1995, **73**, 2774.
14. D. N. Ledgerwood, E. J. DePeters, P. H. Robinson, S. J. Taylor and J. M. Heguy, *Anim. Feed Sci. Technol.*, 2009, **150**, 207.
15. X. F. Li, S. L. Hou, M. Su, M. F. Yang, S. H. Shen, G. M. Jiang, D. M. Qi, H. Y. Chen and G. S. Liu, *Environ. Manage.*, 2010, **46**, 579.
16. I. Lewandowski, J. Scurlock, E. Lindvall and M. Christou, *Biomass Bioenergy*, 2003, **25**, 335.
17. J. H. Yan, Z. Q. Hu, Y. Q. Pu, E. C. Brummer and A. J. Ragauskas, *Biomass Bioenergy*, 2010, **34**, 48.
18. M. Galbe and G. Zacchi, *Adv. Biochem. Eng. Biotechnol.*, 2007, **108**, 41.
19. Y. C. Sun and J. Bioresour., *Technol.*, 2002, **83**, 1.
20. H. Alizadeh, F. Teymouri, T. I. Gilbert and B. E. Dale, *Appl. Biochem. Biotechnol.*, 2005, **121**, 1141.
21. D. S. Powlson, A. B. Riche and I. Shield, *Ann. Appl. Biol.*, 2005, **146**, 193.
22. E. M. Hodgson, S. J. Lister, A. V. Bridgwater, J. Clifton-Brown and I. S. Donnison, *Biomass Bioenergy*, 2010, **34**, 652.
23. Molecular Investigation of Diversity in Wild Source Germplasm to Support a *Miscanthus* Breeding Programme, DEFRA, London, 2002.
24. D. G. Christian, A. B. Riche and N. E. Yates, *Ind. Crops Prod.*, 2008, **28**, 320.
25. C. J. Atkinson, *Biomass Bioenergy*, 2009, **33**, 752.
26. J. C. Clifton-Brown, I. Lewandowski, B. Andersson, G. Basch, D. G. Christian, J. Bonderup Kjeldsen, U. Jørgensen, J. V. Mortensen, A. B. Riche, K.-U. Schwarz, K. Tayebi and F. Teixeira, *Agron. J.*, 2001, **93**, 1013.
27. M. B. Jones and M. Walsh, *Miscanthus for Energy and Fibre*, James & James, London, 2002.
28. K. Jakob, F. S. Zhou and A. H. Paterson, *In Vitro Cell Dev. Biol. Plant*, 2009, **45**, 291.
29. B. Andersson and E. Lindvall, *Canarygrass Breeding in Sweden*, European Commission, Finland, 1999.
30. S. Yaman, *Energy Convers. Manage.*, 2004, **45**, 651.
31. C. N. Hamelinck, G. van Hooijdonk and A. P. C. Faaij, *Biomass Bioenergy*, 2005, **28**, 384.
32. J. Q. Wan, Y. Wang and Q. Xiao, *Bioresour. Technol.*, 2010, **101**, 4577.
33. M. Pauly and K. Keegstra, *Curr. Opin. Plant Biol.*, 2010, **13**, 304.
34. M. E. Himmel, S. Y. Ding, D. K. Johnson, W. Adney, M. Nimlos, J. Brady and T. Foust, *Science*, 2007, **315**, 4.
35. S. Atsumi, A. F. Cann, M. R. Connor, C. R. Shen, K. M. Smith, M. P. Brynildsen, K. J. Chou, T. Hanai and J. C. Liao, *Metab. Eng.*, 2008, **10**, 305.
36. E. J. Steen, Y. S. Kang, G. Bokinsky, Z. H. Hu, A. Schirmer, A. McClure, S. Cardayre and J. D. Keasling, *Nature*, 2010, **463**, 559.

37. G. W. Huber, J. N. Chheda, C. J. Barrett and J. A. Dumesic, *Science*, 2005, **308**, 1446.
38. J. Q. Bond, D. M. Alonso, D. Wang, R. M. West and J. A. Dumesic, *Science*, 2010, **327**, 1110.
39. L. Wu, M. Arakane, M. Ike, M. Wada, T. Takai, M. Gau and K. Tokuyasu, *Bioresour. Technol.*, 2011, **102**, 4793.
40. D. Antoni, V. V. Zverlov and W. H. Schwarz, *Appl. Microbiol. Biotechnol.*, 2007, **77**, 35.
41. P. C. Badger, *Trends in New Crops and New Uses*, ASHS Press, Alexandria, 2002.
42. A. Demirbas, *Energy Sources*, 2005, **27**, 327.
43. S. Chundawat, G. Beckham, M. E. Himmel and B. E. Dale, *Annu. Rev. Chem. Biomol. Eng.*, 2011, **2**, 121.
44. R. L. Howard, E. Abotsi, E. L. Jansen van Rensburg and S. Howard, *Afr. J. Biotechnol.*, 2003, **2**, 602.
45. H. Younesi, G. Najafpour, K. S. Ku Ismail, A. R. Mohamed and A. H. Kamaruddin, *Bioresour. Technol.*, 2008, **99**, 2612.
46. J. R. Regalbuto, *Science*, 2009, **325**, 822.
47. T. M. Carole, J. Pellegrino and M. D. Paster, *Appl. Biochem. Biotechnol.*, 2004, **115**, 871.
48. T. Werpy and G. Petersen, NREL report, Top Value Added Chemicals From Biomass- Volume I: Results of Screening for Potential Candidates from Sugars and Synthesis Gas, *National Renewable Energy Laboratory, Golden*, 2004.
49. S. Fernando, S. Adhikari, C. Chandrapal and N. Murali, *Energy Fuels*, 2006, **20**, 1727.
50. P. Kaparaju, M. Serrano, A. B. Thomsen, P. Kongjan and I. Angelidaki, *Bioresour. Technol.*, 2009, **100**, 2562.
51. L. R. Lynd, E. Larson, N. Greene, M. Laser, J. Sheehan, B. E. Dale, S. McLaughlin and M. Wang, *Biofuels, Bioprod. Biorefin.*, 2009, **3**, 113.
52. M. McCoy, *Chem. Eng. News*, 2007, **85**, 12.
53. D. R. Dodds and R. A. Gross, *Science*, 2007, **318**, 1250.
54. M. Pagliaro and M. Rossi, *The Future of Glycerol: New Usages for a Versatile Raw Material*, RSC Publishing, Cambridge, 2008.
55. T. W. Tan, F. Shang and X. Zhang, *Biotechnol. Adv.*, 2010, **28**, 543.
56. J. van Haveren, E. L. Scott and J. Sanders, *Biofuels, Bioprod. Biorefin.*, 2007, **2**, 41.
57. H. J. Huang, S. Ramaswamy, U. W. Tschirner and B. V. Ramarao, *Sep. Purif. Technol.*, 2008, **62**, 1.
58. Y. P. Zhang, S. Ding, J. R. Mielenz, J. Cui, R. T. Elander and M. Laser, *Biotechnol. Bioeng.*, 2007, **97**, 214.
59. S. A. Nolen, C. L. Liotta, C. A. Eckert and R. Glaser, *Green Chem.*, 2003, **5**, 663.
60. C. A. Eckert, C. L. Liotta, D. Bush, J. S. Brown and J. P. Hallett, *J. Phys. Chem.*, 2004, **108**, 18108.
61. W. A. Farone and J. E. Cuzens, *US Pat.*, 6054 611, 2000.

62. L. A. Lucia, D. S. Argyropoulos, L. Adamopoulos and A. R. Gaspar, *Can. J. Chem.*, 2006, **84**, 960.

63. Y. Zhang, H. Du, X. Qian and E. Y.-X. Chen, *Energy Fuels*, 2010, **24**, 2410.

64. H. Zhao, J. E. Holladay, H. Brown and Z. C. Zhang, *Science*, 2007, **316**, 1597.

65. H. Danner and R. Braun, *Chem. Soc. Rev.*, 1999, **28**, 395.

66. D. M. Rathin, *J. Chem. Technol. Biotechnol.*, 2006, **81**, 1119.

67. Z. Xu, D. Zhang, J. Hu, X. Zhou, X. Ye, K. L. Reichel, N. R. Stewart, R. D. Syrenne, X. Yang, P. Gao, W. Shi, C. Doeppke, R. W. Sykes, J. N. Burris, J. J. Bozell, M. Z. Cheng, D. G. Hayes, N. Labbe, M. Davis, C. N. Stewart Jr and J. S. Yuan, *BMC Bioinf.*, 2009, **10**, 1.

68. C. Lee, Q. Teng, W. Huang, R. Zhong and Z. H. Ye, *Plant Cell Physiol.*, 2009, **50**, 1075.

69. L. E. A. Munoz and M. R. Riley, *Curr. Opin. Biotechnol.*, 2003, **14**, 454.

70. Z. Sun, J. A. Ramsay, M. Guay and B. A. Ramsay, *Appl. Microbiol. Biotechnol.*, 2007, **75**, 475.

71. P. Suriyamongkol, R. Weselake, S. Narine, M. Moloney and S. Shah, *Biotechnol. Adv.*, 2007, **25**, 148.

72. R. E. H. Sims, A. Hastings, B. Schlamadinger, G. Taylor and P. Smith, *Global Clim. Change*, 2006, **12**, 2054.

73. D. Tilman, R. Socolow, J. A. Foley, J. Hill, E. Larson, L. Lynd, S. Pacala, J. Reilly, T. Searchinger, C. Somerville and R. Williams, *Science*, 2009, **325**, 270.

74. M. Abramson, O. Shoseyov and Z. Shani, *Plant Sci.*, 2010, **178**, 61.

75. R. Lal and D. Pimentel, *Science*, 2009, **325**, 1345.

76. P. E. Kauppi and L. Saikku, *Science*, 2009, **326**, 1345.

77. D. E. Akin, *Appl. Biochem. Biotechnol.*, 2007, **136**, 3.

78. C. Nguyen, M. Ladisch and R. Meilan, *Nat. Biotechnol.*, 2007, **25**, 746.

79. T. A. Nguyen, K. R. Kim, S. J. Han, H. Y. Cho and J.W. Kim, *Bioresour. Technol.*, 2010, **101**, 7432.

80. N. Din, N. Gilkes, B. Tekant, R. Miller, R. Warren and D. Kilburn, *Nat. Biotechnol.*, 1991, **9**, 1096.

81. O. Shoseyov, Z. Shani and I. Levy, *Microbiol. Mol. Biol. Rev.*, 2006, **70**, 283.

82. M. Buanafina, T. Langdon, B. Hauck, S. Dalton and P. Morris, *Plant Biotechnol. J.*, 2008, **6**, 264.

83. V. Lionetti, F. Francocci, S. Ferrari, C. Volpi, D. Bellincampi, R. Galletti, R. D'Ovidio, G. Lorenzo and F. Cervone, *Proc. Natl. Acad. Sci. U. S. A.*, 2010, **107**, 616.

84. S. Bhandari, T. Fujino, S. Thammanagowda, D. Zhang, F. Xu and C. P. Joshi, *Planta*, 2006, **224**, 828.

85. H. Z. Wang, U. Avcib, J. Nakashima, M. G. Hahn, F. Chen and R. A. Dixon, *Proc. Natl. Acad. Sci. U. S. A.*, 2010, **107**, 22338.

86. W. D. Reiter, *Curr. Opin. Plant Biol.*, 2002, **5**, 536.

87. J. Vogel, *Curr. Opin. Plant Biol.*, 2008, **11**, 301.

88. W. Zeng, N. Jiang, R. Nadella, T. L. Killen, V. Nadella and A. Faik, *Plant Physiol.*, 2010, **154**, 78.
89. B. S. Dien, H. Jung, K. Vogel, M. Casler, J. Lamb, L. Itena, R. Mitchell and G. Sarath, *Biomass Bioenergy*, 2006, **30**, 880.
90. B. S. Dien, G. Sarath, J. Pedersen, S. Sattler, H. Chen, D. L. Funnell-Harris, N. Nichols and M. Cotta, *Bioenerg. Res.*, 2009, **2**, 153.
91. F. Chen and R. A. Dixon, *Nat. Biotechnol.*, 2007, **25**, 759.
92. C. X. Fu, J. R. Mielenz, X. R. Xiao, Y. X. Ge, C. Y. Hamilton, M. Rodriguez, F. Chen, M. Foston, A. Ragauskas, J. Boutona, R. A. Dixon and Z. Y. Wang, *Proc. Natl. Acad. Sci. U. S. A.*, 2011, **108**, 3803.
93. J. F. Pedersen, K. P. Vogel and D. L. Funnell, *Crop Sci.*, 2005, **45**, 812.
94. H. Liang, C. Frost, X. Wei, N. Brown, J. Carlson and M. Tien, Clean: *Soil, Air, Water*, 2008, **36**, 662.
95. J. H. Grabber, J. Ralph and R. D. Hatfield, *J. Agric. Food Chem.*, 2002, **50**, 6008.
96. J. H. Grabber, J. Ralph, C. Lapierre and Y. Barriere, *C. R. Biol.*, 2004, **327**, 455.
97. X. Xu, J. Lin and P. Cen, *J. Chem. Eng.*, 2006, **14**, 419.
98. J.-C. Leple, R. Dauwe, K. Morreel, V. Storme, C. Lapierre, B. Pollet, A. Naumann, K.-Y. Kang, H. Kim, K. Ruel, A. Lefebvre, J.-P. Joseleau, J. Grima-Pettenati, R. De Rycke, S. Andersson-Gunneras, A. Erban, I. Fehrle, M. Petit-Conil, J. Kopka, A. Polle, E. Messens, B. Sundberg, S. D. Mansfield, J. Ralph, G. Pilate and W. Boerjan, *The Plant Cell*, 2007, **19**, 3669.
99. L. E. Taylor, Z. Dai, S. R. Decker, R. Brunecky, W. S. Adney, S. Y. Ding and M. E. Himmel, *Trends Biotechnol.*, 2008, **26**, 413.
100. M. E. Himmel and E. A. Bayer, *Curr. Opin. Biotechnol.*, 2009, **20**, 316.
101. H. Oraby, B. Venkatesh, B. Dale, R. Ahmad, C. Ransom, J. Oehmke and M. Sticklen, *Trans. Res.*, 2007, **16**, 739.
102. C. Ransom, V. Balan, G. Biswas, B. Dale, E. Crockett and M. Sticklen, *Appl. Biochem. Biotechnol.*, 2007, **137**, 207.
103. G. S. Xie and L. C. Peng, *J. Integr. Plant Biol.*, 2011, **53**, 143.

CHAPTER 7

Technologies to Study Plant Biomass Fermentation Using the Model Bacterium Clostridium Phytofermentans

ANDREW C. TOLONEN,*[a,b,c] ELSA PETIT,[d]
JEFFREY L. BLANCHARD,[e] TOM WARNICK[f] AND
SUSAN B. LESCHINE[f]

[a] CEA, DSV, IG, Genoscope, Evry, France; [b] CNRS-UMR8030, Evry, France; [c] Université d'Evry Val d'Essonne, Evry, France; [d] Amherst College, Amherst MA, USA; [e] Department of Biology, University of Massachusetts, Amherst MA, USA; [f] Department of Veterinary and Animal Sciences, University of Massachusetts, Amherst MA, USA
*Email: atolonen@genoscope.cns.fr

7.1 Introduction

Dwindling oil reserves and climate change have made the development of renewable, carbon-balanced energy resources a global necessity. As cellulosic biomass is the world's most abundant biological energy source,[1] fermentation of inedible, cellulosic feedstocks such as prairie grasses, corn stover, wood chips and agricultural waste represents one of the most promising avenues for replacing petroleum-based fuels and products. Indeed, over 1.3 billion metric tons of cellulosic biomass could be used annually to produce

RSC Energy and Environment Series No. 10
Biological Conversion of Biomass for Fuels and Chemicals: Explorations from Natural Utilization Systems
Edited by Jianzhong Sun, Shi-You Ding and Joy Doran-Peterson
© The Royal Society of Chemistry 2014
Published by the Royal Society of Chemistry, www.rsc.org

fuels and chemicals in North America without affecting food supplies.[2] Fermentation on this scale could provide enough ethanol for 65% of the USA's ground transportation fuel at current levels,[3] which is attractive for USA energy security because it imports 60% of it's petroleum.[4] Encouraged by this opportunity, the US Congress passed the Energy Independence and Security Act (EISA) in 2007 mandating an annual production of 36 billion gallons of biofuels by 2022, of which 16 billion gallons should come from sources such as cellulosic feedstocks.

Presently, the main barrier to the expansion of cellulosic-based fuels and bio-products is the recalcitrance of biomass.[4] While plant biomass is composed primarily of pentose and hexose sugars, they are linked within high-molecular-weight polysaccharides to form a complex network that resists biological and chemical degradation. Furthermore, no single crop can be cultivated in all geographic areas and candidate feedstocks vary widely in their polysaccharide composition: cellulose (24–45%), hemicellulose (18–29%), and pectin (1–24%).[5] Due to the complexity and variability of plant biomass, the cost of breaking it into sugars using purified enzymes and chemical pretreatment currently doubles the price per unit of carbohydrate, which erases the cost savings of biomass relative to corn.[6]

Consolidated bioprocessing (CBP)[7] is a promising strategy to overcome biomass recalcitrance in a cost-effective manner by using a single microbial culture to combine biomass deconstruction and sugar fermentation into one step. The projected costs for converting biomass to ethanol using CBP are four-fold lower than a two-step process based on simultaneous saccharification and co-fermentation (SSCF) using purified cellulases.[8] However, CBP requires a single microbe or stable microbial consortia with five properties: (1) enzyme production to degrade diverse plant polysaccharides; (2) fermentation of 5- and 6-carbon sugars; (3) minimal production of inhibitory side-products; (4) growth in conditions compatible with industrial bioreactors; and (5) the ability to produce a high concentration of fuel product. *Clostridium phytofermentans* is remarkable in possessing all five of these properties, enabling it to ferment biomass with saccharification rates and ethanol yields comparable to SSCF.[9]

C. Phytofermentans, a mesophilic anaerobe isolated from forest soil,[10] which grows on both the soluble and insoluble components of plant biomass by first cleaving plant polysaccharides (cellulose, hemicellulose, starch, and pectin) and then fermenting the resulting pentose and hexose sugars. According to the CAZy database,[11] the *C. phytofermentans* genome encodes 169 carbohydrate-active enzymes (CAZys) to cut diverse plant substrates into sugars, which is the largest number of CAZys among sequenced clostridia. This chapter details how *C. phytofermentans* is unique among the characterized microbes in its broad substrate range and efficient ethanol production, making it an excellent model system for cellulosic biofuels research. Specifically, we will focus on three areas of *C. phytofermentans* research: isolation and physiological characterization; systems biology; and genetic engineering.

7.2   Isolation and Physiology

7.2.1   Isolation

This chapter focuses on the *C. phytofermentans* strain ISDgT ATCC 700394, which was isolated in August 1991 from damp silt in a stream bed up-stream of a mill pond dam near Quabbin Reservoir in central Massachusetts, USA.[10] The strain name "ISDg" refers to the isolation site: "Intermittent Stream behind Dam". Samples were recorded alphabetically, so "g" is the seventh isolate from the ISD location. The "T" means strain ISDg is the "type strain" used for taxonomy.

To isolate strain ISDgT, a spatula of 0.2–0.5 g soil was inoculated into GS-2 medium[12] with $6\,g\,L^{-1}$ ball-milled cellulose as a sole carbon source (GS-2C medium) and incubated anaerobically for 1 month at 30 °C. A 0.1 mL aliquot was twice transferred to 4 mL GS-2C medium and incubated for 2 weeks. The culture was then diluted into tubes containing 4 mL melted GS-2C soft-agar medium and poured onto plates with a basal layer of GS-2 medium with no added carbon. Colonies that produced clear zones in cellulose overlays were twice streaked on GS-2 cellobiose plates and colonies were picked and grown in GS-2 cellobiose liquid medium. Cultures were transferred to GS-2C medium to confirm they were cellulolytic and stocks were frozen at –80 °C in 15% glycerol.

At least 16 different cellulolytic anaerobes were isolated from nine different soil samples during this sampling survey. Four of the eight isolates from the ISD site were cellulolytic and a preliminary characterization of all 16 isolates showed that ISDg produced the highest ethanol titers. Ribosomal sequencing showed that *C. phytofermentans* belongs to clostridia group 14 A, which is phylogenetically distant from well-studied cellulolytic clostridia such as *C. thermocellum* and *C. cellulovorans* (Figure 7.1(a)). *C. phytofermentans* ISDg was thus chosen for further study based on its ability to efficiently ferment cellulose to ethanol and its phylogenetic distance from previously characterized clostridia.

7.2.2   Growth Conditions

C. phytofermentans ISDgT cells are straight rods 0.5–0.8 μm by 3–15 μm in size. Cells growing on soluble substrates are motile, usually with one or two flagella (Figure 7.1(b)). When growing on insoluble plant matter such as cellulose, cells are shorter, non-flagellated, and adhere to the substrate (Figure 7.1(c)). Adhesion to biomass is an adaptation used by some cellulolytic microbes to increase enzyme concentrations near the substrate and to exclude competitors from the liberated sugars.[13] *C. phytofermentans* can bind both cellulose and hemicellulose,[14] even though it lacks the cellulosomes that enable attachment by other clostridia. Cells growing on hemicellulose sometime bear surface nodules, suggesting that hemicellulose particles bind to the cell surface. The mechanism by which *C. phytofermentans* adheres to plant substrates is unknown, but may include cell-surface-binding proteins, secretion of an

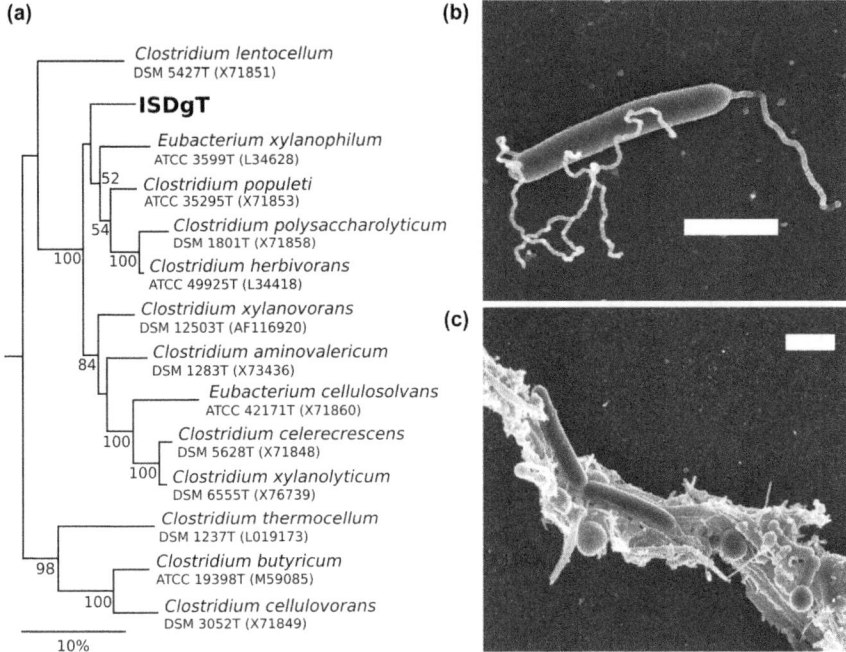

Figure 7.1 (a) Phylogenetic tree based on the maximum likelihood analysis of *C. phytofermentans* and related clostridia 16 S rRNA gene sequences. Clostridia are members of group 14a, except *Clostridium cellulovorans* (cluster 1), *Clostridium thermocellum* (cluster 3) and *Clostridium lentocellum* (cluster 14b) which are included as cluster representatives. Bootstrap values greater than 50%, expressed as percentages of 100 replications, are shown at the branching points. The bar represents 10 substitutions per 100 base pairs. *E. Coli* was used as the outgroup to root the tree (image adapted from ref. 10). (b) Scanning electron microscopy shows a log-phase cell with two flagella. (c) Scanning electron micrograph of cells growing on cellulose fibers. Cells are adhered to cellulose and spores are also visible. White scale bars are 1 μm.

extra-polysaccharide glycocalyx, or cell surface pili that facilitate binding as in *Ruminococcus albus*.[15]

C. phytofermentans grows fastest at 37 °C. It will grow slowly at temperatures as low as 15 °C and as high as 42 °C. Cells grow well at pH 6.0–9.0, but poorly at pH 9.5 and not at all at pH 5.5. The maximum growth rate occurs when the initial pH is 8.0, whereas the highest growth yield is achieved when the initial pH is 8.5. Strain ISDgT ferments numerous carbon sources: pentoses (arabinose, ribose, xylose), hexoses (glucose, glucuronic acid, galactose, galacturonic acid, fructose, mannose), disaccharides (cellobiose, gentiobiose, lactose, maltose), and polysaccharides (cellulose, xylan, pectin, starch, laminarin). ISDgT does not grow on inulin, mannan, glycerol, pyruvate, sucrose, trehalose or tryptone.[10]

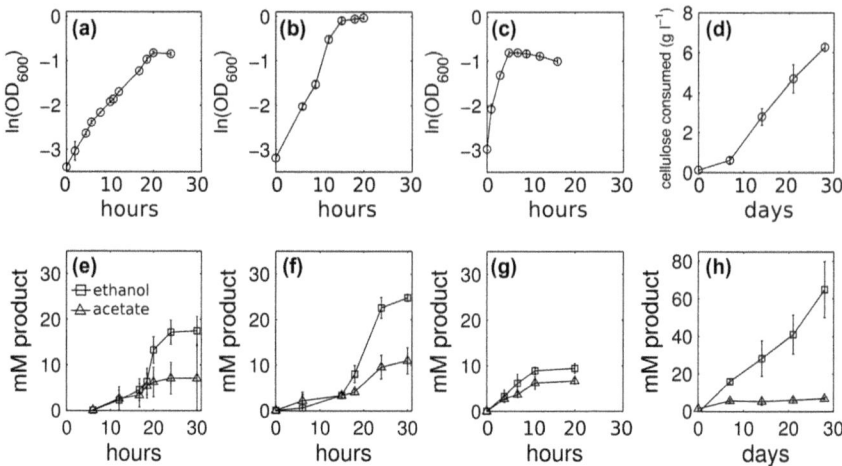

Figure 7.2   (a)–(d) *C. phytofermentans* growth and (e)–(h) accumulation of fermentation products on (a) and (e) $0.3 \, g \, L^{-1}$ xylose, (b) and (f) glucose, (c) and (g) hemicellulose and (d) and (h) $1.2 \, g \, L^{-1}$ cellulose. Data points are means of triplicate cultures. Error bars show one standard deviation and are smaller than the symbols where not apparent. Growth on xylose (a), glucose (b), and hemicellulose (c) was quantified as OD_{600}. Growth on cellulose (d) was measured as dry mass of cellulose in culture. Production of ethanol and acetate, the two most abundant fermentation products, was measured by HPLC as described in Tolonen *et al.*[14]

C. phytofermentans sometimes grows better on polysaccharides than on their constituent sugars. For example, growth on xylose (Figure 7.2(a)) is slower than on hemicellulose xylan (Figure 7.2(c)), even though hemicellulose xylan is a β-1,4-D-xylopyranose polymer that must be cleaved to xylose and isomerized before glycolysis. Faster growth on polysaccharides may result from *C. phytofermentans* having higher affinity transporters for oligosaccharides than for sugar monomers. Also, because ATP for sugar transport is used on a per-molecule basis, uptake of oligosaccharides saves energy relative to sugar monomers.[16] Phosphorolysis of xylan by a yet unidentified phosphorylase might also save ATP relative to hydrolysis by a mechanism similar to cello-dextrin phosphorylases.[17] However, when xylan is hydrolyzed, the resulting xylose is isomerized to xylulose by xylose isomerase (Cphy0200, Cphy1219) before being phosphorylated to xylulose 5-phosphate by xylulokinase (Cphy3419). This suggests that the terminal xylose residues on xylosaccharides would have to be isomerized before phosphorolysis or the phosphorylase would need to isomerize xylose in parallel.

Although standard GS-2 media contains ammonium chloride or urea as a nitrogen source, there is evidence that *C. phytofermentans* can fix nitrogen if grown in a defined medium without nitrogen.[18] Nitrogen fixation is common among clostridia, indeed *C. pasteurianum* was the first free-living, nitrogen-fixing organism described almost 120 years ago.[19] The genomes of nitrogen-fixing clostridia contain *nif* genes: *nifH*, *nifD*, and *nifK* encode nitrogenase proteins and *nifE*, *nifB-N* and *nifV* construct the iron–molybdenum nitrogenase co-factor. *C. phytofermentans* has potential orthologs to *nifH* (*cphy2693*, *cphy3935*), *nifB* (*cphy2391*, *cphy1480*), *nifV* (*cphy3171*), and two genes to provide [Fe–S] clusters for synthesis of the Fe–Mo co-factor, *nifS* (*cphy2907*) and *nifU* (*cphy2907*, *cphy3260*). However, *nifD*, *nifK*, and *nifE* are missing and the putative *nif* genes in *C. phytofermentans* are not co-localized in the genome as they are in *C. pasteurianum*, *C. acetobutylicum*, and *C. beijerinckii*.[20] Further work is needed to characterize nitrogen fixation by *C. phytofermentans*, but it is an interesting opportunity to ferment biomass without nitrogen supplementation and to produce additional hydrogen.[18]

A recent advance in the culturing of *C. phytofermentans* is the formulation of chemically defined media that do not contain the yeast extract present in GS-2 medium. These defined media consist of a carbon source, an inorganic phosphate, inorganic nitrogen, salts ($MgCl_2$, $CaCl_2$, $FeSO_4$), cysteine HCl as a reducing agent, sodium citrate, and seven B-vitamins: thiamine (B1), riboflavin (B2), nicotinate (B3), pantethine (B5), pyridoxal (B6), biotin (B7), and folinic acid (B9).[21] B-Vitamin auxotrophies are common in clostridia. *C. thermocellum* is auxotrophic for pyridoxal, biotin, folate, and cobalamin (B12).[12] Similarly, *C. cellulolyticum* cannot synthesize biotin, folate, nicotinate, riboflavin, panthoteate, thiamine, and cobalamin.[22] An unanticipated benefit of using chemically defined media, at least in *C. cellulolyticum*, is that specific ethanol production rates are 10-fold higher than in complex media.[22] The authors reasoned that in complex media, *C. cellulolyticum* accumulates pyruvate leading to NADH-induced growth arrest, whereas cells in defined media are better able to regulate $NADH$-to-NAD^+ ratios.

Because B-vitamin addition is expensive for large-scale industrial reactors, we want to understand the genetic basis of *C. phytofermentans* B-vitamin auxotrophies (Figure 7.3) to enable the development of prototrophic strains. Four of the seven B-vitamin auxotrophies (thiamine, riboflavin, pantothenate, and biotin) result from *C. phytofermentans* missing entire multi-gene pathways that would be challenging to transfer to the genome. Although *C. phytofermentans* lacks most or all of the genes to make these vitamins, they are each co-factors for many required enzymes. For example, the pyrophosphate derivative (TPP) of thiamine (B1) is a co-enzyme of highly expressed decarboxylase enzymes such as pyruvate ferredoxin oxidoreductase (Cphy3558), oxoglutarate ferredoxin oxidoreductase (Cphy3122-3), and transketolase (Cphy0014). Rather than making thiamine, *C. phytofermentans* scavenges it using a thiamine transporter (Cphy0729). The riboflavin (B2) derivatives flavin mononucleotide (FMN) and flavin adenine dinucleotide (FAD) are co-factors that mediate redox reactions; 10 proteins are annotated as FMN-binding and 42 bind FAD. Riboflavin appears to be taken into the cell with a transporter (Cphy0344) that has a high similarity to the *B. subtilis* riboflavin transporter YpaA. Pantothenate (B5) is a building block of co-enzyme A, which is required to oxidize pyruvate and to build fatty acids. Rather than building pantothenate, *C. phytofermentans* uptakes pantethine, splits it into pantetheine, and converts it to CoA. Biotin (B7) is a co-factor of at least six carboxylase

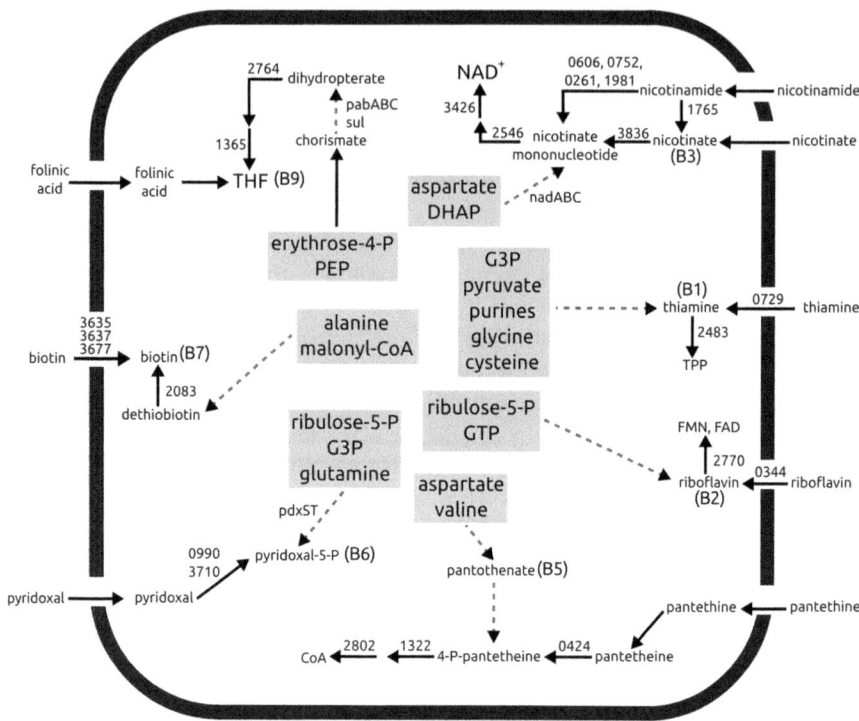

Figure 7.3 Map of the genetic basis of *C. phytofermentans* B-vitamin auxotrophies. The metabolites used for *de novo* synthesis of each vitamin are shown. The arrows represent one or more enzymatic steps: solid arrows are reactions catalyzed in *C. phytofermentans* and dashed arrows are missing reactions. The numbers denote NCBI protein annotations. For vitamins B3, B6, and B9, the *B. subtilis* genes that might complement *C. phytofermentans* auxotrophies are shown. Abbreviations: glyceraldehyde-3-phosphate (G3P), guanosine triphosphate (GTP), phosphoenolpyruvate (PEP), tetra-hydrofolate (THF), flavin mononucleotide (FMN), flavin adenine dinucleotide (FAD), thiamine pyrophosphate (TPP).

enzymes such as acetyl-CoA carboxylase, (Cphy0519, Cphy0521-3) which adds a carboxyl group to acetyl-CoA to form malonyl-CoA for fatty acid synthesis. *C. Phytofermentans* likely uptakes biotin with a tripartite biotin transporter.[23]

Auxotrophies for three other B-vitamins, nicotinate, pyridoxal, and folate, result from *C. phytofermentans* missing only one or a few enzymes, suggesting these auxotrophies could be readily overcome by gene transfer. *C. Phyto-fermentans* requires NAD^+ for various electron-transfer reactions, but needs nicotinate (B3) or nicotinamide for growth because it lacks the *nadABC* genes for *de novo* synthesis of nicotinate mononucleotide from aspartate and dihy-droxyacetone phosphate. *C. phytofermentans* can, however, salvage NAD^+ from nicotinate in three steps: (1) nicotinate and D-ribose-5-phosphate are combined to form nicotinate mononucleotide by Cphy3836; (2) nicotinate adenine dinucleotide is formed from nicotinate mononucleotide and ATP by

Cphy2546; and (3) finally Cphy3426 adds NH_2 to form NAD^+, which can be converted to $NADP^+$ by Cphy2509. According to KEGG, *C. phytofermentans* also has a second pathway to convert nicotinate and nicotinamide to the mononucleotide forms using Cphy0752, Cphy0261, and Cphy0606, which were all up-regulated on cellulose relative to glucose.[14] Although proteins for the other NAD^+ synthesis pathway were not similarly up-regulated, *C. phytofermentans* may have an increased need of NAD^+ during growth on cellulose. Pyridoxal-5-phosphate (B6), PLP, is a putative co-factor for 15 *C. phytofermentans* enzymes including transaminases, amino acid racemases and decarboxylases, and sulfhydrylases. PLP synthesis in *B. subtilis* requires the PdxST complex,[24] also called "YaaDE", which makes PLP from either ribose-5-phosphate or ribulose 5-phosphate, glyceraldehyde 3-phosphate or dihydroxy-acetone phosphate, and glutamine.[25] Because *C. phytofermentans* makes all the substrates of the PdxST complex, it is feasible that transformation of *C. phyofermentans* with *B. subtilis pdxS–pdxT* could rescue the PLP auxotrophy.

Tetrahydrofolate (THF), the active form of folic acid (B9), is required for many single-carbon-transfer reactions, but the pathway to make THF is incomplete in *C. phytofermentans*. It uses erythrose-4-phosphate and PEP to build shikimate which can, in turn, be used to make chorismate. However, the *pabABC* and *sul* genes to convert chorismate to dihydropteroate are missing. Following these three steps, *C. phytofermentans* can use dihydropteroate make dihydrofolate (DHF) using Cphy2764 and can reduce DHF to THF with dihydrofolate reductase (Cphy1365). Experimental evidence supports that the THF auxotrophy can only be subsidized with folinic acid, even though it is not a cofactor in any known reactions.[26] It is unclear if this preference for folinic acid is due to its increased stability,[27] if *C. phytofermentans* has a specific folinic acid requirement, or if the putative folate transporter is more efficient at transporting folinic acid.

A benefit of the alternative and missing enzymes for THF metabolism is that *C. phytofermentans* is resistant to anti-folate antibiotics, which can thus be used to purge contaminant bacteria from mixed cultures. For example, dihydrofolate reductase (Cphy1365) is similar to other clostridial orthologs, but not to those of *E. coli* or *B. subtilis*. As trimethoprim acts by inhibiting dihydrofolate reductase, the variant enzyme likely underlies the natural resistance of *C. phytofermentans* to trimethoprim such that this antibiotic can be used to remove *E. coli* following conjugation.[28] Also, the lack of dihydropteroate synthase to make dihydropteroate from 4-Aminobenzoic acid (PABA) should make *C. phytofermentans* resistant to sulfonamide antibiotics.

7.2.3   Fermentation

Ethanol is the generally the primary fermentation product for *C. phytofermentans* and ethanol yields can approach the theoretical maximum. To calculate ethanol yields, we assume that each glucose molecule entering glycolysis forms two pyruvate molecules that each can be converted to one ethanol molecule, giving a maximum yield of two ethanols formed per glucose consumed. As the initial glucose concentration in cultures was 16 mM (3 g^{-1}), ethanol reaching 24.81 mM in 30 h (Figure 7.2(f)) corresponds to a yield of 77%. After 48 h, ethanol concentrations in glucose cultures reach >95% of the maximum theoretical yield.[14] In cellulose cultures, stable cell densities (10^7–10^8 colony forming units (CFU) mL^{-1}) result in a constant rate of cellulose degradation (Figure 7.2(d)) and ethanol formation (Figure 7.2(h)). Assuming cellulose is composed of polymerized glucose, the conversion of cellulose to ethanol is 68% of the maximum theoretical yield. Furthermore, the ethanol-to-acetate ratio of 9.54 in the cellulose cultures is among the highest yields reported for clostridia.[7]

One xylose can be fermented to 5/3 ethanol equivalents; xylose is converted to xylulose-5-phosphate and fed into the non-oxidative pentose phosphate pathway where three xylulose-5-phosphate molecules are converted into two fructose-6-phosphates (four ethanols) and one glyceraldehyde-3-phosphate (one ethanol). Conversion of 20 mM (3 g^{-1}) xylose to 17.1 mM ethanol in 24 h (Figure 7.2(e)) thus represents a 52% ethanol yield. Assuming that xylan (Sigma X0502) is made entirely of xylose, the initial xylose concentration in the hemicellulose cultures is 22.8 mM and the final ethanol concentration of 10.35 mM (Figure 7.2(g)) gives an ethanol yield of 27.2%. Rapid growth on hemicellulose (Figure 7.2(c)) is thus accompanied by a reduced ethanol-to-acetate ratio (Figure 7.2(g)).

A shift to produce lower ethanol-to-acetate ratios during growth on hemicellulose is reinforced by recent results that acetate can be the most abundant fermentation product when *C. phytofermentans* grows on AFEX-treated corn stover.[9] This study showed that acetate was mostly produced during early growth on AFEX-treated stover when hemicellulose was preferentially consumed; afterwards the fermentation products shifted to ethanol when glucans were metabolized. Acetate could be viewed as the preferred fermentation product because its synthesis makes ATP, whereas ethanol formation wastes potential energy by oxidizing NADH to maintain redox balance. Efficient carbon assimilation during growth on hemicellulose may allow the cell to produce more acetate while still making enough ethanol to maintain redox balance. During growth on cellulose, cells still need to make ethanol to maintain redox balance, but are unable to also produce as much acetate for ATP because it is more expensive to harvest carbon.

7.3   Systems Biology

A comprehensive, molecular-level understanding of how *C. phytofermentans* ferments biomass will enhance our ability to optimize strains for transformation of plant matter into useful biochemicals. Because biomass fermentation by *C. phytofermentans* is such a complex process involving a myriad of enzymes, systems biology is proving to be a vital tool for rapid, high-throughput analysis of changes in cell state under different environmental conditions. Our systems biology studies seek an improved understanding of three areas: genome annotation (defining of genes, regulatory elements, and regulons), assembling proteins into metabolic and regulatory networks, and identifying the function of key enzymes for cellulosic breakdown and fermentation. Initially, we used approaches including genome sequencing and proteomics. These systems-level analyses show that *C. phytofermentans* has a treasure trove of novel, cellulolytic and fermentative enzymes that provide a rational basis to engineer strains for improved biomass fermentation.

7.3.1 Genome Sequence

The *C. phytofermentans* ISDg ATCC 700394 complete genome has been sequenced by the US Department of Energy Joint Genome Institute. The genome has a 35.4% G + C content and consists of a single, circular 4 847 594-bp chromosome encoding 3902 protein coding sequences (CDS). Presently, 2481 (62.2%) of the CDS have a predicted function according to the JGI annotation. The genome also contains 89 non-coding RNAs including eight rRNA operons (16 S, 5 S, 23 S), a 6 S RNA, a tmRNA, and 61 tRNAs. Assuming G–U wobble in the codon–anti-codon pairing, the 61 tRNA can recognize all 61 possible codons. The genome also encodes 33 putative riboswitches that control processes such as amino acid and cobalamin biosynthesis, sodium balance, and perhaps biofilm formation.

The *C. phytofermentans* genome sequence is a basis for understanding the machinery used by this bacterium to ferment biomass.[29] The genome has 169 CAZys including 31 glycosyl transferases, 13 carbohydrate esterases, 9 polysaccharide lyases, and 115 glycoside hydrolases that are spread across 39 sequence families.[11] This abundance and diversity of CAZys enables this bacterium to grow on so many plant substrates. For example, *C. thermocellum* grows on cellulose, but not hemicellulose or pectin, and its genome encodes only 70 glycoside hydrolases in 23 families. Because *C. thermocellum* specializes in cellulose degradation, its CAZys include 16 glycoside hydrolase family 9 (GH9) cellulases. In contrast, *C. phytofermentans* has only one CAZy in GH9, which is required for cellulose degradation.[28] *C. Phytofermentans* CAZys also differ from other cellulolytic clostridia by lacking dockerin domains and the genome does not have a scaffoldin protein to assemble a cellulosome. Cellulolytic enzymes in *C. phytofermentans* are thus either freely secreted or anchored to the cell in a cellulosome-independent manner. The absence of a cellulosome will likely make *C. phytofermentans* cellulolytic enzymes particularly well-suited for over-expression and heterologous expression because they can function without binding a cellulosomal scaffold.

The ability of *C. phytofermentans* cells to uptake numerous substrates is highlighted by the genome encoding 137 genes for ATP-binding cassette (ABC-type) transporters. Along with the transporters, are 53 genes for extra-cellular-solute-binding proteins (ESBs), which are membrane-attached lipoproteins that capture substrates and pass them to the membrane transporter. Forty-seven of the 53 *C. phytofermentans* ESBs are involved in sugar transport and can be very highly expressed. The oligosaccharide-binding ESB Cphy2466 was the third most highly expressed protein in the cell during growth on cellulose.[14]

7.3.2   Gene Expression

We recently completed a proteome-wide analysis of *C. phytofermentans* to identify the hydrolytic and metabolic enzymes used to ferment different cellulosic substrates.[14] Protein quantification by mass spectrometry (LC-MS/MS) remains a challenging research area and we showed that a labeling method we call "ReDi proteomics" gives comprehensive, accurate, low-cost quantification of a microbial proteome and can discern extra-cellular proteins. ReDi labeling is advantageous to alternative approaches for stable isotope incorporation such as ^{15}N labeling and Stable Isotope Labeling by Amino acids in Cell culture (SILAC) in not requiring strains with specific amino acid auxotrophies or optimization of growth in synthetic media. Also, because ReDi isotope labels are added after peptides are isolated, supernatant and cellular proteins can be differentially labeled to identify extra-cellular proteins. Specifically, we applied ReDi proteomics to quantify differences in each protein in cellulosic cultures (hemicellulose and cellulose) relative to protein levels in glucose cultures. We also identified the extra-cellular proteome using ReDi protemics to measure the concentration differences of each protein between the supernatant and culture lysates. To understand the physiological consequences of changes to the proteome, we integrated the proteomics with detailed measurements of growth, fermentation, enzyme activities, and electron microscopy.

Changes in a total of 2567 proteins (65% of the predicted coding sequences) were quantified, including 357 previously hypothetical proteins. Growth on cellulosic substrates entailed numerous proteome changes. Comparing protein expression levels from glucose and hemicellulose cultures showed greater than two-fold differences for 20% of proteins; 49% proteins were expressed at two-fold different levels between glucose and cellulose cultures. As a control, fewer than 4% of proteins differed by two-fold levels when comparing two glucose cultures. Expression changes on hemicellulose were largely confined to pentose transport and assimilation, the pentose phosphate pathway, and CAZys, whereas expression changes on cellulose extended throughout metabolism. Many proteins were up-regulated on cellulose (CAZys, transporters, glycolytic proteins, NADH synthesis); others were repressed (flagella, fatty acid synthesis, housekeeping functions, DNA replication, transcription, translation). Also, the overall fraction of the proteome comprised of hypothetical proteins was much higher on cellulose, suggesting that many proteins of unknown function relate to cellulose metabolism.

Defining the set of secreted proteins "the secretome" was a high priority because enzymes to degrade insoluble biomass must be extra-cellular. We combined ReDi proteomics with localization prediction by PsortB v2.0[30] and SignalP3.0[31] to identify extra-cellular proteins and the cellular mechanisms used for export. The *C. phytofermentans* secretome consists of more than 300 proteins that function mostly in carbohydrate and protein degradation, cell surface and flagellar assembly, and transport. Furthermore, assigning CAZys to intra-cellular and extra-cellular proteins showed which steps in polysaccharide degradation occur inside the cell. Both substrates are initially cleaved by multiple extra-cellular enzymes and transported into the cell as oligosaccharides before being catabolized to sugars, an adaptation that likely conserves energy and prevents sugars from being available to competing microbes.

C. phytofermentans expressed more than 100 CAZys and altered their stoi-chiometries to each substrate. The first step to metabolize hemicellulose is cleavage of the xylan backbone by extra-cellular GH10 and GH11 endo-xylanases. Cphy2108, the most highly expressed extra-cellular xylanase, is predicted to attach to the peptidoglycan wall with an LPXTG motif.[32] Cleavage of hemicellulose by a surface-bound xylanase may help to retain xylosaccharides in close proximity to the cell. In parallel with cleavage of the xylan backbone, additional extra- and intra-cellular enzymes remove the glu-curonic and acetic acid side-chains, which are common in hemicellulose.[33] *C. phytofermentans* likely transport xylosaccharides into the cell before cleavage by two main intra-cellular exo-xylosidases: Cphy3009 cleaves xylosides from the non-reducing end of xylosaccharides and Cphy3207 acts on the reducing end.

Cellulose fibers are depolymerized by the concerted action of endo- and exo-cellulases. Endo-cellulases randomly cleave internal bonds in cellulose chains to disrupt the crystalline structure and expose individual polysaccharide chains. The most highly secreted endo-cellulase was Cphy3367, which is the only GH9 enzyme in *C. phytofermentans*. Cphy3367 is required for cellulose degradation[28] and solubilizes cellulose *in vitro*.[34] *C. phytofermentans* also highly up-regulated 3 GH5 endo-cellulases, which likely have accessory roles in cellulose cleavage. Exo-cellulases channel the freed cellulose chains through an active site tunnel to cleave two to four units from the ends of the exposed chains.[35] Cphy3368 is an exo-cellulase[36] that is likely expressed from the same mRNA as Cphy3367 and these two enzymes synergize to convert cellulose to cellodextrins and glucose.[34]

C. phytofermentans transports cellodextrins into the cell before splitting them into sugar monomers. Cellodextrin cleavage occurs primarily by three cello-dextrin phosphorylases (Cphy3854, Cphy0430, Cphy1929) that remove term-inal glucosides using phosphate as an attacking group. Phosphorolysis avoids ATP hydrolysis for phosphorylation of glucose during import, which is re-quired for glucose transport by the *E. coli* phosphotransferase system. While the ATP savings correlate with the length of the transported cellodextrin, up-take and phosphorolytic cleavage of glucans can reduce cellular ATP re-quirements by 25%.[37]

Sugars are catabolized by the Embden–Meyerhof–Parnas glycolysis pathway using several non-standard enzymes that, similar to the cellulodextrin phos-phorylases described above, highlight how *C. phytofermentans* is pressured to conserve ATP because of the reduced efficiency of fermentation relative to oxidative respiration. For example, ATP is likely conserved during glycolysis by using reversible, pyrophosphate (PPi)-dependent glycolytic enzymes such as PPi-dependent phosphofructokinase and pyruvate phosphate dikinase. *C. phytofermentans* lacks the Ppa phosphatase used by *E. coli* to cleave PPi and instead uses it as a phosphate donor and source of a high-energy bond. PPi-Dependent glycolytic enzymes can increase glycolytic yield from 2 to 5 ATP38. In general, clostridia have low pyrophosphatase activities and high intracellular PPi levels,[39] supporting the general use of PPi as an energy-saving adaptation.

C. phytofermentans likely also conserves energy by using ferredoxin-dependent enzymes in the place of enzymes that use NADH. Pyruvate from glycolysis is converted to acetyl-CoA using a highly expressed pyruvate ferredoxin oxidoreductase (PFOR, Cphy3558) that oxidizes pyruvate and transfers electrons to ferredoxin. In contrast, *E. coli* and *B. subtilis* convert pyruvate to acetyl-CoA using NAD^+-reducing pyruvate dehydrogenase. *C. phytofermentans* also converts 2-oxoglutarate to succinyl-CoA using a ferredoxin-dependent 2-oxoglutarate oxidoreductase (Cphy3122-3) instead of NADH-dependent 2-oxoglutarate dehydrogenase. Reduced ferredoxin ($E_0' \leq -420\,mV$) has a more negative redox potential than NADH ($E_0' = -320\,mV$) and this $100\,mV$ energy difference could be exploited for energy generation. Similar to *C. ljungdahlii*,[40] *C. phytofermentans* highly expresses an Rnf-type NADH ferredoxin oxidoreductase[29] that could use electrons from reduced ferredoxin to pump either sodium or protons across the cell membrane and subsequently deposit the electrons on NAD^+. The resulting electrochemical gradient could then be harnessed to form ATP by a membrane-bound ATPase.

Reducing equivalents in the form of NADH are consumed by reducing acetyl-CoA to ethanol with multiple, highly expressed alcohol dehydrogenase (ADH) enzymes: five Fe-dependent ADHs and one Zn-dependent ADH. The two Fe-dependent ADHs, Cphy3925 and Cphy1029, are among the most highly expressed proteins under all conditions. Cphy3925 shares 54% amino acid identity with *E. coli* AdhE, a bifunctional aldehyde/alcohol dehydrogenase that converts acetyl-CoA directly to ethanol. The other highly expressed ADH, Cphy1029, shares 35% amino acid identity with *Z. mobilis* AdhB, the primary ADH in this organism. This combination of ADH enzymes rapidly consumes excess reducing equivalents and facilitates the high ethanol yields. Another adaptation to consume excess reducing equivalents is the high expression of both monomeric and bifurcating [FeFe]–hydrogenases that re-oxidize ferredoxin and NADH. Monomeric hydrogenases oxidize only ferredoxin, but the bifurcating hydrogenases oxidize both ferredoxin and NADH in a 1 : 1 ratio,[41] thereby saving ferredoxin for use by the Rnf complex. All subunits for two multi-meric, bifurcating hydrogenases were highly expressed on all carbon sources, whereas a monomeric ferredoxin-dependent hydrogenase was expressed at a 10-fold lower level.

In addition to identifying the principle enzymes used to deconstruct and ferment cellulosic substrates (Figure 7.4), other metabolic adaptations likely indirectly enhance cellulosic fermentation. For example, tryptophan biosynthesis enzymes are up-regulated during growth on cellulose. While the tryptophan precursors leading up to chorismate are used for various compounds, the up-regulated enzymes make anthranilate (Cphy3847-8) and are

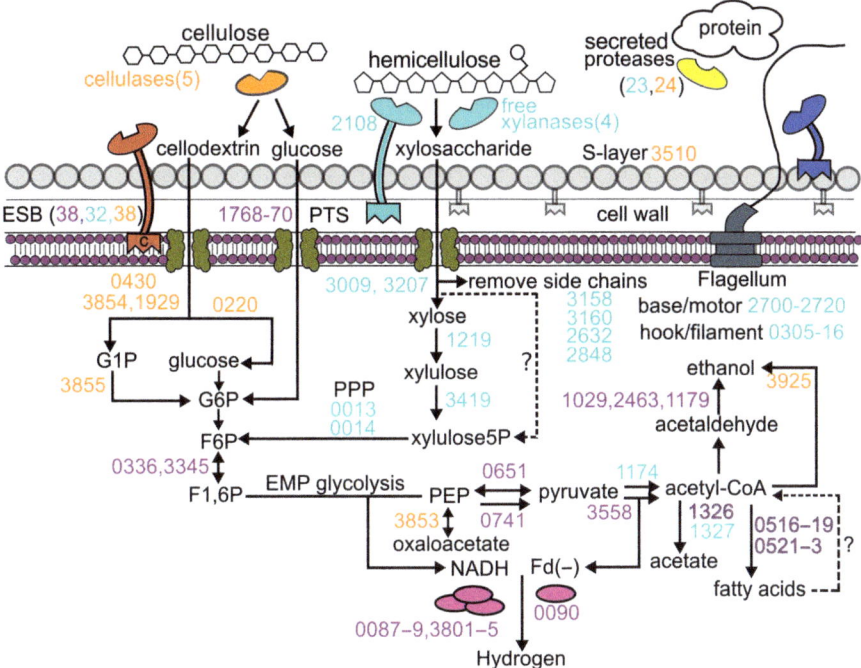

Figure 7.4 Model of the key secreted and intracellular enzymes for the degradation and fermentation of plant biomass. The numbers are NCBI protein annotations and are colored by the highest protein expression on glucose (purple), hemicellulose (turquoise), and cellulose (orange). The numbers in parentheses show the number of proteins of related function. Dashed arrows represent unknown reactions.
Image adapted from ref. 14.

involved in the final steps of tryptophan synthesis (Cphy3842-3, Cphy3845-6), suggesting the expression changes show a direct need for tryptophan. Increased production of tryptophan could increase ethanol tolerance by an unknown mechanism similar to yeast[42,43] and enable translation of CAZys with tryptophan-rich carbohydrate-binding modules (CBMs). CBMs in carbohydratase proteins enhance catalysis by keeping the hydrolytic domain in close association with its substrate[44] using a hydrophobic platform of tryptophan residues[45] that form van der Waals interactions with sugar rings.[46] Thirteen glycoside hydrolases with CBMs were up-regulated on cellulose and the concomitant high expression of tryptophan-synthesis proteins suggests that tryptophan may be an important requirement for cellulolysis.

Fatty acid synthesis was repressed during growth on cellulose relative to glucose, including enzymes for synthesis of malonyl-CoA (Cphy0519, Cphy0521-3) and malonyl-ACP (Cphy 0516), and for fatty acid elongation (Cphy0517. The greater expense required to harvest carbon from cellulose likely caused a metabolic shift to store less carbon and energy as fatty acids, which could complicate efforts to use microbes to produce fatty-acid-based biofuels from cellulosic biomass.[47] Along with repression of fatty acid synthesis, we expected to see increased expression of β-oxidation enzymes to break down fatty acids into acetyl. However, KEGG and homology searches support that *C. phytofermentans* and other clostridia including *C. cellulolyticum* and *C. thermocellum* do not have any of the enzymes for fatty acid β-oxidation, making it unclear if or how these clostridia degrade fatty acids.

A novel means to express cellulolytic enzymes on the cell surface is suggested by the most highly expressed protein in the proteome, Cphy3510, a secreted protein that shares significant sequence similarity with the *Bacillus anthracis* S-layer protein, Sap.[48] The S-layer is a protective outer coat in some bacteria that is generally composed of a single protein that spontaneously assembles into a two-dimensional lattice.[49] Because S-layers of $\sim 5 \times 10^5$ protein molecules are required to cover a microbial cell,[50] the S-layer protein is often the most abundant protein in the proteome. Transmission electron micrographs of *C. phytofermentans* shows a surface layer exterior to the cell wall,[10,14] leading us to propose that *C. phytofermentans* is covered by a Cphy3510-based S-layer. The S-layer has been shown to anchor CAZys in *B. stearothermophilis*[51] and *Thermoanaerobacterium thermosulfurigenes*[52] using C-terminal repeats of about 50 amino acids. Thus, the identification of the S-layer protein in *C. phytofermentans* provides insight into how CAZys may be anchored to the cell surface and suggests a new strategy for engineering polypeptides for cell surface display by binding them to Cphy3510.

In addition to the genomic and proteomic data described above, we are analyzing metabolomic and transcriptomic data with the ultimate goal of integrating these datasets into a comprehensive molecular model of biomass fermentation. These datasets will together reveal genetic regulatory elements, the organization of protein networks, and the dynamics of metabolite pools during cellulosic fermentation. This molecular systems-level understanding of cell state will ultimately enable improved predictions about how *C. phytofermentans* should be genetically engineered for industrial biomass fermentation.

7.4   Genetic Engineering

Although cellulolytic clostridia have been studied for decades, a lack of methods for their genetic manipulation has hindered our ability to learn how they ferment biomass and to create robust strains for industrial bioconversion. Here we discuss progress and future opportunities to develop *C. phytofermentans* as a platform microbe for genetic engineering by focusing on three major prerequisites: (1) delivery of foreign DNA; (2) plasmid origins and resistance markers; and (3) chromosomal integration.

7.4.1   DNA Delivery

Transfer of foreign DNA to clostridia has been achieved by conjugation, electroporation, and natural transformation. We have shown that DNA can be reliably transferred to *C. phytofermentans* by conjugal transfer from *E. coli*,[28] a method that also works in cellulolytic *C. cellulolyticum*[53] and in non-cellulolytic clostridia including *C. acetobutylicum*,[54] *C. perfringens*,[55] and *C. difficile*.[56] Specifically, we used the broad range RP4 conjugal apparatus encoded by pRK24,[57] which can transfer any plasmid that has the 760-bp RP4 conjugal origin of transfer (oriT) sequence.[58]

Following conjugal transfer, it is necessary to eliminate both *E. coli* donors and *C. phytofermentans* cells that did not receive plasmid DNA. As *C. phytofermentans* is naturally resistant to nalidixic acid and trimethoprim, these antibiotics can be used to kill *E. coli* donors. Alternatively, if plates are supplemented with X-gal and IPTG, *lacZ +* *E. coli* strains form blue colonies whereas *C. phytofermentans* colonies are white. The formation of white *C. phytofermentans* colonies on X-gal/IPTG plates is surprising because it grows on lactose and has multiple putative β-galactosidases. Perhaps X-gal is either inefficiently transported into the cell or is not a substrate of *C. phytofermentans* enzymes. A third option to selectively remove *E. coli* from post-conjugation liquid cultures, is inoculation with an *E. coli* phage such as T7.[59] After selection, the absence of residual *E. coli* can be confirmed by plating an aliquot on solid LB medium and incubating aerobically at 37 °C overnight. Selection for *C. phytofermentans* transconjugants that received a resistance marker (see the following section) supports the conjugal delivery of DNA to *C. phytofermentans* with an efficiency of ∼1 transconjugant per 10^6 cells.

Although DNA delivery by electroporation has not yet been shown in *C. phytofermentans*, this method works in various clostridia including *C. acetobutylicum*,[60] *C. tyrobutyricum*,[61] *C. ljungdahlii*,[40] *C. paraputrificum*,[62] and the cellulolytic strains *C. cellulolyticum*[53] and *C. thermocellum*.[63] Electroporation is often inhibited in gram-positive bacteria such as clostridia by their thick peptidoglycan wall, but efficiencies can be improved by weakening the cell wall with muralytic enzymes,[64] glycine,[65] or isoniacin.[66] *C. phytofermentans* is an excellent candidate for electroporation because the genome does not appear to encode any DNA-restriction enzymes, which have plagued electroporation efforts in other clostridia such as *C. acetobutylicum*[67] and *C. perfringens*.[68]

An exciting possibility that could accelerate large-scale genome engineering in *C. phytofermentans* would be to find methods to transfer DNA into the cell by natural transformation. Natural transformation is based on genetic competence, the ability of a cell to uptake naked DNA from the environment.[69] In the first report of natural competence in clostridia, strains of *Thermoanaerobacterium* and *Thermoanaerobacter* were shown to be naturally competent with efficiencies from 1×10^{-3} to 1.9×10^{-6} transformants per CFU.[70] These thermophiles, which ferment xylan and certain sugars to a range of fermentation products, had previously been transformed by electroporation,[71] but now several strains are known to be competent in the early exponential phase in a process that depends upon a TP4 locus, *comEA*, *comEC*, and a *cinA recA* locus.

The genes for natural competence in these thermophilic clostridia all have homologs in *C. phytofermentans*, suggesting it may also be transformable if the conditions for competence were found. *C. phytofermentans* has several genes for assembly of DNA-binding type IV pili and structures resembling these pili are visible in electron micrographs.[14] *C. phytofermentans* has gene homologs of both parts of the ComE DNA transporter: the membrane-bound dsDNA receptor ComEA (*cphy1953*) and the ComEC transmembrane channel (*cphy1957*). Furthermore, the *comEC* homologs are separated in the genome by a putative two-component system (*cphy1954*, *cphy1955*), suggesting competence may be regulated by a sensor kinase/response regulator pair similar to *B. subtilis*.[69] After the transforming DNA has been internalized, it must be delivered to the genome for recombination. In *B. subtilis*, there is a specifically positioned ssDNA binding apparatus that receives ssDNA from the DNA-uptake machinery and processes the DNA for recombination with chromosomal DNA.[72] *C. phytofermentans* has gene homologs of the *B. subtilis* proteins that mediate this process including: DprA, (*cphy2723*), SsbB (*cphy3773*, *cphy2987*, *cphy3483*), CoiA (*cphy0885*), RecA (*cphy2439*), and RecN (*cphy2499*). However, even if *C. phytofermentans* has a complete competence machinery, conditions for natural transformation may be abstruse. *Bacillus megaterium* has a *comE* locus that can complement the *B. subtilis* ComE transporter, but *B. megaterium* has not been found to be naturally competent.[73]

7.4.2   Antibiotic Resistance and Plasmid Origins

In parallel with methods for DNA delivery, we are working to identify antibiotic resistance genes and plasmid origins for clostridial shuttle and suicide vectors. The two most commonly used antibiotics used for clostridia are erythromycin and chloramphenicol. For selections in *C. phytofermentans*, we use a $200\,\mu g\,mL^{-1}$ erythromycin to select for the *erm* gene from *S. pneumoniae* Tn1545 that functions in both gram-negative and gram-positive bacteria.[74] Although we have not had problems with spontaneous resistance to erythromycin, a 1 : 1 combination of erythromycin and lincomycin could potentially give a stronger selection, as with *C. thermocellum*.[63] These antibiotics can be combined because lincomycin and erythromycin resistance are conferred by the same *mls* gene (macrolide–lincosamide–streptogramin).[75] Because the mutations that confer resistance to each antibiotic are independent, the probability that both mutations arise simultaneously is much lower. *C. phytofermentans* is also sensitive to $10\,\mu g\,mL^{-1}$ chloramphenicol and the chloramphenicol acetyltransferase gene *catP* functions as a selectable marker. If desired, thiamphenicol, a methyl-sulfonyl analog of chloramphenicol, can be used instead for selections with *catP*.[76] *C. phytofermentans* is, however, naturally resistant to kanamycin and streptomycin,[10] so these antibiotics should not be used for selection.

   Genetic methods using both replicative and non-replicative plasmid origins have been used in clostridia. Plasmids bearing the *Enterococcus* pAMß1 origin replicate in *C. phytofermentans* during antibiotic selection, but are rapidly lost in the absence of selection.[28] Specifically, we found that 80% of cells can be cured of pAMß1 plasmids by diluting cultures 1 : 100 with a medium lacking antibiotic and growing to late log phase through five serial transfers. The cultures that have been cured of pAMß1 plasmids regain erythromycin sensitivity, permitting further manipulations using the same plasmid and resistance marker. While conditionally replicating plasmids are ideal for many applications, a high-copy-number plasmid that replicates indefinitely without selection would be useful in large, industrial reactors where antibiotic addition is prohibitive. We are currently investigating plasmid origins that may allow stable, high-copy replication of plasmids in *C. phytofementans* without selection, including the pBP1 origin from *C. botulinum*, the pCB101 origin from *C. butyricum*[77] that replicates at high copy in *C. acetobutylicum*,[54] the pIM13 origin from *B. subtilis*,[78] and the pCD6 origin from *C. difficile*.[56]

7.4.3 Chromosomal Insertion

Targeted chromosomal changes in clostridia have traditionally been made by either single or double cross-over homologous recombination. Single cross-over homologous recombination has been used to insert a plasmid into the *pta* genes of *C. acetobutylicum*[79] and of *C. tyrobutyricum*.[61] Similarly, a double cross-over was used to delete the *pta*[80] and *cel48s*[81] genes from *C. thermocellum*. Because the frequencies of chromosomal insertion by homologous recombination initially appeared to be prohibitively low in *C. phytofermentans*, we developed a general system for targeted, chromosomal gene inactivation in *C. phytofermentans* based on group II introns, catalytic RNAs that self-splice into genomic DNA in a site-specific manner.[82]

The group II intron we developed to make targeted chromosomal insertions in *C. phytofermentans* is based on the *Lactococcus lactis* Ll.LtrB group II intron,[83] which is available from Sigma–Aldrich as the "Targetron System". It consists of a 0.9 kb Ll.LtrB-deltaORF intron flanked by short exon sequences and a downstream *ltrA* gene encoding a protein with endo-nuclease and reverse transcriptase activity.[84] Because the DNA-binding specificity of the group II intron is conferred by a short 13–16-bp sequence, the intron can be customized to integrate into a desired DNA target site by a simple two-step cross-over PCR. Group II introns can, in theory, function in any bacterial taxa into which plasmid DNA can be delivered because intron insertion does not require host-supplied factors. Furthermore, the insertion frequencies of group II introns into the genomic DNA of diverse bacteria are quite high: 0.1–22% in *E. coli*,[83] 37–100% in grampositive *Staphylococcus aureus*,[85] and 2.5–100% in clostridia.[86]

When a strong *C. phytofermentans* promoter is used to drive expression of the group II intron, it inserts into the genome with such a high efficiency (often near 100%) that mutants can be easily isolated without selecting for integration. Because the intron insertions are stable in the absence of selection, the intron need not contain a resistance gene. Once the plasmid carrying the intron and *ltrA* gene is cured from *C. phytofermentans*, no antibiotic resistance genes remain. Multiple intron insertions can thus be made in the same strain without the need for independent resistance markers, which is particularly helpful because so few resistance markers are known to function in clostridia.

The elements to make our general system for targeted chromosomal insertions in *C. phytofermentans* are combined in the plasmid, pQint (Figure 7.5(a)). This plasmid has an RP4 conjugal origin of transfer, an

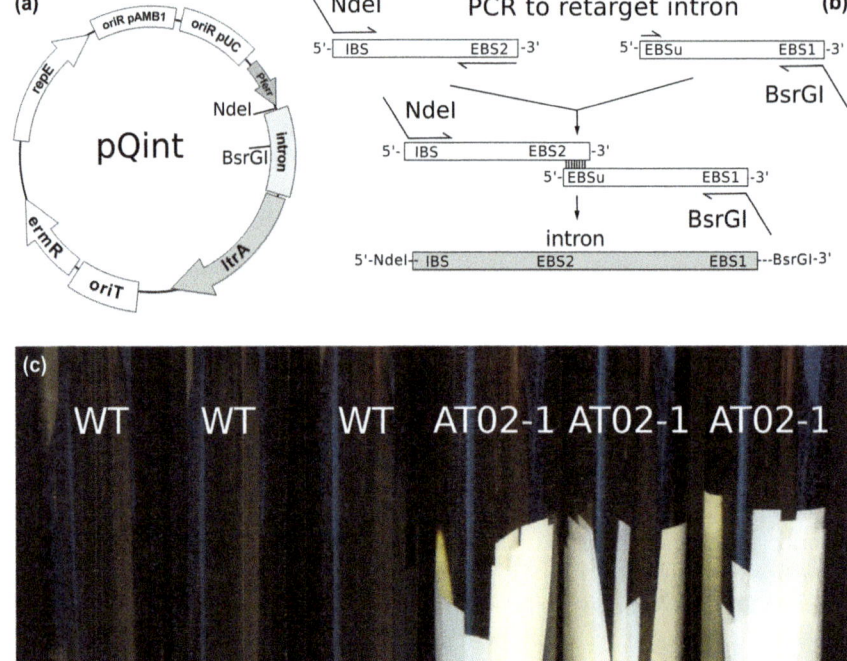

Figure 7.5 The genetic system to make targeted, chromosomal insertions in the *C. phytofermentans* chromsome. (a) Plasmid map of pQint. This plasmid has two origins of replication: a *Enterococcus* pAMß1 origin (for *C. phytofermentans*) and a pUC origin (for *E. coli*). The *C. phytofermentans* pyruvate ferrodoxin oxidoreductase (*cphy3558*) promoter (Pferr) drives the strong expression of the intron cassette and down-stream *ltrA* gene. The oriT facilitates conjugal transfer by an RP4-type conjugal apparatus. The *erm* gene from *S. pneumoniae* Tn1545 functions in *E. coli* and *C. phytofermentans*. (b) The intron cassette can be targeted to insert in nearly any desired DNA sequence by two-step, cross-over PCR. After PCR, the targeted intron is cloned into the NdeI and BsrGI sites of pQint. (c) Insertion of an intron into the gene encoding the sole family 9 hydrolase Cphy3367 results in a strain (AT02-1) that is unable to degrade cellulose. Cellulose degradation was measured as dry mass of cellulose remaining in culture. After four weeks, the cellulose strips in the wild-type tubes had broken down, while the strips in the AT02-1 cultures appeared unchanged. Images adapted from ref. 28.

erythromycin-resistance gene that functions in both gram-negative and gram-positive bacteria, origins of replication for *E. coli* (pUC) and for gram-positive bacteria (*Enterococcus* pAMß1), and a group II intron down-stream of the strong *cphy3558* promoter. This intron can be targeted to insert anywhere in the *C. phytofermentans* chromosome by customizing the targeting sequence using PCR (Figure 7.5(b)). In an initial study, we inserted a group II intron into *cphy3367*, the only family 9 glycoside hydrolase in *C. phytofermentans*.[28] Inactivation of *cphy3367* resulted in a strain (AT02-1) that grew normally on glucose, cellobiose and hemicellulose, but had lost the ability to degrade cellulose (Figure 7.5(c)). These findings reveal a central role played by Cphy3367 in cellulose degradation and show for the first time that a single gene can be required for cellulolysis.[87] We are currently applying this system to dissect the set of cellulolytic enzymes in *C. phytofermentans* in order to identify key genes for the degradation of cellulosic biomass and to develop strains that ferment biomass to specific fermentation products.

7.5 Future Directions

This chapter has highlighted how *C. phytofermentans* is an excellent model system for cellulosic biofuel research because of its unique ability to directly ferment diverse plant polysaccharides to ethanol and its ease of cultivation in the laboratory. Our recent work on *C. phytofermentans* has focused on optimizing growth conditions, high-throughput genomics and proteomics to characterize metabolic networks, and genetic engineering methods. These studies both provide insight into how microbes ferment biomass and reveal unmet challenges that need to be addressed.

The results described here show promising targets for metabolic engineering to make *C. phytofermentans* strains that are more cost-effective for industrial fermentation. For example, *C. phytofermentans* is similar to other cellulolytic clostridia in lacking the ability to make several B-vitamins (Figure 7.3), which are expensive to add to large industrial reactors. Genomic analysis suggests that some of these auxotrophies could be overcome by addition of a few genes, either by putting the genes on a replicating plasmid or by integrating them into the genome using a group II intron. Specifically, these B-vitamin auxotophies might be overcome by adding the following genes: *nadABC* for nicotinamide (B3), *B. subtilis pdxST* for pyridoxal-5-phosphate (B6), and *B subtilis pabABC* and *sul* to convert chorismate to dihydropteroate for folate (B9).

Proteomic identification of highly expressed and secreted enzymes allows us to reduce the 169 *C. phytofermentans* CAZys into a set of high-priority targets to engineer for improved biomass degradation. For example, Cphy3367 is the second-most highly expressed CAZy on cellulose and it is required for cellulose degradation; supporting it would be an ideal gene both to over-express in *C. phytofermentans* and to express in other microbes to improve cellulolysis. Similarly, Cphy2105 and Cphy2108 appear to be the main secreted hemicellulases, suggesting they would be good candidates to manipulate for the degradation of hemicellulose-rich substrates. Along with the secreted enzymes to degrade polysaccharides *in situ*, the principle intra-cellular cellodextrin phosphorylases (Cphy3854, Cphy0430, Cphy1929) and xylosidases (Cphy3207, Cphy3009) should also be over-expressed to cleave oligosaccharides into sugars that can be readily metabolized.

Even though ethanol fermentation by *C. phytofermentans* is quite efficient, metabolic engineering to streamline fermentation could further improve ethanol yields. We observed that the ethanol-to-acetate ratio is much lower on hemicellulose (Figure 7.2(g)) than on other substrates, especially cellulose (Figure 7.2(h)). The proteomics of hemicellulose cultures showed elevated expression of acetate kinase (Cphy1327), supporting that inactivation of this enzyme would improve ethanol yields similar to other clostridia.[79,61,80] The growth rate on hemicellulose (doubling time 1.16 h, Figure 7.2(c)) is also the fastest among the substrates we have tested. It is unknown if high acetate production, which yields ATP by substrate-level phosphorylation, directly enables the rapid growth rate on hemicellulose, but it would be interesting to test if inactivating acetate kinase cripples the growth on hemicellulose.

In addition to altering the expression of native enzymes, methods such as directed evolution and domain shuffling can be used to improve the activities of these enzymes under industrially relevant conditions. For example, the *C. phytofermentans* hydrolase Cphy3202 was amplified by error-prone PCR to create a library of random 20 000 variants, which was then screened to identify a mutant with a two-fold increased half-life at 60 °C relative to the wild-type.[88] Domain shuffling has also been applied to three fungal class II cellobiohydrolases to create themostable chimeric proteins.[89] As an alternative to enzyme screening, a technique called "chemical complementation" allows the selection of highly active cellulases *in vivo*.[90] This method is a reverse yeast three-hybrid assay in which cleavage of a tetrasaccharide substrate decreases expression of a toxic URA3 reporter, thereby selecting for cells with high-activity cellulases. While this assay requires a synthetic saccharide substrate with methotrexate and dexamethasone handles that bears little resemblance to plant biomass, it can analyze more enzyme variants ($\sim 10^8$) than is feasible by *in vivo* enzyme screening and was applied to domain-shuffled cellulases to find a mutant with a six-fold increase in K_{cat}/K_m over the parent enzyme.[90] Chemical complementation could be applied to *C. phytofermentans* CAZys to develop variants with enhanced activities.

Cost-effective biofuel production requires high product titers, which is challenging because ethanol and other prospective biofuels are toxic to microbes at high concentrations. *C. phytofermetans* grows normally in ethanol concentrations up to 4% (40 g l^{-1}). Even though this ethanol tolerance is higher than other clostridia such as *C. thermocellum* (1–2% ethanol tolerance),[91] industrial biofuel production will likely require strains that tolerate even higher ethanol concentrations. Numerous studies have shown that ethanol tolerance can be improved by random mutagenesis[92] or serial transfer.[91,93,94] Two promising recombinant DNA methods have also recently been used to express genes conferring increased biofuel tolerance in clostridia. First, transforming *C. acetobutylicum* with a random genomic library of its own DNA was used to identify genes that improve growth during butanol stress when at increased copy number.[95] Second, heterologous expression in *E. coli* of 43 efflux pumps from various sequenced bacterial genomes was shown to reduce toxicity by exporting the biofuels from the cell.[96] Both these methods are based on heterologous expression of plasmid DNA, which is now tractable in *C. phytofermentans* using the genetic methods described in this chapter.

Plant biomass is an abundant, low-cost feedstock that could enable large-scale production of fuels and biochemicals. In order for cellulosic biofuels to replace a significant fraction of our current petroleum needs, improvements are needed in many areas such as the genetic engineering of bioenergy crops, improved cultivation methods for these plants, and microbes to efficiently convert plant matter into useful chemicals. This chapter has described recent advances in developing *C. phytofermentans* as a model microbe to study biomass fermentation. We hope that these studies on *C. phytofermentans* will help to enable future advances to produce biofuels that both reduce our need for fossil fuels and facilitate sustainable economic development.

References

1. S. B. Leschine, *Annu. Rev. Microbiol.*, 1995, **49**, 399.
2. D. Perlack, L. L. Wright, A. F. Turhollow, R. L. Graham, B. L. Stokes and D. C. Erbach, *A Joint Study by the US Department of Energy and the US Department of Agriculture*, 2005, U.S. Department of Energy (Oak Ridge TN, USA).
3. C. Somerville, *Science*, 2006, **312**, 1277.
4. J. Houghton, S. Weathervax and J. Ferrell, *US Department of Energy Roadmap*, Department of Energy, 2006, U.S. Department of Energy (Oak Ridge TN, USA).
5. J. Doran-Peterson, D. M. Cook and S. K. Brandon, *Plant J.*, 2008, **54**, 582.
6. L. R. Lynd, M. S. Laser, D. Bransby, B. E. Dale, B. Davison, R. Hamilton, M. Himmel, M. Keller, J. D. McMillan, J. Sheehan and C. E. Wyman, *Nat. Biotechnol.*, 2008, **26**, 169.
7. L. R. Lynd, P. J. Weimer, W. H. van Zyl and I. S. Pretorius, *Microbiol. Mol. Biol. Rev.*, 2002, **66**, 506.
8. L. R. Lynd, W. H. van Zyl, J. E. McBride and M. Laser, *Curr. Opin. Biotechnol.*, 2005, **16**, 577.
9. M. Jin, V. Balan, C. Gunawan and B. E. Dale, *Biotechnol. Bioeng.*, 2011, **108**, 1290.
10. T. A. Warnick, B. A. Methé and S. B. Leschine, *Int. J. Syst. Evol. Microbiol.*, 2002, **52**, 1155.
11. B. L. Cantarel, P. M. Coutinho, C. Rancurel, T. Bernard, V. Lombard and B. Henrissat, *Nucleic Acids Res.*, 2009, **37**, D233.
12. E. A. Johnson, A. Madia and A. L. Demain, *Appl. Environ. Microbiol.*, 1981, **41**, 1060.
13. Y. Lu, Y.-H. P. Zhang and L. R. Lynd, *Proc. Natl. Acad. Sci. U. S. A.*, 2006, **103**, 16165.

14. A. C. Tolonen, W. Haas, A. C. Chilaka, J. Aach, S. P. Gygi and G. M. Church, *Mol. Syst. Biol.*, 2011, **7**, 461.
15. M. Morrison and J. Miron, *FEMS Microbiol. Lett.*, 2000, **185**, 109.
16. M. Muir, L. Williams and T. Ferenci, *J. Bacteriol.*, 1985, **163**, 1237.
17. Y.-H. P. Zhang and L. R. Lynd, *J. Bacteriol.*, 2005, **187**, 99.
18. S. Harvey, A. Chambers and P. Zhang, presented at the 26th Army Science Conference, Orlando, 2008.
19. S. Winogradsky, *Arch. Sci. Biol.*, 1895, **3**, 297.
20. J. S. Chen, J. Toth and M. Kasap, *J. Ind. Microbiol. Biotechnol.*, 2001, **27**, 281.
21. S. Leschine, Susan and T. Warnick, *US Pat.*, 7682811 (B2), 2010.
22. E. Guedon, M. Desvaux, S. Payot and H. Petitdemange, *Microbiology*, 1999, **145**, 1831.
23. P. Hebbeln, D. A. Rodionov, A. Alfandega and T. Eitinger, *Proc. Natl. Acad. Sci. U. S. A.*, 2007, **104**, 2909.
24. B. R. Belitsky, *J. Bacteriol.*, 2004, **186**, 1191.
25. M. Strohmeier, T. Raschle, J. Mazurkiewicz, K. Rippe, I. Sinning, T. B. Fitzpatrick and I. Tews, *Proc. Natl. Acad. Sci. U. S. A.*, 2006, **103**, 19284.
26. P. Stover and V. Schirch, *Trends Biochem. Sci.*, 1993, **18**, 102.
27. S. Ogwang, H. T. Nguyen, M. Sherman, S. Bajaksouzian, M. R. Jacobs, W. H. Boom, G.-F. Zhang and L. Nguyen, *J. Biol. Chem.*, 2011, **286**, 15377.
28. A. C. Tolonen, A. C. Chilaka and G. M. Church, *Mol. Microbiol.*, 2009, **74**, 1300.
29. Petit *et al.*, *PLoS One* (2013). Accepted.
30. J. L. Gardy, M. R. Laird, F. Chen, S. Rey, C. J. Walsh, M. Ester and F. S. L. Brinkman, *Bioinformatics*, 2005, **21**, 617.
31. J. D. Bendtsen, H. Nielsen, G. von Heijne and S. Brunak, *J. Mol. Biol.*, 2004, **340**, 783.
32. M. Zhou, J. Boekhorst, C. Francke and R. J. Siezen, *BMC Bioinf.*, 2008, **9**, 173.
33. D. Shallom and Y. Shoham, *Curr. Opin. Microbiol.*, 2003, **6**, 219.
34. X.-Z. Zhang, N. Sathitsuksanoh and Y.-H. P. Zhang, *Bioresour. Technol.*, 2010, **101**, 5534.
35. C. Divne, J. Ståhlberg, T. Reinikainen, L. Ruohonen, G. Pettersson, J. K. Knowles, T. T. Teeri and T. A. Jones, *Science*, 1994, **265**, 524.
36. X.-Z. Zhang, Z. Zhang, Z. Zhu, N. Sathitsuksanoh, Y. Yang and Y.-H. P. Zhang, *Appl. Microbiol. Biotechnol.*, 2010, **86**, 525.
37. Y.-H. P. Zhang and L. R. Lynd, *Proc. Natl. Acad. Sci. U. S. A.*, 2005, **102**, 7321.
38. C. H. Slamovits and P. J. Keeling, *Eukaryotic Cell*, 2006, **5**, 148.
39. J. K. Heinonen and H. L. Drake, *FEMS Microbiol. Lett.*, 1988, **52**, 205.
40. M. Köpke, C. Held, S. Hujer, H. Liesegang, A. Wiezer, A. Wollherr, A. Ehrenreich, W. Liebl, G. Gottschalk and P. Dürre, *Proc. Natl. Acad. Sci. U. S. A.*, 2010, **107**, 13087.

41. G. J. Schut and M. W. W. Adams, *J. Bacteriol.*, 2009, **191**, 4451.
42. T. Hirasawa, K. Yoshikawa, Y. Nakakura, K. Nagahisa, C. Furusawa, Y. Katakura, H. Shimizu and S. Shioya, *J. Biotechnol.*, 2007, **131**, 34.
43. X. Q. Zhao and F. W. Bai, *J. Biotechnol.*, 2009, **144**, 23.
44. P. Tomme, H. Van Tilbeurgh, G. Pettersson, J. Van Damme, J. Vandekerckhove, J. Knowles, T. Teeri and M. Claeyssens, *Eur. J. Biochem.*, 1988, **170**, 575.
45. T. Ponyi, L. Szabó, T. Nagy, L. Orosz, P. J. Simpson, M. P. Williamson and H. J. Gilbert, *Biochemistry*, 2000, **39**, 985.
46. J. Lehtiö, J. Sugiyama, M. Gustavsson, L. Fransson, M. Linder and T. T. Teeri, *Proc. Natl. Acad. Sci. U. S. A.*, 2003, **100**, 484.
47. E. J. Steen, Y. Kang, G. Bokinsky, Z. Hu, A. Schirmer, A. McClure, S. B. Del Cardayre and J. D. Keasling, *Nature*, 2010, **463**, 559.
48. I. Etienne-Toumelin, J. C. Sirard, E. Duflot, M. Mock and A. Fouet, *J. Bacteriol.*, 1995, **177**, 614.
49. U. B. Sleytr and P. Messner, *Annu. Rev. Microbiol.*, 1983, **37**, 311.
50. U. B. Sleytr and P. Messner, *J. Bacteriol.*, 1988, **170**, 2891.
51. E. Egelseer, I. Schocher, M. Sára and U. B. Sleytr, *J. Bacteriol.*, 1995, **177**, 1444.
52. M. Matuschek, G. Burchhardt, K. Sahm and H. Bahl, *J. Bacteriol.*, 1994, **176**, 3295.
53. K. C. Jennert, C. Tardif, D. I. Young and M. Young, *Microbiology*, 2000, **146**, 3071.
54. D. R. Williams, D. I. Young and M. Young, *J. Gen. Microbiol.*, 1990, **136**, 819.
55. D. Lyras and J. I. Rood, *Plasmid*, 1998, **39**, 160.
56. D. Purdy, T. A. T. O'Keeffe, M. Elmore, M. Herbert, A. McLeod, M. Bokori-Brown, A. Ostrowski and N. P. Minton, *Mol. Microbiol.*, 2002, **46**, 439.
57. R. Meyer, D. Figurski and D. R. Helinski, *Mol. Gen. Genet.*, 1977, **152**, 129.
58. D. G. Guiney and E. Yakobson, *Proc. Natl. Acad. Sci. U. S. A.*, 1983, **80**, 3595.
59. A. C. Tolonen, G. B. Liszt and W. R. Hess, *Appl. Environ. Microbiol.*, 2006, **72**, 7607.
60. S. Nakotte, S. Schaffer, M. Böhringer and P. Dürre, *Appl. Microbiol. Biotechnol.*, 1998, **50**, 564.
61. Y. Zhu, X. Liu and S.-T. Yang, *Biotechnol. Bioeng.*, 2005, **90**, 154.
62. K. Sakka, M. Kawase, D. Baba, K. Morimoto, S. Karita, T. Kimura and K. Ohmiya, *J. Biosci. Bioeng.*, 2003, **96**, 304.
63. M. V. Tyurin, S. G. Desai and L. R. Lynd, *Appl. Environ. Microbiol.*, 2004, **70**, 883.
64. P. T. Scott and J. I. Rood, *Gene*, 1989, **82**, 327.
65. H. Holo and I. F. Nes, *Appl. Environ. Microbiol.*, 1989, **55**, 3119.
66. M. V. Tyurin, C. R. Sullivan and L. R. Lynd, *Appl. Environ. Microbiol.*, 2005, **71**, 8069.

67. L. D. Mermelstein and E. T. Papoutsakis, *Appl. Environ. Microbiol.*, 1993, **59**, 1077.
68. C. K. Chen, C. M. Boucle and H. P. Blaschek, *FEMS Microbiol. Lett.*, 1996, **140**, 185.
69. D. Dubnau, *Microbiol. Rev.*, 1991, **55**, 395.
70. A. J. Shaw, D. A. Hogsett and L. R. Lynd, *Appl. Environ. Microbiol.*, 2010, **76**, 4713.
71. V. Mai and J. Wiegel, *Appl. Environ. Microbiol.*, 2000, **66**, 4817.
72. D. Kidane and P. L. Graumann, *Cell*, 2005, **122**, 73.
73. M. Lammers, H. Nahrstedt and F. Meinhardt, *J. Basic Microbiol.*, 2004, **44**, 451.
74. P. Trieu-Cuot, C. Carlier, C. Poyart-Salmeron and P. Courvalin, *Gene*, 1991, **102**, 99.
75. M. Monod, C. Denoya and D. Dubnau, *J. Bacteriol.*, 1986, **167**, 138.
76. J. R. O'Connor, D. Lyras, K. A. Farrow, V. Adams, D. R. Powell, J. Hinds, J. K. Cheung and J. I. Rood, *Mol. Microbiol.*, 2006, **61**, 1335.
77. J. G. Morris and N. P. Minton, *J. Gen. Microbiol.*, 1981, **127**, 325.
78. H. Azeddoug, J. Hubert and G. Reysset, *J. Gen. Microbiol.*, 1992, **138**, 1371.
79. E. M. Green, Z. L. Boynton, L. M. Harris, F. B. Rudolph, E. T. Papoutsakis and G. N. Bennett, *Microbiology*, 1996, **142**, 2079.
80. S. A. Tripathi, D. G. Olson, D. A. Argyros, B. B. Miller, T. F. Barrett, D. M. Murphy, J. D. McCool, A. K. Warner, V. B. Rajgarhia, L. R. Lynd, D. A. Hogsett and N. C. Caiazza, *Appl. Environ. Microbiol.*, 2010, **76**, 6591.
81. D. G. Olson, S. A. Tripathi, R. J. Giannone, J. Lo, N. C. Caiazza, D. A. Hogsett, R. L. Hettich, A. M. Guss, G. Dubrovsky and L. R. Lynd, *Proc. Natl. Acad. Sci. U. S. A.*, 2010, **107**, 17727.
82. A. M. Lambowitz and S. Zimmerly, *Annu. Rev. Genet.*, 2004, **38**, 1.
83. M. Karberg, H. Guo, J. Zhong, R. Coon, J. Perutka and A. M. Lambowitz, *Nat. Biotechnol.*, 2001, **19**, 1162.
84. H. Guo, M. Karberg, M. Long, J. P. Jones 3rd, B. Sullenger and A. M. Lambowitz, *Science*, 2000, **289**, 452.
85. J. Yao, J. Zhong, Y. Fang, E. Geisinger, R. P. Novick and A. M. Lambowitz, *RNA*, 2006, **12**, 1271.
86. J. T. Heap, O. J. Pennington, S. T. Cartman, G. P. Carter and N. P. Minton, *J. Microbiol. Methods*, 2007, **70**, 452.
87. D. B. Wilson, *Mol. Microbiol.*, 2009, **74**, 1287.
88. W. Liu, X.-Z. Zhang, Z. Zhang and Y.-H. P. Zhang, *Appl. Environ. Microbiol.*, 2010, **76**, 4914.
89. P. Heinzelman, C. D. Snow, I. Wu, C. Nguyen, A. Villalobos, S. Govindarajan, J. Minshull and F. H. Arnold, *Proc. Natl. Acad. Sci. U. S. A.*, 2009, **106**, 5610.
90. P. Peralta-Yahya, B. T. Carter, H. Lin, H. Tao and V. W. Cornish, *J. Am. Chem. Soc.*, 2008, **130**, 17446.
91. T. I. Williams, J. C. Combs, B. C. Lynn and H. J. Strobel, *Appl. Microbiol. Biotechnol.*, 2007, **74**, 422.

92. P. Tailliez, H. Girard, J. Millet and P. Beguin, *Appl. Environ. Microbiol.*, 1989, **55**, 207.
93. A. A. Herrero and R. F. Gomez, *Appl. Environ. Microbiol.*, 1980, **40**, 571.
94. X. Shao, B. Raman, M. Zhu, J. R. Mielenz, S. D. Brown, A. M. Guss and L. R. Lynd, *Appl. Microbiol. Biotechnol.*, 2011, **92**, 641.
95. J. R. Borden and E. T. Papoutsakis, *Appl. Environ. Microbiol.*, 2007, **73**, 3061.
96. M. J. Dunlop, Z. Y. Dossani, H. L. Szmidt, H. C. Chu, T. S. Lee, J. D. Keasling, M. Z. Hadi and A. Mukhopadhyay, *Mol. Syst. Biol.*, 2011, **7**, 487.

CHAPTER 8

Lignocellulose Degradation in Termite Symbiotic Systems

LEI XIE, NING LIU AND YONGPING HUANG*

Key Laboratory of Insect Developmental and Evolutionary Biology, Institute of Plant Physiology and Ecology, Shanghai Institute for Biological Sciences, Chinese Academy of Sciences, Shanghai 200032, P. R. China
*Email: yongping@sippe.ac.cn

8.1 Introduction

As a principal component of plant cell walls, lignocellulose constitutes the most abundant renewable resource on Earth, and the chemical components of lignocellulosic biomass make them a substrate of enormous biotechnological value. Thus, the development of methods to utilize such a tremendous bioresource in industry, especially in biofuel production, has been attracting increasing scientific attention. There are two main steps involved in the conversion process: depolymerization of celluloses to simple sugars, and fermentation of simple sugars to ethanol, of which the first step has proven to be the most intractable and critical. With the complex and recalcitrant structures that are inherent in plant cell walls, lignocellulosic materials are extremely resistant to thermochemical pretreatments as well as enzymatic attacks, increasing the cost of cellulolytic enzymes and producing unaddressed bottlenecks in the biofuel industry.[1,2] One strategy to solve this problem is to study analogous systems in nature in which lignocellulose is efficiently decomposed.

RSC Energy and Environment Series No. 10
Biological Conversion of Biomass for Fuels and Chemicals: Explorations from Natural Utilization Systems
Edited by Jianzhong Sun, Shi-You Ding and Joy Doran-Peterson
© The Royal Society of Chemistry 2014
Published by the Royal Society of Chemistry, www.rsc.org

8.2 Primary Lignocellulose-Degrading Enzymes

Lignocellulose is a matrix of cellulose, hemicellulose and lignin, and therefore, enzymes involved in its effective degradation mainly include cellulases, hemicellulases and lignin-degrading enzymes.

8.2.1 Cellulases

Cellulose is the principal constituent of lignocellulose, which is an unbranched polysaccharide consisting of several hundred to over 10 000 glucose residues, and serves as the backbone of plant cell walls. Enzymes that can hydrolyze cellulose into glucose are termed "cellulases". These include three main types of enzymes: endo-β-1,4-glucanases (EGs; EC 3.2.1.4), cellobiohydrolases (CBHs; EC 3.2.1.91) and β-glucosidases (BGs; EC 3.2.1.21). EGs randomly shear internal glycosidic bonds at the surface of cellulose, CBHs cut off cellobiose units from either the non-reducing or reducing end of a cellulose chain, and BGs further convert the derived cello-oligosaccharide into glucose.[3,4]

8.2.2 Hemicellulases

Unlike cellulose, hemicelluloses are a structurally heterogeneous group of plant-cell-wall polysaccharides that have an average degree of polymerization between 70 and 200.[5] They can be generally classified into xylan, arabinoxylan, glucuronoxylan, arabinoglucuronoxylan, glucomannan and galactoglucomannan, according to their backbone composition and side-chains.[6] Because of this high heterology, efficient hydrolysis of hemicellulose requires the concerted action of an equally diverse set of enzymes, including xylanases (EC 3.2.1.8), β-xylosidases (EC 3.2.1.37), β-mannanases (EC 3.2.1.78), α-L-arabinofuranosidases (EC 3.2.1.55), α-L-arabinanases (EC 3.2.1.99), α-glucuronidases (EC 3.2.1.131), acetyl xylan esterases (EC 3.1.1.72), feruloyl esterases (EC 3.1.1.73) *etc.*

8.2.3 Lignases

In lignocellulosic materials, lignin is the most abundant non-polysaccharide embedded in the carbohydrate polymer matrix of cellulose and hemicellulose. It is a high-molecular-mass complex that is composed of radical couplings of randomly organized phenylpropene units, including guaiacyl units (G, from the precursor coniferyl alcohol), syringyl units (S, from the precursor sinapyl alcohol) and *p*-hydroxyphenyl units (H, from the precursor *p*-coumaryl alcohol).[5] Typical lignolytic enzymes include phenol-oxidase (laccase), lignin peroxidase (LiP) and manganese peroxidase (MnP).[7] Note that any enzymatic breakdown or modification of lignin could potentially improve the accessibility of glycosyl hydrolases to cellulose and hemicelluloses and would thus enhance the efficiency of lignocellulose digestion.

8.3   Termite Symbiotic Systems as Efficient Lignocellulose-Degrading Systems

Termites, an influential group of social insects notorious for the tremendous damage they can cause to human property, are now considered an informative bioreactor that can efficiently degrade lignocellulose. In fact, they play an important role in global carbon recycling in natural ecosystems. They can dissimilate 74–99% of the cellulose and 65–87% of the hemicellulose that they ingest and use it as a primary energy source. Some fungus-growing species even consume over 90% of dry wood in some tropical regions, and mineralize as much as 20% of the net primary production in wetter savannas.[8] The ability of termites to flourish on recalcitrant plant biomass and its ubiquitous distribution make them an ideal biological model to improve the current biorefinery processing of cellulosic biomass.

To sustain this expertise, termites have evolved delicate alimentary tracts and versatile symbiotic microbial systems to constitute small-scale, yet highly efficient, bioreactors, which resemble industrial biomass-conversion systems but consist of a grinding machine (mandible and proventriculus), a reaction chamber (digestive tract), microbial flora and their enzymes. Within this microscale bioreactor, mechanical and enzymatic action work together to achieve extensive lignocellulose degradation.

8.4   The Origin of Lignocellulose-Degrading Enzymes in Termite Symbiotic Systems

With thousands of years of evolution, termites have established obligate mutualistic relationships with diverse and unique sets of microbial populations. For lower termites, which show general xylophagy, dense and diverse populations of protists are recruited in the gut which are largely responsible for the degradation of plant matter. In higher termites, according to the variation in their feeding behavior, they can be classified as "wood-feeders", "soil-feeders" or "fungus-cultivators". Among these, the first two groups exclusively develop symbiotic relationships with gut microbes, which mainly include prokaryotic bacteria. Members of the latter group distinguish themselves by their peculiar ability to cultivate and consume a basidiomycete fungus of the genus *Termitomyces* on their nests, introducing a third symbiont in addition to the gut microorganisms found in other termite guilds. Cellulases from termites themselves, together with those derived from their microbial symbionts, work cooperatively to achieve the extensive xylophagy of this insect group.

8.4.1   Termite Hosts

Termites were once widely believed to depend entirely on microbial symbionts for their xylophagy. However, in 1998, Watanabe *et al.*[9] cloned a cellulase gene from the salivary gland tissue of *Reticulitermes speratus*. Now termites are generally acknowledged to possess endogenous cellulase systems of their own, typically consisting of a few paralogous EG copies from GH9 and one or a few BGs from GH1, while with no CBHs are included. For lower termites, the salivary glands were the only observed expression site of endogenous cellulases (both EGs and BGs), while in higher termites (family Termitidae), the expression of both EG and BG shifted from the salivary glands to the midgut sometime during their evolution.[10] For example, although EG was still observed to be expressed only in the salivary glands of the fungus-growing termite, *Odontotermes formosanus*, it changed dramatically and was only observed in the midgut of the wood-feeding termite, *Nasutitermes takasagoensis* and the soil-feeder *Sinocapritermes mushae*.[11] Multiple BG copies have also been found in midgut cDNA preparations, in addition to copies from salivary glands in a wood-feeding termite, *N. takasagoensis*.[12] This contrasts with the exclusive expression in the salivary gland that is observed in lower termites. Remarkably, although termites can produce cellulases on their own,[12] the relative importance of these enzymes in sustaining the regular metabolism of termites may be limited given the obvious absence of CBHs as well as cellulose-binding domains (CBDs) for EGs.

8.4.2 Symbiotic Protists

The symbiotic protists are a typically found resident in the guts of lower termites, as listed in Table 8.1. Numerous protists, as shown in Figure 8.1,[13,14] inhabit the enlarged portion of the hindgut and establish a close symbiotic relationship with the host termites. These protists are critical to the survival of the termite host because of their indispensable assistance in lignocellulose degradation. Cleveland[15] first demonstrated that *Reticulitermes flavipes* was incapable of consuming cellulose and died within 10–20 days after its gut protists were defaunated. The protists engulf wood particles to gain a rich cellulase diet and produce their own cellulases so that they can digest and decompose the cellulose components and turn them into nutrients.

Protists produce a variety of cellulose-degrading enzymes. Recently, several meta-transcriptomic studies aimed at describing this system identified a set of genes related to lignocellulolytic processing.[16–19] These genes belong to several Glycoside Hydrolase Families (GHFs), including cellulases and hemicellulases. One of 10 expressed genes in the protist community was estimated to be related to lignocellulose decomposition after several representative species of lower termites were examined. The concurrence of GHF5 and GHF7 cellulases in all of these protist communities suggests that they represent the 'core enzyme set'. Also note that the expression levels of the GHFs are not even. The most highly expressed genes belong to GHF7. An environmental Expressed Sequence Tags (EST) study on *R. flavipes* protists analyzed over 1000 clones and found that 6.2% of the sequences belonged to GHF7. Two kinds of cellulases (CBH and EG) exist in this family. GHF7 CBHs account for 4.1% of the total ESTs while EGs account for 2.1%. The high expression levels of these enzymes suggest that they are very important for cellulose decomposition. Protist cellulases belonging to GHF45 were also found.[20,21]

Table 8.1 Flagellate species in representative termite lineages.[13,14]

Termite host		Symbiotic flagellate				Number of species
Family	*Genus*	*Class*	*Order*	*Family*	*Genus*	
Rhinotermitidae	*Coptotermes*	Parabasalia	Spirotrichonymphida	Holomastigotoididae	*Holomastigotoides*	5
			Trichonymphida	Teranymphidae	*Pseudotrichonympha*	2
				Trichonymphidae	*Spirotrichonympha*	2
	Reticulitermes	Parabasalia	Spirotrichonymphida	Holomastigotoididae	*Holomastigotoides*	1
			Trichonymphida	Trichomonadidae	*Trichomonas*	1
				Teranymphidae	*Teranympha*	1
				Trichonymphidae	*Spirotrichonympha*	1
					Trichonympha	2
			Tritrichomonadida	Monocercomonadidae	*Monocercomonas*	1
		Preaxostyla	Oxymonadida	Pyrsonymphidae	*Dinenympha*	9
					Pyrsonympha	3
	Rhinotermes	Parabasalia	Cristamonadida	Lophomonadidae	*Gigantomonas*	1
	Schedorhinotermes		Trichonymphida	Teranymphidae	*Pseudotrichonympha*	3
Kalotermitidae	*Cryptotermes*	Parabasalia	Cristamonadida	Lophomonadidae	*Stephanonympha*	1
					Devescovina	1
	Epicalotermes	Parabasalia	Trichonymphida	Staurojoeninidae	*Staurojoenina*	1
	Glyptotermes	Parabasalia	Cristamonadida	Lophomonadidae	*Devescovina*	1
					Macrotrichomonas	2
	Incisitermes	Parabasalia	Cristamonadida	Lophomonadidae	*Coronympha*	1
				Staurojoeninidae	*Staurojoenina*	1
			Trichonymphida	Trichonymphidae	*Trichonympha*	1
	Kalotermes	Parabasalia	Cristamonadida	Lophomonadidae	*Calonympha*	1
					Devescovina	1
					Joenia	1
					Stephanonympha	1
			Tritrichomonadida	Monocercomonadidae	*Monocercomonas*	1

	Neotermes	Parabasalia	Cristamonadida	Lophomonadidae	*Devescovina*	3
					Foaina	1
		Preaxostyla	Oxymonadida	Oxymonadidae	*Oxymonas*	1
Termopsidae	*Archotermopsis*	Parabasalia	Honigbergiellida	Honigbergiellidae	*Ditrichomonas*	1
			Trichonymphida	Teranymphidae	*Pseudotrichonympha*	1
	Hodotermopsis	Parabasalia	Spirotrichonymphida	Holomastigotoididae	*Spirotrichonymphella*	1
			Trichomonadida	Trichomonadidae	*Trichomonas*	1
				Hoplonymphidae	*Hoplonympha*	1
				Teranymphidae	*Eucomonympha*	1
			Trichonymphida	Trichonymphidae	*Spirotrichonympha*	1
					Trichonympha	1
	Porotermes	Parabasalia	Cristamonadida	Lophomonadidae	*Joenina*	
			Spirotrichonymphida	Holomastigotoididae	*Spirotrichonymphella*	1
					Pseudotrypanosoma	1
			Trichonymphida	Teranymphidae	*Pseudotrichonympha*	1
			Trichonymphida	Trichonymphidae	*Trichonympha*	1
	Zootermopsis	Parabasalia	Hypotrichomonadida	Hypotrichomonadidae	*Trichomitus*	1
			Trichonymphida	Trichonymphidae	*Trichonympha*	1
		Preaxostyla	Oxymonadida	Streblomastigidae	*Streblomastix*	1
Hodotermitidae	*Hodotermes*	Parabasalia	Cristamonadida	Lophomonadidae	*Devescovina*	1
					Foaina	1
					Gigantomonas	1
					Stephanonympha	1
Mastotermitidae	*Mastotermes*	Parabasalia	Trichonymphida	Spirotrichosomidae	*Leptospironympha*	1
			Cristamonadida	Lophomonadidae	*Deltotrichonympha*	2
					Koruga	1
					Metadevescovina	1
					Mixotricha	1
			Trichomonadida	Trichomonadidae	*Pentatrichomonoides*	1

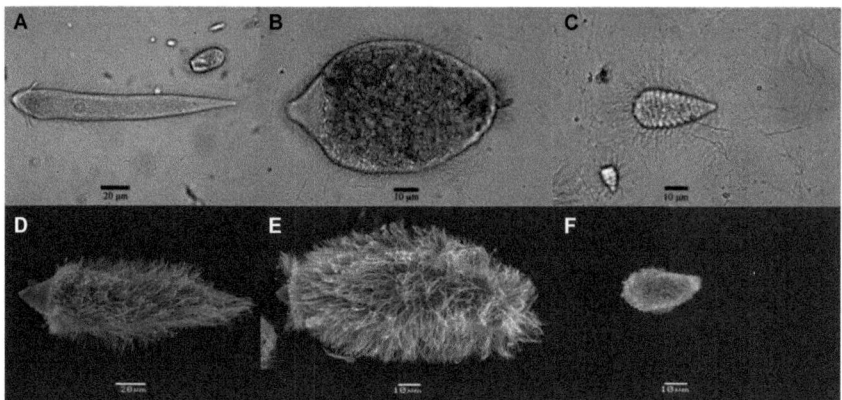

Figure 8.1 Morphological identification of flagellates in the gut of the lower termite, *Coptotermes formosanus.*[14] (a)–(c) Optical microscope images, and (d)–(f) scanning electron micrographs. (a) and (d) *Pseudotrichonympha grassii*, (b) and (e) *Holomastigotoides mirabile*, (c) and (f) *Spirotrichonympha leidyi*.

In addition, various hemicellulases were identified from the protists. They consisted of xylanases from GHF8, 10, 11, 43 and 62, xylosidases from GHF5, mannanases from GHF26, α-mannanases from GHF47 and β-galactosidases from GHF42.[16–18] Hemicellulose accounts for 15–35% of the dry weight of plant cell walls and wraps around cellulose to form a tight structure. Therefore, the decomposition of hemicellulose is necessary for increasing the efficiency of cellulose digestion. Protists have evolved elaborate glycosyl hydrolases to obtain energy and carbon from recalcitrant food material.

Although the protist community was found to be a reservoir of novel cellulose and hemicellulose hydrolases, no lignin-modification or -degradation enzyme was identified. The host's mechanical and enzymatic pretreatment may compensate for the lack of this type of enzyme in the protists.[22,23]

8.4.3 Symbiotic Bacteria

In lower termites, as protists predominantly occupy the gut and proved to be pivotal for host xylophagy, the role of gut bacteria in host lignocellulose degradation has always been a matter of debate. However, the isolation of cellulolytic bacteria from gut homogenates of various lower termite species has been attempted and has been successful in quite a few cases.

In the case of two lower drywood termites: *Incisitermes tabogae* and *Neotermes castaneus*, two novel spirochaetes, *Treponema isoptericolens* SPIT5T (Figure 8.2)[24] and *Spirochaeta coccoides* SPN1 (Figure 8.3),[25] were isolated from the hindgut contents, respectively. Both strains proved to be obligately anaerobic and possessed various enzyme activities such as β-D-glucosidase, α-L-arabinosidase and β-D-xylosidase. In the case of two lower subterranean termites: *Mastotermes darwiniensis* and *Coptotermes curvignathus*, a novel actinobacteria: *Cellulosimicrobium variabile* MX5 (Figure 8.4),[26] a facultative

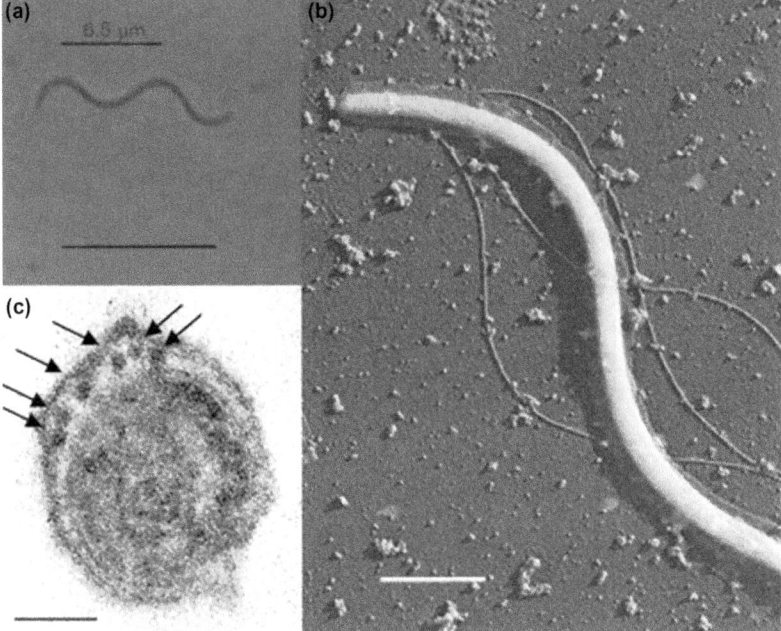

Figure 8.2 Morphology of the strain SPIT5T (*Treponema isoptericolens* sp. nov.).[24]
(a) Phase-contrast micrograph of a single cell with the wavelength indicated above the cell. The scale bar is 10 μm. (b) Scanning electron micrograph of a cell with three flagella released from the periplasmic space. The scale bar is 10 μm. (c) Transmission electron micrograph of an ultra-thin cross-section showing six flagella in the periplasmic space. The scale bar is 0.5 μm.
Reproduced from ref. 24 with permission. © International Union of Microbiological Societies, 2008.

anaerobe with both cellulolytic and xylanolytic activity, was isolated from the hindgut contents of the former, while three isolates, including one actino-bacteria (*Clavibacter agropyri*) and two proteobacteria (*Enterobacter aerogenes* and *Enterobacter cloacae*),[27] were isolated from the whole gut homogenate of the latter and all could grow on carboxymethyl cellulose (CMC) and cellobiose substrates.

In the case of a lower dampwood termite, *Z. angusticollis*, through a culture-enrichment method, a total of 119 cellulolytic strains were isolated under aerobic conditions.[28] These strains belonged to 23 groups according to the restriction fragment pattern of their 16S rDNA. Among these groups, the gram-positive isolates belonged to both the high-GC-content subdivision (order actinomycetales) and the low-GC-content subdivision (order bacillales), whereas the gram-negative isolates belonged to α-proteobacteria or flex-ibacteriaceae. As these cellulolytic bacteria were isolated from lower termite gut, this indicated that they may also play a role in cellulose digestion in the termite gut in addition to the cellulolytic flagellates and termites themselves.

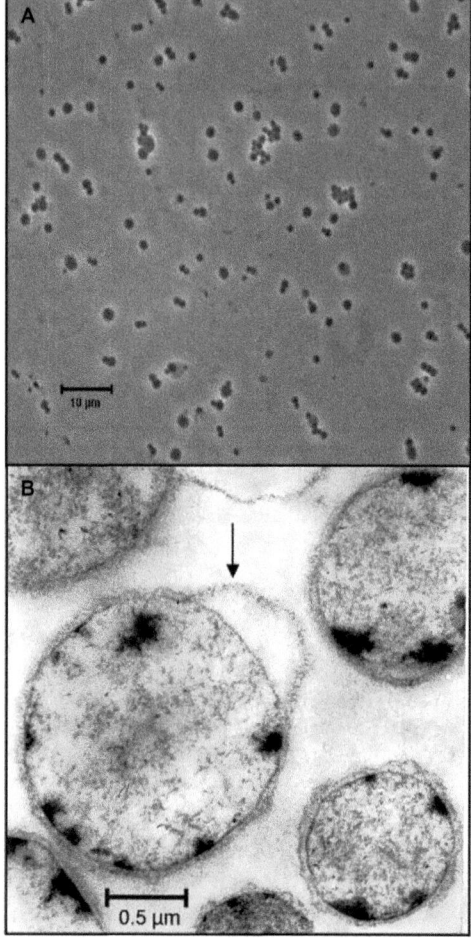

Figure 8.3 Phase-contrast micrograph of whole coccoid cells.[25] (a) Electron micrograph of an ultra-thin section. (b) The thin section shows the outer envelope (arrow) surrounding the coccoid forms with loose cytoplasma and small condensed areas.
Reproduced from ref. 25 with permission. © American Society for Microbiology, 2006.

Unlike the case of lower termites, higher termites typically lack flagellated protists and depend mainly on gut bacteria for plant matter degradation. The potential of this ability is largely related to the feeding behavior of the higher termite hosts. Conceivably, gut microbiota in wood-feeders of those higher termites should be the most lignocellulolytic. This was first supported by the isolation of cellulolytic actinomycetes (*Streptomyces* sp. *and Micromonospora* sp.) from the hindgut contents of *Armitermes*, and *Microcerotermes* species,[29] and then by the isolation of a novel anaerobic cellulolytic clostridium: *Clostridium termitidis* CT1112 (Figure 8.5), obtained from a *Nasutitermes* species.[30]

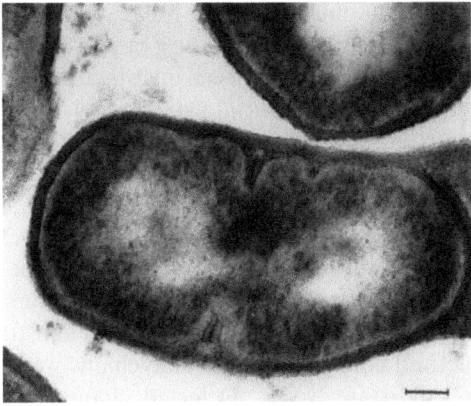

Figure 8.4 Electron micrograph of thin sections of the strain MX5T. The scale bar is 0.06 μm.[26]
Reproduced from ref. 26 with permission. © International Union of Microbiological Societies, 2002.

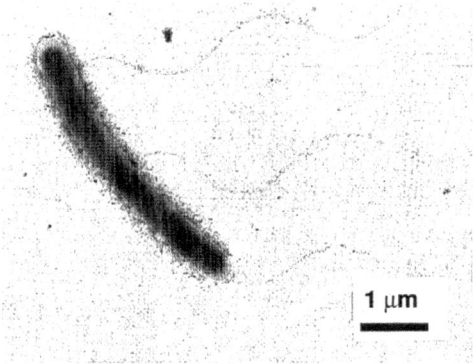

Figure 8.5 Morphology of the strain CT1112; a transmission electron micrograph showing phosphotungstate negative staining.[30]
Reproduced from ref. 30 with permission. © Gustav Fischer Verlag, Stuttgartmew York, 1992.

The lignocellulolytic potential of gut bacteria in wood-feeders was further supported by the cloning of a GH11 xylanase gene from the gut microbiota of a *Nasutitermitidae* species[31] and then firmly testified by a massive meta-genomic sequencing and functional screening of the hindgut meta-genome of a *Nasutitermes* species,[32] in which batches of plant polysaccharide-degrading genes as well as their putative taxonomic origins were also identified.

In fungus-growers of the wood-feeding termites, which have orthodox cellulolytic fungal symbionts and gut bacterial symbionts, were neglected as a potential contributor to host lignocellulose degradation until recently, when one GH11 xylanase gene and a batch of β-glucosidase-positive clones were first reported from a fosmid library of *Macrotermes annandalei*.[33] This investigation directly indicated that gut microbiota in fungus-growing termites can also contribute to the hosts' overall lignocellulose degradation. This hypothesis was further verified by a recent culture-based investigation of gut microflora in *O. formosanus*, during which eight *Bacillus* species isolated from the termite gut showed enzymatic activities against xylan and/or carboxymethyl cellulose.[34] The relative importance of fungus *vs.* gut bacteria for plant matter digestion in fungus-growing termites requires further investigation, but they likely work synergistically and/or complementarily with each other. While the lignocellulolytic potential of gut bacteria in soil-feeders may be far lower than that of their counterparts in wood-feeders or fungus-growers, this is probably largely due to the low level of cellulose in their food substrates.

In addition, structural changes in lignin side-chains were detected in several termite species by solid-state NMR. Detected changes included cleavage of ether linkages in lignin side-chains in *Cryptotermes brevis*,[35] and oxidation along with demethylation, and even some ring hydroxylation in lignin side-chains in *Zootermopsis angusticollis*.[36] Moreover, considerable numbers of gut isolates that were obtained from various termite species were proven to be capable of rapidly mineralizing lignin-derived monoaromatic and dimeric model compounds *in vivo* and in gut homogenates that were incubated in air.[37–40]

8.4.4 Symbiotic Fungi

Higher termites of the subfamily macrotermitine live in peculiar symbiosis with a *Termitomyces* fungus. These fungus-growing termites constitute one of the most influential groups of social insects in the Asian and African tropics.[41] The fungal symbiont is broadly accepted as being largely responsible for the extensive decomposition of lignocellulose for fungus-growing termites, which is implemented in two main ways. First, *Termitomyces* can degrade lignin in the nest for its termite hosts; this was confirmed by the biochemical detection of an apparently increased C-to-N ratio and higher nitrogen quality in some fungus combs,[42,43] and also by the genetic identification of laccases, one of the typical lignolytic enzymes produced by the symbiotic *Termitomyces* spp., in several fungus-growing termites.[44,45] Apparently, destruction of the lignin barrier improves the accessibility of glycosyl hydrolases to plant cellulose and hemicellulose, enhancing overall degrading efficiency. Second, *Termitomyces* secrete a series of cellulolytic or hemicellulolytic enzymes to process plant polysaccharides in the nests, and this conclusion was supported by a subtractive EST analysis of the cultivated *Termitomyces* of *Macrotermes gilvus*,[44] where high expression levels of genes involved in cellulose (6 endo-glucanase and 12 cellobiohydrolase ESTs), hemicellulose (5 endo-1,4-β-xylanase and 5 β-mannanase ESTs) and pectin (17 endo-polygalacturonase, 9 exo-polygalacturonase, 4 pectate lyase and 2 rhamnogalacturonan lyase ESTs) degradation were observed in its cDNA library. These enzymes are distributed throughout the typical fungal CAZy families of GH6, 7 and 61 for cellulases, GH11 for xylananse and PL2 and 4 for pectinases. Partially processed plant matter derived from fungus combs could then be subjected to gut microbiota for further digestion after being ingested by worker termites.

8.5 The Discovery of Lignocellulose-Degrading Enzymes from Termites

8.5.1 Culture-Dependent Strategies

Attempts to isolate cellulolytic bacteria from termite gut have been made for various termite species including both higher termites (*Nasutitermes, Armitermes, Microcerotermes, Macrotermes,* and *Odontotermes*)[29,30] and lower termites (*Reticulitermes, Coptotermes, Neotermes, Hodotermopsis, Zootermopsis and Incisitermes*).[24-28] While compared to the large number of as-yet-unknown microbes in the termite gut, only a limited number of them were isolated, indicating an urgent demand for the improvement of the traditional culture strategy. Moreover, further research is required to better understand the ecology of these microbes and to investigate their potential applications in industry, where one possible problem may be that the conditions adopted in industrial processes may not sustain the normal growth of these strains.

8.5.2 Culture-Independent Strategies

The termite gut features distinct biological, physical and chemical conditions, which have been specifically adapted for gut symbionts, and are difficult to establish in pure cultures. However, because of the rapid development of culture-independent molecular approaches, we have increasingly greater access to the naturally occurring but "as-yet uncultured" termite gut microbial communities, including both microbial diversity and phylogeny, and their metabolic properties and capabilities. New approaches have been applied to a few termite species, including transcriptome analyses of symbiotic protist communities in *R. flavipes*,[46-48] *R. speratus*,[16,17] *Hodotermopsis sjostedti, Neotermes koshunensis, Mastotermes darwiniensis*[16] and *Coptotermes formosanus*,[19] and an analysis of the symbiotic *Termitomyces* fungus of *M. gilvus*. Complete genome analyses were conducted for uncultured Termite Group 1 bacteria in a single host protist cell in *R. speratus*[49] and uncultured bacteroidale endosymbionts of the cellulolytic protist in *C. formosanus*.[50] Metagenomic and functional analyses examined the hindgut microbiome of a higher wood-feeding *Nasutitermes* species.[32] These comprehensive transcriptomic, genomic and meta-genomic studies have shed light on the molecular processes underlying termite polyphenism, caste differentiation and particularly the mutualistic relationships between termites and their microbial symbionts, including both the roles of various symbionts in host cellulolytic processes and the energy flow between them. Moreover, a broad set of lignocellulose-degrading genes and enzymes that may have promising industrial applications have been obtained, highlighting the importance of applying culture-independent strategies to explore the otherwise inaccessible microbial genome resources found in termite symbiotic systems. Environmental microbiologists will surely make sustained efforts in the future, based on culture-independent approaches, to obtain new insights into the not-yet solved mysteries of termite symbiotic systems.

8.6 Potential Applications in Industry

Lignocellulose is a promising feedstock for the biofuel industry because of its sustainability and vast distribution. One of the major impediments to lignocellulose bioconversion is the high cost of the process of solubilization.[51] The production of biofuels such as ethanol can be divided into three main processes: pretreatment, hydrolysis, and fermentation.[52] Termite symbiotic systems have acquired their own counterparts through evolution. Symbionts produce enzymes to remove/modify lignin and hemicelluloses in the pretreatment, then hydrolyse cellulose into sugar monomers and finally ferment them into acetate. Unlike industrial processes requiring heat or chemicals, these reactions are accomplished by consortia of living organisms and carried out under mild conditions. Thus study of the termite symbiotic system could help us alleviate the high cost in industrial utilization. There have been some successful examples of using *Escherichia coli* to express cellulases from termite symbionts.[53–55] Nowadays more and more attention is paid to eukaryotic expression systems such as yeast and insect cell lines, for some enzymes may need post-translational modification to function properly. Sasagawa *et al.*[56] developed a eukaryotic expression system which used homologous recombination to insert exogenous genes into the genome of *Pichia pastoris*. Several endo-β-1,4-glucanases and xylanases were expressed and secreted into the medium. Todaka *et al.*[57] expressed 11 endo-β-1,4-glucanases, belonging to GH5, 7, and 45 endo-glucanases in *Saccharomyces cerevisiae* and obtained a recombinant yeast strain SM2042B24 EG I which was more active than *Trichoderma reesei* EG I. Apart from cellulases, xylanases also have significant applications in industry, including in chemicals, fuels, food, breweries, pulp and paper. Pentoses in hemicelluloses account for 20–40% of lignocellulose biomass, making it another valuable resource for the bioethanol conversion industry. Recent studies regarding ethanol fermentation from pentose sugars shed light on its future application. An analysis conducted by scientists at the Solar Energy Research Institute indicated that the production cost can be reduced by 25%, about $1.23 per gallon, if xylose is completely fermented into ethanol.[58]

The termite symbiotic system is a rich resource to discover new genes and enzymes. The latest progress in -omics, such as meta-genomics, meta-transcriptomics and meta-proteomics, has been gradually applied to reveal the nature of this system. High-throughput technology, such as next-generation sequencing and fosmid library screening provides unprecedented tools to exploit functional genes. The combination of high-throughput technology and bioinformatic analysis will provide us with information which will allow the in-depth study of both the microbial diversity and life processes of biomass degradation. Future research will increasingly focus on the characterization of these enzymes and genetic engineering to make them more suitable for industrial use. In order to achieve efficient utilization of biomass resources, the synergistic collaboration in lignocellulose digestion between enzymes also deserves attention.[59] Novel enzymes that have high activity and stability could reduce the amount of enzyme used, thus lower the cost of the product. In addition, a thorough understanding of the mechanisms involved in the symbiotic system would allow us to develop optimized procedures that imitate the contributions and collaborations within this multi-species symbiotic system.[60]

References

1. M. E. Himmel, S. Y. Ding, D. K. Johnson, W. S. Adney, M. R. Nimlos, J. W. Brady and T. D. Foust, *Science*, 2007, **315**, 804.
2. M. Galbe and G. Zacchi, *Appl. Microbiol. Biotechnol.*, 2002, **59**, 618.
3. C. Oppert, W. E. Klingeman, J. D. Willis, B. Oppert and J. L. Jurat-Fuentes, *Comp. Biochem. Physiol., Part B: Biochem. Mol. Biol.*, 2010, **155**, 145.
4. H. Watanabe and G. Tokuda, *Cell Mol. Life Sci.*, 2001, **58**, 1167.
5. D. W. Fengel and G. Wegener, *Wood: Chemistry, Ultrastructure, Reactions*, Walter de Gruyter, Berlin, 1983.
6. H. V. Scheller and P. Ulvskov, *Annu. Rev. Plant Biol.*, 2010, **61**, 263.
7. D. O. Krause, S. E. Denman, R. I. Mackie, M. Morrison, A. L. Rae, G. T. Attwood and C. S. McSweeney, *FEMS Microbiol. Rev.*, 2003, **27**, 663.
8. C. Rouland-Lefèvre, in *Termites: Evolution, Sociality, Symbioses, Ecology*, ed. T. Abe, D. E. Bignell and M. Higashi, Kluwer Academic Publishers, Dordrecht, 2000, p. 289.
9. H. Watanabe, H. Noda, G. Tokuda and N. Lo, *Nature*, 1998, **394**, 330.
10. N. Lo, G. Tokuda and H. Watanabe, in *Biology of Termites: A Modern Synthesis*, ed. D. E. Bignell, Y. Roisin and N. Lo, Springer Netherlands, 2011, p. 51.
11. G. Tokuda, N. Lo, H. Watanabe, G. Arakawa, T. Matsumoto and H. Noda, *Mol. Ecol.*, 2004, **13**, 3219.
12. G. Tokuda, H. Saito and H. Watanabe, *Insect Biochem. Mol. Biol.*, 2002, **32**, 1681.
13. M. Slaytor, A. Sugimoto, J. Azuma, K. Murashima and T. Inoue, *J. Insect Physiol.*, 1997, **43**, 235.
14. L. Xie, N. Liu, Y. Huang and Q. Wang, *Acta Entomol. Sin.*, 2011, **54**, 1140.
15. L. R. Cleveland, *Biol. Bull. U. S. A.*, 1924, **46**, 178.
16. N. Todaka, T. Inoue, K. Saita, M. Ohkuma, C. A. Nalepa, M. Lenz, T. Kudo and S. Moriya, *PLoS One*, 2010, **5**, e8636.
17. N. Todaka, S. Moriya, K. Saita, T. Hondo, I. Kiuchi, H. Takasu, M. Ohkuma, C. Piero, Y. Hayashizaki and T. Kudo, *FEMS Microbiol. Ecol.*, 2007, **59**, 592.
18. A. Tartar, M. M. Wheeler, X. Zhou, M. R. Coy, D. G. Boucias and M. E. Scharf, *Biotechnol. Biofuels*, 2009, **2**, 25.

19. L. Xie, L. Zhang, Y. Zhong, N. Liu, Y. Long, S. Wang, X. Zhou, Z. Zhou, Y. Huang and Q. Wang, *Genomics*, 2012, **99**, 246.
20. L. Li, J. Frohlich, P. Pfeiffer and H. Konig, *Eukaryotic Cell*, 2003, **2**, 1091.
21. K. Ohtoko, M. Ohkuma, S. Moriya, T. Inoue, R. Usami and T. Kudo, *Extremophiles*, 2000, **4**, 343.
22. S. L. Chen, J. Ke, D. D. Laskar and D. Singh, *Biotechnol. Biofuels*, 2011, **4**, 17.
23. M. R. Coy, T. Z. Salem, J. S. Denton, E. S. Kovaleva, Z. Liu, D. S. Barber, J. H. Campbell, D. C. Davis, G. W. Buchman, D. G. Boucias and M. E. Scharf, *Insect Biochem. Mol. Biol.*, 2010, **40**, 723.
24. S. Droge, R. Rachel, R. Radek and H. Konig, *Int. J. Syst. Evol. Microbiol.*, 2008, **58**, 1079.
25. S. Droge, J. Frohlich, R. Radek and H. Konig, *Appl. Environ. Microbiol.*, 2006, **72**, 392.
26. A. Bakalidou, P. Kampfer, M. Berchtold, T. Kuhnigk, M. Wenzel and H. Konig, *Int. J. Syst. Evol. Microbiol.*, 2002, **52**, 1185.
27. M. Ramin, A. R. Alimon, N. Abdullah, J. M. Panandam and K. Sijam, *Res. J. Microbiol.*, 2008, **3**, 288.
28. M. Wenzel, I. Schonig, M. Berchtold, P. Kampfer and H. Konig, *J. Appl. Microbiol.*, 2002, **92**, 32.
29. M. B. Pasti and M. L. Belli, *FEMS Microbiol. Lett.*, 1985, **26**, 107.
30. P. Hethener, A. Brauman and J. L. Garcia, *Syst. Appl. Microbiol.*, 1992, **15**, 52.
31. Y. Brennan, W. N. Callen, L. Christoffersen, P. Dupree, F. Goubet, S. Healey, M. Hernandez, M. Keller, K. Li, N. Palackal, A. Sittenfeld, G. Tamayo, S. Wells, G. P. Hazlewood, E. J. Mathur, J. M. Short, D. E. Robertson and B. A. Steer, *Appl. Environ. Microbiol.*, 2004, **70**, 3609.
32. F. Warnecke, P. Luginbuhl, N. Ivanova, M. Ghassemian, T. H. Richardson, J. T. Stege, M. Cayouette, A. C. McHardy, G. Djordjevic, N. Aboushadi, R. Sorek, S. G. Tringe, M. Podar, H. G. Martin, V. Kunin, D. Dalevi, J. Madejska, E. Kirton, D. Platt, E. Szeto, A. Salamov, K. Barry, N. Mikhailova, N. C. Kyrpides, E. G. Matson, E. A. Ottesen, X. N. Zhang, M. Hernandez, C. Murillo, L. G. Acosta, I. Rigoutsos, G. Tamayo, B. D. Green, C. Chang, E. M. Rubin, E. J. Mathur, D. E. Robertson, P. Hugenholtz and J. R. Leadbetter, *Nature*, 2007, **450**, 560.
33. N. Liu, X. Yan, M. Zhang, L. Xie, Q. Wang, Y. Huang, X. Zhou, S. Wang and Z. Zhou, *Appl. Environ. Microbiol.*, 2011, **77**, 48.
34. G. M. Mathew, Y. M. Ju, C. Y. Lai, D. C. Mathew and C. C. Huang, *FEMS Microbiol. Ecol.*, 2012, **79**, 504.
35. K. S. Katsumata, Z. F. Jin, K. Hori and K. Iiyama, *J. Wood Sci.*, 2007, **53**, 419.
36. S. M. Geib, T. R. Filley, P. G. Hatcher, K. Hoover, J. E. Carlson, M. D. Jimenez-Gasco, A. Nakagawa-Izumi, R. L. Sleighter and M. Tien, *Proc. Natl. Acad. Sci. U. S. A.*, 2008, **105**, 12932.
37. A. Brune, E. Miambi and J. A. Breznak, *Appl. Environ. Microbiol.*, 1995, **61**, 2688.

38. T. Kuhnigk, E. M. Borst, A. Ritter, P. Kampfer, A. Graf, H. Hertel and H. Konig, *Syst. Appl. Microbiol.*, 1994, **17**, 76.
39. T. Kuhnigk and H. Konig, *J. Basic Microbiol.*, 1997, **37**, 205.
40. D. K. Ngugi, M. K. Tsanuo and H. I. Boga, *J. Basic Microbiol.*, 2007, **47**, 87.
41. D. E. Bignell and P. Eggleton, in *Termites: Evolution, Sociality, Symbioses, Ecology*, ed. T. Abe, D. E. Bignell and M. Higashi, Kluwer Academic Publishers, Dordrecht, 2000, p. 363.
42. F. Hyodo, T. Inoue, J. I. Azuma, I. Tayasu and T. Abe, *Soil Biol. Biochem.*, 2000, **32**, 653.
43. F. Hyodo, I. Tayasu, T. Inoue, J. I. Azuma, T. Kudo and T. Abe, *Funct. Ecol.*, 2003, **17**, 186.
44. T. Johjima, Y. Taprab, N. Noparatnaraporn, T. Kudo and M. Ohkuma, *Appl. Microbiol. Biotechnol.*, 2006, **73**, 195.
45. Y. Taprab, T. Johjima, Y. Maeda, S. Moriya, S. Trakulnaleamsai, N. Noparatnaraporn, M. Ohkuma and T. Kudo, *Appl. Environ. Microbiol.*, 2005, **71**, 7696.
46. M. E. Scharf, D. Wu-Scharf, B. R. Pittendrigh and G. W. Bennett, *Genome Biol.*, 2003, **4**, R62.
47. M. E. Scharf, D. Wu-Scharf, X. Zhou, B. R. Pittendrigh and G. W. Bennett, *Insect Mol. Biol.*, 2005, **14**, 31.
48. X. G. Zhou, J. A. Smith, F. M. Oi, P. G. Koehler, G. W. Bennett and M. E. Scharf, *Gene*, 2007, **395**, 29.
49. Y. Hongoh, V. K. Sharma, T. Prakash, S. Noda, T. D. Taylor, T. Kudo, Y. Sakaki, A. Toyoda, M. Hattori and M. Ohkuma, *Proc. Natl. Acad. Sci. U. S. A.*, 2008, **105**, 5555.
50. Y. Hongoh, V. K. Sharma, T. Prakash, S. Noda, H. Toh, T. D. Taylor, T. Kudo, Y. Sakaki, A. Toyoda, M. Hattori and M. Ohkuma, *Science*, 2008, **322**, 1108.
51. J. R. Mielenz, *Curr. Opin. Microbiol.*, 2001, **4**, 324.
52. S. T. Merino and J. Cherry, *Adv. Biochem. Eng. Biotechnol.*, 2007, **108**, 95.
53. H. Watanabe, K. Nakashima, H. Saito and M. Slaytor, *Cell Mol. Life Sci.*, 2002, **59**, 1983.
54. K. I. Nakashima, H. Watanabe and J. I. Azuma, *Cell Mol. Life Sci.*, 2002, **59**, 1554.
55. T. Inoue, S. Moriya, M. Ohkuma and T. Kudo, *Gene*, 2005, **349**, 67.
56. T. Sasagawa, M. Matsui, Y. Kobayashi, M. Otagiri, S. Moriya, Y. Sakamoto, Y. Ito, C. C. Lee, K. Kitamoto and M. Arioka, *Plasmid*, 2011, **65**, 65.
57. N. Todaka, R. Nakamura, S. Moriya, M. Ohkuma, T. Kudo, H. Takahashi and N. Ishida, *Biosci. Biotech. Biochem.*, 2011, **75**, 2260.
58. N. Hinman, D. Schell, J. Riley, P. Bergeron and P. Walter, *Appl. Biochem. Biotechnol.*, 1992, **34–35**, 639.
59. M. E. Scharf, Z. J. Karl, A. Sethi, R. Sen, R. Raychoudhury and D. G. Boucias, *Commun. Integr. Biol.*, 2011, **4**, 761.
60. J. Z. Sun and M. E. Scharf, *Insect Sci.*, 2010, **17**, 163.

CHAPTER 9

Functional Gene Resources from Cellulose-Feeding Insects for Novel Catalysts

HIROFUMI WATANABE*[a] AND JINFENG NI*[b]

[a] National Institute of Agrobiological Sciences, Owashi, Tsukuba, Ibaraki 305-8634, Japan; [b] State Key Laboratory of Microbial Technology, Shandong University, Shandong 250100, P. R. China
*Email: hinabe@affrc.go.jp; jinfgni@sdu.edu.cn

9.1 The Cellulolytic Systems of Xylophagous Insect Species

Cellulolytic processes in xylophagous insects likely consist of: (1) extra-corporeal digestion by salivary enzymes ejected from the mouth; (2) physical breakdown by masticating organs (mandibles and proventriculus); (3) digestion in the alimentary canal by endogenous enzymes; and (4) symbiotic digestion by gut microbes or extra-corporeal symbionts.[1] In these four categories of cellulolytic processes, (1) remains to be elucidated, (2) is starting to gather interest in relation to body size and cellulolytic efficiency in termites,[1,2] and (3) and (4) are growing fields as next-generation sequencing technology leads to an accumulation of sequence information. Since the purification/recombinant production of 'hypothetical' enzymes has not yet caught up with sequencing analysis, the majority of primary structural (deduced amino acid sequence) information has

RSC Energy and Environment Series No. 10
Biological Conversion of Biomass for Fuels and Chemicals: Explorations from Natural Utilization Systems
Edited by Jianzhong Sun, Shi-You Ding and Joy Doran-Peterson
© The Royal Society of Chemistry 2014
Published by the Royal Society of Chemistry, www.rsc.org

not been correlated with enzymatic properties, with the exception of pioneer enzymes in each homologous group.[1]

In this chapter, we summarize reports on the plant cell wall degrading enzymes of insects based on glycoside hydrolase family classification.[3] We mainly consider reports combined with genomic information, since possible industrial uses of the reported enzymes will be based on recombinant production *via* their corresponding genes (cDNAs) in expression hosts (usually fungi or bacteria), not merely on their acquirement from native hosts (cellulolytic insects and their symbionts).

9.2   Historical Background

Termites and wood-feeding cockroaches were the initial insect model systems used in cellulase studies. In the early research into animal cellulose digestion, there was no concept of symbiotic cellulose digestion, as seen in the study of snails. However, after a series of studies into the termite *Reticulitermes flavipes*, and the wood-feeding cockroach, *Cryptocercus punctulatus*, by L. R. Cleveland, symbiotic cellulose digestion became known as one of the iconic phenomena in nature. The presence of endogenous cellulolytic enzymes was claimed in the middle of 20th Century, and demonstrated in 1998 in the termite *R. speratus*.[4,5]

9.3   Host Origins of Glycoside Hydrolases

9.3.1   Glycoside Hydroxylase 9 Endo-β-1,4-Glucanase

Genes of the glycoside hydroxylase family 9 (GH9) are widely distributed among insects (especially hemimetabola), from silverfish (Thysanura) to honey bees (Hymenoptera). The function of proteins translated from GH9 genes of non-cellulolytic species is still unknown, although some of them exhibit hydrolyzing activity against synthetic substrates such as carboxymethyl cellulose (CMC).[6–8] With the exception of several studies on GH9 members of termite and wood-feeding cockroach origin, there are few studies describing their enzymatic properties, although there have been many reports and deposited sequences of insect GH9 genes and cDNA in GenBank (Table 9.1). All GH9 group members in insects and other invertebrate species show higher homology to each other than to other members of non-insect origins (plants and bacteria), indicative of the potential presence of a common ancestral gene.[9,10] Particularly in termites, all reported endo-β-1,4-glucanase (EG) genes encode 448 amino acids (432 amino acids in mature form), and their intron positions, as well as their related phases are highly conserved (introns can be divided into three different phases according to their position relative to the codon; a highly conserved intron position and its associated phases could be indicative of the potential presence of a common ancestor).[4,9–12]

Recombinant production (over-expression) of insect GH9 members was difficult for many years until the first report of mutant termite EG over-expression in *Escherichia coli*.[13] The first over-expressed mutant termite GH9

Table 9.1   List of insects reported with GH9 cellulase genes and homologs.

Insect species	Common name	Family	GenBank accession no.	Reference
Orthoptera				
Teleogryllus emma	Emma cricket	Gryllidae	EU126927	64
Blattaria				
Polyphaga aegyptiaca	Cockroach	Polyphagidae	AF220583-5	65
Blattella germanica	Cockroach	Blattellidae	AF220595	65
Panesthia angustipennis	Cockroach	Blaberidae	AB438950-2	66
Panesthia cribrata	Cockroach	Blaberidae	AF220596-7	65
Salganea esakii	Cockroach	Blaberidae	AB438946-8	66
Periplaneta americana	Cockroach	Blattidae	AF220586&7	65
Cryptocercus clevelandi	Cockroach	Cryptocercidae	AF220588-90	65
Mastotermes darwiniensis	Termite	Mastotermitidae	AJ511339-43, AF220593&4	67,65
Hodotermopsis sjoestedti (synonym. *H. japonica*)	Termite	Termopsidae	AF220592, AB118662, AB118794-6	65,68
Neotermes koshunensis	Termite	Kalotermditae	AB118797-9, AF220591	68,65
Reticulitermes speratus	Termite	Rhinotermitidae	AB008778, AB019095	5,11
Reticulitermes flavipes	Termite	Rhinotermitidae	AY572862	69
Coptotermes acinaciformis	Termite	Rhinotermitidae	AF336120	Unpublished
Coptotermes formosanus	Termite	Rhinotermitidae	AB058669-71, ADB12483	44 Unpublished
Odontotermes formosanus	Termite	Termitidae	AB118800-2, ADB82658	68,70
Nasutitermes takasagoensis	Termite	Termitidae	AB013272&3, AB019585, AB118803	11,68
Nasutitermes walkeri	Termite	Termitidae	AB013273	11
Sinocapritermes mushae	Termite	Termitidae	AB118804-6	68
Phthiraptera				
Pediculus humanus corporis	Human body louse	Pediculidae	XP_002426465, XM_002426420	71,72
Hemiptera				
Acyrthosiphon pisum	Pea aphid	Aphididae	XM_001944739	73[a]
Coleoptera				
Tribolium castaneum	Rust-red flour beetle	Tenebrionidae	XM_001810641, CM000281	74[a]
Hymenoptera				
Camponotus floridanus	Carpenter ant	Formicidae	GL437663	75
Apis mellifera	Honey bee	Apoidae	XM_396791	76
Nasonia vitripennis	Jewel wasp	Pteromalidae	XM_001606404	Unpublished[a]

[a]Derived from a genomic sequence annotated using the gene-prediction method GNOMON of NCBI.

(A18) was a shuffled cDNA of four parental native cDNAs from *Coptotermes formosanus*, *C. lacteus*, *Nasutitermes takasagoensis*, and *R. speratus* that was adapted for the host *E. coli* JM109.[13] Following this experiment, a thermo-stable over-expression clone (PA68) was screened from the shuffled clones of the same parental cDNAs.[14] Recently, a native form of termite EG (RsEG and NtEG) was recombinantly produced by over-expression in the traditional Japanese fermentation fungus (sake-, miso-, and shoyu-producing), *Aspergillus oryzae*.[15] Recombinant expression of the native form of termite EG from *C. formosanus* (CfEG), using the over-expression vector pET28α in *E. coli*, was also recently reported.[12]

Some members of insect GH9 exhibit high specific activity against CMC. A GH9 endo-glucanase (NkEG) purified from a higher termite species (no protozoan symbionts in its hindgut), *N. takasagoensis*, showed 2222 units per mg (one unit is defined as the amount of enzyme which produces 1 μM of reducing sugar per min),[7] while a mutant termite EG (made by family shuffling of four GH9 homologues from different species, which shows 93% of identity in its amino acid sequence with RsEG from *R. speratus*) exhibited 588 units per mg.[13] Another recombinant termite EG (CfEG5) showed 325 units per mg.[12] RsEG and NtEG expressed in *A. oryzae* reached 1200 and 1392 units per mg.[15] Purified termite EGs mainly produce cellobiose (G2) and a trace of glucose (G1) from native cellulose chains (crystalline cellulose and/or amorphous cellulose chains), while recombin-ant EGs produce cellotriose (G3) in addition to G1 and G2.[6,7,12,13,15]

9.3.2 Glycoside Hydroxylase 5 Endo-β-1,4-Glucanase

Glycoside hydroxylase 5 (GH5) endo-β-1,4-glucanase genes have been found in several beetle and nematode species (Table 9.2). A GH5 EG of insect origin was first purified from the longhorn beetle, *Psacothea hilaris* (PhEG).[8] Distribution of the GH5 group is limited in the Coleoptera order,[8,16–19] and a common ancestor of coleopterans may have been formed by the possible horizontal transfer from bacteria (bacterial genes do not have introns in general, and the *PhEG* gene does not have introns[8]). GH5 genes have also been reported in many plant-parasitic nematodes,[20] but nematode GH5 members did not directly relate to those of insect origin.[9]

The properties of insect GH5 EGs have been reported for longhorn beetles, *P. hilaris* and *Apriona germari*.[8,17] Purified PhEG showed a specific activity of 150 units per mg of protein against CMC, and produced G2 and G3, but no G1 from cellodextrins. The *PhEG* gene encodes 325 amino acids ($M_w = 33\,800$ Da), and the deduced sequence belongs to subgroup 2 of GH5. The actual molecular weight is larger (47 000 Da) than the deduced value, thus, the possible addition of sugar chains is expected.[8] The cDNA of GH5 EG from *A. germari* also encodes 325 amino acids, and the protein sequence has a high similarity (89%) to that of PhEG. The purified recombinant Ag-EGase III had an activity of approximately 1037 units per mg of protein and showed the highest activity at 55 °C and pH 6.0. In addition, the *N*-glycosylation of Ag-EGase III was proved

Table 9.2 List of beetles and nematodes reported with GH5 cellulase genes.

Beetles and nematodes	Common name	Family	Protein name	GenBank accession no.	Reference
Aphelenchus avenae	Fungivouous nematode	Aphelenchoididae	Endo-β-1,4-glucanase	BAI44493-96	77
Apriona germari	Mulberry longhorn beetle	Cerambycidae	Endo-glucanase III	AAX18655	Unpublished
Gastrophysa viridula	Leaf beetle	Chrysomelidae	Glycoside hydrolase family protein 5	ADU33333	18
Globodera rostochiensisRO-1	Golden nematode	Heteroderidae	Endo-β-1,4-glucanase	AAN03645-48	78
			Endo-glucanase 1	AAC48325	79
			Endo-glucanase 2	AAC48341	79
			Endo-glucanase 1, 2	AAC63988-89	Unpublished
Globodera tabacum solanacearum	Nematode	Heteroderidae	Endo-β-1,4-glucanase	AAD56392-93	Unpublished
Heterodera avenae	Oat cyst nematode	Heteroderidae	Endo-β-1,4-glucanase	ACO55952	108
Heterodera glycines	Soybean cyst nematode	Heteroderidae	Endo-glucanase 4	AAK85303	80
			Cellulase	AAM50039	81
			Cellulase	AAN32884	81
			Endo-glucanase 1,2	AAC15707-08	Unpublished
			Endo-glucanase 3	AAC33848	Unpublished
Heterodera schachtii	Beet cyst nematode	Heteroderidae	Endo-β-1,4-glucanase	CAC12958-59	Unpublished
Meloidogyne incognita	Southern root-knot nematode	Heteroderidae	Endo-β-1,4-glucanase	AAD45868	82
			Cellulase	AAR37374-75	83
			Endo-β-1,4-glucanase	AAK21881-82	109
			β-1,4-Endo-glucanase	AAK21883.2	109
Meloidogyne javanicaHYN	Root-knot nematode	Heteroderidae	Cellulase	ACA66271, CAJ77137	Unpublished
Oncideres albomarginate chamela	Beetle	Cerambycidae	Endo-β-1,4-glucanase	ADI24131	19
Pratylenchus penetrans	Root-lesion nematode	Pratylenchidae	Endo-β-1,4-glucanase	BAB68522-23	84
Psacothea hilaris	Yellow-spotted longhorn beetle	Cerambycidae	Endo-glucanase/cellulase	BAB86867	8

and the result suggested that the carbohydrate moieties are not necessary for enzyme activity.[17]

9.3.3 Glycoside Hydroxylase 45 Endo-β-1,4-Glucanase

Glycoside hydroxylase 45 (GH45) EG genes of insect origin were first reported in a gut-tissue-expressed sequence tag (EST) analysis for the leaf beetle species, *Phaedonco chleariae*.[21] The cDNA encodes a putative endo-glucanase with 242 amino acid residues. The presence of cellulase activity was demonstrated by zymogram analysis using CMC as the substrate, but there was no direct proof showing that cDNA of GH45 was responsible for its enzymatic activity. The first characterized GH45 EG was the Ag-EGase identified in the midgut of a longhorn beetle species, *A. germari*. The cDNA of Ag-EGase encoding 237 amino acids was functionally expressed as a 29 kDa polypeptide in the baculovirus-infected insect Sf9 cells. The enzyme activity of the purified recombinant protein reached about 992 units per mg of protein, and the recombinant Ag-EGase exhibited its highest enzymatic activity at 50 °C and pH 6.0.[16] About 28 nucleotide sequences are listed as insect GH45 genes in GenBank (Table 9.3). Considering the fact that, in general, beetles express plural GH members (GH5, 45, and at least 48) in their guts,[18] the enzymatic properties of insect GH45 enzymes should be further studied since many cellulolytic enzymes degrade cellulose synergistically.[22,23]

9.3.4 Glycoside Hydroxylase 48 Genes

Glycoside hydroxylase 48 (GH48) is a family consisting of endo-glucanase (EC 3.2.1.4), chitinase (EC 3.2.1.14), cellobiohydrolase (EC 3.2.1.91), endo-

Table 9.3 List of insects reported with GH45 cellulase genes.

Insect species	Common name	Family	Protein name	GenBank accession no.	Reference
Apriona germari	Mulberry longhorn beetle	Cerambycidae	Cellulase 1 Endo-glucanase 2	AAN78326, AAU44973 AAR22385	16,85 85
Chrysomela tremulae	Beetle	Chrysomelidae	Endo-β-1,4-glucanase	ADU33285-86	18
Dendroctonus ponderosae	Mountain pine beetle	Curculionidae	Endo-β-1,4-glucanase	ADU33288-95	18
Leptinotarsa decemlineata	Colorado potato beetle	Chrysomelidae	Endo-β-1,4-glucanase	ADU33351	18
Oncideres albomarginate chamela	Beetle	Cerambycidae	Endo-β-1,4-glucanase	ADI24132	19
Phaedonco chleariae	Mustard leaf beetle	Chrysomelidae	Endo-β-1,4-glucanase	CAA76931	Unpublished
Sitophilus orzyae	Beetle rice weevil	Curculionidae	Endo-β-1,4-glucanase	ADU33246-50	18
Reticulitermes speratus (hindgut protist)	Termite	Rhinotermitidae	Endo-glucanase	BAA98048, BAG71490	86,87

Table 9.4   List of insects reported with GH48 cellulase/chitinase genes.

Insect species	Common name	Family	Protein name	GenBank accession no.	Reference
Chrysomela tremulae	Beetle	Chrysomelidae	Glycoside hydrolase family protein 48	ADU33283-84	18
Dendroctonus ponderosae	Mountain pine beetle	Curculionidae	Glycoside hydrolase family protein 48	ADU33297-301	18
Gastrophysa atrocyanea	Leaf beetle	Chrysomelidae	Chitinase	BAE94320	25
Gastrophysa viridula	Leaf beetle	Chrysomelidae	Glycoside hydrolase family protein 48	ADU33335-37	18
Leptinotarsa decemlineata	Colorado potato beetle	Chrysomelidae	Glycoside hydrolase family protein 48	ADU33352-54	18
Otiorhynchus sulcatus	Black vine weevil	Curculionidae	Cellulose 1,4-β-cellobiosidase	CAH25542-43	Unpublished
Sitophilus orzyae	Rice weevil	Curculionidae	Glycoside hydrolase family protein 48	ADU33251-52	18

processive cellulases (EC 3.2.1.-), and [reducing end] cellobiohydrolase (EC 3.2.1.-) (http://www.cazy.org/GH48.html). The first insect GH48 cDNA is registered as "cellulose 1,4-β-cellobiosidase" (cellobiohydrolase) in GenBank (Table 9.4). It was reported in 2002 that the GH48 cDNA, isolated from a black vine weevil, *Otiorhynchus sulcatus*, encodes 637 amino acids with a signal peptide of 17 amino acids,[24] but no related reports have subsequently been published with this insect species. In 2006, two proteins of the family GH48 were isolated from the active adults of the leaf beetle species, *Gastrophysa atrocyanea*, and the following enzyme assays showed that one of the proteins has a chitinase activity but no cellobiohydrolase activity. The cDNA of GH48 chitinase encodes 645 amino acids.[25] Recently, meta-transcriptome analysis using a pyro DNA sequencer, elucidated that several different beetle species express GH48 members in their gut tissues (Table 9.4),[18] but no real function of the translated products of these members has been reported. Exploration of authentic substrates to test insect GH48 members and their enzymatic properties is needed.

9.3.5   Glycoside Hydroxylase 28 Polygalacturonase

Pectin makes up about one third of the dry cell-wall substance of higher plants, and is a complex mixture of polysaccharides, consisting mainly of D-galacturonic acid units.[26] It would be useful if polygalacturonase were present in the digestive systems of herbivorous species. As expected, insect polygalacturonase was first purified from a weevil,[27] *Sitophilus oryzae*, in 1996 and in the following investigations, its cDNA was further isolated and classified as a GH28 member in 2003. The purified enzyme drastically reduced the viscosity of

Table 9.5   List of beetle and nematode reported with GH28 genes.

Insect species	Common name	Family	Protein name	GenBank accession no.	Reference
Chrysomela tremulae	Beetle	Chrysomelidae	Endo-polygalacturonase	ADU33275-82	18
Dendroctonus ponderosae	Mountain pine beetle	Curculionidae	Endo-polygalacturonase	ADU33302-9	18
Gastrophysa viridula	Leaf beetle	Chrysomelidae	Endo-polygalacturonase	ADU33338-44	18
Leptinotarsa decemlineata	Colorado potato beetle	Chrysomelidae	Endo-polygalacturonase	ADU33355-64	18
Phaedon cochleariae	Mustard leaf beetle	Chrysomelidae	Polygalacturonase	CAA76930	Unpublished
Sitophilus orzyae	Rice weevil	Curculionidae	Endo-polygalacturonase	ADU33253-56	18
Meloidogyne incognita	Southern root-knot nematode	Meloidogynidae	Polygalacturonase	AAM28240	88

the substrate solution (0.3% pectic acid in 0.1 M, pH 5.5 acetic acid buffer), but it only produced a relatively small amount of reducing sugars (10% of the final value by the time the enzymatic mixture had decreased to 90% of its initial viscosity), therefore, this enzyme was considered to act with endo-action.[27] Since this GH28 member was found as an orphan in the insect order at the time of discovery, it was assumed that this gene could be obtained by a possible horizontal transfer event from symbionts.[28] However, as with the later cases investigated, several more GH28 members have been identified using meta-analysis from different beetle genomes (Table 9.5), thus, the origin of insect GH28 has become a more complex issue.[18]

9.3.6 β-Glucosidases

"β-Glucosidase" (BG) is a general term for enzymes that hydrolyze not only the β-1,4, but also the β-1,6 and β-1,3 linkages of polysaccharides. In addition to cellulose, chitin and xylan could be potential substrates for β-glucosidases for many cellulose-feeding insects because of insect extra-skeleton body shells (chitin) and their feeding habits (plant materials). As a digestive β-glucosidase of termites, NkBG (60 kDa) identified in the lower termite, *Neotermes koshunesis*, was the first purified enzyme followed by the isolation of its corresponding cDNA.[29] NkBG showed high activities against laminaribiose and *p*-nitrophenyl-β-D-fucopyranoside (*p*NPβFuc), in addition to a high activity against cellobiose (G2, a possible native substrate) and very low activities against lactose, *p*-nitrophenyl-β-D-glucopyranoside (*p*NPβGlu), and *p*-nitrophenyl-β-D-galactopyranoside (pNPβGal). The K_m values for G2, laminaribiose, and pNPβFuc were 2.5, 1.6, and 0.8 mM, respectively, thus, NkBG may have multi-digestive functions in the host digestive system, in addition to hydrolyzing the β-1,4 linkages of cellodexitrins. NkBG cDNA encodes 498 amino acids, including a signal peptide of 20 amino acids, is only expressed in the salivary glands and belongs to GH1. The deduced molecular weight is lower (54 400) than that (60 000) on SDS-PAGE. As a result, it is considered to be slightly glycosylated.[29] More recently, BG cDNAs were cloned from the higher termites, *N. takasagoensis* and *Macrotermes barneyi*. In *N. takasagoensis*, two different sets of orthologs exist; one set (sgNtEGs) is expressed in the salivary glands, while the other set (mgNtEGs) is expressed in the midgut tissue, consisting of 489 and 490 amino acids, respectively.[30] While in *M. barneyi*, a single BG cDNA encoding 493 amino acids (MbmgBG1) was isolated and is dominantly expressed only in the midgut.[31] All BGs of the higher termites have high similarity to GH1.[30,31]

NkBG cDNA was first heterologously expressed in *E. coli*, and the enzymatic properties of its recombinant enzyme (rNkBG) were investigated. rNkBG hydrolyzes both native (*e.g.*, G2–G4, laminaribiose, cellulose) and synthetic (*e.g.*, pNPβGlu, pNPβGal, pNPβFuc) substrates, indicating that it has similar properties to those of native BG. It has also been reported that rNkBG is the most active on laminaribiose among the native substrates investigated, suggesting that it preferentially hydrolyzes β-1,3 glycosidic bonds over β-1,4 glycosidic linkages.[32] Recently, NkBG was also expressed in *A. oryzae*, and the properties of the recombinant enzyme (G1NkBG) were consistent with those previously described. An unusual finding was that G1NkBG showed an increase in activity in the presence of 200 mM glucose (1.3× the activity observed in the absence of glucose), implying its high potential for industrial applications.[33] In 2012, one of the β-glucosidases from termite *N. takasagoensis* (mgNtBG1 [AB508958]) was also expressed in *Pichia pastoris* and further characterized. The recombinant enzyme had an optimum temperature of 65 °C and was thermostable for 5 h at 60 °C, with little inhibition by glucose ($K_i = 600$ mM).[34] Another β-glucosidase (CfBG, GH1) from a wood-feeding termite species, *C. formosanus*, was cloned and successfully expressed in *E. coli*.[35] Recombinant CfBG produces glucose from filter paper substrates when combined with recombinant CfEG.[12,36] Following these successful expressions of termite BGs, MbmgBG1 from *M. barneyi* was also recombinantly expressed in *E. coli*, and showed several times higher activity (100%) against cellobiose than on other substrates (lactose [33%], G3–G6 [28–32%], pNPFuc [19%], and pNPG [15.8%]).[31]

In addition to termites, GH1 BG genes have also been found in other kinds of insects, such as the beetle, cockroach, and silkworm (Table 9.6). It is noteworthy that cDNA encoding a member of GH3 (a family consisting of β-glucosidases) was identified using an EST analysis from the organs of the termite, *Hodotermopsis sjostedti*.[37] At this stage, it is the only report on GH3 of termite origin. It will be interesting to know whether cockroaches (including termites) or other insects have GH3 members in their genomes, and if they regularly function as enzymes.

Table 9.6 List of insect GH1 members and their homologs. mRNAs/genes for GH1 members of insect families with the top 100 scores (E-values) against NkBG (AB073638) are listed.

Species	Common name	Family	Protein name	GenBank accession no. (as protein)	Reference
Blattodea					
Leucophaea maderae	Madeira cockroach	Blaberidae	Male-specific β-glycosidase	AY064214	89
Cryptotermes secundus	Termite	Kalotermitidae	Female neotenic-specific protein 2	EF029055	90
Neotermes koshunensis	Termite	Kalotermitidae	NkBG	AB073638	29
Reticulitermes flavipes	Eastern subterranean termite	Rhinotermitidae	BGluc-1	HM152540	91
Coptotermes formosanus	Formosan subterranean termite	Rhinotermitidae	β-Glucosidase	GQ911585, JN565078, JN565079	36,52
Nasutitermes takasagoensis	Termite	Termitidae	sgNtBG1-4, mgNtBG1-3,	AB508954-60	30
Macrotermes barneyi	Termite	Termitidae	β-Glucosidase	AFD33364.1	31
Odontotermes formosanus	Termite	Termitidae	β-Glucosidase	GU591172	Unpublished
Psocoptera					
Pediculus humanus corporis	Human body louse	Pediculidae	β-Glucosidase, putative	XM_002426164	Unpublished
Coleoptera					
Tribolium castaneum	Rust-red flour beetle	Tenebrionidae	Similar to β-glucosidase, Similar to AGAP006424-PA, Similar to glycoside hydrolases, *etc.*	XM_961239, XM_965131, XM_963225, XM_966939, XM_966989, XM_967041, XM_967089, XM_970572&3, XM_967138, XM_967192, XM_967249, XM_967293, XM_967344, XP_972134	Unpublished

Table 9.6 (*Continued*)

Species	Common name	Family	Protein name	GenBank accession no. (*as protein*)	Reference
Tenebrio molitor	Yellow mealworm	Tenebrionidae	β-Glucosidase precursor	AF312017	92
Lepidoptera					
Bombix mori	Domestic silkworm	Bombycidae	Digestive β-glucosidase	NM_001043608	93
Danaus plexippus	Monarch butterfly	Nymphalidae	Glycoside hydrolase, Hypothetical proteins, β-glucosidase precursor *etc.*	AGBW01008972, AGBW01008644, AGBW0100361, AGBW01006917, AGBW01011607, AGBW01002850, AGBW01005630 (EHJ71114), AGBW01007409 (EHJ68775), AGBW01006917 (EHJ69449)	94
Papilio xuthus	Butterfly	Papilionidae	Similar to CG9701-PA	AB264665, AB264710	95
Chilo suppressalis	Striped rice-borer	Crambidae	Seminal fluid protein	JN033714, JN033718, JN033720-8	Unpublished
Spodoptera frugiperda	Fall armyworm	Noctuidae	β-Glucosidase precursor	AF052729	96,97
Hymenoptera					
Acromyrmex echinatior	Panamanian leafcutter ant	Formicidae	lactase-phlorizin hydrolase	GL888336 (EGI62504)	Unpublished
Nasonia vitripennis	jewel wasp	Pteromalidae	glycoside hydrolase-like protein, myrosinase 1-like	NM_001142860, XM_001600058	98
Bombus impatiens	common eastern bumble bee	Apidae	myrosinase 1-like	XP_003489656, XM_003402875	

Diptera

Species	Common name	Family	Protein/gene name	Accession	Reference
Culex quinquefasciatus	southern house mosquito	Culicidae	glycoside hydrolase, non-cyanogenic beta-glucosidase	XM_001850267-9, XM_001864430	Unpublished
Aedes aegypti	yellow fever mosquito	Culicidae	glycoside hydrolases	XM_001659803-6, XM_001650290&1, XM_001650354	99
Anopheles gambiae s.str	African malaria mosquito	Culicidae	AGAP006422-PA, AGAP006424-PA, AGAP006426-PA, AGAP000481-PA,	XM_001237812, XM_557100, XM_316460&1, XM_310611	100,101,
Anopheles darlingi	American malaria mosquito	Culicidae	hypothetical protein	ADMH01001807, ADMH01001454 (EFR26975)	110
Drosophila ananassae	Fruit fly	Drosophilidae	Dana/GF23929-PA	XM_001957596	102,103
Drosophila erecta	Fruit fly	Drosophilidae	Dere/GG13575	XM_001972980	102,103
Drosophila grimshawi	Fruit fly	Drosophilidae	Dgri\GH16221	XM_001984009	102,103
Drosophila melanogaster	Fruit fly	Drosophilidae	CG9701-PA	NM_140661	111
Drosophila mojavensis	Fruit fly	Drosophilidae	Dmoj\GI13367-PA	XM_002008185	102,103
Drosophila persimilis	Fruit fly	Drosophilidae	GL13287-PA	XM_002026724	102
Drosophila pseudoobscura pseudoobscura	Fruit fly	Drosophilidae	GA21974-PA	XM_001352791	102
Drosophila sechellia	Fruit fly	Drosophilidae	Dsec\GM25655	XM_002030781	102
Drosophila simulans	Fruit fly	Drosophilidae	Dsim\GD14660	XM_002085152	102
Drosophila yakuba	Fruit fly	Drosophilidae	GE19872-PA	XM_002095053	102

9.3.7 Hemicellulases and Other Lignocellulolytic Enzymes

The first xylanase cDNA of insect origin was cloned from the phytophagous beetle, *P. cochleariae*. The cDNA encodes 217 amino acid residues and belongs to GH11.[21] Recently, xylanase cDNA from a coffee-berry-borer, *Hypothenmus hampei*, belonging to GH10, was also reported in GenBank (ADN94682.1) and its cDNA encodes 316 amino acids with a predicted molecular weight of 34 923 Da.[38]

9.4 Symbiotic Origins of Glycoside Hydrolases

9.4.1 Lignocellulolytic Enzymes Identified from Insect Symbionts

Due to recent advances in bulk-sequencing technologies, there have been many reports on the genes (or cDNAs) of lignocellulolytic enzymes identified from hindgut symbionts (protozoa and prokaryotes) by meta-genomic analysis methods.[39–42] The first cellulase genes, which have high similarities to GH7, were isolated from the parabasalian symbionts in the hindgut of *C. formosanus*. Through construction of a cDNA library from an as yet uncultivated symbiotic protist community of the lower termite, *R. speratus*, multiple glycosyl hydrolase enzymes, including GH3, 5, 7, 8, 10, 11, 26, 43, 45, and 62 were found. Recently, meta-genomic and functional analyses of the bacterial flora in the gut of a higher termite revealed a huge, diverse set of bacterial genes essential to the degradation of plant materials. In the cellulolytic system of hindgut protozoan fauna, the key enzymes for cellulose digestion are the GH7 members.[42–45] To the best of our knowledge, there have been no reports on stable recombinant production of GH7 enzymes. Establishing a recombinant production system of GH7 enzymes is virtually required for further study on protozoan cellulolytic systems.

In the process of enzymatic hydrolysis of lignocellulosic substrates, xylan is one of the major obstacles that limits the accessibility of cellulase to cellulose. The addition of xylanase can enhance enzyme accessibility through increasing fiber swelling and fiber porosity, and thus significantly improve hydrolysis yields.[46,47] Several xylanase genes have been identified from the gut microbiota of termites (Table 9.7).[48–50] The xylanase gene, encoding a 64.5 kDa multidomain endo-1,4-β-xylanase (Xyl6E7), which consists of a GH11 family catalytic domain and other domains, was obtained by functional screening of the fosmid library of a fungus-growing termite. Xyl6E7 was a highly active and endo-acting alkaline enzyme with wide pH tolerance and stability.[48] Three functional xylanases were also purified from the Formosan subterranean termite, *C. formosanus*,[50] and furthermore, the corresponding cDNAs were cloned and the amino acid sequence showed high homology to the GH11 family xylanase. Reverse-transcription PCR revealed that the xylanase was produced by the symbiotic flagellates residing in the hindgut of *C. formosanus*. In addition to GH11 xylanase, GH8 and GH30 xylanase genes were also found in the

Table 9.7   List of xylanase genes and fragment from insects and nematodes sources.

Insect and nematodes (source)	Common name	Family	GH family	GenBank accession no.	Reference
Hypothenemus hampei	Beetle (coffee-berry-borer)	Curculionidae	GH10	ADN94682	38
Phaedon cochleariae	Mustard leaf beetle	Chrysomelidae	GH11	CAA76932	Unpublished
Reticulitermes speratus (symbiotic protist)	Japanese termite	Rhinotermitidae	GH11	BAF57316–BAF57319	Unpublished
Macrotermes annandalei (uncultured bacterium)	Termite	Termitidae	GH11	HM483387	48
Coptotermes formosanus(symbiotic flagellate)	Formosan subterranean termite	Rhinotermitidae	GH11	AB469372–AB469376	104
Cotton bollworm (*Paenibacillus sp.*)	Corn earworm	Noctuidae	GH11	HQ657204	49
Nasutitermitidae (uncultured bacterium)	Termite	Termitidae	GH11	AY542135	51
Caterpillar lepidopteran (uncultured bacterium)	Larvae	Saturniidae	GH11	AY542136	51
Adult lepidopteran (uncultured bacterium)	Moth	Saturniidae	GH11	AY542137	51
Adult lepidopteran (uncultured bacterium)	Moth	Saturniidae	GH8	AY542134	51
Radopholus similis	Banana-root nematode	Pratylenchidae	GH30	ABZ78968	105
Meloidogyne incognita	Southern root-knot nematode	Meloidogynidae	GH30	AAF37276	106
Meloidogyne incognita	Southern root-knot nematode	Meloidogynidae	GH30	AAC77826-27 (fragment)	107

environmental DNA libraries from Lepidoptera (moths) and the gut of nematodes.[51]

9.4.2   Meta-Genome and Meta-Transcriptome Analyses for Gut Microbiota

Recently, meta-genomic and meta-transcriptomic analytical methods have been applied to research on wood-feeding termites and their symbiotic fauna. Plural large-scale meta-analyses have also been used for investigations into termite gut systems (host tissue and microbial contents), such as *C. formosanus*.[52] As a result, a large number of carbohydrate-active proteins, have been identified.[52–54]β-Glucosidases were also expressed and characterized from the transcriptome library.[59] A feruloyl esterase (FAE) was identified from a thermophile from the gut fauna of *C. formosanus*, which is thought to split ester linkages between xylose and phenolic acids in the plant cell wall matrix to separate the lignin matrix from hemicellulose.[55] From aerobic free-living bacterial genomes of the gut system (position unspecified) of the termite *R. santonensis*, eight β-glucosidases (GH1, 2, and 3) were successfully isolated and characterized by activity screening of an expressive genomic library constructed in *E. coli*.[56,57] A thermostable xylanase produced from a bacterium, *Paenibacillus macerans*, found in the termite gut, was isolated on beechwood xylan plates and the xylanase (205 kDa, putative pI of 5.38, and a half-life of 6 h at 60 °C and 2 h at 90 °C) was then purified and also characterized.[58] A trans-species comparison between cellulase and xylanase activity, which lead to a trans-meta-genomic comparative study, was also recently reported.[60] This growing number of reports demonstrates the importance of wood-feeding termites and their gut microbiota as resources of novel carbohydrate enzymes and proteins. On the other hand, it is important to recognize that the functional and nutritional contributions to the host termites from many of those symbiotic microbes have not been clearly elucidated. Some reports in the last century have been quite careful about using the term "symbiont", preferring to refer to them as "facultative gut-luminal microbes with carbohydrase activity".[61–63] Functional experiments, *i.e.* knock-out, replacement, and mutation of the microbes and their digestive genes, and elucidation of trans-generational and social transmission of the microbes, are truly needed to elucidate their symbiotic functions, and further evaluate their importance as cellulolytic agents in termite cellulolytic systems.

9.5   Future Perspectives of Xylophagous Insects as Novel Enzyme Resources

As discussed in this chapter, many plant-cell-wall degrading enzymes with extremely high specific activities have been isolated from various herbivorous and xylophagous insect species that only ingest cellulosic leaves or wood with low protein contents. Thus, these species would develop a quite efficient cellulolytic system during the course of evolution, which must be highly adapted to their native substrates (feeds). If we define the roles of "hypothetical proteins" or "homologs" by only relying on amino acid sequence homology, we may lose the opportunity to identify novel enzymes with high potential. Therefore, post-genomic applications of plant-cell-wall degrading enzymes from insects should be the focus of future research.

References

1. H. Watanabe and G. Tokuda, *Annu. Rev. Entomol.*, 2010, **55**, 609.
2. C. A. Nalepa, *Evol. Biol.*, 2011, **38**, 243.
3. B. Henrissat, *Biochem J.*, 1991, **280**, 309.
4. H. Watanabe and G. Tokuda, *Cell Mol. Life Sci.*, 2001, **58**, 1167.
5. H. Watanabe, H. Noda, G. Tokuda and N. Lo, *Nature*, 1998, **394**, 330.
6. H. Watanabe, M. Nakamura, G. Tokuda, I. Yamaoka, A. M. Scrivener and H. Noda, *Insect Biochem. Mol. Biol.*, 1997, **27**, 305.
7. G. Tokuda, H. Watanabe, T. Matsumoto and H. Noda, *Zool. Sci.*, 1997, **14**, 83.

8. M. Sugimura, H. Watanabe, N. Lo and H. Saito, *Eur. J. Biochem.*, 2003, **270**, 3455.
9. N. Lo, H. Watanabe and M. Sugimura, *Proc. R. Soc. London, Ser. B*, 2003, **270**, S69.
10. A. Davison and M. Blaxter, *Mol. Biol. Evol.*, 2005, **22**, 1273.
11. G. Tokuda, N. Lo, H. Watanabe, M. Slaytor, T. Matsumoto and H. Noda, *Biochim. Biophys. Acta*, 1999, **1447**, 146.
12. D. Zhang, A. R. Lax, J. M. Bland and A. B. Allen, *Insect Biochem. Mol. Biol.*, 2011, **41**, 211.
13. J. Ni, M. Takehara and H. Watanabe, *Biosci. Biotechnol. Biochem.*, 2005, **69**, 1711.
14. J. Ni, M. Takehara, M. Miyazawa and H. Watanabe, *Protein Eng., Des. Sel.*, 2007, **20**, 535.
15. K. Hirayama, H. Watanabe, G. Tokuda, K. Kitamoto and M. Arioka, *Biosci. Biotechnol. Biochem.*, 2010, **74**, 1680.
16. S. J. Lee, S. R. Kim, H. J. Yoon, I. Kim, K. S. Lee, Y. H. Je, S. M. Lee, S. J. Seo, H. D. Sohn and B. R. Jin, *Comp. Biochem. Physiol., Part B: Biochem. Mol. Biol.*, 2004, **139**, 107.
17. Y. D. Wei, K. S. Lee, Z. Z. Gui, H. J. Yoon, I. Kim, G. Z. Zhang, X. Guo, H. D. Sohn and B. R. Jin, *Comp. Biochem. Physiol., Part B: Biochem. Mol. Biol.*, 2006, **145**, 220.
18. Y. Pauchet, P. Wilkinson, R. Chauhan and R. H. ffrench-Constant, *PLoS One*, 2010, **5**, e15635.
19. N. Calderon-Cortes, H. Watanabe, H. Cano-Camacho, G. Zavala-Páramo and M. Quesada, *Insect Mol. Biol.*, 2010, **19**, 323.
20. M. Mitreva, G. Smant and J. Helder, *Methods Mol. Biol.*, 2009, **532**, 517.
21. C. Girard and L. Jouanin, *Insect Biochem. Mol. Biol.*, 1999, **29**, 1129.
22. M. Qi, H. S. Jun and C. W. Forsberg, *Appl. Environ.Microbiol.*, 2007, **73**, 6098.
23. T. Jeoh, D. B. Wilson and L. P. Walker, *Biotechnol. Prog.*, 2002, **18**, 760.
24. M. G. Edwards, PhD thesis, Durham University, 2002.
25. K. Fujita, K. Shimomura, K. Yamamoto, T. Yamashita and K. A. Suzuki, *Biochem. Biophys. Res. Commun.*, 2006, **345**, 502.
26. K. H. Caffall and D. Mohnen, *Carbohydr. Res.*, 2009, **344**, 1879.
27. Z. C. Shen, J. C. Reese and G. R. Reeck, *Insect Biochem. Mol. Biol.*, 1996, **26**, 427.
28. Z. Shen, M. Denton, N. Mutti, K. Pappan, M. R. Kanost, J. C. Reese and G. R. Reeck, *J. Insect Sci.*, 2003, **3**, 24.
29. G. Tokuda, H. Saito and H. Watanabe, *Insect Biochem. Mol. Biol.*, 2002, **32**, 1681.
30. G. Tokuda, M. Miyagi, H. Makiya, H. Watanabe and G. Arakawa, *Insect Biochem. Mol. Biol.*, 2009, **39**, 931.
31. Y. Wu, S. Chi, C. Yun, Y. Shen, G. Tokuda and J. Ni, *Insect Mol. Biol.*, 2012, **21**, 604.
32. J. Ni, G. Tokuda, M. Takehara and H. Watanabe, *Appl. Entomol. Zool.*, 2007, **42**, 457.

33. C. A. Uchima, G. Tokuda, H. Watanabe, K. Kitamoto and M. Arioka, *Appl. Microbiol. Biotechnol.*, 2011, **89**, 1761.
34. C. A. Uchima, G. Tokuda, H. Watanabe, K. Kitamoto and M. Arioka, *Appl. Environ. Microbiol.*, 2012, **78**, 4288.
35. D. Zhang, A. B. Allen and A. R. Lax, *J. Insect Physiol.*, 2012, **58**, 205.
36. D. H. Zhang, A. R. Lax, J. M. Bland, J. Yu, N. Fedorova and W. C. Nierman, *Insect Sci.*, 2010, **17**, 245.
37. M. Yuki, S. Moriya, T. Inoue and T. Kudo, *Zool. Sci.*, 2008, **25**, 401.
38. B. Padilla-Hurtado, C. Florez-Ramos, C. Aguilera-Galvez, J. Medina-Olaya, A. Ramirez-Sanjuan, J. Rubio-Gomez and R. Acuna-Zornosa, BMC Res. *Notes*, 2012, **5**, 23.
39. X. Zhou, J. A. Smith, F. M. Oi, P. G. Koehler, G. W. Bennett and M. E. Scharf, *Gene*, 2007, **395**, 29.
40. M. E. Scharf, *Biofuels Bioprod. Biorefin.*, 2008, **2**, 540.
41. F. Warnecke, P. Luginbühl, N. Ivanova, M. Ghassemian, T. H. Richardson, J. T. Stege, M. Cayouette, A. C. McHardy, G. Djordjevic, N. Aboushadi, R. Sorek, S. G. Tringe, M. Podar, H. G. Martin, V. Kunin, D. Dalevi, J. Madejska, E. Kirton, D. Platt, E. Szeto, A. Salamov, K. Barry, N. Mikhailova, N. C. Kyrpides, E. G. Matson, E. A. Ottesen, X. Zhang, M. Hernández, C. Murillo, L. G. Acosta, I. Rigoutsos, G. Tamayo, B. D. Green, C. Chang, E. M. Rubin, E. J. Mathur, D. E. Robertson, P. Hugenholtz and J. R. Leadbetter, *Nature*, 2007, **450**, 560.
42. N. Todaka, S. Moriya, K. Saita, T. Hondo, I. Kiuchi, H. Takasu, M. Ohkuma, C. Piero, Y. Hayashizaki and T. Kudo, *FEMS Microbiol. Ecol.*, 2007, **59**, 592.
43. K. Nakashima, H. Watanabe and J. I. Azuma, *Cell Mol. Life Sci.*, 2002, **59**, 1554.
44. K. Nakashima, H. Watanabe, H. Saitoh, G. Tokuda and J. I. Azuma, *Insect Biochem. Mol. Biol.*, 2002, **32**, 777.
45. H. Watanabe, K. Nakashima, H. Saito and M. Slaytor, *Cell Mol. Life Sci.*, 2002, **59**, 1983.
46. O. Garcia-Kirchner, M. Muñoz-Aguilar, R. Pérez-Villalva and C. Huitrón-Vargas, *Appl. Biochem. Biotechnol.*, 2002, **98–100**, 1105.
47. J. Hu, V. Arantes and J. N. Saddler, *Biotechnol. Biofuels*, 2011, **4**, 36.
48. N. Liu, X. Yan, M. Zhang, L. Xie, Q. Wang, Y. Huang, X. Zhou, S. Wang and Z. Zhou, *Appl. Environ. Microbiol.*, 2011, **77**, 48.
49. N. Adlakha, R. Rajagopal, S. Kumar, V. S. Reddy and S. S. Yazdani, *Appl. Environ. Microbiol.*, 2011, **77**, 4859.
50. G. Arakawa, H. Watanabe, H. Yamasaki, H. Maekawa and G. Tokuda, *Biosci. Biotechnol. Biochem.*, 2009, **73**, 710.
51. Y. Brennan, W. N. Callen, L. Christoffersen, P. Dupree, F. Goubet, Healey, M. Hernandez, M. Keller, K. Li., N. Palackal, A. Sittenfeld, G. Tamayo, S. Wells, G. P. Hazlewood, E. J. Mathur, J. M. Short, D. E. Robertson and B. A. Steer, *Appl. Environ. Microbiol.*, 2004, **70**, 3609.
52. D. H. Zhang, A. R. Lax, B. Henrissat, P. Coutinho, N. Katiya, W. C. Nierman and N. Fedorova, *Insect Mol. Biol.*, 2012, **21**, 235.

53. L. Xie, L. Zhang, Y. Zhong, N. Liu, Y. Long, S. Wang, X. Zhou, Z. Zhou, Y. Huang and Q. Wang, *Genomics*, 2012, **99**, 246.
54. Y. J. Han and H. Z. Chen, *Prog. Chem.*, 2007, **19**, 1153.
55. M. Chandrasekharaiah, A. Thulasi, M. Bagath, D. P. Kumar, S. S. Santosh, C. Palanivel, V. L. Jose and K. T. Sampath, *BMB Rep.*, 2011, **44**, 52.
56. C. Matteotti, F. Francis, E. Haubruge, J. Destain, C. Brasseur, J. Bauwens, E. De Pauw, D. Portetelle and M. Vandenbol, *Microbiol. Res.*, 2011, **166**, 629.
57. C. Matteotti, E. Haubruge, P. Thonart, F. Francis, E. D. Pauw, D. Portetelle and M. Vandenbol, *FEMS Microbiol. Lett.*, 2011, **314**, 147.
58. P. Dheeran, N. Nandhagopal, S. Kumar, Y. K. Jaiswal and D. K. Adhikari, *J. Ind. Microbiol. Biotechnol.*, 2012, **39**, 851.
59. D. H. Zhang, A. B. Allen and A. R. Lax, *J. Insect Physiol.*, 2012, **58**, 205.
60. W. B. Shi, S. Y. Ding and J. S. Yuan, *Bioenergy Res.*, 2011, **4**, 1.
61. D. W. Thayer, *J. Gen. Microbiol.*, 1976, **95**, 287.
62. D. W. Thayer, *J. Gen. Microbiol.*, 1978, **106**, 13.
63. M. L. Eutick, R. W. O'Brien and M. Slaytor, *Appl. Environ. Microbiol.*, 1978, **35**, 823.
64. N. Kim, Y. M. Choo, K. S. Lee, S. J. Hong, K. Y. Seol, Y. H. Je, H. D. Sohn and B. R. Jin, *Comp. Biochem. Physiol., Part B: Biochem. Mol. Biol.*, 2008, **150**, 368.
65. N. Lo, G. Tokuda, H. Watanabe, H. Rose, M. Slaytor, K. Maekawa, C. Bandi and H. Noda, *Curr. Biol.*, 2000, **10**, 801.
66. K. Shimada and K. Mekawa, *Sociobiology*, 2008, **52**, 417.
67. L. Li, J. Fröhlich, P. Pfeiffer and H. König, *Eukaryotic Cell*, 2003, **2**, 1091.
68. G. Tokuda, N. Lo, H. Watanabe, G. Arakawa, T. Matsumoto and H. Noda, *Mol. Ecol.*, 2004, **13**, 3219.
69. M. E. Scharf, D. Wu-Scharf, X. Zhou, B. R. Pittendrigh and G. W. Bennett, *Insect Mol. Biol.*, 2005, **14**, 31.
70. T. F. Deng, C. Chen, M. Cheng, C. Pan, Y. Zhou and J. Mo, *Sociobiology*, 2008, **51**, 697.
71. E. F. Kirkness, B. J. Haas, W. Sun, H. R. Braig, M. A. Perotti, J. M. Clark, S. H. Lee, H. M. Robertson, R. C. Kennedy, E. Elhaik, D. Gerlach, E. V. Kriventseva, C. G. Elsik, D. Graur, C. A. Hill, J. A. Veenstra, B. Walenz, J. M. C. Tubío, J. M. C. Ribeiro, J. Rozas, J. S. Johnston, J. T. Reese, A. Popadic, M. Tojo, D. Raoult, D. L. Reed, Y. Tomoyasu, E. Kraus, O. Mittapalli, V. M. Margam, H.-M. Li, J. M. Meyer, R. M. Johnson, J. Romero-Severson, J. P. VanZee, D. Alvarez-Ponce, F. G. Vieira, M. Aguadé, S. Guirao-Rico, J. M. Anzola, K. S. Yoon, J. P. Strycharz, M. F. Unger, S. Christley, N. F. Lobo, M. J. Seufferheld, N. Wang, G. A. Dasch, C. J. Struchiner, G. Madey, L. I. Hannick, S. Bidwell, V. Joardar, E. Caler, R. Shao, S. C. Barker, S. Cameron, R. V. Bruggner, A. Regier, J. Johnson, L. Viswanathan, T. R. Utterback, G. G. Sutton, D. Lawson, R. M. Waterhouse, J. C. Venter,

R. L. Strausberg, M. R. Berenbaum, F. H. Collins, E. M. Zdobnov and B. R. Pittendrigh, *Proc. Natl. Acad. Sci. U. S. A.*, 2010, **107**, 12168.

72. E. F. Kirkness, B. J. Haas, W. Sun, H. R. Braig, M. A. Perotti, J. M. Clark, S. H. Lee, H. M. Robertson, R. C. Kennedy, E. Elhaik, D. Gerlach, E. V. Kriventseva, C. G. Elsik, D. Graur, C. A. Hill, J. A. Veenstra, B. Walenz, J. M. C. Tubío, J. M. C. Ribeiro, J. Rozas, J. S. Johnston, J. T. Reese, A. Popadic, M. Tojo, D. Raoult, D. L. Reed, Y. Tomoyasu, E. Kraus, O. Mittapalli, V. M. Margam, H.-M. Li, J. M. Meyer, R. M. Johnson, J. Romero-Severson, J. P. VanZee, D. Alvarez-Ponce, F. G. Vieira, M. Aguadé, S. Guirao-Rico, J. M. Anzola, K. S. Yoon, J. P. Strycharz, M. F. Unger, S. Christley, N. F. Lobo, M. J. Seufferheld, N. Wang, G. A. Dasch, C. J. Struchiner, G. Madey, L. I. Hannick, S. Bidwell, V. Joardar, E. Caler, R. Shao, S. C. Barker, S. Cameron, R. V. Bruggner, A. Regier, J. Johnson, L. Viswanathan, T. R. Utterback, G. G. Sutton, D. Lawson, R. M. Waterhouse, J. C. Venter, R. L. Strausberg, M. R. Berenbaum, F. H. Collins, E. M. Zdobnov and B. R. Pittendrigh, *Proc. Natl Acad. Sci. U. S. A.*, 2011, **108**, 6335.

73. The International Aphid Genomics Consortium, *PLoS Biol.*, 2010, **8**, 1.

74. *Tribolium* Genome Sequencing Consortium, *Nature*, 2008, **452**, 949.

75. R. Bonasio, G. Zhang, C. Ye, N. S. Mutti, X. Fang, N. Qin, G. Donahue, P. Yang, Q. Li, C. Li, P. Zhang, Z. Huang, S. L. Berger, D. Reinberg, J. Wang and J. Liebig, *Science*, 2010, **329**, 1068.

76. M. Solignac, L. Zhang, F. Mougel, B. Li, D. Vautrin, M. Monnerot, J.-M. Cornuet, K. C. Worley, G. M. Weinstock and R. A. Gibbs, *Genome Biol.*, 2007, **8**, 403.

77. N. Karim, J. T. Jones, H. Okada and T. Kikuchi, *BMC Genomics*, 2009, **10**, 525.

78. S. Rehman, P. Butterbach, H. Popeijus, H. Overmars, E. L. Davis, J. T. Jones, A. Goverse, J. Bakker and G. Smant, *Phytopathology*, 2009, **99**, 194.

79. G. Smant, G. Smant, J. P. W. G. Stokkermans, Y. Yan, J. M. de Boer, T. J. Baum, X. Wang, R. S. Hussey, F. J. Gommers, B. Henrissat, E. L. Davis, J. Helder, A. Schots and J. Bakker, *Proc. Natl. Acad. Sci. U. S. A.*, 1998, **95**, 4906.

80. B. Gao, R. Allen, T. Maier, E. L. Davis, T. J. Baum and R. S. Hussey, *J. Nematol.*, 2002, **34**, 12.

81. B. Gao, R. Allen, T. Maier, E. L. Davis, T. J. Baum and R. S. Hussey, *Mol. Plant-Microbe Interact.*, 2003, **16**, 720.

82. M. N. Rosso, B. Favery, C. Piotte, L. Arthaud, J. M. De Boer, R. S. Hussey, J. Bakker, T. J. Baum and P. Abad, *Mol. Plant-Microbe Interact.*, 1999, **12**, 585.

83. G. Huang, R. Dong, T. Maier, R. Allen, E. L. Davis, T. J. Baum and R. S. Hussey, *Mol. Plant Pathol.*, 2004, **5**, 217.

84. T. Uehara, A. Kushida and Y. Momota, *Nematology*, 2001, **3**, 335.

85. S. J. Lee, K. S. Lee, S. R. Kim, Z. Z. Gui, Y. S. Kim, H. J. Yoon, I. Kim, P. D. Kang, H. D. Sohn and B. R. Jin, *Comp. Biochem. Physiol., Part B: Biochem. Mol. Biol.*, 2005, **140**, 551.

86. K. Ohtoko, M. Ohkuma, S. Moriya, T. Inoue, R. Usami and T. Kudo, *Extremophiles*, 2000, **4**, 343.

87. S. Sasaguri, J. Maruyama, S. Moriya, T. Kudo, K. Kitamoto and M. Arioka, *J. Gen. Appl. Microbiol.*, 2008, **54**, 343.

88. S. Jaubert, J. B. Laffaire, P. Abad and M. N. Rosso, *FEBS Lett.*, 2002, **522**, 109.

89. R. Cornette, J. Farine, D. Abed-Viellard, B. Quennedey and R. Brossut, *Biochem. J.*, 2003, **372**, 535.

90. T. Weil, M. Rehli and J. Korb, *BMC Genomics*, 2007, **8**, 198.

91. M. E. Scharf, E. S. Kovaleva, S. Jadhao, J. H. Campbell, G. W. Buchman and D. G. Boucias, *Insect Biochem. Mol. Biol.*, 2010, **40**, 611.

92. A. H. Ferreira, S. R. Marana, W. R. Terra and C. Ferreira, *Insect Biochem. Mol. Biol.*, 2001, **31**, 1065.

93. G. M. Byeon, K. S. Lee, Z. Z. Gui, I. Kim, P. D. Kang, S. M. Lee, H. D. Sohn and B. R. Jin, *Comp. Biochem. Physiol., Part B: Biochem. Mol. Biol.*, 2005, **144**, 418.

94. S. Zhan, C. Merlin, J. L. Boore and S. M. Reppert, *Cell*, 2011, **147**, 1171.

95. R. Futahashi and H. Fujiwara, *Dev. Genes Evol.*, 2008, **218**, 491.

96. S. R. Marana, E. H. P. Andrade, C. Ferreira and W. R. Terra, *Eur J. Biochem.*, 2004, **271**, 4169.

97. S. R. Marana, M. Jacobs-Lorena, W. R. Terra and C. Ferreira, *Biochim. Biophys. Acta*, 2001, **1545**, 41.

98. The Nasonia Genome Working Group, *Science*, 2010, **327**, 343.

99. V. Nene, J. R. Wortman, D. Lawson, B. Haas, C. Kodira, Z. J. Tu, B. Loftus, Z. Xi, K. Megy, M. Grabherr, Q. Ren, E. M. Zdobnov, N. F. Lobo, K. S. Campbell, S. E. Brown, M. F. Bonaldo, J. Zhu, S. P. Sinkins, D. G. Hogenkamp, P. Amedeo, P. Arensburger, P. W. Atkinson, S. Bidwell, J. Biedler, E. Birney, R. V. Bruggner, J. Costas, M. R. Coy, J. Crabtree, M. Crawford, B. deBruyn, D. DeCaprio, K. Eiglmeier, E. Eisenstadt, H. El-Dorry, W. M. Gelbart, S. L. Gomes, M. Hammond, L. I. Hannick, J. R. Hogan, M. H. Holmes, D. Jaffe, J. S. Johnston, R. C. Kennedy, H. Koo, S. Kravitz, E. V. Kriventseva, D. Kulp, K. LaButti, E. Lee, S. Li, D. D. Lovin, C. Mao, E. Mauceli, C. F. M. Menck, J. R. Miller, P. Montgomery, A. Mori, A. L. Nascimento, H. F. Naveira, C. Nusbaum, S. O'Leary, J. Orvis, M. Pertea, H. Quesneville, K. R. Reidenbach, Y.-H. Rogers, C. W. Roth, J. R. Schneider, M. Schatz, M. Shumway, M. Stanke, E. O. Stinson, J. M. C. Tubio, J. P. VanZee, S. Verjovski-Almeida, D. Werner, O. White, S. Wyder, Q. Zeng, Q. Zhao, Y. Zhao, C. A. Hill, A. S. Raikhel, M. B. Soares, D. L. Knudson, N. H. Lee, J. Galagan, S. L. Salzberg, I. T. Paulsen, G. Dimopoulos, F. H. Collins, B. Birren, C. M. Fraser-Liggett and D. W. Severson, *Science*, 2007, **316**, 1718.

100. R. A. Holt, G. M. Subramanian, A. Halpern, G. G. Sutton, R. Charlab, D. R. Nusskern, P. Wincker, A. G. Clark, J. M. C. Ribeiro, R. Wides, S. L. Salzberg, B. Loftus, M. Yandell, W. H. Majoros, D. B. Rusch, Z. Lai, C. L. Kraft, J. F. Abril, V. Anthouard, P. Arensburger, P. W. Atkinson, H. Baden, V. de Berardinis, D. Baldwin, V. Benes, J. Biedler, C. Blass, R. Bolanos, D. Boscus, M. Barnstead, S. Cai, A. Center, K. Chatuverdi, G. K. Christophides, M. A. Chrystal, M. Clamp, A. Cravchik, V. Curwen, A. Dana, A. Delcher, I. Dew, C. A. Evans, M. Flanigan, A. Grundschober-Freimoser, L. Friedli, Z. Gu, P. Guan, R. Guigo, M. E. Hillenmeyer, S. L. Hladun, J. R. Hogan, Y. S. Hong, J. Hoover, O. Jaillon, Z. Ke, C. Kodira, E. Kokoza, A. Koutsos, I. Letunic, A. Levitsky, Y. Liang, J.-J. Lin, N. F. Lobo, J. R. Lopez, J. A. Malek, T. C. McIntosh, S. Meister, J. Miller, C. Mobarry, E. Mongin, S. D. Murphy, D. A. O'Brochta, C. Pfannkoch, R. Qi, M. A. Regier, K. Remington, H. Shao, M. V. Sharakhova, C. D. Sitter, J. Shetty, T. J. Smith, R. Strong, J. Sun, D. Thomasova, L. Q. Ton, P. Topalis, Z. Tu, M. F. Unger, B. Walenz, A. Wang, J. Wang, M. Wang, X. Wang, K. J. Woodford, J. R. Wortman, M. Wu, A. Yao, E. M. Zdobnov, H. Zhang, Q. Zhao, S. Zhao, S. C. Zhu, I. Zhimulev, M. Coluzzi, A. D. Torre, C. W. Roth, C. Louis, F. Kalush, R. J. Mural, E. W. Myers, M. D. Adams, H. O. Smith, S. Broder, M. J. Gardner, C. M. Fraser, E. Birney, P. Bork, P. T. Brey, J. C. Venter, J. Weissenbach, F. C. Kafatos, F. H. Collins and S. L. Hoffman, *Science*, 2002, **298**, 129.
101. E. Mongin, C. Louis, R. A. Holt, E. Birney and F. H. Collins, *Trends Parasitol.*, 2004, **20**, 49.
102. Drosophila 12 Genomes Consortium, *Nature*, 2007, **450**, 203.
103. A. V. Zimin, D. R. Smith, G. Sutton and J. A. Yorke, *Assem. Reconciliation Bioinf.*, 2008, **24**, 42.
104. G. Arakawa, H. Watanabe, H. Yamasaki, H. Maekawa and G. Tokuda, *Biosci. Biotechnol. Biochem.*, 2009, **73**, 710.
105. J. Jacob, M. Mitreva, B. Vanholme and G. Gheysen, *Mol. Genet. Genomics*, 2008, **280**, 1.
106. M. Mitreva-Dautova, E. Roze, H. Overmars, L. de Graaff, A. Schots, J. Helder, A. Goverse, J. Bakker and G. Smant, *Mol. Plant-Microbe Interact*, 2006, **19**, 521.
107. C. Schultz, P. Gilson, D. Oxley, J. Youl and A. Bacic, *Trends Plant Sci.*, 1998, **3**, 426.
108. H. Long, H. Peng, W. Hiang, G. Wang, B. Gao, M. Moens and D. Peng, *Eur. J. Plant Pathol.*, 2012, **134**, 391.
109. T. N. Ledger, S. Jaubert, N. Bosselut, P. Abad and M. N. Rosso, *Gene*, 2006, **382**, 121.
110. N. D. Mendes, A. T. Freitas, A. T. Vasconcelos and M. F. Sagot, *BMC Genomics*, 2010, **11**, 529.
111. R. A. Hoskins, J. W. Carlson, C. Kennedy, D. Acevedo, M. Evans-Holm, E. Frise, K. H. Wan, S. Park, M. Mendez-Lago, A. Villasante, P. Dimitri, G. H. Karpen and S. E. Celniker, *Science*, 2007, **316**, 1625.

CHAPTER 10

Biological Pretreatment of Biomass in Wood-Feeding Termites

JING KE AND SHULIN CHEN*

Department of Biological Systems Engineering, Washington State University, Pullman, WA 99164, USA
*Email: chens@wsu.edu

10.1 Effect of Lignin on the Saccharification of Cellulose and Hemicelluloses

Lignin, the second most abundant organic substance in nature after cellulose, is covalently linked to hemicellulose and cross-linked with different plant polysaccharides to provide mechanical strength to plant cell walls (PCWs) and protect other structures from being damaged by external factors.[1,2] Unlike most natural biopolymers, lignin is heterogeneously and irregularly constructed, and is comprised of various carbon-to-carbon and ether linkages.[3,4] The network of the lignin polymer consists primarily of p-hydroxyphenyl (H), guaiacyl (G), and syringyl (S) units, derived from the dehydrogenation and polymerization of three different hydroxycinnamyl alcohols (monolignols), namely p-coumaryl alcohol, coniferyl alcohol, and sinapyl alcohol, respectively.[5,6] These monomers are incorporated in the lignin primary structure in the form of various substructures and/or inter-unit linkages, such as β–O–4′, β–5′, β–β′, etc., each as a result of an oxidative coupling process.

RSC Energy and Environment Series No. 10
Biological Conversion of Biomass for Fuels and Chemicals: Explorations from Natural Utilization Systems
Edited by Jianzhong Sun, Shi-You Ding and Joy Doran-Peterson
© The Royal Society of Chemistry 2014
Published by the Royal Society of Chemistry, www.rsc.org

The phenylpropanoid polymer resists chemical, enzymatic and microbial degradation,[7] and is covalently linked at various points with hemicellulose to form a matrix surrounding the orderly cellulose microfibrils.[8] This arrangement results in resistance to saccharification, thus severely limiting biochemical conversion of lignocellulosic biomass to fermentable sugars.[9] In addition, the lignin framework inhibits several different aspects of cellulase function: (1) competitive adsorption effects, during which lignin binds non-covalently to cellulases *via* hydrophobic interactions.[10] This competitive adsorption effect is considered to be reversible, and lignin generally does not undergo chemical reactions when bound to the cellulases, but can be easily removed by dilution or dialysis; (2) chemical inhibition effects with which lignin covalently modifies cellulase, and this is quite often due to the reactive functional groups of lignin, such as phenolic hydroxyl groups;[11] these electrophilic groups react with amino acid side-chains on cellulase to form covalent adducts, decreasing the activity or deactivating the cellulase;[12] and (3) steric hindrance effects which arise from the fact that each atom within a molecule occupies a certain amount of space, and the effects occur when the large sizes of functional groups within lignin prevent the chemical reactions of related groups on smaller molecules. The existence of more methyl and methoxy groups leads to a more branched lignin structure, thus hindering diffusion and productive binding of cellulases to cellulose.[13,14] It has been demonstrated that the extent of pure cellulose hydrolysis could be reduced by 14 – 60% by the addition of up to 15% lignin to the substrate. The inhibitory effect of lignin addition was only partially overcome by a 10-fold increase in cellulase loading, suggesting inhibitory interactions between lignin and the substrate and/or enzyme.[15] Unmodified lignins from various sources were more detrimental to cellulose hydrolysis than modified lignins.[16] Various lignin–cellulase interactions make releasing sugars from the polysaccharides of intact lignocellulosics more difficult. Thus, overcoming the recalcitrance of lignin to obtain cellulosic sugars from PCWs has been a major challenge in the development of cellulose-based biofuels and biochemicals. There are two general approaches to accomplish this, the first one is reducing lignin content or modifying its structure in plants, and the second is to chemically remove lignin from PCWs. Ideally, genetic lignin modification in future energy crops or modifying existing plants should be done in a way so as to avoid negative effects on plant productivity. It is, therefore, critical to understand to what extent lignin content and its composition can be modified without causing deleterious consequences to plant growth and development.

In bioprocessing lignocellulosic biomass, the process removing of lignin from PCWs to facilitate the subsequent saccharification is called "pretreatment". Different pretreatment strategies/approaches are aimed at alleviating the different lignin inhibition factors described above. However, all the existing pretreatment schemes suffer from various inherent drawbacks. Consequently, pretreatment is not only currently among the most costly steps but also has a significant impact on the cost of upstream as well as down-stream operations.[18,19] Ideal pretreatment technologies have the potential to increase the net yield of reduced sugar and eventually increase the yield of biofuel. Realizing commercial success in the production of lignocellulosic fuels, such as cellulosic ethanol, will require more advanced pretreatment methods that efficiently remove protective lignin from cellulose.[17]

The existing saccharification of cellulose and hemicellulose typically relies on a combination of thermochemical pretreatments that require high temperatures and pressures, thus consume a great deal of energy and require chemicals. A major drawback of these chemical and physical processes is the lack of selectivity. Although effective, these processes tend to damage the basic units of biopolymers *via* reactions that form new and often unwanted compounds, such as inhibitors and pollutants.[20] For example, dilute acid and alkaline pretreatments have been demonstrated to be highly efficient with fast reaction rates. The removal of hemicelluloses by acid, or removal of lignin by alkaline will result in increased accessibility of cellulase to the cellulose substrate. However, in addition to the requirement of chemicals, heat, pressure and corrosion/heat/pressure-resistant reactors, these pretreatment processes often lead to production of inhibitors including furfurals or phenols, which adversely affect the subsequent cellulose hydrolysis or sugar fermentation.[21] Meanwhile, additional processes are needed for recycling wastewater and remnants. Another representative thermochemical pretreatment process is ammonia fiber expansion (AFEX). AFEX can either modify or effectively reduce the lignin fraction of lignocellulosic materials, while keeping the hemicellulose and cellulose fractions intact. Besides its high efficiency, the two major advantages of AFEX are the formation of few inhibitory by-products and short reaction times. However, even with high temperatures and pressures, AFEX is only effective for the lignocellulosics with low lignin content, such as herbaceous crops and grasses. It has been proved to be much less effective for hardwood and softwood, which suggests that AFEX effectiveness decreases with increasing lignin contents. Biological pretreatments, such as those by white- and brown-rot fungi, have been proposed as alternative processes, which, however, are often too slow to be economically feasible.[22] Consequently, there is currently no single pretreatment technology that is perfectly acceptable for the conversion of biomass into biofuels or chemicals.

10.2 Termite Gut Physiochemical Characteristics and their Association with Lignin Pretreatment

Nature has created other types of pretreatment process. For example, animal or insect guts show morpho-anatomical and physiological adaption to diet, and this is especially obvious in the primary consumers of lignocelluloses, such as wood-feeding termites.[23] Each gut section plays a particular role in wood digestion, as Figure 10.1 shows. Understanding the morpho-anatomical and physiological conditions in the intestinal tract of wood-feeding termites is believed to provide insight into the biochemical process involved in pretreatment of PCWs. The process starts in the mandible where wood biomass is reduced to small particle sizes. As the particles are moved along the digestion tract

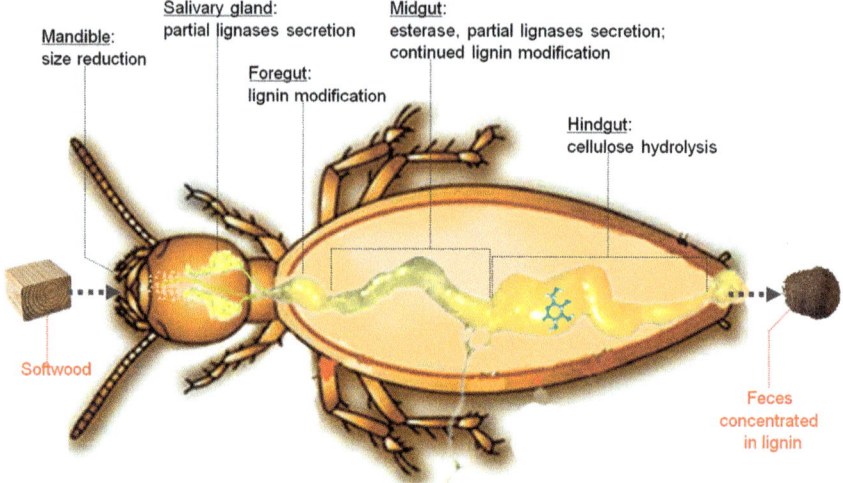

Mandible:
size reduction

Salivary gland:
partial lignases secretion

Midgut:
esterase, partial lignases secretion;
continued lignin modification

Foregut:
lignin modification

Hindgut:
cellulose hydrolysis

Softwood

Feces
concentrated
in lignin

Figure 10.1   The function of each segment of the termite gut system for wood
digestion.

including the foregut, midgut and the hindgut, various enzymes are added in
sequence and the substrates are treated under different physical, chemical and
biochemical conditions that function synergistically to effectively deconstruct
the PCWs.

10.2.1   The Physiology of each Compartment in a Termite Gut System

The passage of wood along the gut of the worker termite, *Microcerotermes
edentates* Wasmann, was investigated by Kovoor under good feeding conditions at temperatures between 25 and 30 °C. The total duration of wood
passage was reported as around 24 h.[24,25] The retention time of the food in the
foregut was from 1 to 2 h and the emptying of the crop was continuous. Food
passage through the midgut was around 1 to 1.5 h, but much slower in the
hindgut segment (14 to 20 h). In addition, Kovoor demonstrated that there was
no reflux of food from the hindgut to the midgut.[25]

10.2.1.1   *The Mouth Parts: the Mandible for Size Reduction*

The main functions of termite mouth parts are the physical ingestion and
chewing of the wood particles. Mechanical digestion is the fragmentation of
food aided by specialized physical structures, such as the mandible, prior to
extra-cellular digestion. The termite mandibles in particular have a very conservation dentition,[26] and the molar region presents the mastication ridge.[27]
The mobility of the labrum and the clypeus are demonstrated to play an important role in mastication, ingestion and regurgitation of wood diets.[28] The
mechanical fragmentation of food helps digestive enzymatic reactions because

it provides a larger total area for the contact between enzymes and their substrates.

10.2.1.2 Labial or Salivary Glands for the Secretion of Digestive Enzymes

The glands beside the termite alimentary canal are mainly responsible for the secretion of digestive enzymes. Each salivary gland of termites consists of a series of lobes, or acini, connected by a salivary canal, which opens symmetrically at the base of the labium slightly before the margin of the buccal cavity. Saliva lubricates the food bolus and initiates the enzymatic extracellular digestion of food. The termite saliva, which is used for the claying of nesting materials, was reported in previous studies to contain various digestive enzymes. Fujita and his colleagues have already cloned two cDNAs that encode premature lysozyme peptides (Rs-Lys1 and Rs-Lys2) from workers of a Japanese dampwood termite, *Reticulitermes speratus*.[29] Moreover, they demonstrated that the total digestive lysozyme activity in termites predominates in the salivary glands, and to a lesser extent in the digestive tract. At the same time, Nakashima *et al.*[30] found that 80.8% of the total cellulase activity of *C. formosanus* originated in the salivary glands. Cellobiase was also reported to exist in the salivary glands, midgut and hindgut of *Mastotermes darwiniensis* and *Neotermes koshenensis*.[31,32] It has been proposed that the cellulose is first degraded to some extent by carboxymethyl cellulase (CMCase) produced in the salivary glands of termites, and then ingested by the protozoa, which finally decompose cellulose to glucose using their own endo-β-1,4-glucanases, exocellobiohydrolase, and β-D-glucosidase.[33,34] For the digestion of hemicelluloses, Inoue *et al.* demonstrated endo-β-1,4-xylanase activity in the salivary glands of the lower termite *Reticulitermes speratus*, however, no β-xylosidase activity was documented.[35] For lignin degradation, laccase and phenol-oxidase gene expression, as well as pyrogallol oxidation activity, were confirmed to occur in the salivary glands of *R. flavipes*.[36] The labial glandular ducts open into the oral cavity and from there termites release the watery labial gland secretion onto the wood particles.[37] Labial gland secretion, which contributes to the saliva, is reported to have various species-specific functions in nest construction and social nutrition, and is a source of digestive enzymes.[31,37–40] Also, during the communal food exploitation of termites, the labial gland secretion is released onto the food by feeding workers, as demonstrated in the Africa termite, *Schedorhinotermes lamanianus*, and the French species *Reticulitermes santonensis*,[41–43] thereby aiding efficient food exploitation.

10.2.1.3 The Foregut Performs as the Valve System (Crop-Filter Chamber), Storage and Initial Digestion Organ

The foregut of wood-feeding termites may play a critical role in biomass pretreatment. The termite foregut is generally considered to consist of four sections: the pharynx, esophagus, crop, and proventriculus. The pharynx is a tube that connects the interior of the buccal cavity with the more inward parts of the gut. The esophagus is a narrow tube extending as far as the middle or posterior part of the thorax, followed by the crop.[38] The crop is simply a storage area and is the proventriculus with a muscular extension known as the "gizzard". In wood-feeding termites it is used to grind the wood into smaller particles, and also serves as a filter to keep over-sized particles out of the main digestive tract and as a valve controlling the flow of food into the midgut. Cookson also suggested that the termite gizzard fragments some lignin to low-molecular-weight components before cellulose digestion.[44] The foregut termi-nates with a cardiac valve which penetrates deeply into the midgut. Besides meshing and storage of wood particles, the main function of this digestive organ is lignin pretreatment. The laccase and phenol-oxidase secreted by sal-ivary glands pass into and start to function on ingested wood substrates in the foregut.[36] It is also worthwhile mentioning here that it has recently been demonstrated that the lignin pretreatment starts during the termite chewing process, since the enzymes are first secreted into the buccal cavity and pass into the foregut with the wood particles. In other words, the initial lignin pre-treatment starts in the buccal cavity during the termite chewing process,[45] which was detailed in Section 10.2.1.1.

10.2.1.4 The Midgut Epithelial Cell for the Secretion of Digestive Enzymes

The midgut of wood-feeding lower termites is longer than that of higher ter-mites.[38] It is a slender tube of uniform diameter with a remarkably constant histological structure, which runs from the gastric caeca to just before the Malpighian tubules. The gastric caeca serve to increase the surface area of the midgut, and thus increase both its ability to secrete digestive enzymes and to extract byproducts from the partially digested wood. The digestive enzymes are released across the midgut intestinal wall into the body cavity. The midgut is lined with a semi-permeable peritrophic membrane composed of protein and chitin, like the cuticle, which allows the passage of liquids and dissolved sub-stances to the midgut intestinal cell while preventing the passage of solid wood particles. The midgut and the hindgut are separated by the proctodeal valve. The termite midgut is the main place for esterase secretion[46] and lignin–hemicellulose matrix modification, such as lignin–hemicellulose dissociation, lignin phenolic dehydroxylation and β–5′, β–β′ linkage cleavage.[47]

10.2.1.5 The Hindgut Microbial Niche for Synergistic Cellulose Hydrolysis

The hindgut in termites is the most important structure as it is the main place for cellulose hydrolysis.[30,48,49] The rectum which compresses the undigested wood and waste products as feces, such as undigested or modified lignin residuals, extracts more water from this region before the waste is passed out through the anus. The hindgut is also considered as a microbial niche, where the deconstruction of lignin monomers and dimers resulting from lignin modification in the midgut occurs to alleviate the inhibitory effect on cellulose hydrolysis.[50] It has been reported that the symbiotic microorganisms in termite guts are responsible for aromatic lignin model degradation.[51–54] It has also been clarified that oligomer lignin is degraded in the termite hindgut.[44,55,56] In another study regarding the degradation of monomeric lignin and related aromatic compounds, Kuhnigk *et al.* found that aromatic compounds were only degraded aerobically by the hindgut flora of xylophagous termites.[57]

10.2.2 Microenvironments in the Termite Gut System

10.2.2.1 Oxygen and its Role in Lignin Modification

Lignin degradation by wood-feeding termites is intriguing and although much evidence regarding it has been gathered recently, some important processing characteristics remain ambiguous. A termite gut system is generally considered to be an anaerobic environment due to the large community of strictly anaerobic microorganisms present in the hindgut.[58–60] On the other hand, high oxygen levels are virtually required for lignin degradation.[61,62] In an anaerobic environment, polymeric lignin will not be degraded, and wood may persist in its non-degraded form for several hundreds or thousands of years.[63] The presence of molecular oxygen is a prerequisite for the initiation of lignin depolymerization since oxygen acts as a co-substrate in the modification/disruption of lignin by oxidative enzymes.[62,64] There is no conclusive evidence that the intermonomeric linkages in insoluble, highly polymeric lignins are attacked in the absence of oxygen.[65,66]

The oxygen concentrations in each gut compartment of *Coptotermes Formosanus* Shiraki, (Figure 10.2) demonstrated that morpho-anatomical differentiation in the termite intestinal tract corresponds to different oxygen levels, which will most likely create different physicochemical microenvironments.[47,67] Various gut segments have been found to show different oxygen gradients, indicating different oxygen penetration/consumption rates in the gut compartments of the foregut, midgut and hindgut. Among which, the foregut of *C. formosanus* showed a slow decline in radial oxygen concentration, the

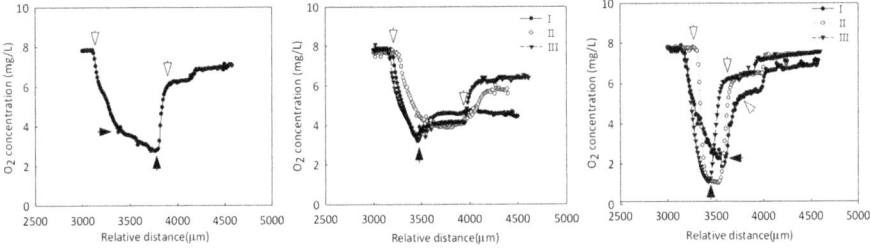

Figure 10.2 Radial oxygen profiles along the gut of *C. formosanus*.[47]

midgut contained a significant concentration of oxygen, despite the small diameters, and it exhibited a more gradual loss of oxygen than the hindgut did. The anterior and posterior portions of the midgut revealed steeper oxygen gradients than the middle region. In the three hindgut regions of *C. formosanus*, there were linear decreases and then increases in oxygen concentrations from the ambient concentration, directed from the gut wall towards the gut center. Only the central region of the paunch was relatively anoxic, with an oxygen concentration of approximately 1 mg L^{-1}. This was most obvious in the second region of the paunch.

10.2.2.2 The Redox Potential for Electron Transfer and Lignin Oxidation

Redox potential (E_m), which is a measure of the tendency of a chemical to acquire electrons and therefore be reduced, is another important factor to indicate the mechanism of lignin oxidation in termites. The redox potential is also an indicator of the affinity of a substance for electrons compared with hydrogen, which is set at 0 V. Substances more strongly electronegative (*i.e.*, more capable of oxidizing) than hydrogen have positive redox potentials. Molecules less electronegative (*i.e.*, more capable of reducing) than hydrogen have negative redox potentials. The enzymatic reactions involved in lignin pretreatment are oxidation–reduction reactions in which the lignin framework is oxidized and the oxidant, such as oxygen or H_2O_2, is reduced. The ability of an enzyme to carry out oxidation–reduction reactions depends on the oxidation–reduction state of the environment, or its redox potential (E_m). For microbes, strictly aerobic microorganisms are generally active at positive E_m values, whereas strict anaerobes are generally active at negative E_m values. Redox also affects the solubility of nutrients, especially metal ions. Thus, a positive redox potential is speculated to exist in termite guts, especially in the midgut, which has been proven to be the main site for lignin modification.

The redox potential is the combined result of the pH, the concentrations of O_2, H_2, and N_2, as well as other redox-active compounds (*i.e.*, lignin, lignolytic enzymes, or iron minerals), and decides the oxidation rate. To be specific, a higher redox potential will lead to the quicker oxidation of the substrate. The redox potential of the whole gut has been proven to be positive, indicating aerobic conditions in the termite gut. This is in accordance with the results from oxygen measurements (Section 10.2.2.1). The midgut exhibited a much higher redox potential than the foregut and hindgut. Based on the definition of redox potential, this means that midgut fluid possess a higher tendency to acquire electrons and therefore be reduced, with a concurrently higher tendency for lignin oxidation. The importance of iron(II) as a potential electron donor in the termite gut is increased by the fact that lignin can act as an electron donor for iron-oxidizing lignolytic enzymes, and may also serve as an electron sink for hindgut fermenting bacteria.[68] Since lignin can act as an electron shuttle between lignolytic enzymes and iron(II) in a purely chemical reaction, the mediation by lignins relieves the kinetic limitation of microbial iron oxidation.[69] The sharp decrease in redox potential between the midgut and hindgut is speculated not only to be attributed to the pH difference, but also the lignolytic enzymes in the midgut. The redox potential difference between the foregut and midgut is then only because the midgut is the main region where the lignases function. The redox potential in different regions of the gut may play a significant role in the digestion and assimilation of food materials, in detoxification reactions, and in the production of toxic metabolites from ingested food materials.

In addition to the axial dynamics, there is also the radial dynamics of the redox potential. The redox potential of ~ 0.25 V at the foregut wall increased within the gut periphery, reaching values above 0.275 V at the gut center. In the midgut, the redox potential steeply increased from 0.25 to 0.325 V. All the redox transition zones basically coincide with the oxygen profiles in the gut periphery as detailed in Section 10.2.2.1.

The apparent redox potential is an indicator of the prevailing redox processes in a given environment, although it has to be considered that in biological systems, redox processes are usually not in equilibrium. The resulting mixed potential depends on the thermodynamic status of the system and the ability of the electro-active components to react with the electrode surface.[70] In the hindgut, the gas atmosphere has little impact on the redox potential, which is explained by the large size of these compartments (O_2 penetrates only into the periphery) and the significant amounts of H_2 present.[71] In the midgut, the redox potential is controlled by O_2, and this actually agrees with the overall oxic conditions in this region, caused by its small diameter, and is also expected to be controlled by the complete oxidation–reduction of the electro-active component(s) under all other conditions. These electro-active components are most likely to be lignin, lignases, and iron in the termite midgut.[13] In addition, O_2-dependent lignin oxidation processes in the microoxic periphery are also possible in the midgut. This ability of termites to adjust their gut redox potential could be one of the ways that termites adapt in the co-evolutionary race with plants in combating the lignins in PCWs.[72]

10.2.2.3   The pH and its Role in Lignase Function

The pH value in a termite gut system is another important factor for enzymatic pretreatment of lignin. It has been reported that the pH of a reaction system influences the structure and activity of enzymes. On one hand, the dissolved H^+ will influence the state of ionization of both the acidic and basic amino acids, any change of which will subsequently alter the ionic bonds within protein structure that determine the three-dimensional shape of the enzymes. On the other hand, changes in pH may also affect the charge properties of an enzyme so that it can undergo neither substrate binding nor catalysis. Within a narrow pH fluctuation, the resulting changes in the shape of an enzyme and its substrates may be reversible. But for a significant change in pH level, an enzyme may undergo denaturation. In general, there is an optimal pH for a specific enzyme, which varies for different enzymes. All the currently reported fungal lignases work optimally at acidic pH, and neutral pH or basic environments will influence the enzyme activity or even deactivate the enzymes. For example, the optimal working pH range for laccase, lignin peroxidase (LiP), as well as manganese peroxidase (MnP) identified from fungi is 3.0–5.0, 3.7–6.0, and 4.5–5.5, respectively. However, the acidic working conditions of these fungal lignases make them very challenging for industrial applications due to the requirement for special bioreactors. Hence, enzymes with a neutral optimal pH range will be more desirable.

The pH in a termite gut segment greatly influences the action of any enzyme secreted into that location. In addition, gut pH may also influence the solubility of ingested components, the toxicity of some generated toxins, and the gut symbiotic microorganisms. As demonstrated in the gut of worker termites, *C. formosanus*, both the anterior and posterior regions of hindgut exhibit acidic pH values ranging from 5.0 to 7.0. This is due to the production of acetic acid, which is employed by termites as the direct energy source for gut cell adsorption, as well the absence of buffering agents that could alter the pH. Also, the carbohydrate-digesting enzymes usually prefer slightly acidic working conditions.[72,73] The pH in the termite foregut ranges from 7.5 to 8.0, and from 6.5 to 7.0 in the midgut. For most insects, the crops tend to be slightly acidic because of the production of glycolytic cycle acids and the existence of few buffering agents there. Moreover, the pH level in the termite midgut is slight basic, revealing a tolerance of termite-origin lignolytic enzymes for basic conditions. As discussed, the neutral pH microenvironment in the termite midgut compartment, where the main lignin modification occurs, indicates mild working conditions for termite lignases compared to the reported fungal lignases. In other words, if termite-origin lignases are successfully bioengineered and expressed in other host media, these enzymes will be highly promising for industrial pretreatment processing due to their neutral pH working conditions.

10.2.2.4   Hydrogen Peroxide and its Role in Peroxidase and Non-Enzymatic Lignin Modification

The overall performance of an enzyme depends not only on its reaction temperature and pH, but also on some co-factors. Termite digestive systems are highly evolved for efficient lignin-unlocking processes, which specifically target unique functional groups, such as the ester bonds between hemicelluloses and lignin, and the covalent bonds within the lignin network. Since the midgut of a termite has been demonstrated to be the main place for lignin modification, it is also speculated to be the most likely site for LiP function. In addition, the existence of co-factor H_2O_2 is also required for lignin modification. In fungal systems, H_2O_2 serves as the oxidant for LiP to achieve high oxidation states, and then peroxidases utilize aromatic compounds as their reducing substrates; Fenton's reaction. In addition, laccase could be effective for the depolymerization of milled wood lignin in the presence of H_2O_2.

Our preliminary data support the existence of H_2O_2 in the midgut of *C. formosanus* when termite workers were fed with native pinewood. The same investigation also found that H_2O_2 was actually concentrated in the anterior and central regions of the midgut, ranging in concentration from 5 to 56 μM, but H_2O_2 was not detected in the foregut and hindgut of *C. formosanus*. Furthermore, H_2O_2 was not detectable in the termite gut system when they were fed with cellulosic substrate for 15 days, indicating the potential relevance of H_2O_2 to lignin components and their processing in termite gut systems. Further research is need to determine how the H_2O_2 is produced in the termite gut system and what function it may have. Firstly, based on another finding for converting aromatic alcohols to aldehydes during wood digestion by termites,[45] it has been speculated that an enzyme functioning as a glyoxal oxidase may catalyze the conversion of the lignin aliphatic hydroxyl group to a carbonyl group, with the production of H_2O_2 during the reaction.[74] It is also well known that H_2O_2 usually serves as a co-factor for peroxidases, and also affects the selectivity and reactivity of laccase.[36,75,76] More importantly, Vu *et al.* reported 1.3 mM of iron(II) to be present in termite gut,[68] which is just within the same order of magnitude as the H_2O_2 detected (approximately 4.3 mM). These investigations may shed light on the existence of a non-enzymatic lignin-modifying system initiated by H_2O_2, which may be responsible for the selective lignin modification in termite gut systems. During wood pretreatment processing in the termite gut system, this non-enzymatic processing system is speculated to easily penetrate inside the PCWs because of the small molecular size of H_2O_2. However, the speculated non-enzymatic lignin-modifying system in termite gut systems is believed to differ from the Fenton's reaction and alkaline H_2O_2 (AHP) pretreatment. In other words, it is a more advanced process than the existing free radical reaction, as it is a selective process of lignin structural modification, rather than a severe degradation. The primary data we obtained may infer that the electron-transfer mechanism of this non-enzymatic lignin modifying system is a partial reversal process of plant lignin synthesis. This processing needs to take place first, prior to the action of the lignases, and may accomplish around 10–15% of the lignin degradation in the termite gut system.[77]

10.3 The Lignin-Unlocking Mechanism in Termites for Maximum Sugar Release

10.3.1 Step-Wise Pretreatment for Sugar Release

10.3.1.1 The Chewing Process

In exploring the mechanisms of PCW degradation by termites, it has always been speculated that termite chewing action will play an important role in initial pretreatment. Fujita *et al.* have already demonstrated one of the characteristics of wood degradation by termites to be the mechanical grinding of food by the mandibles to increase the surface area of the substrates.[78]

The termite workers chew wood blocks into small particles during the early stage of its digestion process for better access to carbohydrates. During the mechanical grinding of wood particles by the termites, the cellulases appended to the wood particles or mixed with them to initiate the catalysis process.[79] Yoshimura demonstrated that the lower termite, *C. formosanus*, crushes and grinds the wood to produce sharp edges on surface for better accessibility with the functional enzymes.[33] It has been reported that the termite workers in the Rhinotermitidae family excavate wood for nesting.[80] When they were moved to new environment with only wood blocks, they continued their chewing behavior to acquire a large amount of small wood particles for nesting. The nest consists of ground wood particles, pasted together with a special secretion from the salivary glands.[80,81] A large proportion of these particles might then be redigested. During the chewing process, the workers will release a secretion, which consists of various cellulolytic and lignolytic enzymes, from the salivary and labial glands for initial digestion and community feeding stimulation activities.[37]

It is important to note that the current research on lignin modification by termites is mainly focused on the gut processes of digestion. Such studies are centered around the understanding of symbiont evolution, and the transcriptomics, meta-transcriptomics, and meta-genomics of these wood-feeding termites, which are necessary for understanding the lignocellulose-degradation process. However, attention has been drawn towards understanding the combined process of biological lignin modification and physical chewing, for cellulose release during the chewing process in the termite worker *C. formosanus*, which is essential for targeting the fundamental barrier in the enzymatic hydrolysis of cellulose.[82] The results showed that the termite chewing process induces selective pretreatment on the lignin counterpart and/or the lignin–hemicellulose association. Consequently, structural modification within the lignin matrix during the chewing process resulted in a much better exposure of the carbohydrate counterpart within the PCWs. These results could significantly contribute to the current understanding of specific lignin structure modification by the lower termites for efficient cellulose utilization, as well as promote existing biomass pretreatment technologies for effective disintegration of the PCW structure to produce bioenergy-derived products and fuels. Combining the achievements of nature with those of mankind should lead to new breakthroughs in science and technology that are critical for converting lignocelluloses to biofuels and bioproducts.

10.3.1.2 Alimentary Canal Metabolism

Lignin modification in each segment of a termite gut system involves side-chain oxidation/cleavage and demethylation of the ring methoxyl groups of wood lignin, which usually starts in the foregut and continues in the midgut. Ring demethylation makes the substrate more suitable for carbon–carbon bond cleavage. Side-chain oxidation will cause depolymerization of lignin polymers or release of lignin units from the lignin–carbohydrate complex. Both of these reactions are believed to require oxygen, as previous experiments have demonstrated.[83,84] Our findings support that a similar system of catalysis might exist in the salivary glands/foregut and midgut of wood-feeding lower termites.[45,47,50,85,86] Furthermore, lignin modification may not only occur in the foregut and midgut, but also in the hindgut as it might also participate in the processes dealing with simpler products. As the lignin depolymerizes, it liberates phenolic groups, breaks away from hemicellulose,[87] and perhaps releases low-molecular-weight lignin fragments,[86,88] increasing its hydrophilia. It is speculated that when passing through the hindgut, the easily digested byproducts, such as ester and monomer lignin units, might be taken up through the gut epithelium and oxidized aerobically by termite tissue or gut symbionts. However, further investigations are needed to clarify this. This process may be attributed to the function of esterase and phenol-oxidizing enzymes.[89] The results acquired in these studies have provided new insights into our understanding of lignin degradation by wood-feeding termites. Lignocellulosic degradation occurs throughout the entire gut system, with initiation of lignin modification during the chewing process. It is likely that wood-feeding termites have evolved optimal processes for efficient lignocellulose degradation. Further understanding of the exact reactions and mechanisms that modify lignin components in the termite gut system will contribute to our understanding of the biochemical roles that these insects play in lignin modification, and promote industrial utilization of these processes for faster and better access of enzymes to polymer carbohydrates.

Termites have proven to be highly efficient wood digesters. The unique *in vivo* degradation/modification of lignin-related aromatic compounds and three-ring polycyclic hydrocarbons also provides strong evidence for lignin pretreatment in the termite gut.[50,90] The conjugated structures of these aromatic compounds are metabolized by side-chain addition, ring hydroxylation, ring/side-chain oxidation, ring mineralization, and β–*O*–4′ and 5–5′ linkage cleavage. All of the modifications begin in the foregut but mainly happen in the midgut, which indicates that the enzyme(s) responsible for aromatic compound degradation/modification may be excreted in the fore- and/or midgut of the termite itself. We cannot exclude the possibility that there may be microbe(s) collaborating with the termite origin enzyme(s) for degradation of these aromatic compounds. These reactions would likely be similar to lignin modification during natural wood digestion by wood-feeding termites. Further elucidation of the functional mechanisms of lignin modifying systems in termite guts will contribute to our understanding of the biochemical routes of lignin degradation/modification in termite gut systems.

10.3.2 Enzymatic/Non-Enzymatic and Microbial Unlocking Reactions on Lignin and its Analogs

Currently, the results obtained on lignin modification by wood-feeding termites may provide some insight into a unique lignin-unlocking mechanism for PCW deconstruction, which will indeed assist in developing novel enzymatic biomass pretreatment processes. To be specific, termite-induced lignin-unlocking processes for cellulose release are not achieved by lignin degradation, rather by selective functional group change and covalent linkage cleavage. The three-dimensional structure of lignin is then altered. and cellulose fibers are exposed away from lignin protection. In addition, the inhibition effects of the lignin framework on cellulases are alleviated with specific structural modifications. For example, phenolic demethoxylation and demethylation are considered to reduce lignin steric hindrance effects on cellulases, as the existence of $-OCH_3$ and $-CH_3$ in lignin increases its branch overlapping, thus, inhibits cellulolytic enzyme diffusion, as well as blocking the enzymes from accessing the cellulose substrate. Also, phenolic/aliphatic dehydroxylation and carboxylation on lignin will reduce its non-reversible chemical inhibition effect on cellulose. Since lignin hydroxyl groups are reported to react with cellulase to induce enzyme inactivation, the introduction of more carboxyl groups to lignin has been demonstrated to have a positive inhibitory influence on pure cellulose hydrolysis. Furthermore, the increased solubility of lignin, which is speculated to be a result of carboxylation, demethylation and demethoxylation, leads to a reduction in lignin's competitive and unproductive binding effects on cellulases. Hydrophobic interactions between lignin and cellulase are also considered to be one of the main substrate–enzyme hindering factors of cellulose enzymatic hydrolysis. Cellulose is considered to be largely exposed by selective dissociation of β–β′, β–5′ and 5–5′ lignolytic ether linkages. The alleviation of physical and chemical inhibition effects on cellulase and the release of cellulose achieved by all these selective lignin modifications are regarded as the 'recipe' used by wood-feeding termites for highly efficient wood-cellulose utilization. Notably, the lignolytic catalyzing system in termite gut systems leaves the lignin backbone alone, thus consuming less energy and producing negligible lignin-derived inhibitors for subsequent cellulose hydrolysis and sugar fermentation. On this basis, it is possible that termite-inspired pretreatment technologies can be developed by mimicking the whole lignin-unlocking process in a termite gut system. It is highly desirable that such technologies will perform lignin pretreatment as efficiently and robustly as termites do. Continued research in these related areas is needed, and will make major contributions to the fundamental science that is critical for filling the existing knowledge and technology gaps currently holding back sustainable and economical biofuel production from lignocellulosic biomass.

10.4 Summary

The efficient wood-degrading process that has evolved in termite gut systems provides support for a new concept that complete degradation or removal of lignin may not be essential for enhanced cellulose hydrolysis, but that the selective modification of lignin functional groups and ether linkages, which play critical roles in lignin recalcitrance to enzymatic cellulose hydrolysis, will result in an energy-efficient pretreatment process. Compared to the existing pretreatment technologies, this nature-inspired pretreatment scheme is more advanced, as the scheme can be implemented under ambient conditions with no heat, no pressure, and few chemicals, and results in lignin modification. Additionally, the knowledge obtained regarding wood-feeding termites could be used to improve and modify existing biomass pretreatment technologies for effective disintegration of the PCW structure to release sugars. Combining the achievements of nature with those of mankind should lead to new breakthroughs in science and technology that are critical for using lignocellulosic biomass as renewable sources for producing bioproducts and biofuels.

References

1. W. Böerjan, J. Ralph and M. Baucher, *Annu. Rev. Plant Biol.*, 2003, **54**, 519.
2. M. Chabannes, K. Ruel, A. Yoshinaga, B. Chabbert, A. Jauneau, J. Joseleau and A. M. Boudet, *Plant J.*, 2001, **28**, 271.
3. B. Goodell, D. D. Nicholas and T. P. Schultz, *Wood Deterioration and Preservation*, American Chemical Society, Washington DC, 2003.
4. R. R. Sederoff, J. J. MacKay, J. Ralph and R. D. Hatfield, *Curr. Opin. Plant Biol.*, 1999, **2**, 145.
5. J. Ralph, K. Lundquist, G. Brunow, F. Lu, H. Kim, P. F. Schatz, J. M. Marita, R. D. Hatfield, S. A. Ralph, J. H. Christensen and W. Boerjan, *Phytochem. Rev*, 2004, **3**, 29.
6. L. B. Davin and N. G. Lewis, *Curr. Opin. Biotechnol.*, 2005, **16**, 398.
7. M. Tuomela, M. Vikman, A. Hatakka and M. Itavaara, *Bioresour. Technol.*, 2000, **72**, 169.
8. T. W. Jeffries, *Biodegradation*, 1990, **1**, 163.
9. F. Chen and R. A. Dixon, *Nat. Biotechnol.*, 2007, **25**, 759.
10. R. Sutcliffe and J. Saddler, *Biotechnol. Bioeng.*, 1986, **17**, 749.
11. X. Pan, *J. Biobased Mater. Bioenergy*, 2008, **2**, 25.
12. R. L. Lundblad, Chemical Reagents for Protein Modification, 3rd Edition, CRC Press, Boca Raton, 2004.
13. A. Kappler and A. Brune, *Soil Biol. Biochem.*, 2002, **34**, 221.
14. V. M. Chernoglazov, O. V. Ermolova and A. A. Klyosov, *Enzyme Microb. Technol.*, 1988, **10**, 503.
15. V. J. H. Sewalt, W. G. Glasser and K. A. Beauchemin, *J. Agric. Food Chem.*, 1997, **45**, 1823.
16. X. Pan, D. Xie, N. Gilkes, D. J. Gregg and J. N. Saddler, *Appl. Biochem. Biotechnol.*, 2005, **121–124**, 1069.
17. N. Mosier, C. Wyman, B. Dale, R. Elander, Y. Y. Lee, M. Holtzapple and M. Ladisch, *Bioresour. Technol.*, 2005, **96**, 673.
18. R. Wooley, M. Ruth, D. Glassner and J. Sheehan, *Biotechnol. Prog.*, 1999, **15**, 794.
19. L. R. Lynd, R. T. Elamder and C. E. Wyman, *Appl. Biochem. Biotechnol.*, 1996, **57–58**, 741.
20. Y. Pu, N. Jiang and A. J. Ragauskas, *J. Wood Chem. Technol.*, 2007, **27**, 23.

21. F. Parisi, *Adv. Biochem. Eng. Biotechnol.*, 1989, **38**, 53.
22. T. R. Filley, P. G. Hatcher, W. C. Shortle and R. T. Praseuth, *Org. Geochem.*, 2000, **31**, 181.
23. M. Charrier and A. Brune, *Can. J. Zool.*, 2003, **81**, 928.
24. J. Kovoor, Contribution à l'étude de la digestion chez un termite supérieur (Microcerotermes edentatus, Was., Isopters, Termitidae), Faculté des Sciences, Paris, 1966.
25. J. Kovoor, *Cell. Mol. Life Sci.*, 1967, **23**, 820.
26. M. Ahmad, PhD thesis, University of Chicago, 1950.
27. J. Deligne, *Compte-Rendu Acad. Sci.*, 1966, **263**, 1323.
28. H. S. Vishnoi, *J. Zool. Soc. India*, 1956, **8**, 1.
29. A. Fujita, T. Minamoto, I. Shimizuand and T. Abe, *Insect Biochem. Mol. Biol.*, 2002, **32**, 1615.
30. K. Nakashima, H. Watanabe, H. Saitoh, G. Tokuda and J. I. Azuma, *Insect Biochem. Mol. Biol.*, 2002, **32**, 777.
31. P. C. Veivers, A. M. Musca, R. W. O'Brien and M. Slaytor, *Insect Biochem*, 1982, **12**, 35.
32. G. Tokuda, H. Saito and H. Watanabe, *Insect Biochem. Mol.*, 2002, **32**, 1681.
33. T. Yoshimura, *Wood Res.*, 1995, **82**, 68.
34. N. Todaka, C. M. Lopez, T. Inoue, K. Saita, J. Maruyama, M. Arioka, K. Kitamoto, T. Kudo and S. Moriya, *Appl. Biochem. Biotech.*, 2010, **160**, 1168.
35. T. Inoue, K. Murashima, J. I. Azuma, A. Sugumoto and M. Slaytor, *J. Insect Physiol.*, 1997, **43**, 235.
36. M. R. Coy, T. Z. Salem, J. S. Denton, E. S. Kovaleva, Z. Liu, D. S. Barber, J. H. Campbell, D. C. Davis, G. W. Buchman, D. G. Boucias and M. E. Scharf, *Insect Biochem. Mol. Biol.*, 2010, **40**, 723.
37. J. Reinhard and M. Kaib, *J. Chem. Ecol.*, 2001, **27**, 175.
38. C. Noirot, *Biology of the Termites*, Academic Press, New York, 1969.
39. M. Hogan, P. C. Veivers, M. Slaytor and R. T. Czolij, *J. Insect Physiol.*, 1988, **34**, 891.
40. T. K. Mednikova, *Zh. Evol. Biokhim. Fiziol.*, 1991, **27**, 86.
41. M. Kaib and J. Ziesmann, *Insectes Soc.*, 1992, **39**, 373.
42. J. Reinhard and M. Kaib, *Physiol. Entomol.*, 1995, **20**, 266.
43. J. Reinhard, H. Hertel and M. Kaib, *J. Chem. Ecol.*, 1997, **23**, 2371.
44. L. J. Cookson, *J. Insect Physiol.*, 1988, **34**, 409.
45. J. Ke, D. Singh, S. Chen and X. Yang, *Biomass Bioenergy*, 2010, **35**, 3617.
46. M. M. Wheeler, M. R. Tarver, M. R. Coy and M. E. Scharf, *Arch. Insect Biochem. Physiol.*, 2010, **73**, 30.
47. J. Ke, J. Sun, H. D. Nguyen, D. Singh, K. C. Lee, H. Beyenal and S. Chen, *Insect Sci.*, 2010, **17**, 277.
48. F. Warnecke, P. Luginbühl, N. Ivanova, M. Ghassemian, T. H. Richardson, J. T. Stege, M. Cayouette, A. C. McHardy, G. Djordjevic, N. Aboushadi, R. Sorek, S. G. Tringe, M. Podar, H. G. Martin, V. Kunin, D. Dalevi, J. Madejska, E. Kirton, D. Platt, E. Szeto, A. Salamov, K. Barry,

N. Mikhailova, N. C. Kyrpides, E. G. Matson, E. A. Ottesen, X. Zhang, M. Hernández, C. Murillo, L. G. Acosta, I. Rigoutsos, G. Tamayo, B. D. Green, C. Chang, E. M. Rubin, E. J. Mathur, D. E. Robertson, P. Hugenholtz and J. R. Leadbetter, *Nature*, 2007, **450**, 560.

49. A. Brune, *Trends Biotechnol.*, 1998, **16**, 16.
50. J. Ke, D. Singh and S. Chen, *Int. Biodeter. Biodegr.*, 2011, **65**, 744.
51. K. Harazono, N. Yamashita, N. Shinzato, Y. Watanabe, T. Fukatsu and R. Kurane, *Biosci. Biotech. Biochem.*, 2003, **67**, 889.
52. T. Kuhnigk and H. König, *J. Basic Microb.*, 1997, **37**, 205.
53. K. Kato, S. Kozaki and M. Sakuranaga, *Biotechnol. Lett.*, 1998, **20**, 459.
54. M. B. Pasti, A. L. Pometto, M. P. Nuti and D. L. Crawford, *Appl. Environ. Microbiol.*, 1990, **56**, 2213.
55. J. H. A. Butler and J. C. Buckerfield, *Soil Biol. Biochem.*, 1979, **11**, 507.
56. S. C. Mishra and P. K. Sen-Sarma, *Mater. Org.*, 1980, **15**, 119.
57. T. Kuhnigk, E. M. Borst, A. Ritter, P. Kampfer, A. Grad and H. Herte, *Syst. Appl. Microbiol.*, 1994, **17**, 76.
58. M. L. Eutick, R. W. O'Brien and M. Slaytor, *J. Insect Physiol.*, 1976, **22**, 1377.
59. J. A. Breznak, *Annu. Rev. Microbiol.*, 1982, **36**, 323.
60. K. S. Katsumata, Z. F. Jin, K. Hori and K. Iiyama, *J. Wood Sci.*, 2007, **53**, 419.
61. R. Ten Have and P. J. M. Teunissen, *Chem. Rev.*, 2001, **101**, 3397.
62. M. E. Scharf and A. Tartar, *Biofuels, Bioprod. Biorefin.*, 2008, **2**, 540.
63. R. A. Blanchette, *Can. J. Bot.*, 1995, **73**, 999.
64. J. A. Breznak and A. Brune, *Annu. Rev. Entomol.*, 1994, **39**, 453.
65. P. J. Colberg, *Biology of Anaerobic Microorganisms*, John Wiley & Sons, New York, 1988.
66. L. Y. Young and A. C. Frazer, *Geomicrobiol. J.*, 1987, **5**, 261.
67. A. Brune, D. Emerson and J. A. Breznak, *Appl. Environ. Microbiol.*, 1995, **61**, 2681.
68. A. T. Vu, N. C. Nguyen and J. R. Leadbetter, *Geobiology*, 2004, **2**, 239.
69. D. R. Lovley, J. L. Fraga, E. L. Blunt-Harris, L. A. Hayes, E. J. P. Phillips and J. D. Coates, *Acta Hydrochim. Hydrobiol.*, 1998, **26**, 152.
70. W. Stumm and J. J. Morgan, *Aquatic Chemistry*, Wiley, New York, 1981.
71. A. Ebert and A. Brune, *Appl. Environ. Microbiol.*, 1997, **63**, 4039.
72. J. L. Nation, *Insect Physiology and Biochemistry*, CRC Press, Boca Raton, 2008.
73. R. L. Howard, E. Abotsi, E. L. Jansen van Rensburg and S. Howard, *Afr. J. Biotech.*, 2003, **2**, 602.
74. P. J. Kersten, *Proc. Natl. Acad. Sci. U. S. A.*, 1990, **87**, 2936.
75. C. Srinivasan, T. M. D'Souza, K. Boominathan and C. A. Reddy, *Appl. Environ. Microbiol.*, 1995, **61**, 4274.
76. A. J. Damle and S. R. Shukla, *Clean: Soil, Air, Water*, 2010, **38**, 663.
77. J. Z. Sun and X. G. Zhou, *Recent Advances in Entomological Research*, Higher Education Press, Beijing, 2010.

78. A. Fujita, M. Hojo, T. Aoyagi, Y. Hayashi, G. Arakawa, G. Tokuda and H. Watanabe, *J. Wood Sci.*, 2010, **56**, 222.
79. L. Li, J. Fröhlich, P. Pfeiffer and H. König, *Eukaryotic Cell*, 2003, **2**, 1091.
80. A. E. Emerson, *Ecol. Monogr.*, 1938, **8**, 247.
81. M. Oshima, *Philos. J. Sci.*, 1919, **15**, 319.
82. J. Ke, D. D. Laskar and S. Chen, *Biotechnol. Biofuels*, 2012, **5**, 11.
83. G. L. Kedderis and P. F. Hollenberg, *J. Biol. Chem.*, 1983, **258**, 8129.
84. R. Ten Have, J. A. Field and M. C. R. Franssen, *Biochem. J.*, 2000, **347**, 585.
85. J. Ke, D. D. Laskar, D. Singh and S. Chen, *Biotechnol. Biofuels*, 2011, **4**, 17.
86. A. Tartar, M. M. Wheeler, X. Zhou, M. R. Coy, D. G. Boucias and M. E. Scharf, *Biotechnol. Biofuels*, 2009, **2**, 25.
87. T. Higuchi, *Wood Sci. Technol.*, 1990, **24**, 23.
88. H. Janshekar and A. Fiechter, *Adv. Biochem. Eng./Biotechnol.*, 1983, **27**, 119.
89. T. K. Kirk, T. Higuchi and H. Chang, *Lignin Biodegradation: Microbiology, Chemistry, and Potential Applications*, CRC Press, Boca Raton, 2nd edn, 1980.
90. J. Ke, D. Singh and S. Chen, *J. Agric. Food Chem.*, 2012, **60**, 1788.

Lignocellulolytic Systems of Insects and their Potential for Viable Biofuels

JIANZHONG SUN*[a] AND XUGUO JOE ZHOU[b]

[a] Biofuels Institute, Jiangsu University, Zhenjiang 212013, P. R. China;
[b] Department of Entomology, University of Kentucky, Lexington,
KY 40546, USA
*Email: jzsun1002@hotmail.com

11.1   Introduction

Much information has accumulated in the past decade on cellulose-feeding insects and their potential application in the biofuels industry, which has encouraged the emergence of a new area of entomological science that interacts with other relevant disciplines, such as bioengineering for fuels and chemicals. Most insects are unable to use plant cell walls as their main food sources, but some insects subsist on lignocellulosic biomass, from agricultural crops to forest woody substrates, as their only food, for example, termites, wood-feeding roaches, wood-feeding beetles, wood wasps, leaf-shredding aquatic insects, silver fish, and leaf-cutting ants.[1]

In spite of the presence of different types of carbohydrolytic activities identified in these insects, their lignocellulolytic mechanisms and the relevant system characteristics are poorly understood. The ability of these insects to feed on wood, foliage and detritus has recently stimulated extensive investigation into

RSC Energy and Environment Series No. 10
Biological Conversion of Biomass for Fuels and Chemicals: Explorations from Natural Utilization Systems
Edited by Jianzhong Sun, Shi-You Ding and Joy Doran-Peterson
© The Royal Society of Chemistry 2014
Published by the Royal Society of Chemistry, www.rsc.org

the mechanisms of how these cellulose-feeding insects digest the structural and recalcitrant lignocellulose in their food. With this knowledge, scientists could possibly advance biofuel technologies using the discovery of novel lignocellulolytic enzymes and a better understanding of the bioconversion mechanisms that deconstruct plant cell walls inside insect digestive systems.[2]

Biofuels derived from lignocellulosic biomass provide the potential to reduce our dependence on traditional fossil fuels, mitigate global climate change, and further secure world economic growth. However, breakthrough technologies are still needed to overcome barriers in the development of cost-effective processes for converting biomass to fuels and chemicals.[3,4] The demand for the identification of novel biomass-degrading enzymes at higher efficiencies has catalyzed an interest in revealing some unusual animal lignocellulolytic systems, such as the wood-feeding termites.[2,5,6] It is believed that these unique insect lignocellulolytic systems could potentially shed light on an efficient, low-cost, cellulose-based biofuel production system. This is because a variety of insect lignocellulolytic systems have capabilities, efficiency, and advantages with respect to wood deconstruction that are generally lacking in most microbial lignocellulolytic systems. The challenges we currently face in the biorefinery of fuels are to acquire a deeper understanding of biomass recalcitrance and to develop an efficient conversion system.[2] Thus, given the relative novelty of this field and its interdisciplinary nature, this chapter will address various insect lignocellulolytic systems, their characteristics, potential value, and opportunities that exist for current biofuels industries.

11.2   The Diversity of Cellulose-Feeding Insects and their Lignocellulolytic Systems

Although lignocellulolytic activities were originally thought to be restricted to plants, bacteria, and fungi, evidence has accumulated in recent years for the existence of animal lignocellulases (such as cellulases, hemicellulases, and lignases), especially in cellulose-feeding insects.[1,2,6,7] Indeed, these insects belong to a group of potential candidates for exploring novel lignocellulolytic catalysts, due to their diverse and highly adapted lignocellulolytic systems that can efficiently digest various lignocellulosic foods.[7] Numerous reports on cellulolytic activities in insects have illustrated their unique mechanisms and advantages in biomass deconstruction, which are quite different from most microbial degradation systems, such as bacteria and fungi. Recent studies using advanced molecular biotechnologies, including meta-genomics, proteomics and transcriptomics, have brought new insight into the mechanisms of biomass deconstruction within these small, but complicated digestive gut systems. Furthermore, with the systems biology method, we may potentially reach a fundamental understanding of how wood-feeding insects, such as termites, degrade wood substrates in such a quick and efficient way by their highly adapted digestive systems. These systems are more successful at biomass processing than most single microbial systems in various environments.

In what follows, we will examine some representative cellulose-feeding insects, especially wood-feeding termites, and discuss the relevant lignocellulolytic systems for the efficient breakdown of recalcitrant biomass in their digestive systems.

11.2.1   Cellulose-Feeding Insects and their Digestibility

It is believed that most of the phytophagous insects are cytoplasm consumers that could not utilize cellulosic biomass in their diets, and indeed, cellulolytic capability is uncommon in insects as well as in most animals, simply because the traits required to overcome the recalcitrance of the plant cell walls are rarely advantageous to possess. However, cellulose digestion has been demonstrated in a variety of insect species from many diverse taxonomic groups including: various wood-feeding insects (woodroaches – Dictyoptera, lower and higher termites – Isoptera, various beetles – Coleoptera, and wood wasps – Hymenoptera), detritus/litter-feeding insects (*e.g.*, the immature stages of leaf-shedding aquatic insects – Trichoptera, Diptera, and Plecoptera, silverfish – Thysanura, and crickets – Orthoptera), and forage-feeding insects (beetles – Coleoptera). It has been reported that cellulose digestion has been demonstrated in more than 20 families representing 10 distinct insect orders, *e.g.*, Thysanura, Plecoptera, Dictyoptera, Orthoptera, Isoptera, Coleoptera, Trichoptera, Hymenoptera, Phasmida, and Diptera.[1] Table 11.1 lists most of reported lignocellulose-feeding insects, representing various insect taxa in order and family, as well as the evidence that has been presented for the utilization of lignocellulose *via* various means.

Insects are an important group of animals that play a significant role in recycling woody and grassy lignocellulosic biomass on Earth, although a large variation in cellulose-degradation capability exists among different insect taxa and species (11–99%, Table 11.1[7,8]). Among these cellulose-feeding insects, termites, as the key decomposers in numerous ecosystems in the tropics and beyond, are the most efficient cellulose digesters in terms of not only their efficiency in cellulose degradation, but also the total amount of lignocellulose consumed in each year (Figure 11.1[9–11]).[1] It has been reported that wood-feeding termites consume at least $3–7 \times 10^{15}$ g of lignocellulose annually[12,13] using an unbelievably large population that has been estimated to be 2.4×10^{17} individuals.[14] Regardless of the small size of a single termite (<5 mg), the biomass density of termites in tropical and subtropical regions can be so large (>50 g M^{-2}) that their impact often surpasses that of grazing mammals (0.013–17.5 g M^{-2}).[12,15]

Most people are aware of termites, either directly or indirectly because termites are voracious eaters of houses, crops, as well as almost all lignocellulosic substrates. Thus, in the public's perception, wood-feeding termites are commonly recognized as one of the pests of wooden structures because of their economic impact, and their super capability in wood digestion. However, only <4% of total termite species (~2700) in the world attack man-made products and structural materials. The large majority of termite species are not pests

Table 11.1   Lignocellulose-feeding insects.[a]

Order	Family	Representative species	Evidence for capability to digest cellulose[b]
Thysanura	Lepismatidae	Ctenolepisma lineata	AD (72-87), ^{14}C, EN
		Lepisma sp.	EN
Orthoptera	Gryllidae	Acheta domesticus	AD (41)
		Gryllus pennsylvanicus	EN
		Allonemobius socius	EN
	Acrididae	Schistocerca gregaria	EN
		Melanoplus differentialis	EN
		M. femurrubrum	EN
		Schistocerca Americana	EN
		S. damnifica	EN
		Dicromorpha viridis	EN
		Syrbula admirabilis	EN
		Chortophaga viridifasciata	EN
		Hippiscus ocelote	EN
		Spharagemon bolli	EN
	Tettigoniidae	Conocephalus strictus	EN
		Orchelimum vulgare	EN
	Copiphorinae	Neoconocephalus triops	EN
		Microcentrum retinerve	EN
		Scudderia curvicauda	EN
		S. furcata	EN
Dicyoptera	Cryptocercidae	Cryptocercus punctulatus	EN
	Blattidae	Periplaneta americana	[14]C
		P. australasia	EN
		Pycnoscelus surinamensis	EN
	Blaberidae	Gromphadorrhina portentosa	EN
Phasmida	Phasmatidae	Eurycantha calcarata	EN
Isoptera	Mastotermitidae	Mastotermes darwiniensis	EN, [14]C
	Kalotermitidae	Kalotermes flavicollis	AD (74-91)
		Cryptotermes brevis	AD (86), [13]C
		Neotermes bosei	EN
	Hodotermitidae	Zootermopsis angusticollis	AD (82), EN, [13]C
	Rhinotermitidae	Coptotermes formosanus	[14]C, EN, AD (75), DNA
		C. acinaciformis	[14]C
		C. lacteus	[14]C
		Reticulitermes flavipes	AD (91), ^{14}C, EN, DNA
		R. lucifugus	AD (96-99), EN
		R. speratus	EN, DNA
		R. santonensis	[14]C
		R. hangeni	EN
		Heterotermes indicola	AD (78-89)
	Termitidae	Macrotermes natalensis	EN
		M. bellicosus	EN
		M. subhyalinus	EN
		M. michaelseni	EN
		M. gilvus	[13]C, DNA
		Trinervitermes trinervoides	EN

Table 11.1   (*Continued*)

Order	Family	Representative species	Evidence for capability to digest cellulose[b]
		Odontotermes formosanus	EN, DNA
		Nasutitermes ephratae	AD (91-97)
		N. exitiosus	EN, [14]C
		N. nigriceps	[14]C
		Nasutitermes sp.	DNA (hindgut)
		Cubitermes orthognathus	[14]C
Plecoptera	Pteronarcyidae	*Pteronarcys proteus*	AD (11), [14]C
Trichoptera	Limnephilidae	*Pycnopsyche luculenta*	AD (12), [14]C
Hymenoptera	Siricidae	*Sirex cyaneus*	EN
		S. gigas	AD (22)
		S. phantoma	AD (31)
	Formicidae	*Atta colombica*	EN
		A. cephalotes	EN
		A. sexdens	EN
		Acromyrmex octospinosus	EN
		A. crassispinosus	EN
		A. echinator	EN
		A. subterraneus	EN
	Tenthredinidae	*Allantus cinctus*	EN
		Macremphytus tarsatus	EN
		Cladius difformis	EN
Coleoptera	Anobiidae	*Anobium punctatum*	AD (33), EN
		A. striatum	AD (31)
		Ernobius mollis	EN
		Ptilinus pectinicornis	EN
		Xestobium rufovillosum	AD (68), [14]C
	Buprestidae	*Agrilus fsucicollis*	EN
		Anthaxia corinthia	EN
		Capnodis miliaris	EN
		Capnodis sp.	EN
		Chalcophora mariana	EN
		Melanophila picta	EN
		Chrysobothris sp.	EN
	Cerambycidae	*Acanthocyinus aedilus*	EN
		Aegosoma scabricornae	EN
		Anoplophora glabripennis	EN, [13]C
		Callidium sanguineum	EN
		Cerambyx cerdo	EN
		Derobrachus brunneus	EN
		Ergates faber	EN
		Gracilia minuta	AD (33)
		Hylotrupes bajulus	AD (12-21), EN
		Isotomus speciosus	EN
		Leptura rubra	AD (33)
		Leptura sp.	EN
		Macrotoma palmata	AD (14-47)
		Monochamus marmorator	AD (27), ^{14}C, EN
		Morimus funereus	EN
		Oxymirus cursor	AD (49), EN
		Phymatodea testaceus	EN

Table 11.1   (*Continued*)

Order	Family	Representative species	Evidence for capability to digest cellulose[b]
		Physocnemum brevilineum	EN
		Plagionotus detritus	EN
		Rhagium bifasciatum	EN
		R. inquisitor	EN
		R. mordax	EN
		Saperda populnea	EN
		Smodicum cucujiforme	EN
		Stromatium barbatum	AD (30-57), EN
		S. fulvum	EN
		Xylotrechus rusticus	EN
		Apriona germari	EN, DNA
		Elaphidion mucronatum	EN
		Neoclytus a. acuminatus	EN
		Cerambycid sp.	EN
	Scarabaeidae	*Oryctes nasicornis*	AD (68), [14]C
		Sericesthis geminata	EN
		Pachnoda marginata	EN, DNA
		Phyllophaga sp.	
	Coccinellidae	*Epilachna varivestis*	AD (17-47), ^{14}C, EN
	Passalidae	*Odontotaenius disjunctus*	EN (by AFM and HPLC)
	Curculionidae	*Pissodes notatus*	EN
		Graphognathus leucoloma	EN
	Scolytinae	*Scolytus rugulosus*	EN
	Diprionidae	*Neodiprion lecontei*	EN
	Lyctidae	*Lyctus planicollis*	EN
	Tenebrionidae	*Tenebrio molitor*	EN
		Tribolium castaneum	EN
Diptera	Tipulidae	*Tipula abdominalis*	AD (19), EN, DNA

[a]This table has been modified from refs. 7 and 8. Some data for the larvae in Lepidoptera are not included in this table because of the general low enzymatic activities on cellulosic substrates that have been reported.

[b]AD = cellulose digestion demonstrated by the determination of assimilation efficiency (value given in parentheses) of ingested cellulose. [13]C or [14]C = cellulose digestion demonstrated by detection of the incorporation of carbon-14 or 13 into tissue or respiratory CO_2 following ingestion of [U-^{14}C or δ^{13}C] cellulose. EN = gut fluid demonstrated to possess enzymatic capability to degrade filter paper, cotton, Avicell, or some other form of insoluble crystalline cellulose. DNA = cellulolytic genes identified from different symbiont meta-genomes/transcriptomes and termite genomes.

under any circumstances, instead, they carry out wholly beneficial activities. The extent of their benefits is only just gaining recognition and has scarcely been economically evaluated.[1]

Wood-feeders, especially wood-feeding termites and beetles, dominate the majority of biomass consumed by wood-feeding insects listed in Table 11.1. Specifically, the direct and indirect wood-feeding termites account for approximately >80% of total termite species, leading to a distinct role in recycling

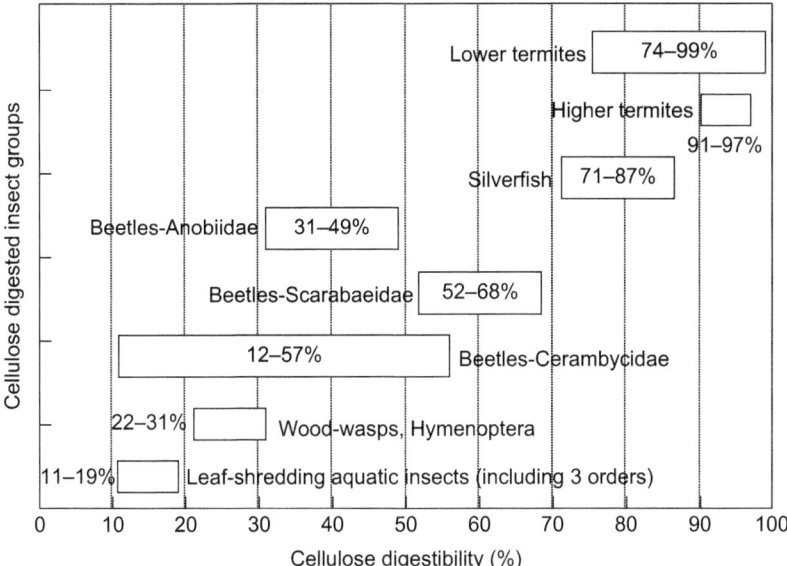

Figure 11.1 The capability of cellulose digestion in some insects. Approximate cellulose digestibility determined by comparing the cellulose contents of food and feces.
Modified from refs. 9–11. Reproduced from ref. 1 with permission.

wood or other lignocellulosic biomass on our planet.[13] Currently, seven termite families in the order Isoptera are classified, which includes 283 genera with approximately 2700 species.[16] Six of these families belong to the phylogenetically lower termites (Figure 11.2[16–18]). One of the main characteristics that separates lower termites from the phylogenetically higher termites is the presence of symbiotic protozoa in their enlarged hindgut, which helps them digest lignocellulose.

Lignin in plant cell walls is a source of useful chemicals and also the major barrier to the saccharification of lignocellulosic biomass by cellulose-feeding insects for energy. It is noteworthy that most lignocellulose-feeding insects are unable to deconstruct the aromatic polymer lignin from woody or herbaceous biomass. Without an efficient modification or disruption of lignin structure, most of the cellulose-feeding insects would not be able to biodegrade the polysaccharide components of lignocellulose substrates, such as lepidopteran which only possess a weak capability to use cellulose food,[7] simply because lignin structures serve to protect cellulose and hemicellulose from biodegradation. However, significant pretreatment activity on lignin components seems to exist in some of the cellulose-feeding insects, such as the wood-feeding termites and beetles, which allow a large quantity of cellulose or hemicellulose carbohydrates to be converted into a variety of monosaccharides for insect use.[19–23] A study by Itakura *et al.*[11] of the lower termite, *Coptotermes formosanus*, indicated that this species assimilates 25–30% of total lignin polymers

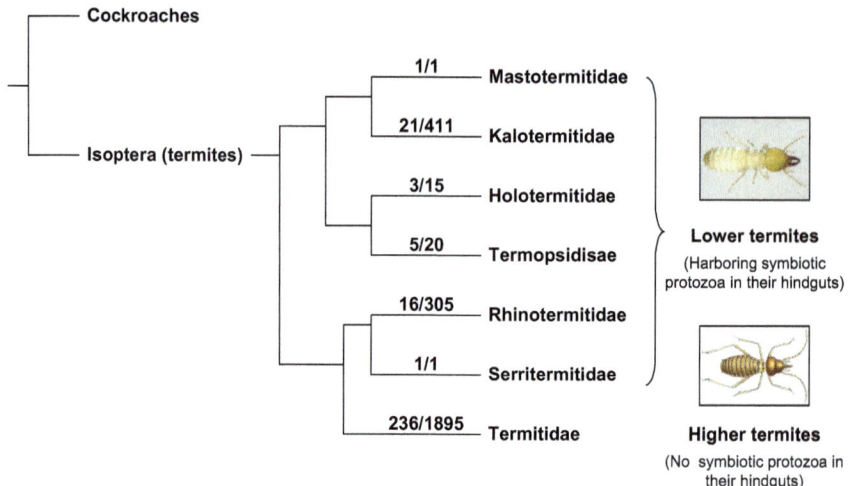

Figure 11.2 Phylogenic scheme of termite evolution showing the presumed relationships among the seven termite families. The numbers superimposed on
the single branches represent the currently recognized genera/species of
the respective families.
Adapted from refs. 16–18. Reproduced from ref. 1 with permission.

(Figure 11.3[11]). However, in contrast to this data, the fungus-growing termites
(Termitidae family) can completely degrade lignin polymers and the polysaccharides with help from an exo-symbiotic fungus, *Termitomyces*, cultured in
their nest.[13] Further discussion of lignin degradation in termite gut systems as a
unique pretreatment process for lignocellulosic biomass is available in Chapter
10 of this book, as well as in the review by Scharf and Boucias.[24]

11.2.2 Lignocellulolytic Systems in Cellulose-Feeding Insects

Lignocellulose digestion in a biological system is a complex process, involving a
suite of enzymes with diverse modes of action. Although the production of a
complete cellulolytic system is common among wood-rotten bacteria and fungi,
it is unusual among insects and other animals. In general, cellulose digestion in
animals is most often mediated by symbiotic cellulolytic microorganisms.[25]
However, recent investigations have also indicated that invertebrates, such as
arthropods and gastropods, can produce their own active cellulases.[6,26] It is
believed that each cellulose-feeding insect has to acquire or develop a lignocellulolytic system to cope with the biomass recalcitrance, which may extend
beyond cellulases to lignases and phenol-oxidases. Indeed, cellulose-feeding
insects require sophisticated cooperation between their hosts and the gut symbiotic microorganisms either inhabiting their guts (endo-symbiosis) or their
habitats (exo-symbiosis), to efficiently break down lignocellulosic biomass.[1,27,28]
    The intestinal tract of cellulose-feeding insects harbors a variety of symbiotic
microorganisms, from bacteria, archaea, fungi to protista, many of which are

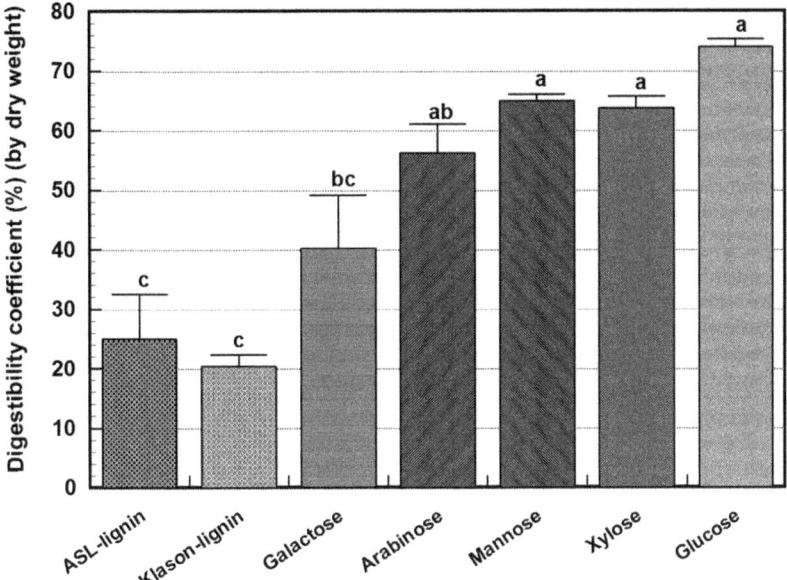

Wood components consumed by Coptotermes formosanus

Figure 11.3 Wood component digestibility (%) by the lower termite species, *Copto-termes formosanus*. The letters above each bar represent the mean separation, and bars with same letter did not differ significantly at $\alpha = 0.05$, Tukey HSD (Honestly Significant Difference). The error bars indicate 1 SEM (Standard Error of the Mean).
Adapted from ref. 11. Reproduced from ref. 1 with permission.

believed to be engaged in the processing of cellulosic food.[29] It seems that cellulose-feeding insects must have direct or indirect associations with symbiotic microorganisms, in the form of endo-symbiosis or exo-symbiosis, to deconstruct lignocellulosic biomass. These symbiotic relationships between host insects and microbiota are often obligatory. For nearly a century, the idea that intestinal symbionts were the sole contributors to insect lignocellulose digestion has remained popular and compelling. However, growing evidence, at both the gene and protein levels, has further confirmed that cellulose-feeding animals can produce their own catalysts, and that these are significantly involved in biomass degradation.[2,26,30–32] Thus, the hypothesis that insect-derived enzymatic activities must play an important or indispensible role in cellulose degradation has been proposed. From another perspective, there is no evidence of an insect species that demonstrates a complete set of host-derived lignocellulases independently capable of degrading plant cell walls. In fact, a realistic estimate regarding the relative contributions of a host insect and its gut/nest symbionts to the efficient deconstruction of recalcitrant biomass food has not been made.

In nature, some unusual cellulolytic animal systems *e.g.*, some crustaceans (such as marine isopod and crayfish) do not rely on symbiotic microbes to help digest cellulosic food. It has been reported that terrestrial wood-boring isopods and a marine wood-inhabiting amphipod, a member of the phylum Crustacea, maintain their guts completely free of microbial symbionts.[32,33]

Clearly, cellulose-feeding animals have evolved various lignocellulolytic systems and strategies to deal with the recalcitrance of biomass. Based on the origins of lignocellulases and the relationship between insect hosts and the microorganisms hosted by insects, Sun and Zhou[1] have proposed three types of lignocellulolytic systems for cellulose-feeding insects: (1) symbiont-dependent (microorganism-affiliated, *e.g.*, gut endo-symbiosis) systems; (2) symbiont-independent (insect tissue-affiliated, *e.g.*, termite salivary gland) systems; and (3) systems that acquire the necessary enzyme system *via* food ingestion (*e.g.*, from fungus *via* exo-symbiosis).[1] Two different lignocellulolytic systems can be incorporated into an animal's digestive system, such as wood-feeding termites, to synergize the conversion of rigid plant cell walls (dual lignocellulolytic systems).[28] Numerous investigations have indicated that both host insect tissues, such as salivary glands and midgut, and the symbiotic microbiota in insect guts, produce an array of different complex cellulases that differ in their functions, but complement one another, for the efficient and synergistic degradation of cellulose.[30,34–42] In an interesting investigation, Scharf *et al.*[28] quantified the synergistic collaboration between a host insect and its gut symbionts for the wood-feeding lower termite, *Reticulitermes flavipes*, on pine wood lignocellulose. Additional investigations on termite gut cellulases and their encoding genes have supported the hypothesis that both gut symbiont-dependent and independent lignocellulolytic systems in an animal's digestive system work together through a well-coordinated collaboration.[1,2,3,6,28,37,43,44] It has been confirmed in recent years that wood-feeding termites possess a dual cellulolytic system: in lower termites the cellulases are produced by both the insect host and its gut flagellates, whereas in higher termites, both host cellulases and hindgut bacteria are involved in wood digestion.[1,27]

The third important lignocellulolytic system in some insects is referred to as an "acquired enzyme system", where insects regularly encounter fungi in their environments or culture symbiotic fungi in their nests and commonly consume fungal tissues to obtain essential lignocellulases (an exo-symbiotic association). This unique pattern, although certain arguments still persist, is generally demonstrated by the fungus-growing animals either as an obligate association with the symbiotic fungi cultured in their nests (*i.e.*, termites, Attine ants, and the siricid woodwasps) or as a facultative association with the fungi encountered in their environments (*e.g.*, cerambycid beetles, the detritus feeders).[1,8] With this exo-symbiosis system between hosts and fungi, cellulose-feeding animals sustainably obtain a diverse array of active and lignocellulolytic enzymes by ingesting the fungal fruiting body and the comb substrate,[8,45–47] combining with insect-derived cellulolytic enzymes in their guts to accomplish the complete degradation of lignocellulose. In recent years, less attention and effort has been paid to this lignocellulolytic system due to the relative difficulty in sampling as well as the methods employed to identify/track the enzyme origins.

In general, lignocellulose-feeding insects have at least one of these three lignocellulolytic systems to deal with cellulosic diets. Indeed, there is no evidence that any cellulose-feeding insect is completely independent from symbionts for cellulose degradation. However, it is not always clear which system prevails over another, or how well a dual lignocellulolytic system is being coordinated, or how well each process complements each other, for each essential process performed in insect guts.[1,2]

11.3 Lignocellulose-Degradation Machinery

11.3.1 Depolymerization of Lignocelluloses

Plants capture solar energy and convert CO_2 and H_2O to polysaccharides $(CH_2O)_n$. Plants store thermochemical energy predominantly in plant cell walls, which eventually become a source of biomass following the death of the plant. The resultant biomass is primarily composed of three types of polymer, cellulose, hemicellulose and lignin, all of which link together and form complex compounds called "lignocelluloses".[1,6,48]

The embedding of celluloses in hemicelluloses and lignin *via* cross-linking to form a complex three-dimensional structure makes lignocelluloses completely insoluble in water and well protected from enzymatic and/or chemical attacks. Although it is chemically stable and resistant to decomposition, many organisms can thrive on this planet using lignocelluloses as their sole or main food source, including protozoa, bacteria, fungi, herbivores, and wood-feeding insects. More than 100 insect species have been reported to possess lignocellulose-degradation capabilities, and they have developed different strategies involving a sophisticated symbiosis between insects and symbiotic microorganisms either in their guts (endo-symbiosis) or in their habitats (exo-symbiosis) to break down the recalcitrant plant cell wall.[1,49–51]

Lignocellulases, both symbiotic and endogenous, are lignocellulolytic enzymes involved in the entire process of lignocellulose decomposition, starting with the depolymerization of hydrophobic lignin polymers, proceeding to hemicellulose degradation, and finally releasing fermentable single unit sugars. During the past century, scientists have devoted enormous amounts of time and effort to unlocking the mystery of lignocellulose degradation and to search for the 'holy grail' of enzymatic digestion machinery.[6] Most animals, however, cannot digest cellulose independently without the assistance of cellulolytic microorganisms. As we have discussed above, there are three lignocellulolytic systems: symbiont-dependent, symbiont-independent, and an acquired system using lignocellulolytic food sources.[1] Termites, the most efficient biomass decomposers, in terms of their efficiency and the total amount of lignocellulose consumed each year, possess convoluted machineries involving cellulolytic enzymes from all three systems.

The enzymatic breakdown of lignocellulose, typically, involves three steps. The heterogeneity and stereo-irregularity of lignin polymers protect celluloses and hemicelluloses from chemical and/or enzymatic degradation,[1,52] therefore, the detachment of non-carbohydrate lignin polymers occurs first,[19,53] followed by the depolymerization of hemicelluloses, and thereafter the decomposition of celluloses to eventually release glucose monomers. For the critical first step of lignocellulose decomposition, the lignin-degradation machinery in insects is mainly composed of oxidative enzymes such as peroxidases, oxidases and laccases. Other enzymes involved in xenobiotic metabolism could contribute to lignin degradation, including alcohol dehydrogenase, catalase, superoxide dismutase, methyltransferases, cytochrome P450, cytochrome P450 co-factor, epoxide hydrolase, reductase, glutathione-*S*-transferase, and esterase.[1,54]

The second step in the overall lignocellulose-digestion process, hemicellulose degradation, requires the combined effects of endo-β-1,4-xylanases, exo-β-1, 4-xylanases, β-xylosidases, and several accessory enzymes, including α-arabinofuranosidases, α-glucuronidases, acetylxylan esterase, ferulic acid esterase, and *p*-coumaric acid esterase.[1,55,56] Specifically, the conversions accomplished by each of these enzymes are: (1) endo-xylanases hydrolyze the β-1,4-xylose linkages in the xylan backbone; (2) exo-xylanases hydrolyze reduced β-1, 4-xylan linkages releasing xylobiose; (3) β-xylosidases act on xylobiose to liberate xylose and other short-chain oligosaccharides; (4) α-arabinofuranosidases hydrolyze terminal non-reducing α-arabinofuranose from arabinoxylans; (5) α-uronidases release α-glucuronic, α-mannuronic and α-galacturonic acids; and (6) esterases hydrolyze the phenolic ester linkages between the xylose units of the xylan and acetic acid or between arabinose side-chain residues and phenolic acids, such as ferulic acid and *p*-coumaric acid.[1,55]

The final step is the depolymerization of the celluloses after lignocellulose deconstruction and the generation of glucose monomers. Three primary enzymes are involved in cellulose depolymerization including endo-β-1,4-glucanases, exo-β-1,4-glucanases and β-glucosidases. It is believed that these enzymes work synergistically to efficiently breakdown the cellulose polymer into glucose monomers in the following sequence: (1) endo-glucanases attack the feasible amorphous regions of cellulose and cleave the β-1,4-glycosyl linkages of the primary cellulose backbone thus randomly releasing glucose, cellobiose, cellotriose and other longer oligomers, in particular, generating more termini on cellulose fibers for the exo-glucanses to target; (2) exo-glucanses act on the non-reducing or reducing terminal regions of polysaccharides to liberate either glucose or cellobiose, and additionally to loosen the tightly assembled crystalline regions of celluloses for endo-glucanase degradation; and (3) β-glucosidases hydrolyze cellobiose and cellotriose from the non-reducing ends to produce glucose monomers.[1,49]

11.3.2 The Distribution of Lignocellulose-Degradation Machinery

It was previously believed that most animals could not utilize lignocellulose independently without the aid of lignocellulolytic microorganisms. During the past two decades, evidence has indicated the existence of endogenous lig-nocellulases in a wide range of invertebrates.[57,58] Several insect species from Isoptera, Thysanura, Coleoptera, Blattodea, and Hemiptera can digest woody materials using lignocellulolytic enzymes secreted from their salivary glands and/or midgut.[59,60]

Insect digestive systems typically include the mouth, oesophagus, salivary glands, foregut, midgut and hindgut (Figure 11.4).[6] First, the woody material is chewed and masticated by mandibles and ground into an assembly of fine particles called a "bolus". This mechanical degradation process increases the surface area of the food to allow more efficient breakdown by the lig-nocellulolytic enzymes.[6,60] The foregut is composed of the oesophagus, crop and salivary glands, which secrete endogenous lignocellulolytic enzymes mainly for lignin depolymerization. The midgut is a slender tube of uniform diameter distally connected to 4–12 Malpighian tubules.[62] Peritrophic membranes

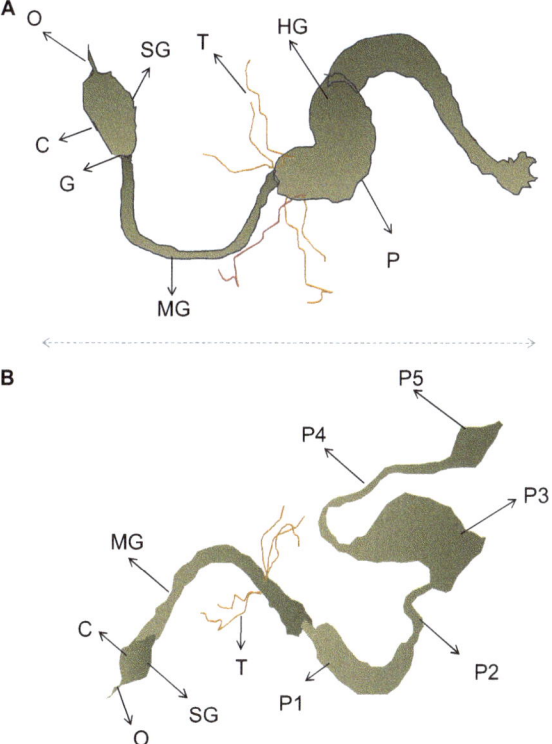

Figure 11.4 A schematic drawing of the termite digestive tract. (a) *Reticulitermes flavipes*, a lower termite (b) *Nasutitermes costalis*, a higher termite. The hindgut of higher termites can be divided into five successive segments following ref. 61. O = oesophagus, SG = salivary gland, C = crop, G = gastric cecum, MG = midgut, MX = mixed gut, T = Malpighian tubules, HG = hindgut, P = paunch, and P1–P5 = proctodeal segments.

surrounding the food material are produced by the midgut, in which limited lignocellulose degradation occurs. Analogous to the vertebrate kidney, Malpighian tubules connect at the junction of the midgut and hindgut and participate in waste excretion and nitrogen recycling.[49] All higher termites possess a prominent hindgut which can be divided into five successive segments: the proctodeal segment (P1), the enteric valve (P2) controlling the entrance of food, the fermentation chamber or "paunch" (P3) harboring symbiotic flora, as well as the colon (P4) and rectum (P5) (Figure 11.4).[61] The paunch is generally anaerobic, but does possess a microoxic zone around its periphery. The hindgut, housing the majority of gut symbionts, is the primary tissue where most cellulose degradation as well as fermentation occurs.[49]

11.3.3 The Discovery of Insect Lignocellulases and their Industrial Potential

Cleveland[63] demonstrated that the lower termites *Reticulitermes flavipes* rely on intestinal protists for wood consumption, suggesting a mutualistic relationship between termites and their gut microorganisms. Yokoe,[64] however, showed that termites retain partial cellulase activity even after the removal of gut fauna, implying that the termite itself can produce its own cellulases. Yamaoka and Nagatani[65] also documented the participation in cellulose digestion of both endogenous and protozoan cellulases, and notably they proposed the possible dual origin of digestive enzymes in lower termites. Recently, digestome analysis of *R. flavipes* identified an endo-glucanase, named "Cell-1", from its salivary glands. Cell-1, a glycoside hydrolase family 9 (GHF9) member, has significant sequence similarity to cellulases identified from termite, cockroach and bivalve genomes.[41,66] While, the other putative cellulase, Cell-2, which belongs to GHF7, has strong sequence similarity to endo-glucanases previously identified from termite gut symbionts.[41] Lo and colleagues[67] proposed that GHF9 subgroup E2 genes from termites, abalone, and sea squirt have a common ancient origin. Davison and Blaxter[57] later acknowledged that GHF9 genes are present in at least five metazoan phyla with a single ancient origin, suggesting that eukaryote GHF9 genes are probably derived from an ancient eukaryote gene, not from horizontal gene transfer. This contradicts the results of a previous phylogenetic analysis, which showed that termites have acquired their cellulolytic endo-glucanases by horizontal gene transfer from prokaryotes.[68]

Using Carbohydrate-Active EnZyme (CAZy, http://www.cazy.org) and Fungal Oxidative Lignin Enzyme (FOLys, http://foly.esil.univ-mrs.fr) databases, lignocellulose-degrading enzymes can be categorized into glycoside hydrolase (GH), polysaccharide lyase (PL), carbohydrate esterase (CE), lignin oxidase (LO), and lignin-degrading auxiliary enzyme (LDA) families according to their sequence and structural homology.[69–71] There are 14 families of glycosyl hydrolases (GHFs) with cellulositic catalytic capability, of which 5 families are present in the Animalia kingdom (GHF5, GHF6, GHF9, GHF10, and GHF45).[57,72] Family members share not only structural motifs and catalytic machinery, but also an evolutionary origin.[6,72] Most of the insect-derived endo-glucanases are members of the GHF9 family. The GHF9 homologs have been identified in the genomes and transcriptomes of various wood-feeding insects including termites and cockroaches.[6,41,60] The cellulase genes in gut protists of termites belong to several GHFs. From meta-transcriptome analysis of the entire protistan gut microbiota, nearly 10% of the 910 expressed sequence tags (ETSs) were identified as putative cellulases,[60,73] of which, nearly half were homologs of GHF7 exo-glucanases, the remaining half belonging to the families of GHF5, GHF7, GHF45, GHF8, GHF11, GHF3, GHF62 and GHF26.[60,73] Transcriptome analysis of the salivary glands of the lower termite *Hodotermopsis sjoestedti* identified that more than 10% of the 851 ETSs were transcripts for endo-glucanases.[60,74] Similar results were obtained from tissue-specific transcriptome studies of *R. flavipes*.[60,66]

Lignocellulosic biomass, including agricultural and forestry by-products, has the potential to become a major source of fermentable sugars for biofuels. Biofuel, an environmental friendly 'green' energy, is a promising solution to the current problem of the remaining finite amount of fossil fuel and the continuously increasing energy demand from human consumption as well as industrial development. With the dawn of the "peak oil era" (the era with the maximum rate of oil production), there is a greater need for the development of sustainable and renewable energies.[75] Crocker and Andrews[76] predicted that biofuels would become the largest single energy source, supplying about 50% of all sources by 2030.

Currently, there are two primary routes to break down lignocellulosic biomass and to liberate sugar monomers, including thermochemical pathways, such as pyrolysis and gasification, and biochemical pathways involving cellulose hydrolysis.[77] The thermochemical approach requires intensive energy input, which partially negates the energy output generated from the fermentable sugars. On the other hand, enzymatic hydrolysis of sugars requires much less energy input, which renders the biochemical pathway attractive as long as cost-effective and highly efficient lignocellulases can be developed. The lignin-preconditioning procedure (pretreatment) is a critical first step for the complete degradation of lignocellulose. Despite a number of limitations, including low specific activity and being susceptible to product inhibition and environmental factors, cellulases derived from fungal and bacterial systems have been developed and are widely used in industrial applications.[78] Lignocellulolytic enzymes, which can be used for the depolymerization of lignins and hemicelluloses, however, have yet to be developed. Phytophagous insects represent the best available reservoir for the identification of novel lignocellulolytic enzymes for the enzymatic pretreatment of biomass.

The dawn of the "genomic era" has provided scientists with new tools and insights into lignocellulose degradation.[59,79] Meta-omics has started to unveil the mystery of some symbiotic systems associated with wood-feeding insects,[60] including the higher termites *Nasutitermes ephratae*[80] and lower termites *Reticulitermes speratus*.[73] To reduce costs and increase yields in lignocellulosic ethanol production, more than 100 phytophagous species belonging to 10 taxonomic orders have been investigated and screened for industrial application.[7] The success of heterologous expression of termite and termite-associated symbiotic cellulases in multiple systems, including bacterial, yeast, and insect cultures, has shed light on the development of efficient and cost-effective bioreactors and biorefineries for biofuel production.[81,82]

The first insect cellulase-encoding gene, RsEG, was identified as an endo-β-1,4-glucanase (EC 3.2.1.4) from a lower termite *R. speratus*.[68] To date, the majority of these insect-derived lignocellulases are secreted from salivary glands and/or the midgut, and function as either endo-β-1,4-glucanases, which belong to GHF9, or as β-glucosidases, which belong to GHF1 (Table 11.2[41,66,68,83–94]). Specifically, NtEG, an endo-β-1,4-glucanase from *Nasutitermes takasagoensis*, was found predominantly in the midgut, whereas, RsEG2, the same endo-β-1,4-glucanase from *R. speratus*, was exclusively secreted from the salivary glands,[83] which is consistent with Cell-1, a homolog in *R. flavipes*. The recombinant Cell-1, which was heterologously expressed in both *E. coli* and insect culture, maintained endo-β-1,4-glucanase activity, which was comparable to the native enzyme, with optimal activities at pH 6.5–7.5 and 50–60 °C.[81] Recently, Uchima *et al.*[94] reported the expression and characterization of an endogenous β-glucosidase, G1mgNtBG1, from *N. takasagoensis*. As a member of GHF1, G1mgNtBG1 has a relatively high temperature optimum of 65 °C and a high level of glucose tolerance. In comparison to Novozym 188, the most common commercially available β-glucosidase produced by *Aspergillus niger*, G1mgNtBG1 was more thermostable after 5 h of incubation at 60 °C and exhibited better tolerance to glucose inhibition.[94] The other successfully expressed and characterized β-glucosidase was G1NkBG, which was secreted from the salivary glands of *Neotermes koshunensis*. G1NkBG was most active toward laminaribiose and pnitrophenyl-β-D-fucopyranoside between pH 5.0 and 9.0. Furthermore, β-glucosidase activity was stimulated by the addition of glucose, which makes G1NkBG a gene-of-interest for industrial applications, especially for bioethanol production.[92,95] To better understand the role of specific amino acid residues, Ni *et al.*[96] applied site-directed mutagenesis and found that G147, a glycine residue, played a significant role in maintaining the endo-β-1,4-glucanase activity. In addition, heterologous over-expression of a mutant termite cellulase in *E. coli* yielded fold higher endo-β-1,4-glucanase activity than the wild type.[97]

11.4 The Potential of Cellulose-Feeding Insects for Viable Biofuels

The potential attractiveness of bioconversion technologies for liquid fuels is related to the idea that bioconversion strategies are relatively scale-neutral and may require lower capital and energy inputs than thermal conversion pathways.[98] However, the current state of technology with respect to the biological conversion of biomass is still far away from being mature for large-scale

Table 11.2 Endogenous lignocellulases derived from termites.

Gene name	Description	GHF[a]	Termite species	Distribution	Accession no.	Reference
RsEG	Endo-β-1,4-glucanase	GHF9	*Reticulitermes speratus*	Salivary gland	AB008778	68
RsEG2	Endo-β-1,4-glucanase	GHF9	*Reticulitermes speratus*	Salivary gland	AB019095	83
NtEG	Endo-β-1,4-glucanase	GHF9	*Nasutitermes takasagoensis*	Midgut	AB013272	83
NwEG	Endo-β-1,4-glucanase	GHF9	*Nasutitermes walkeri*	Midgut	AB013273	83
CfEG1a	Endo-β-1,4-glucanase	GHF9	*Coptotermes formosanus*	Salivary gland/midgut	AB058667	84
CfEG1b	Endo-β-1,4-glucanase	GHF9	*Coptotermes formosanus*	Salivary gland/midgut	AB058668	84
CfEG2	Endo-β-1,4-glucanase	GHF9	*Coptotermes formosanus*	Salivary gland/midgut	AB058669	84
CfEG3	Endo-β-1,4-glucanase	GHF9	*Coptotermes formosanus*	Salivary gland/midgut	AB058670	84
CfEG4	Endo-β-1,4-glucanase	GHF9	*Coptotermes formosanus*	Salivary gland/midgut	AB058671	84
Cel1	Endo-β-1,4-glucanase	GHF9	*Mastotermes darwinensis*	Salivary gland	AJ511339	85
Cel2	Endo-β-1,4-glucanase	GHF9	*Mastotermes darwinensis*	Salivary gland	AJ511340	85
Cel3	Endo-β-1,4-glucanase	GHF9	*Mastotermes darwinensis*	Salivary gland	AJ511341	85
Cel4	Endo-β-1,4-glucanase	GHF9	*Mastotermes darwinensis*	Salivary gland	AJ511342	85
Cel5	Endo-β-1,4-glucanase	GHF9	*Mastotermes darwinensis*	Salivary gland	AJ511343	85
NtEG2[b]	Endo-β-1,4-glucanase	GHF9	*Nasutitermes takasagoensis*	Midgut	AB118803	86
CelI-1	Endo-β-1,4-glucanase	GHF9	*Reticulitermes flavipes*	Salivary gland/foregut	AY572862	41
CaEG3(CfEG3a)	Endo-β-1,4-glucanase	GHF9	*Coptotermes formosanus*	Salivary gland/midgut	ACI45756	87
CaEG5(CfEG3b)	Endo-β-1,4-glucanase	GHF9	*Coptotermes formosanus*	Salivary gland	ADB12483	88
Cf β-glucosidase	β-Glucosidase	GHF1	*Coptotermes formosanus*	Salivary gland	GQ911585	89, 90
G1NkBG	β-Glucosidase	GHF1	*Neotermes koshunensis*	Salivary gland	AB073638	91, 92
RfBGluc-1	β-Glucosidase	GHF1	*Reticulitermes flavipes*	Salivary gland/foregut	ADK12988	66, 93
RfBGluc-2[b]	β-Glucosidase	GHF1	*Reticulitermes flavipes*	Salivary gland	FL635576	66, 93
G1mgNtBG1	β-Glucosidase	GHF1	*Nasutitermes takasagoensis*	Midgut	AB508958	94

[a] GHF = glycoside hydrolase family.
[b] Denotes partial cDNA sequences. The RfBGluc-2 partial cDNA sequence (FL635576) was based on a single clone composed of 496 nucleotides.[66]

applications due to its efficiency and processing economics. To improve upon our current strategy and technologies, it appears that we need to explore/learn from other sound cellulolytic systems, instead of insisting on traditional bio-catalysis and fermentation technologies derived from some limited bacterial and fungal cellulolytic systems.[1,2] Wood-feeding insects surviving on lig-nocellulosic biomass as their only food source demonstrate some unique ad-vantages, and can degrade lignocellulose much more efficiently than most single microorganisms due to their special and advanced lignocellulolytic systems, which may offer possible options for realizing biomass conversion in an efficient and economical way.[6,99] Here, we will outline the potential of cellulose-feeding insects for viable biofuels with a focus on wood-feeding termites.

11.4.1 A Reservoir of Functional Genes and other Resources

Most cellulose-feeding insects harbor diverse and unique microbial populations in their gut systems that are critical for cellulose deconstruction processing. Molecular studies have disclosed that the diversity, especially of bacteria, is far greater than previously expected and that the gut community consists of mostly novel and unique microbial species.[100] Except for some types of cellulose-feeding animals, such as wood-feeding termites, silverfish, and a few species of earthworm, animals, including most insects, cannot directly utilize lig-nocellulosic biomass as their food resource because they produce an incomplete set of lignocellulases.[101] These species must have direct or indirect associations with microorganisms to consume this food,[1,101] and these microorganisms may be present either in their gut systems or in their nest environments. Termites, as a typical type of cellulose-feeding insects, have developed a unique hindgut flora consisting of bacteria, Archaea, archaezoa, yeasts and probably fungi. It has also been reported that fungus-growing termites are very successful in the almost complete deconstruction of lignocellulose in a sophisticated cooperation with basidiomycetes fungi cultivated in their nests. Although the wood-feeding termite gut provides only a tiny, microscale habitat, it is a reservoir of mostly novel and unique microbial complexity that may have some special functions, found nowhere else in nature.[100] Wood-feeding termites harbor diverse and unique microbial symbionts, with densities reaching up to 10^{11} cells per mL, for the digestion and utilization of their food, the highly recalcitrant lignocellulose. To date, the accumulated evidence has confirmed that microbial symbionts are present in each of the cellulose-feeding insects, where they have various func-tions in biomass degradation.

It is noteworthy that, along with the application of various classical enrichment culture techniques and recently meta-genomics, as well as other-omics approaches, many cellulose-feeding insects and their symbionts are po-tentially unique resources of functional genes for industrial applications. Culture-independent molecular approaches in the past two decades have en-hanced our ability to assess natural microbial communities, and have been successfully applied to understanding the diversity, structure, and evolution of the microbiota in termite guts, as well as the roles of individual symbionts.[100]

The first meta-genomic analysis of the inhabitants of a *Nasutitermes* termite gut provided insight into the feat of biomass-to-energy conversion, revealing a rich diversity of bacterial genes encoding hitherto unknown glycosyl hydrolases that consist of more than 100 families of proteins relevant to breaking the glycosidic bonds of carbohydrates or non-carbohydrate entities.[80] Based on this milestone investigation, the authors' sequence analysis suggested that many of the gene products fall within the GHFs specializing in the degradation of carbohydrates, where there are more than 500 genes related to the enzymatic deconstruction of cellulose and hemicellulose. Many of the genes identified by this investigation were assigned to one of two groups of bacteria—the fibrobacters, relatives of the cellulose-degrading bacteria found in the intestines of cows and other ruminant animals, and spirochetes, primarily relevant to sugar fermentation and nitrogen-fixation functions.[102] Recent advances in -omics technologies have enabled scientists to explore microbial symbionts in the gut systems of cellulose-feeding animals in an unprecedented way. The high-throughput meta-genome, meta-transcriptome, and meta-proteome analysis of microbial populations will allow the molecular-, organism-, and population-level investigation of how the chemical and biological processes are enabled, and controlled, and how they evolved.[79,103] Lignocellulose degradation in the termite symbiotic system is further discussed in Chapter 8 of this book. For more information, additional reviews of molecular studies on the intestinal microbial community, especially for wood-feeding termites, can be found in refs. 100, 101 and 104–109.

In addition to the formidable unculturability of most of gut symbionts in the gut systems of cellulose-feeding insects, a variety of cultivable microbial communities (another set of potentially good resources of functional genes for industrial applications) also generally co-exist with those culture-independent microbiota in gut systems that assist the cellulose-degradation processing or other important metabolic processes. Advanced molecular analyses and other comprehensive and elaborate culture-dependent studies in the past decade have gradually unveiled the complex nature of the intestinal microbiota. It has been reported that the dense gut microbiota in termites can include a variety of microorganisms from the domains Bacteria, Archaea, flagellates, and also yeasts and fungi.[110] Although several hundred microbial strains have been obtained in pure culture from cellulose-feeding insect guts, the majority of cultivable microbial symbionts have not yet been isolated.[111] Most of the bacterial isolates obtained in pure culture from termite guts have been identified as being gram-positive bacteria and members of the Proteobacteria, and Bacteroides/Fusobacterium branches, as reviewed by König *et al.*[111]

In the past decade, many investigations have demonstrated that the termite gut is a rich reservoir of mostly novel and valuable microorganisms. It has been reported that the bacterial communities in the gut of *Reticulitermes* termites are extraordinarily diverse, and can be categorized into 11 divisions in the bacterial domain, most of them from as-yet uncultured symbionts.[112] A recent survey of host and symbiont transcriptomes from the *R. flavipes* gut revealed over 175 genes encoding cellulase, hemicellulase, candidate lignase, and other potentially relevant digestive enzymes.[49] Since termites have developed a sophisticated and efficient lignocellulolytic system, as described in Section 11.2.2, their intestinal tracts are not only a model community for digestive symbiosis, but also an attractive microbe and gene source for potential application in viable biofuels.[113] Although a large proportion of the symbionts may be uncultivable without special -omics tools and a great deal of effort, recent advances in culture-independent molecular techniques would make it possible to retrieve and use some functional genes buried in termite guts as well as in their nest environments.[79,80,104,114]

11.4.2 The Advantages of Insect Lignocellulolytic Systems

In addition to cellulosic material decomposition, which is achieved by most cellulose-feeding animals, termite lignocellulolytic systems can also uniquely cope with the deconstruction of lignin-derived compounds, lignin being a significant component of lignocellulose. The lignin content of ingested cellulosic diets is particularly important for efficient cellulose and hemicellulose utilization by wood-feeding insects because it is tightly associated with cellulose and other plant polysaccharides. Lignin breakdown in the termite gut requires oxygen,[115] a fact that has been supported by evidence that various compartments in termite gut systems are not completely anaerobic environments, especially the foregut and midgut.[53] This is a significant advantage over ruminants' digestive systems because the cow's stomach is generally an anaerobic environment that cannot support lignin-degrading processing. No mechanism is known for the cleavage of the non-hydrolyzable bonds in the lignin polymer in the absence of oxygen.

In general, the chemical disruption of lignin-derived compounds is an oxidative process by lignases (*e.g.*, laccase, peroxidase, esterase) requiring molecular oxygen, which is in conflict with the process of microbial fermentation of polysaccharides in the hindgut because high titers of oxygen are harmful to the anaerobic microorganisms, such as cellulolytic protozoa, methanogenic and acetogenic bacteria.[116] Amazingly, termite lignocellulolytic systems are well designed to cope with this challenge, and they integrate the following strategies: (1) the spatial separation of lignin (in the fore- and midgut) and cellulose degradation (primarily in the hindgut) into oxic and anoxic gut regions, respectively; (2) using oxygen-consuming bacteria (facultative and strictly anaerobic) to maintain an anoxic microhabitat in the hindgut;[117,118] and (3) having an anaerobic gradient system in the hindgut, where oxygen penetrates 50–200 mm into the periphery of the hindgut lumen, leaving only the central portion of the dilated compartments anoxic.[53,108] Clearly, the termite gut structure and its microhabitat, as well as the mechanisms of the lignocellulolytic system on biomass processing are the result of extensive evolution to cope with the challenges of efficient and quick biomass conversion. For most wood-feeding termites, the food substrates have to pass rapidly through their gut systems to obtain adequate energy due to the small scale of their gut size (0.5–10 mL gut fluid), Thus, the mean retention time of digesta (wood particles) in termite gut systems is estimated to be 24–26 h.[15] These facts have a significant influence on the function/mechanism of termite lignocellulolytic systems as well as the composition/contribution of symbiotic microbial flora in termite gut systems.[1,13] In recent years, significant research effort has been directed towards understanding the particular strategy of processing lignin compounds by termite lignocellulolytic systems,[1,19–21,24,53,119] which will potentially promote advanced technology development in biomass pretreatment for industrial applications (see Chapter 10 for details).

As discussed above, a so-called "dual lignocellulose-digesting system" (a symbiont-dependent and symbiont-independent lignocellulolytic system) is generally observed in the wood-feeding termites[1] and it synergistically operates to cope with the deconstruction processing of lignocellulosic food, which presents some unique system advantages to enable efficient biomass conversion.[28] The use of integrative molecular biology and biochemistry studies has provided considerable evidence of host digestion capabilities (symbiont-independent) and their contributions have accumulated, now extending beyond cellulases to lignases and phenol-oxidases for potentially multiple levels of synergistic action among various enzymes or between two lignocellulolytic systems (symbiont-dependent and symbiont-independent).[66,120,121] In recent years, Scharf and his colleagues have confirmed that the host termite and its gut symbionts collaborate synergistically during wood processing, rather than additively for lignocellulose digestion,[121] which suggests that the lignocellulolytic system in wood-feeding termites has some unique system advantages and can potentially serve as a model system to improve our current biomass bioconversion technology for fuels and chemicals.

As another important advantage regarding termite lignocellulolytic systems, recent papers and patents have demonstrated that wood-feeding termites and their gut symbionts have not only cellulolytic, hemicellulolytic, and lignin-modification activity, but also aromatic hydrocarbon degradation activity.[19,56,114] It has been reported that *Pseudomonas cepacia*, isolated from the gut of *Nasutitermes takasagoensis* (a higher termite species), can degrade phenol, cresol, furane, and trichloroethylene, which suggests that it could be used for soil remediation for soil polluted with such compounds.[114,122,123]

In addition, the gut microbiota of wood-feeding termites can theoretically convert a sheet of A4 paper into two liters of hydrogen,[102] an ideal energy format without carbon in its molecular structure. The bioconversion of cellulose to molecular H_2 by the subterranean termites tested could reach as high as 3858 ± 294 µmol g^{-1} cellulose, suggesting that the termite gut system is unique and efficient in H_2 conversion from cellulosic substrates.[124]

11.4.3 Nature-Inspired Technologies from Cellulose-Feeding Insects

The cellulose-feeding termite gut systems, combined with other important insect digestive systems, may provide unique and attractive model systems to develop nature-inspired technologies for viable biofuels. Many scientists in various disciplines have long focused their attention on termite guts as an excellent source of microorganisms, and unique catalyst systems for the production of biofuels.[102] After millions of years of trial and error, the evolution of lignocellulolytic systems in some cellulose-feeding insects has come up with amazing solutions/mechanisms to cope with the challenges of biomass recalcitrance. In recent years, the trend of "biomimicry", copying or adapting those answers, has emerged leading to innovations in designing novel reactor systems that mimic the biomass pretreatment functions (see Chapter 10 of this book), as well as the hydrolysis/fermentation processes in cellulose-feeding animals, such as wood-feeding termites[2,28] and cows (see Chapter 14). It has been reported that wood-feeding termites harbor some host lignases, such as laccase and peroxidase, with other accessory oxidases (*e.g.*, phenol-oxidase and esterase) that are integrated to modify lignin's structure[2] and improve the accessibility to, and functionality of, saccharifying enzymes. Current knowledge about the processing strategies of cellulosic diets in wood-feeding termites and their unique lignocellulolytic systems has provided the required critical insight for designing novel bioreactors to break down lignocellulosic biomass. A nature-inspired lignocellulose-processing system can be potentially developed from wood-feeding termites by building upon a stepwise lignin/cellulose/hemicellulose degrading catalytic process, and by mimicking their gut structure and unique physiochemical microenvironment, as well as their thermal dynamic processing system. The largely unexplored biodiversity and biochemistry of the termite gut system is a promising source of novel catalytic capacities.

Exploring various insect lignocellulolytic systems will definitely lead to the discovery of a variety of novel biocatalysts and the genes that encode them, as well as associated unique mechanisms for developing nature-inspired technologies. This new and evolving multi-disciplinary area has emerged in recent years, and no doubt, with more attention from the academic and industrial communities, will pave the way for future breakthroughs and innovations in biomass conversion for fuels and chemicals.

Acknowledgements

Special thanks go to Linghua Xu and John Obrycki (Department of Entomology, University of Kentucky) for their suggestions and comments on an earlier draft of this chapter. This work was supported by a project funded by the Priority Academic Program Development (PADP) of the Jiangsu Higher Education Institutions, China, a start-up fund from the University of Kentucky, a hatch project from USDA-NIFA (KY008053), USA, and a pilot project at the Kentucky Tobacco Research and Development Center (KTRDC), University of Kentucky, USA. The granting agencies had no role in study design, data collection and analysis, the decision to publish, or preparation of this manuscript.

References

1. J. Z. Sun and X. G. Zhou, in *Recent Advances of Entomological Research: from Molecular Biology to Pest Management*, ed. T. X. Liu and L. Kang, Springer, Berlin, 2011, p. 434.
2. J. Z. Sun and M. E. Scharf, *Insect Sci.*, 2010, **17**, 163.
3. M. E. Himmel and S. K. Picataggio in *Biomass Recalcitrance: Deconstructing the Plant Cell Wall for Bioenergy*, ed. M. E. Himmel, Blackwell Publishing, New Delhi, 2008.
4. L. Olsson, in *Biofuels*, Springer, New York, 2007.
5. N. Lo, G. Tokuda and H. Watanabe, in *Biology of Termites: A Modern Synthesis*, ed. D. E. Bignell, Y. Roisin and N. Lo, Springer, New York, 2011, p. 51.
6. H. Watanabe and G. Tokuda, *Annu. Rev. Entomol.*, 2010, **55**, p. 609.
7. C. Oppert, W. E. Klingeman, J. D. Willis, B. Oppert and J. L. Hurat-Futentes, *Comp. Biochem. Physiol.*, 2010, **155**, 145.
8. M. M. Martin, *Invertebrate–Microbial Interactions: Ingested Fungal Enzymes in Arthropod Biology*, Cornell University Press, Ithaca, 1987.
9. R. A. Prins and D. A. Kreulen, *Anim. Feed Sci. Technol.*, 1991, **32**, 101.
10. M. M. Martin, *Comp. Biochem. Physiol., Part A: Mol. Integr. Physiol.*, 1983, **75**, 313.
11. S. Itakura, K. Ueshima, H. Tanaka and A. Enoki, *Mokuzaigakkaishi*, 1995, **41**, 580.
12. J. A. Breznak and A. Brune, *Annu. Rev. Entomol.*, 1994, **39**, 453.
13. H. König, J. Frohlich and H. Hertel, in *Intestinal Microorganisms of Termites and other Invertebrates*, ed. H. König and A. Varma, Springer, Berlin, 2006, p. 271.
14. P. R. Zimmerman, J. P. Greenberg and J. P. E. C. Darlington, *Science*, 1984, **224**, 84.
15. J. A. Breznak, in *Invertebrate Microbial Interactions*, ed. J. M. Anderson, A. D. M. Rayner and D. W. H. Walton, Cambridge University Press, Cambridge, 1984, p. 173.
16. T. Abe, D. E. Bignell and M. Higashi, *Termites: Evolution, Sociality, Symbiosis, Ecology*, Kluwer Academic Publishers, Dordrecht, 2000.
17. H. Higashi and T. Abe, in *Biodiversity – An Ecological Perspective*, ed. T. Abe, S. A. Levin and M. Higashi, Springer, New York, 1997, p. 83.
18. M. Pester, PhD thesis, Max Planck Institute for Terrestrial Microbiology, 2006.
19. J. Ke, D. D. Laskar, D. Singh and S. Chen, *Biotechnol. Biofuels*, 2011, **4**, 17.
20. J. Ke, D. Singh, X.-W. Yang and S. Chen, *Biomass Bioenergy*, 2011, **35**, 3617.
21. J. Ke, D. D. Laskar and S. Chen, *Biomacromolecules*, 2011, **12**, 1610.
22. K. S. Katsumata, Z. Jin, K. Hori and K. Iiyama, *J. Wood Sci.*, 2007, **53**, 419.

23. S. M. Geib, T. R. Filley, P. G. Hatcher, K. Hoover, J. E. Carlson, M. D. M. Jimenez-Gasco, A. Nakagawa-lzumi, R. L. Sleighter and M. Tien, *Proc. Natl. Acad. Sci. U. S. A.*, 2008, **105**, 12932.
24. M. E. Scharf and D. G. Boucias, *Insect Sci.*, 2010, **17**, 166.
25. M. M. Martin, *Philos. Trans. R. Soc., B.*, 1991, **333**, 281.
26. H. Watanabe and G. Tokuda, *Cell. Mol. Life Sci.*, 2001, **58**, 1167.
27. K. Nakashima, H. Watanabe, H. Saitoh, G. Tokuda and J.-I. Azuma, *Insect Biochem. Mol. Biol.*, 2002, **32**, 777.
28. M. E. Scharf, Z. J. Karl, A. Sethi and D. G. Boucias, *PLoS One*, 2011, **6**, e21709.
29. J. A. Breznak and A. Brune, *Annu. Rev. Entomol.*, 1994, **39**, 453.
30. H. Watanabe, H. Noda, G. Tokuda and N. Lo, *Nature*, 1998, **394**, 330.
31. X. M. Xue, A. J. Anderson, N. A. Richardson, A. J. Anderson, G. P. Xue and P. B. Mather, *Aquaculture*, 1999, **180**, 373.
32. A. J. King, S. M. Cragg, Y. L. J. Dymond, M. J. Guille, D. J. Bowles, N. C. Bruce, I. A. Graham and S. J. McQueen-Mason, *Proc. Natl. Acad. Sci. U. S. A.*, 2010, **107**, 5345.
33. P. J. Boyle and R. Mitchell, *Science*, 1978, **200**, 1157.
34. G. W. O'Brien, P. C. Veivers, S. E. McEwen, M. Slaytor and R. W. O'Brien, *Insect Biochem.*, 1979, **9**, 619.
35. S. Itakura, H. Tanaka and A. Enoki, *Mater. Org.*, 1997, **31**, 17.
36. K. Nakashima and J.-I. Azuma, *Biosci. Biochem.*, 2000, **64**, 1500.
37. K. Nakashima, H. Watanabe, H. Saitoh, G. Tokuda and J.-I. Azuma, *Insect Biochem. Mol. Biol.*, 2002, **32**, 777.
38. K. Nakashima, H. Watanabe and J.-I. Azuma, *Cell. Mol. Life Sci.*, 2002, **59**, 1554.
39. L. Li, J. Fröhlich and H. König, in *Intestinal Microorganisms of Termites and other Invertebrates*, ed. H. König and A. Varma, Springer, Berlin, 2006, p. 221.
40. M. Slaytor, in *Intestinal Microorganisms of Termites and other Invertebrates*, ed. H. König and A. Varma, Springer, Berlin, 2006, p. 307.
41. X. G. Zhou, J. A. Smith, F. M. Oi, P. G. Koehler, G. W. Bennett and M. E. Scharf, *Gene*, 2007, **395**, 29.
42. A. E. Douglas, *Funct. Ecol.*, 2009, **23**, 38.
43. T. Inoue, S. Moriya, M. Ohkuma and T. Kudo, *Gene*, 2005, **349**, 67.
44. G. Tokuda and H. Watanabe, *Biol. Lett.*, 2007, **3**, 336.
45. N. Abo-Khatwa, *J. King Abulaziz Univ. Sci.*, 1989, **1**, 51.
46. N. Abo-Khatwa, *Experientia*, 1978, **34**, 559.
47. M. Ohkuma, Y. Maeda, T. Johjima and T. Kudo, *Riken Rev.*, 2001, **42**, 39.
48. I. C. Prentice, G. D. Farquhar, M. J. R. Fasham, M. L. Goulden, M. Heimann, H. S. Kheshi, L. C. Quere, R. J. Scholes, D. W. R. Wallace, D. Archer, M. R. Ashmore, O. Aumont, D. Baker, M. Battle, M. Bender, L. P. Bopp, P. Bousquet, K. Caldeira, P. Ciais, W. Cramer, F. Dentener, I. G. Enting, C. B. Field, E. A. Holland, R. A. Houghton, J. I. House, A. Ishida, A. K. Jain, I. Janssens, F. Joos, T. Kaminski, C. D. Keeling, D. W. Kicklighter, K. E. Kohfeld, W. Knorr, R. Law, T. Lenton, K. Lindsay,

E. Maier-Reimer, A. Manning, R. J. Matear, A. D. McGuire, J. M. Melillo, R. Meyer, M. Mund, J. C. Orr, S. Piper, K. Plattner, P. J. Rayner, S. Sitch, R. Slater, S. Taguchi, P. P. Tans, H. Q. Tian, M. F. Weirig, T. Whorf and A. Yool, The carbon cycle and atmospheric carbon dioxide, in *Climate Change 2001: the Scientific Basis: Contribution of Working Group I to the Third Assessment Report of the Intergovernmental Panel on Climate Change*, ed. J. T. Houghton, Cambridge University Press, Cambridge, 2001, p. 183.

49. M. E. Scharf and A. Tartar, *Biofuels, Bioprod. Biorefin.*, 2008, **2**, 540.

50. G. Tokuda, H. Watanabe, M. Hojo, A. Fujita, H. Makiya, M. Miyagi, G. Arakawa and M. Arioka, *J. Insect Physiol.*, 2012, **58**, 147.

51. L. R. Cleveland, *Proc. Natl. Acad. Sci. U. S. A.*, 1923, **9**, 424.

52. I. D. Reid, *Enzyme Microbiol. Technol.*, 1989, **11**, 786.

53. J. Ke, J. Sun, H. D. Nguyen, D. Singh, K. C. Lee, H. Beyenal and S. Chen, *Insect Sci.*, 2010, **17**, 277.

54. D. Cullen and P. J. Kersten, *The Mycota III: Biochemistry and Molecular Biology*, Springer, New York 2004, p. 249.

55. B. C. Saha, *J. Ind. Microbiol. Biotechnol.*, 2003, **30**, 279.

56. X. Su and J. Sun, MSc thesis, Jiangsu University, 2012.

57. A. Davison and M. Blaxter, *Mol. Biol. Evol.*, 2005, **22**, 1273.

58. L. Tokuda and H. Watanabe, *Biology of Termites: a Modern Synthesis*, ed. D. E. Bignell, Y. Roisin and N. Lo, Springer, The Netherlands, 2011, p. 51.

59. J. D. Willis, C. Oppert and J. L. Jurat-Fuentes, *Insect Sci.*, 2010, **17**, 184.

60. Y. Hongoh, *Cell Mol. Life Sci.*, 2011, **68**, 1311.

61. C. Noirot and C. Noirot-Timothée, in *Biology of Termites*, ed. K. Krishna and F. M. Weesner, Academic Press, New York, 1969, p. 49.

62. D. A. Waller and J. P. La Fage, in *The Nutritional Ecology of Insects, Mites and Spiders and Related Invertebrates*, ed. E. Slansky and J. G. Rodriguez, Wiley, New York, 1987, p. 487.

63. (a) L. R. Cleveland, *Biol. Bull.*, 1924, **46**, 178; (b) L. R. Cleveland, *Biol. Bull.*, 1924, **46**, 203.

64. Y. Yokoe, *Sci. Pap. Coll. Gen. Educ., Univ. Tokyo*, 1964, **14**, 115.

65. I. Yamaoka and Y. Nagatani, *Zool. Mag.*, 1977, **3**, 175.

66. A. Tartar, M. M. Wheeler, X. Zhou, M. R. Coy, D. G. Boucias and M. E. Scharf, *Biotechnol. Biofuels*, 2009, **2**, e2.

67. N. Lo, H. Watanabe and M. Sugimura, *Proc. R. Soc. London, Ser. B*, 2003, **270**, S69.

68. H. Watanabe, H. Noda, G. Tokuda and N. Lo, *Nature*, 1998, **394**, 330.

69. M. D. Sweeney and F. Xu, *Catalysts*, 2012, **2**, 244.

70. B. L. Cantarel, P. M. Coutinho, C. Rancurel, T. Bernard, V. Lombard and B. Henrissat, *Nucleic Acids Res.*, 2009, **37**, D233.

71. A. Levasseur, F. Piumi, P. M. Coutinho, C. Rancurel, M. Asther, M. Delattre, B. Henrissat, P. Pontarotti, M. Asther and E. Record, *Fungal Genet. Biol.*, 2008, **45**, 638.

72. B. Henrissat, *Biochem. J.*, 1991, **280**, 309.

73. N. Todaka, S. Moriya, K. Saita, T. Hondo, I. Kiuchi, H. Takasu, M. Ohkuma, C. Piero, Y. Hayashizaki and T. Kudo, *FEMS Microbiol. Ecol.*, 2007, **59**, 592.

74. M. Yuki, S. Moriya, T. Inoue and T. Kudo, *Zool. Sci.*, 2008, **25**, 401.

75. B. E. Dale and R. G. Ong, *Biotechnol. Prog.*, 2012, **28**, 893.

76. M. Crocker and R. Andrews, *Thermochemical Conversion of Biomass to Liquid Fuels and Chemicals*, ed. M. Crocker, Springer, New York, 2010, p. 532.

77. J. P. Lange, *Biofuels Bioprod. Biorefin.*, 2007, **1**, 39.

78. L. R. Lynd, P. J. Weimer, W. H. van Zyl and I. S. Pretorius, *Microbiol. Mol. Biol. Rev.*, 2002, **66**, 506.

79. W. Shi, R. Syrenne, J. Z. Sun and J. S. Yuan, *Insect Sci.*, 2010, **17**, 199.

80. F. Warnecke, P. Luginbühl, N. Ivanova, M. Ghassemian, T. H. Richardson, J. T. Stege, M. Cayouette, A. C. McHardy, G. Djordjevic, N. Aboushadi, R. Sorek, S. G. Tringe, M. Podar, H. G. Martin, V. Kunin, D. Dalevi, J. Madejska, E. Kirton, D. Platt, E. Szeto, A. Salamov, K. Barry, N. Mikhailova, N. C. Kyrpides, E. G. Matson, E. A. Ottesen, X. Zhang, M. Hernández, C. Murillo, L. G. Acosta, I. Rigoutsos, G. Tamayo, B. D. Green, C. Chang, E. M. Rubin, E. J. Mathur, D. E. Robertson, P. Hugenholtz and J. R. Leadbetter, *Nature*, 2007, **450**, 560.

81. X. Zhou, E. S. Kovaleva, D. Wu-Scharf, J. H. Campbell, G. W. Buchman, D. G. Boucias and M. E. Scharf, *Arch. Insect Biochem. Physiol.*, 2010, **74**, 147.

82. E. E. Brevnova, V. Rajgarhia, M. Mellon, A. Warner, J. Mcbride, C. Gandhi and E. Wiswall, *US Pat.* 2012 0003701, 2012.

83. G. Tokuda, N. Lo, H. Watanabe, M. Slaytor, T. Matsumoto and H. Noda, *Biochim. Biophys. Acta*, 1999, **1447**, 146.

84. K. Nakashima, H. Watanabe, H. Saitoh, G. Tokuda and J. I. Azuma, *Insect Biochem. Mol. Biol.*, 2002, **32**, 777.

85. L. Li, J. Frohlich, P. Pfeiffer and H. Konig, *Eukaryotic Cell*, 2003, **2**, 1091.

86. G. Tokuda, N. Lo, H. Watanabe, G. Arakawa, T. Matsumoto and H. Noda, *Mol. Ecol.*, 2004, **13**, 3219.

87. D. Zhang, A. R. Lax, A. K. Raina and J. M. Bland, *Insect Biochem. Mol. Biol.*, 2009, **39**, 516.

88. D. Zhang, A. R. Lax, J. M. Bland and A. B. Allen, *Insect Biochem. Mol. Biol.*, 2011, **41**, 211.

89. D. Zhang, A. R. Lax, J. M. Bland, J. Yu, N. Fedorova and W. C. Nierman, *Insect Sci.*, 2010, **17**, 245.

90. D. Zhang, A. B. Allen and A. R. Lax, *J. Insect Physiol.*, 2012, **58**, 205.

91. G. Tokuda, H. Saito and H. Watanabe, *Insect Biochem. Mol. Biol.*, 2002, **32**, 1681.

92. C. A. Uchima, G. Tokuda, H. Watanabe, K. Kitamoto and M. Arioka, *Appl. Microbiol. Biotechnol.*, 2011, **89**, 1761.

93. M. E. Scharf, E. S. Kovaleva, S. Jadhao, J. H. Campbell, G. W. Buchman and D. G. Boucias, *Insect Biochem. Mol. Biol.*, 2010, **40**, 611.

94. C. A. Uchima, G. Tokuda, H. Watanabe, K. Kitamoto and M. Arioka, *Appl. Environ. Microbiol.*, 2012, **78**, 4288.
95. G. Tokuda, H. Saito and H. Watanabe, *Insect Biochem. Mol. Biol.*, 2002, **32**, 1681.
96. J. Ni, M. Takehara and H. Watanabe, *Bioresour. Technol.*, 2010, **101**, 6438.
97. J. Ni, M. Takehara and H. Watanabe, *Biosci. Biotechnol. Biochem.*, 2005, **69**, 1711.
98. A. Carroll and C. Somerville, *Annu. Rev. Plant Biol.*, 2009, **60**, 165.
99. A. Sethi and M. E. Scharf, in *eLS*, John Wiley & Sons, Chichester, 2013, doi: 10.1002/9780470015902.a0020374.
100. M. Ohkuma and A. Brune, in *Biology of Termites: A Modern Synthesis*, ed. D. E. Bignell, Y. Roisin and N. Lo, Springer, New York, 2011, p. 413.
101. T. Abe and M. Higashi, *Okos*, 1991, **60**, 127.
102. A. Brune, *Nature*, 2007, **450**, 487.
103. E. E. Allen and J. F. Banfield, *Nat. Rev. Microbiol.*, 2005, **3**, 489.
104. A. Brune and M. Friedrich, *Curr. Opin. Microbiol.*, 2000, **3**, 263.
105. M. Ohkuma, in *Symbiosis: Mechanisms and Model Systems*, ed. J. Seckbach, Kluwer Academic Publishers, Dordrecht, 2002, p. 715.
106. M. Ohkuma, *Appl. Microbiol. Biotechnol.*, 2003, **61**, 1.
107. M. Ohkuma, *Trends Microbiol.*, 2008, **16**, 345.
108. A. Brune and U. Stingl, in *Molecular Basis of Symbiosis*, ed. J. Overmann, Springer, Berlin, 2005, p. 39.
109. A. Brune, in *Prokaryotes*, ed. M. Dworkin, S. Falkow and E. Rosenberg, Springer, New York, 3rd edn, 2006, vol. 1, p. 439.
110. M. Ohkuma, Y. Hongoh and T. Kudo, in *Intestinal Microorganisms of Termites and other Invertebrates*, ed. H. König and A. Varma, Springer, Berlin, 2006, p. 303.
111. H. König, J. Frohlich and H. Hertel, in *Intestinal Microorganisms of Termites and other Invertebrates*, ed. H. König and A. Varma, Springer, Berlin, 2006, p. 271.
112. Y. Hongoh, M. Ohkuma and T. Kudo, *FEMS Microbiol. Ecol.*, 2003, **44**, 231.
113. A. Tartar, M. M. Wheeler, X. G. Zhou, M. R. Coy, D. G. Boucias and M. E. Scharf, *Biotechnol. Biofuels*, 2009, **2**, 25.
114. T. Matsui, G. Tokuda and N. Shinzato, *Recent Pat. Biotechnol.*, 2009, **3**, 10.
115. J. A. Breznak and A. Brune, *Annu. Rev. Entomol.*, 1994, **39**, 453.
116. A. Brune and M. Ohkuma, in *Biology of Termites: A Modern Synthesis*, ed. D. E. Bignell, Y. Roisin and N. Lo, Springer, New York, 2011, p. 439.
117. P. C. Veivers, R. W. O'Brien and M. Slaytor, *J. Insect Physiol.*, 1982, **28**, 947.
118. L. Adams and R. Boopathy, *Bioresour. Technol.*, 2005, **96**, 1592.
119. S. M. Geib, T. R. Filley, P. G. Hatcher, K. Hoover, J. E. Carlson, M. D. M. Jimenez-Gasco, A. Nakagawa-lzumi, R. L. Sleighter and M. Tien, *Proc. Natl. Acad. Sci. U. S. A.*, 2008, **105**, 12932.

120. M. R. Coy, T. Z. Salem, J. S. Denton, E. Kovaleva, Z. Liu, D. S. Barber, J. H. Campbell, D. C. Davis, G. W. Buchman, D. G. Boucias and M. E. Scharf, *Insect Biochem. Mol. Biol.*, 2010, **40**, 723.
121. M. E. Scharf, Z. J. Karl, A. Sethi, R. Sen, R. Raychoudhury and D. G. Boucias, *Commun. Integr. Biol.*, 2011, **4**, 761.
122. D. Kamanda Ngugi, M. Khamis Tsanuo and H. Iddi Boga, *J. Basic Microbiol.*, 2007, **47**, 87.
123. T. Matsui, Y. Tanaka, K. Maruhashi and R. Kurane, *Curr. Microbiol.*, 2003, **46**, 39.
124. Y.-Q. Cao, J. Z. Sun, M. Rodriguez and K. C. Lee, *Insect Sci.*, 2010, **17**, 237.

CHAPTER 12

The Lignocellulolytic Wood-Feeding Cockroach— A Forgotten Treasure

XIANGRUI LI[a,b] AND XUGUO ZHOU*[a]

[a] Department of Entomology, University of Kentucky, Lexington, KY 40546, USA; [b] Institute of Plant Protection, Chinese Academy of Agricultural Science, Beijing 100193, P. R. China
*Email: xuguozhou@uky.edu

12.1 Dictyoptera and Biofuels

12.1.1 Wood-Feeding Dictyoptera

As stores of fossil fuels deplete, attention has turned to the most abundant carbohydrate energy resource on Earth; lignocelluloses, a suite of naturally occurring plant-derived materials that includes hydrophilic sugar polymers, cellulose and hemicellulose, and a hydrophobic organic 'glue' known as "lignin". However, to release the chemical energy stored in lignocelluloses (approximately 10 times the current world energy consumption[1]), an array of enzymes referred to as "lignocellulases" is required to disassociate the matrix of cellulose, hemicellulose, pectins, lignin, and glycosidic linkages.[2] Lignocellulosic biomass, including residues and by-products from forest, agriculture, and industry, can be converted to ethanol by hydrolysis and down-stream fermentation processes.

RSC Energy and Environment Series No. 10
Biological Conversion of Biomass for Fuels and Chemicals: Explorations from Natural Utilization Systems
Edited by Jianzhong Sun, Shi-You Ding and Joy Doran-Peterson
© The Royal Society of Chemistry 2014
Published by the Royal Society of Chemistry, www.rsc.org

However, the complexity and crystallinity of the lignocellulosic biomass present major challenges in the biomass conversion processes.[1]

Wood-feeding dictyopterans including termites and woodroaches, are the most efficient lignocellulose-digesting bioreactors on Earth, and have intrigued scientists for decades because of their unique capacity to rapidly break down recalcitrant lignocelluloses. Typically, lignocellulose depolymerization in wood-feeding dictyopterans is a complex enzymatic process that involves an intricate interaction between host organisms and their pro- and eukaryotic symbionts. Recent genomic and meta-genomic analyses have discovered hundreds of lignocellulolytic enzymes derived from wood-feeding termites and their symbiotic partners.[3–6] While the pressing needs for bioenergy and biomass conversion have redirected termite research to focus more on lignocellulose digestion and degradation,[2,7] there is virtually no information available on the wood-feeding cockroach *Cryptocercus,* a sister group to Isoptera which shares many traits with the lower termites.

12.1.2 A Key Reference for Tracing the Lignocellulose-Digestion Machinery

The wood-feeding cockroach *Cryptocercus punctulatus* and its obligate gut symbionts have co-evolved into an efficient mini-bioreactor that involves a suite of specialized enzymes which synergistically break down the matrix of cellulose, hemicelluloses, and lignin in plant cell walls. Within 24 h, both termites and woodroaches can efficiently convert over 90% of the recalcitrant lignocellulose into fermentable sugars in their hindguts. Such efficiency and low energy input make wood-feeding dictyopterans an ideal resource to identify novel lignocellulolytic enzymes that are specifically adapted to hydrolyze each type of lignocellulose, namely lignin, hemicellulose, and cellulose. One of the unique aspects of the dictyopteran lignocellulases is that they give us the option to integrate biocatalyst-based approaches with current thermochemical-based biomass pretreatments.

Because of its ancestral status, and the diverse cellulolytic microbial fauna shared with extant termites, *C. punctulatus* represent the best available taxonomic reference to identify evolutionarily conserved lignocellulases within wood-feeding dictyopterans. Of the most relevance will be the discovery of lignocellulose-digestion machinery, a group of enzymes fundamentally important to the degradation of woody materials.

12.2 The Wood-Feeding Cockroaches *Cryptocercus*

12.2.1 Ecological Significance

Wood-feeding insects have intrigued scientists for decades because of their unique capability of breaking down the lignocellulose matrix of cellulose, hemicelluloses, and lignin in plant cell walls. Similar to the lower termite, *Reticulitermes flavipes,* the wood-feeding cockroach, *C. punctulatus,* and their obligate gut symbionts have co-evolved into an efficient mini-bioreactor including a suite of specialized enzymes that synergistically break down lignocelluloses, making them the most abundant and sustainable solar batteries on Earth. Lignocellulose degradation/depolymerization is a complex enzymatic process that involves an intricate collaboration between the host organism and its symbionts. *Cryptocercus punctulatus* progressively degrades the logs it inhabits by not only ingesting the wood, but also shredding it without consumption when excavating tunnels. Their abundant feces line galleries, pack side-chambers, and are pushed to the outside of the logs, no doubt influencing local populations of bacteria, fungi and microfauna.[8]

Cockroaches potentially influence biogeochemical cycles *via* two known pathways: nitrogen fixation and methane production. *Cryptocercus* is the only cockroach currently known to harbor gut microbes capable of fixing atmospheric nitrogen,[9] but spirochetes found in the hindgut of other species may also fix nitrogen.[5,10] Acetylene reduction assays indicate that adults and juveniles of *Cryptocercus* fix nitrogen at rates comparable to those in termites on a body weight basis (0.01–0.12 μg nitrogen per day per g of wet weight).[9,11,12] The process provides a mechanism for nitrogen cycling in the ecosystem and may have a significant ecological impact,[13] particularly in the food chains of the montane mesic forests where *Cryptocercus* is the dominant macroarthropod feeding on rotting logs. Recently, several studies have shown that endosymbionts in the abdominal fat body of cockroaches are involved in nitrogen recycling and the provision of amino acids.[14–16]

The social behavior of *C. punctulatus* and their substantial body size also contribute to their ecological impact. Adults are typically about 3 cm in length and they live in sizable biparental family groups. They may pulverize logs on a timescale comparable to, if not faster than, termites. The *C. punctulatus* species complex dominates the saproxylic guild in the Southern Appalachian Mountains, and occupies the same niche as the subterranean termite *Reticulitermes* at lower elevations.[17]

Like many cockroaches, *Cryptocercus* harbor methanogenic bacteria in the hindgut and consequently emit methane.[9] On a global scale, estimates of methane production by cockroaches vary widely and are debatable, given first, the paucity of data on biomass estimates of field populations, and second, the finding that methane production varies with cockroach age and the fiber content of their diet.[18,19] It has been suggested that cockroaches contribute significantly to global methane, particularly in the tropics.[20,21] However, methane oxidation by methanotropic bacteria in the soil may buffer the atmosphere from methane production by gut Archaea, and although cockroaches may be a source of methane, little of it may be escaping into the atmosphere.[22] The sink capacity of the soil may exceed methane production by cockroaches, just as it does for termites. Nonetheless, their typically large body size (relative to termites), and the tendency of many species to live in aggregations in enclosed spaces (*e.g.*, tree-holes, caves, logs) may engender substantial atmospheric changes at a local level. For example, hundreds of *C. punctulatus* may live in a log of the appropriate stage of moisture and rot. On a per-weight basis methane production by this species is comparable to the termite *R. flavipes* and may surpass levels emitted by ruminants.[9,12]

12.2.2 Phylogenetic Relationship with Termites

Morphological, molecular, symbiotic, and genomic evidence support the contentions that: (1) cockroaches are the evolutionary ancestor of termites; and (2) phylogenetically speaking, the wood-feeding cockroach *Cryptocercus* is a sister group of termites. *Trichonympha acuta*, a symbiotic protist, resides in both woodroach and termite hindguts. The common possession of hindgut flagellates in *Cryptocercus* and lower termites is often a focal point in discussions of the evolutionary origins of termites. These eukaryotic symbionts are unique; most are found only in the hindguts of these two groups. The goal for the comparative genomic analyses is to identify the highly conserved lignocellulases among the wood-feeding cockroach, *Cryptocercus,* and the lower termites, *Reticulitermes* and *Coptotermes*.

The classic view regarding termite origins is that they either branched directly from cockroaches, or that cockroaches and termites are sister groups.[23] In the past decade, there has been overwhelming evidence that termites are a monophyletic clade embedded within cockroaches and a sister group to the cockroach genus *Cryptocercus*.[24–27] This placement is strongly supported, and reflected in phylogenetic analyses based on morphology, molecular markers including mitochondria DNA, nuclear DNA,[28,29] and functional groups like GHF9 endogenous endo-glucanases, the phylogeny of bacteroids in the fat body, genome size[30] and more recently, the phylogeny of hindgut protozoa[31] and the bacterial ectosymbionts of hindgut protozoa.[32] Although it has not been settled as to how these findings affect the taxonomic nomenclature of termites and cockroaches,[33,34] the fundamental message is clear: cockroaches are the ancestors of termites (Figure 12.1).

12.2.3 Microbiota in *Cryptocercus*

Digestion in *Cryptocercus* is comparable to that in lower termites. The hindgut is a fermentation chamber filled to capacity with a community of interacting symbionts, including eukaryotic flagellates, and prokaryotic spirochetes and bacteria that are free in the digestive tract, attached to the gut wall, and symbiotic with resident protozoans. Included are uricolytic bacteria, cellulolytic bacteria, methanogens, and those capable of nitrogen fixation, as well as bacteria that participate in the biosynthesis of volatile fatty acids.[9,35,36] The metabolic capabilities of this symbiotic system reflect the combined contributions of the host, the eukaryotic symbionts and the bacterial symbionts, which can occur in the gut lumen or within the cytoplasm of symbiotic protozoans. Preliminary evidence from termites and cockroaches suggests that both carbon and nitrogen cycling rely on contributions from both the host and one or more symbionts.

The possession of oxymonad and hypermastigid hindgut flagellates in both *Cryptocercus* and lower termites is often a focal point in discussions of the

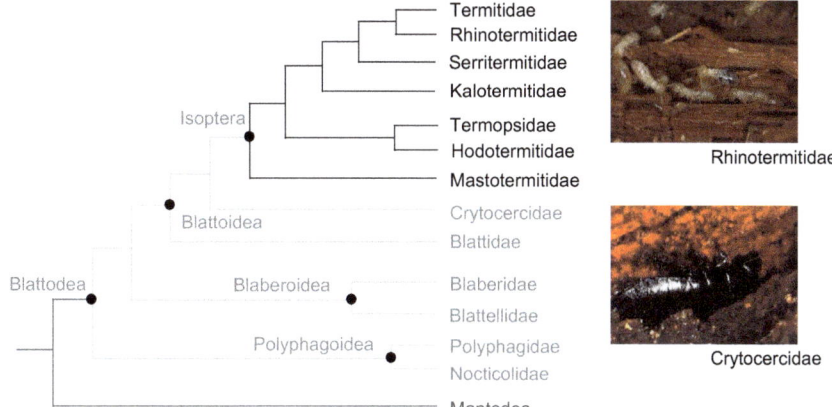

Figure 12.1 The phylogenetic relationships of Dictyoptera (adapted from refs. 28 and 29) based on morphology and molecular markers including mitochondria and nuclear DNA.

evolutionary origins of termites. These protozoans are unusually large, making them good subjects for a variety of experimental investigations; selected species in the gut of *Cryptocercus* are 0.3 mm in length and visible to the unaided eye.[23] They are unusually intricate, with singular morphological structures and a complex of bacterial symbionts of their own.[37] They are unique; most are found nowhere in nature but the hindguts of these two sister groups.[38] Patrick Keeling and his colleagues have characterized eukaryotic flagellates in the *C. punctulatus* hindgut using light microscopy (LM), transmission electron microscopy (TEM), and scanning electron microscopy (SEM), and have developed molecular markers for most of the eukaryotic species.[32,39] A comparative study of gut microbiome using the 16S rDNA maker suggested that 45, 19, and 24 new phylotypes of symbiotic bacteria were common constituents of the intestinal microbiota of lower termites and *Cryptocercus* under three physiological conditions, including active, fasting, and dead, respectively.[40] Neef et al.[14] sequenced the genome of the symbiotic Blattabacterium sp. from *C. punctulatus*, and compared its genome with two other cockroaches, *Blattella germanica*[15] and *Periplaneta Americana*.[16,41] These results suggested that the bacterial genome of *C. punctulatus* consists of a 605 745 bp chromosome and a 3816 bp plasmid. Although the chromosome contains 585 genes, of which 545 are protein-coding, only 3 genes, *dut*, *nrdF*, and a hypothetical protein, are conserved across these three cockroach species.

12.3 Lignocellulose Degradation in *Cryptocercus*

12.3.1 Enzymatic Depolymerization of Lignocellulose

Among the major components of lignocelluloses, cellulose is the most abundant biopolymer on Earth and a principal cell-wall constituent of higher plants.[42] Hemicellulose, also referred to as "xylan", is composed of shorter β-l,4-linked polymers of mixed sugars.[43] Lignin is not a carbohydrate, but a three-dimensional polymer of phenolic compounds that are linked to each other and to hemicellulose by ester bonds.[44] The naturally occurring enzymatic degradation of lignocellulose is a comprehensive three-step process that requires an array of highly specialized biocatalysts. As a first step in the process, depolymerization of hydrophobic lignin polymers is extremely important to enable hemicellulose degradation. Relevant enzymes involved in lignin degradation primarily include laccases and peroxidases.[44] As a second step, hemicellulose degradation is equally important for making cellulose accessible for depolymerization. Complete biodegradation of hemicellulose requires the combined activity of endo-β-1,4-xylanases, exo-β-1,4-xylanases, β-xylosidases, α-arabinofuranosidases, α-uronidases, and esterases.[43] As the third and final step, cellulose depolymerization requires the combined action of endo-β-1,4-glucanases, exo-β-1,4-glucanases, and β-glucosidases.[42,45]

12.3.2 Lignocellulose Digestibility in Wood-Feeding Dictyopterans

In addition to termites and woodroaches, a diverse group of insects is capable of digesting plant materials (Figure 12.2[46–48]). Over 100 lignocellulose-feeding insects have been documented using various lignocellulolytic systems, including symbiont-dependent, symbiont-independent, and acquired-enzymatic systems.[49] However, in terms of efficiency in cellulose degradation and consumption, termites and woodroaches are the most efficient cellulose digesters (Figure 12.2). The advent of the "genomic era" has paved the way for the comprehensive survey of lignocellulases at the genomic scale. Recent -omics studies, including transcriptome, meta-transcriptome, and meta-genome analyses, have corroborated the long-standing hypothesis of termites and their microbial symbionts contributing synergistically to lignocellulose degradation.[5,6,35,50]

Typically, lignocellulose depolymerization in wood-feeding dictyopterans is a complex enzymatic process that involves an intricate collaboration between host insects and their pro- and eukaryotic symbionts. In an effort to better understand the relative digestive contributions of host and symbiont in *R. flavipes*, a normalized symbiont-free gut library and a non-normalized hindgut symbiont library were sequenced. The genomic organization of lignocellulases between host and symbionts is as expected: a non-overlapping complement of host and symbiont glycohydrolase genes that participate in α-carbohydrate, cellulose, and hemicellulose digestion.[6]

In comparison to termites, lignocellulose digestion in the wood-feeding cockroaches has yet to be investigated. So far, only a handful of cellulases have been documented including an endogenous endo-glucanase, a glycoside hydrolase family 9 (GHF9) enzyme from *C. clevelandi*,[24] and a symbiotic endo-glucanase GHF5 (GenBank accession numbers: BAF57456–BAF57461), GHF7 (BAF57462 BAF57470), GHF10 (BAF57471–BAF57478), GHF45 (BAF57479) and GHF23[14] from *C. punctulatus*. Nevertheless, this limited information supports the hypothesis that wood-feeding cockroaches retain similar dual lignocellulose-digestion systems (Figure 12.3), to those which have

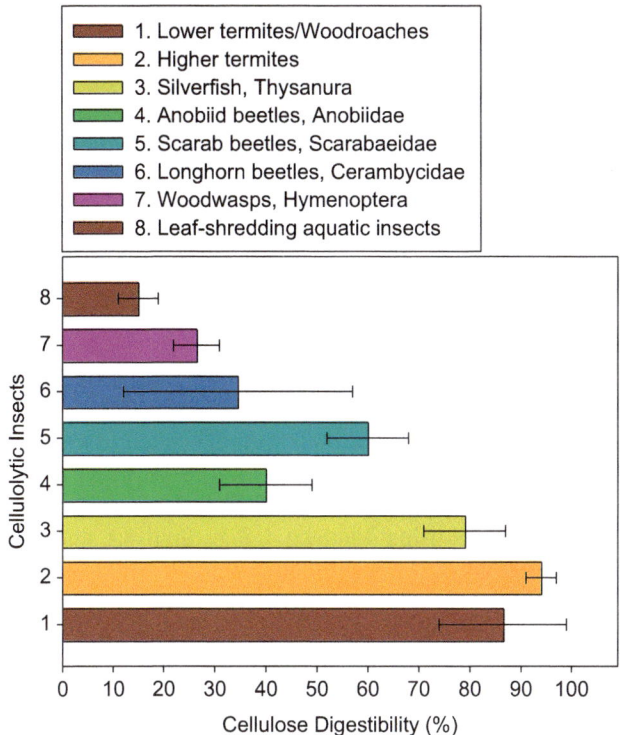

Figure 12.2 Cellulose digestibility among different cellulolytic insect groups. The approximate cellulose digestibility was determined by comparing the cellulose content of food and feces (adapted from refs. 46–48).

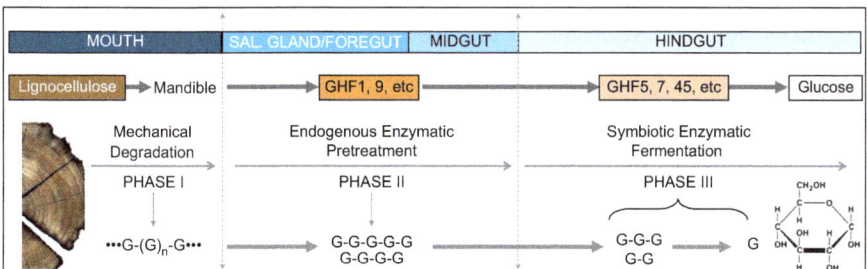

Figure 12.3 Schematic drawing of the dual lignocellulose-digestion system in wood-roaches and termites. The lignocellulose digestion in these two Dictyoptera species involves both hosts and symbionts, suggesting a dual system. Typically, the degradation of lignocelluloses starts with mechanical breakdown of woody materials into micro-scale particles (Phase I), followed by enzymatic pretreatments with host lignocellulases to dis-associate the matrix of lignin, hemicellulose, and cellulose (Phase II). The entire process with fermentation using symbiotic enzymes (Phase III).

been widely documented in the lower termites.[3,6,51,52] The most recent digestome sequencing efforts identified several endogenous endo-glucanases (GHF9)
and β-glucosidases (GHF1 and 3) in *C. punctulatus* salivary glands, and exo-
glucanases (GHF5 and 7) from the woodroach hindgut which have the highest
sequence similarity with GHF5 and 7 genes, respectively, from the uncultured
symbiotic protists in the termite hindgut.[56]

12.4 Current Progress

12.4.1 Genomic Resources

The genomic information regarding wood-feeding cockroaches substantially
lags behind its sister group, termites. As of July 31, 2012, the number of nucleotides and proteins from *Cryptocercus* spp. deposited in GenBank was 1379
and 1887, respectively, with the majority of the sequences generated from
C. punctulatus (Table 12.1). The genome size of *C. punctulatus* has been estimated to be $\sim$1320 Mb using Feulgen image analysis and 6-diamidino-2-
phenylindole (DAPI)-based flow cytometry.[30] The genome size of female *C.*
punctulatus ($\sim$1340 Mb) is slightly larger than that of the male ($\sim$1300 Mb). In
comparison to the subsocial *C. punctulatus,* the eusocial lower termite, *R. fla-*
vipes (Kollar) (Isoptera: Rhinotermitidae) has an estimated $\sim$1070 Mb
genome size.

Tissue-specific transcriptomes including salivary glands, foregut, midgut,
and hindgut, and the meta-transcriptome representing *C. punctulatus* hindgut
microbiota have been pyrosequenced (Figure 12.4). Preliminary analysis shows

Table 12.1 Genomic information for *Cryptocercus* in GenBank.

| *Cryptocercus spp.* | *Sequence* | *Insect origin* | *Symbiotic origin* | | *Total* |
			Eukaryote	*Prokaryote*	
C. punctulatus	Nucleotide	167	312	376	855
	Protein	59	340	1313	1712
C. darwini	Nucleotide	92	0	15	107
	Protein	5	0	0	5
C. wrighti	Nucleotide	60	0	12	72
	Protein	2	0	0	2
C. kyebangensis	Nucleotide	33	1	12	46
	Protein	8	0	0	8
C. garciai	Nucleotide	37	0	9	46
	Protein	2	0	0	2
C. clevelandi	Nucleotide	34	0	4	38
	Protein	6	0	0	6
C. relictus	Nucleotide	19	0	13	32
	Protein	28	0	0	28
C. primarius	Nucleotide	8	0	5	13
	Protein	2	0	0	2
C. matilei	Nucleotide	2	0	0	2
	Protein	0	0	0	0

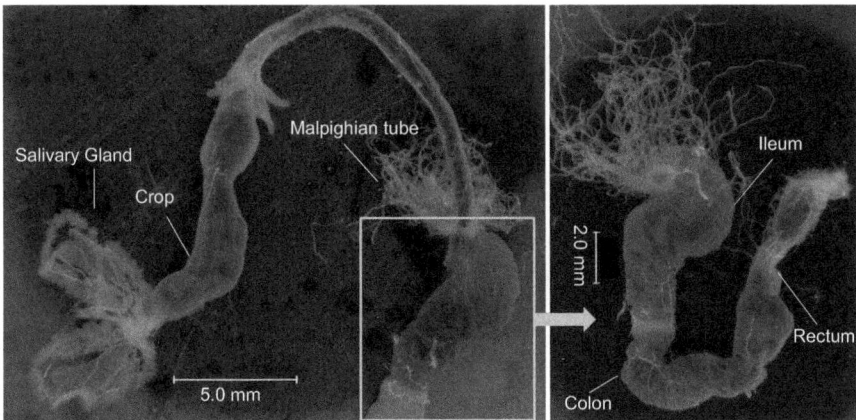

Figure 12.4   A schematic showing of the entire digestive tract of *C. punctulatus*. The alimentary canal is divided into the foregut (*stomodeum*), midgut (*mesenteron*) and hindgut (*proctodeum*). The foregut consists of the undifferentiated *pharynx* and *esophagus*. The paired salivary glands are seen wrapped around the esophagus with delicate connective tissue. The esophagus leads to a distensible "crop" that terminates in the *proventriculus*. The finger-like blind pouches are located at the junction of the foregut and midgut. The midgut is a simple tube. Numerous coiled Malpighian tubules are seen at the junction of the mid- and hindgut. The hindgut can be divided into an *ileum*, a *colon*, and a *rectum*.

that endogenous lignocellulolytic enzymes from *C. punctulatus* are predominantly located in the salivary glands, which include a suite of biocatalysts critical for the enzymatic pretreatment of woody biomasses.[56]

12.4.2   Meta-Proteomic Profiling of Hindgut Microbiota in *C. punctulatus*

The luminal contents collected from *C. punctulatus* adults have been used for both meta-proteomics and meta-transcriptomic analyses. The total protein that was extracted from luminal contents was subjected to trypsin digestion, followed by peptide identification using MALDI-TOF/TOF mass spectrometry. A peptide database was created by translating gut and symbiont Expressed sequence tag (EST) datasets from *R. flavipes*,[6] and was used for protein identification. The initial analysis using the ProteinPilot Paragon algorithm used thousands of peptide sequences collected from MS/MS analysis to identify hundreds of significant protein matches in all MS analyses. These matches were then annotated by searching the corresponding ESTs against the NCBI database to provide annotation and putative classification of the peptides (Figure 12.5).

A total of 694 and 565 proteins were identified from *C. punctulatus* hindgut microbiota using peptide libraries from termite symbionts and guts, respectively. Among them, 194 and 169 sequences were unclassified. The sequence

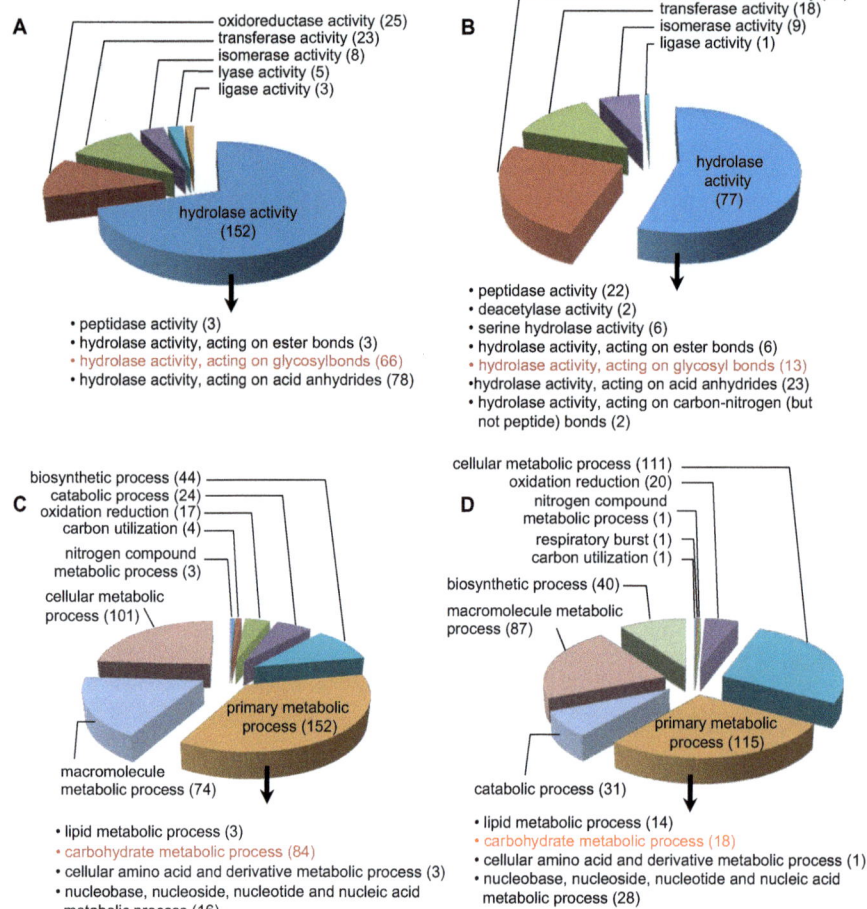

A

oxidoreductase activity (25)
transferase activity (23)
isomerase activity (8)
lyase activity (5)
ligase activity (3)

hydrolase activity
(152)

• peptidase activity (3)
• hydrolase activity, acting on ester bonds (3)
• hydrolase activity, acting on glycosylbonds (66)
• hydrolase activity, acting on acid anhydrides (78)

B

oxidoreductase activity (34)
transferase activity (18)
isomerase activity (9)
ligase activity (1)

hydrolase
activity
(77)

• peptidase activity (22)
• deacetylase activity (2)
• serine hydrolase activity (6)
• hydrolase activity, acting on ester bonds (6)
• hydrolase activity, acting on glycosyl bonds (13)
•hydrolase activity, acting on acid anhydrides (23)
• hydrolase activity, acting on carbon-nitrogen (but
  not peptide) bonds (2)

C

biosynthetic process (44)
catabolic process (24)
oxidation reduction (17)
carbon utilization (4)
nitrogen compound
metabolic process (3)
cellular metabolic
process (101)

primary metabolic
process (152)

macromolecule
metabolic process (74)

• lipid metabolic process (3)
• carbohydrate metabolic process (84)
• cellular amino acid and derivative metabolic process (3)
• nucleobase, nucleoside, nucleotide and nucleic acid
  metabolic process (16)

D

cellular metabolic process (111)
oxidation reduction (20)
nitrogen compound
metabolic process (1)
respiratory burst (1)
carbon utilization (1)
biosynthetic process (40)
macromolecule metabolic
process (87)

primary metabolic
process (115)

catabolic process (31)

• lipid metabolic process (14)
• carbohydrate metabolic process (18)
• cellular amino acid and derivative metabolic process (1)
• nucleobase, nucleoside, nucleotide and nucleic acid
  metabolic process (28)

Figure 12.5 Distribution of the gene ontology annotation based on the termite
symbiont and gut library. Based on the GO-molecular function, 66
putative glycoside hydrolases are nested within the 152-member hydrolase activity gene family in the symbiont library (a), and 13 putative
glycoside hydrolases are nested within the 77-member hydrolase activity
gene family in the gut library (b). Similarly, according to the GO-
biological process, 84 peptides potentially involved in the carbohydrate
metabolic process belong to the 152-member primary metabolic process
groups in the symbiont library (c), and 18 peptides potentially involved in
the carbohydrate metabolic process belong to the 115-member primary
metabolic process groups in the gut library (d).

distributions according to their GO (molecular function, biological process and
cellular component) are summarized in Figure 12.5. Based on the GO term
"molecular function", 79 peptide sequences were putatively classified as
glycoside hydrolases in which majority of them belonged to GHF7, 8, 2, 3, 5,
and 18 (66 derived from the symbiont library and 13 came from the gut library,

respectively). Hindgut microbiota from *C. punctulatus* and *R. flavipes* apparently share some lignocellulolytic enzymes (*e.g.*, GHF7), however the low degree of overlapping GHFs in the meta-proteomics study suggests that *C. punctulatus* is a rich reservoir for novel lignocellulases.

12.5 Future Perspectives

Highly active and chemically/thermally stable lignocellulolytic biocatalysts are a major bottleneck for the large-scale production of biofuels from lignocellulosic biomass. The wood-feeding cockroach and its hindgut microbiota provide an irreplaceable resource for the identification of core lignocellulolytic enzymes conserved across the woodroach–termite lineage. The distribution of lignocellulase genes among different insects has yet to be explored in great detail. However, there has been a recent influx of sequence information from cellulolytic insect taxa.[7,53] Given the fact that *Cryptocercus* and lower termites share many flagellate symbionts such as oxymonadid and hypermastigid in their hindguts,[54] a survey of the homogeneity of lignocellulases across Dictyoptera with basal *C. punctulatus* as a reference will be an important next step. Evolutionarily conserved lignocellulases will shed light on the discovery of lignocellulose-digestion machinery.

The low degree of overlapping GHFs, especially in the endogenous lignocellulolytic enzymes, from the initial meta-proteomic profiling suggests that *C. punctulatus* is a rich reservoir for novel lignocellulases. In addition, transcriptomic and meta-genomic sequencing will reveal the complete suite of genes that maintain the unique microenvironment of the cockroach gut, which supports cellulolytic symbionts and enables the entire lignocellulose-digestion process to take place. Consequently, future studies will have two major components: (1) physical and functional genomics; and (2) protein engineering. The former is the continuation of the initial gene discovery efforts, and includes: (1) the whole genome sequencing of multiple species of woodroaches with *C. punctulatus* as the model species; (2) meta-genome sequencing of *C. punctulatus* hindgut microbiota; (3) transcriptomic/meta-transcriptomic profiling of the *C. punctulatus* digestive tract; and (4) functional characterization of newly discovered lignocellulolytic enzymes from previous steps. To exploit their industrial potential, candidate genes will need to be subject to protein engineering such as directed evolution and DNA recombination to improve their enzymatic properties and meet the need for biomass conversion at an industrial scale.[55] Meanwhile, mass production of insect lignocellulolytic enzymes through transgenic plants or microbial-mediated protein production systems can also be the focal point for down-stream application research.

Acknowledgements

Special thanks go to John Obrycki (Department of Entomology, University of Kentucky) for his suggestions and comments on an earlier draft of this chapter. This work was supported by a start-up fund from the University of Kentucky, a hatch project from USDA-NIFA (KY008053), USA, and a pilot project at the Kentucky Tobacco Research and Development Center (KTRDC), University of Kentucky, USA. The granting agencies had no role in study design, data collection and analysis, the decision to publish, or preparation of this manuscript.

References

1. R. Kumar, S. Singh and O. V. Singh, *J. Ind. Microbiol. Biotechnol.*, 2008, **35**, 377.
2. E. M. Rubin, *Nature*, 2008, **454**, 841.
3. X. Zhou, J. A. Smith, F. M. Oi, P. G. Koehler and M. E. Scharf, *Gene*, 2007, **395**, 29.
4. X. Zhou, E. Kovaleva, D. Wu-Scharf, J. H. Campbell, G. W. Buchman, D. G. Boucias and M. E. Scharf, *Arch. Insect Biochem. Physiol.*, 2010, **74**, 147.
5. F. Warnecke, P. Luginbtlhl, N. Ivanova, M. Ghassemian, T. H. Richardson, J. T. Stege, M. Cayouette, A. C. McHardy, G. Djordjevic, N. Aboushadi, R. Sorek, S. G. Tringe, M. Podar, H. G. Martin, V. Kunin, D. Dalevi, J. Madejska, E. Kirton, D. Platt, E. Szeto, A. Salamov, K. Barry, N. Mikhailova, N. C. Kyrpides, E. G. Matson, E. A. Ottesen, X. Zhang, M. Hernández, C. Murillo, L. G. Acosta, I. Rigoutsos, G. Tamayo, B. D. Green, C. Chang, E. M. Rubin, E. J. Mathur, D. E. Robertson, P. Hugenholtz and J. R. Leadbetter, *Nature*, 2007, **450**, 560.
6. A. Tartar, M. M. Wheeler, X. Zhou, M. R. Coy, D. G. Boucias and M. E. Scharf, *Biotechnol. Biofuels*, 2009, **2**, 25.
7. H. Watanabe and G. Tokuda, *Annu. Rev. Entomol.*, 2010, **55**, 609.
8. W. J. Bell, L. M. Roth, C. A. Nalepa, *Cockroaches: Ecology, Behavior, and Natural History,* The Johns Hopkins University Press, Baltimore, 2007.
9. J. A. Breznak, J. W. Mertins and H. C. Coppel, *Univ. Wis. For. Res. Notes*, 1974, **184**, 1.
10. T. G. Lilburn, K. S. Kim, N. E. Ostrom, K. R. Byzek, J. R. Leadbetter and J. A. Breznak, *Science*, 2001, **292**, 2495.
11. J. A. Breznak, W. J. Brill, J. W. Mertins and H. C. Coppel, *Nature*, 1973, **244**, 577.
12. J. A. Breznak, *Symp. Soc. Exp. Biol.*, 1975, **29**, 559.
13. J. B. Nardi, R. I. Mackie and J. O. Dawson, *J. Insect Physiol.*, 2002, **48**, 751.
14. A. Neef, A. Latorre, J. Peretó, F. J. Silva, M. Pignatelli and A. Moya, *Genome Biol. Evol.*, 2011, **3**, 1437.
15. M. J. López-Sánchez, A. Neef, J. Peretó, R. Patiño-Navarrete, M. Pignatelli, A. Latorre and A. Moya, *PLoS Genet.*, 2009, **5**, e1000721.
16. Z. L. Sabree, S. Kambhampati and N. A. Moran, *Proc. Natl. Acad. Sci. U. S. A.,* 2009, **106**, 19521.

17. C. A. Nalepa, P. Luykx, K. D. Klass and L. L. Deitz, *Ann. Entomol. Soc. Am.*, 2002, **95**, 276.
18. M. D. Kane and J. A. Breznak, *Appl. Environ. Microbiol.*, 1991, **57**, 2628.
19. H. J. Gijzen, C. A. M. Broers, M. Barughare and C. K. Strumm, *Appl. Environ. Microbiol.*, 1991, **57**, 1630.
20. H. J. Gijzen and M. Barugahare, *Appl. Environ. Microbiol.*, 1992, **58**, 2565.
21. J. H. P. Hackstein and C. K. Strumm, *Proc. Natl. Acad. Sci. U. S. A.*, 1994, **91**, 5441.
22. P. Eggleton, R. D. T. Homathevi, J. A. Jones, D. MacDonald, D. E. Jeeva, R. G. Bignell, Davies and M. Maryati, *Philos. Trans. R. Soc. London*, 1999, **354**, 1791.
23. L. R. Cleveland, S. R. Hall, E. P. Sanders and J. Collier, *Mem. Am. Acad. Arts Sci.*, 1934, **17**, 185.
24. N. Lo, M. S. Engel, S. Cameron, C. A. Nalepa, G. Tokuda, D. Grimaldi, O. Kitade, K. Krishna, K. D. Klass, K. Maekawa, T. Miura and G. J. Thompson, *Biol. Lett.*, 2000, **3**, 562.
25. D. G. Inward, G. Beccaloni and P. Eggleton, *Biol. Lett.*, 2007, **3**, 331.
26. D. G. Inward, A. P. Vogler and P. Eggleton, *Mol. Phylogenet. Evol.*, 2007, **44**, 953.
27. J. L. Ware, J. Litman, K. D. Klass and L. A. Spearman, *Syst. Entomol.*, 2008, **33**, 429.
28. F. Legendre, M. F. Whiting, C. Bordereau, E. M. Cancello, T. A. Evans and P. Grandcolas, *Mol. Phylogenet. Evol.*, 2008, **48**, 615.
29. Y. Zhang, W. Xuan, J. Zhao, C. Zhu and G. Jiang, *Mol. Biol. Rep.*, 2010, **37**, 3509.
30. S. Koshikawa, S. Miyazaki, R. Cornette, T. Matsumoto and T. Miura, *Naturwissenschaften*, 2008, **95**, 859.
31. M. Ohkuma, S. Noda, Y. Hongoh, C. A. Nalepa and T. Inoue, *Proc. Biol. Sci.*, 2009, **276**, 239.
32. K. J. Carpenter, L. Chow and P. J. Keeling, *J. Eukaryot. Microbiol.*, 2009, **56**, 305.
33. N. Lo, M. S. Engel, S. Cameron, C. A. Nalepa, G. Tokuda, D. Grimaldi, O. Kitade, K. Krishna, K. D. Klass, K. Maekawa, T. Miura and G. J. Thompson, *Biol. Lett.*, 2007, **3**, 562.
34. P. Eggleton, G. Beccaloni and D. Inward, *Biol. Lett.*, 2007, **3**, 564.
35. J. A. Breznak in *Invertebrate-Microbial Interactions*, ed. J. M. Anderson, A. D. M. Rayner and D. W. H. Walton, Cambridge University Press, Cambridge, 1982, p. 173.
36. C. Noirot, *Ann. Soc. Entomol. Fr.*, 1995, **31**, 197.
37. S. Noda, T. Inoue, Y. Hongoh, M. Kawai, C. A. Nalepa, C. Vongkaluang, T. Kudo and M. Ohkuma, *Environ. Microbiol.*, 2006, **8**, 11.
38. B. M. Honigberg in *Biology of Termites*, ed. K. Krishna and F. M. Weesner, Academic Press, New York, 1970.
39. K. J. Carpenter, A. Horak, L. Chow and P. J. Keeling, *J. Eukaryot. Microbiol.*, 2011, **58**, 426.

40. M. Berlanga, B. J. Paster and R. Guerrero, *Int. Microbiol.*, 2009, **12**, 227.
41. Z. L. Sabree, P. H. Degnan and N. A. Moran, *Appl. Environ. Microbiol.*, 2010, **76**, 4076.
42. A. J. Ragauskas, C. K. Williams, B. H. Davison, G. Britovsek, J. Cairney, C. A. Eckert, W. J. Jr. Frederick, J. P. Hallett, D. J. Leak, C. L. Liotta, J. R. Mielenz, R. Murphy, R. Templer and T. Tschaplinski, *Science*, 2006, **311**, 484.
43. B. C. Saha, *J. Ind. Microbiol. Biotechnol.*, 2003, **30**, 279.
44. W. F. Anderson and D. E. Akin, *J. Ind. Microbiol. Biotechnol.*, 2008, **35**, 355.
45. J. A. Breznak and A. Brune, *Annu. Rev. Entomol.*, 1995, **39**, 453.
46. R. A. Prins and D. A. Kreulen, *Anim. Feed Sci. Technol.*, 1991, **32**, 101.
47. M. M. Martin, Comp. Biochem. *Physiol., Part A: Mol. Integr. Physiol.*, 1983, **75**, 313.
48. S. Itakura, K. Ueshima, H. Tanaka and A. Enoki, *Mokuzai gakkaishi*, 1995, **41**, 580.
49. J. Z. Sun and X. G. Zhou, in *Recent Advances of Entomological Research: from Molecular Biology to Pest Management*, ed. T. X. Liu and L. Kang, Springer, Berlin, 2011, p. 434.
50. M. E. Scharf, Z. J. Karl, A. Sethi and D. G. Boucias, *PLoS One*, 2011, **6**, e21709.
51. K. Nakashima, H. Watanabe, H. Saitoh, G. Tokuda and J. I. Azuma, *Insect Biochem. Mol. Biol.*, 2002, **32**, 777.
52. X. Zhou, M. R. Tarver and M. E. Scharf, *Development*, 2007, **134**, 601.
53. C. Oppert, W. E. Klingeman, J. D. Willis, B. Oppert and J. L. Jurat-Fuentes, *Comp. Biochem. Physiol., Part B: Biochem. Mol. Biol.*, 2010, **155**, 145.
54. K. D. Klass, C. A. Nalepa and N. Lo, *Mol. Phylogenet. Evol.*, 2008, **46**, 809.
55. M. E. Scharf and A. Tartar, *Biofuels Bioprod. Biorefin.*, 2008, **2**, 540.
56. X. Li and X. Zhou, unpublished data.

CHAPTER 13

Reverse Design of Natural Biomass Utilization Systems for Biomass Conversion

RYAN D. SYRENNE,[a,b,c] SHANGXIAN XIE,[a,b] JIANZHONG SUN[d] AND JOSHUA S. YUAN*[a,b]

[a] Department of Plant Pathology and Microbiology, Texas A&M University, College Station, TX 77843, USA; [b] Institute for Plant Genomics and Biotechnology, Texas A&M University, College Station, TX 77843, USA; [c] Program in Molecular Environmental Plant Sciences, Texas A&M University, College Station, TX 77843, USA; [d] Biofuels Institute, Jiangsu University, Zhengjiang, Jiangsu 212013, P. R. China
*Email: syuan@tamu.edu

13.1 The Challenges of Biofuels

13.1.1 Introduction

As early as the late 1800s, agriculturally derived bioethanol has been used as a substitute for gasoline to some extent by every current industrialized nation.[1] However, significant effort was not invested in bioethanol as a complete substitute for gasoline until the late-1970s' oil crisis, and more recently in 2008. The high demand for petroleum to support a massive transportation industry has driven biotechnology research towards utilizing first- and second-generation biofuel feedstocks for the development of ethanol-based liquid fuels.

RSC Energy and Environment Series No. 10
Biological Conversion of Biomass for Fuels and Chemicals: Explorations from Natural Utilization Systems
Edited by Jianzhong Sun, Shi-You Ding and Joy Doran-Peterson
© The Royal Society of Chemistry 2014
Published by the Royal Society of Chemistry, www.rsc.org

First-generation biofuel research focused on the conversion of starch and sucrose molecules into ethanol through a fermentative process, and therefore generally focused on crops that naturally accumulate these compounds such as maize, soybean, and sugarcane. However, concerns over first-generation bioethanol, such as the increase in global food prices, sustainability, and relatively small decreases in greenhouse gas emissions, have increased the attention on second-generation biofuels derived from lignocellulosic feed-stocks. Biofuels produced from lignocellulosic materials such as grass, wood, and herbaceous plants have the potential to supply domestic ethanol demands.[2] However, a cost-competitive model for second-generation biofuels is currently the Achilles' heel of lignocellulose-derived bioethanol. Reducing the cost of biomass-conversion technologies by increasing their efficiency has been the focus of tremendous effort, with various technologies emerging in recent years.[3] To understand the direction of current research and development in biomass-conversion technologies, a bottom-up approach is necessary.

13.1.2   Structure of Lignocellulosic Biomass

It is important to understand that cellulose, hemicelluloses, and lignin represent the major chemical constituents of the plant cell wall (PCW). Equally important, however, are the chemical differences between these components. Cellulose is a linear polymer of glucose subunits joined by repeating β-1,4-glycosidic bonds. The linear structure of cellulose allows it to form microfibrils of multiple cellulose molecules, which create difficulties during saccharification. Saccharification, also called "hydrolysis", of the cellulose polymer breaks the β-1,4-glycosidic bonds to yield individual glucose monomers. Compared to cellulose, hemicelluloses have an amorphous (highly branched, non-crystalline) polymer structure composed of xylose and arabinose (C5 sugars) and galactose, glucose, and mannose (C6 sugars).[4,5] The shorter branched polymers of hemicelluloses undergo saccharification more readily than cellulose due to their non-crystalline structure. Lignin is one of the primary non-carbohydrate fractions of the PCW, contributing significantly to cell-wall rigidity, and remains a major technical barrier to cost-competitive lignocellulosic biofuel. Lignin is a highly complex cross-linked polymer composed of phenolic monomers, providing structural support between hemicelluloses and celluloses in the primary cell wall.[4] Achieving increased conversion efficiency for the production of specific fuel types requires further understanding of chemical composition variations among the different lignocellulosic feedstocks.[5]

13.1.3   The Biomass Conversion Process

Lignocellulosic biomass conversion follows three general steps; pretreatment, saccharification, and fermentation. Ethanol and other biofuels can be produced by fermentation, which converts various monomer or oligomer sugars into energy *via* microorganisms. For fermentation to occur, lignocellulosic biomass, in particular cellulose and hemicelluloses, must first be hydrolyzed into fermentable sugars, which is also known as "saccharification". In this step, biocatalytic enzymes from microbes, such as cellulases and xylanases, catalyze the hydrolysis of the glycosidic linkages connecting sugar monomers. However, efficient enzymatic saccharification is not straightforward because lignocellulosic biomass is naturally resistant to deconstruction—also referred to as "recalcitrance".[6,7] As a result, pretreatment methods are introduced as a key initial step for biomass conversion; breaking the covalent linkages within the lignin–carbohydrate complexes to reduce the recalcitrance.[7–9] A pretreatment step is necessary to increase the accessibility of hydrolytic components to cellulase, increasing the overall efficiency and yield of ethanol from lignocellulosic biomass.

Pretreatment strategies are classified based by their mode of action to disrupt the cell wall and include: (1) physical or mechanical destruction; (2) physiochemical (AFEX); (3) chemical (dilute acid, alkali); and (4) biological (addition of hydrolytic enzymes).[4] Regardless of the method used, all pretreatment strategies must satisfy the following criteria. They should: (1) improve the availability of sugars; (2) avoid the loss of fermentable sugars; (3) avoid the formation of inhibitory compounds that interfere with hydrolysis or fermentation; and (4) be cost-effective.[4] Regardless of the pretreatment strategy used, the three-step biomass conversion represents the major cost in biofuel production. The conversion process needs to be optimized based on the mechanisms in natural biomass-utilization systems.

13.2 Reverse-Design of Natural Biomass-Utilization Systems for Biorefineries

13.2.1 Introduction

Millions of years of evolution and adaptation to dynamic environmental conditions have resulted in highly efficient biological systems. Land plants have evolved extensively complicated biomass macrostructures—namely the PCW—to conquer the terrestrial environment. The contributing factors to cell-wall recalcitrance are extensive; from the intricate arrangements of vascular bundles to the extent of lignification and the heterogeneity of PCW components.[10] The exact composition of these components varies among plant species; grass feedstocks like switchgrass contain less lignin then their woody counterparts for example.[11] Gymnosperms, mainly conifers, are highly lignified with large amounts of cellulose, whereas angiosperms are less lignified but have more hemicelluloses.[10] Microalgae have vastly different cell-wall compositions that are less well understood than those of land plants, and appear to lack lignin altogether.[12] Variations in the complex structures and chemicals between plant species likely evolved as adaptations for resisting attack from fungi, microbes and insects.

13.2.2 Natural Biomass-Utilization Systems

Natural factors believed to contribute to the recalcitrance of lignocelluloses include: (1) the epidermal tissue of the plant body; (2) the arrangement and density of the vascular bundles; (3) the amount of thick wall tissue; (4) the extensive lignification between cellulose and hemicelluloses; (5) the structural heterogeneity and complexity of the cell constituents; (6) the challenges for enzymes acting on insoluble substrates; and (7) the inhibitors to subsequent fermentations that exist naturally in cell walls or are generated during the conversion process.[7,10,13] These features all affect enzyme accessibility to cellulose, reducing the efficiency and increasing the conversion cost of lignocelluloses to ethanol. These features, ironically, drove the evolution of highly efficient biological systems capable of degrading plant material. Plant biomass is rich in potential chemical energy for those organisms able to access it.

Characteristic of all naturally occurring biomass-utilization systems is the extraordinarily high efficiency and extent to which the lignocellulose is degraded. Insects are among the most diverse group of organisms, and their guts represent unique natural biomass-utilization systems for novel enzyme discovery and biomass-deconstruction mechanism studies.[14–16] Wood-feeding termites are among one of the most efficient wood-degrading organisms, and play a crucial role in carbon turnover in the environment, thus representing the classic example of a natural biomass-utilization system.[17,18] Ruminant animals, such as cattle, also possess efficient biomass-utilization systems. However, the conversion efficiency is lower than those found in insects, and thus do not represent a model system for the reverse-design of natural biomass-utilization systems.[19] In addition to the above examples, system microorganisms such as white-rot fungi and many biomass-converting bacteria represent simpler natural biomass-utilization systems.

13.2.3   Insect Gut Symbiotic Microbiota

"Symbiosis" can be defined as the long-term mutually beneficial interactions between two or more organisms. "Endo-symbionts" are those beneficial microbes that simply reside within the host.[18–20] Specifically, natural biomass-utilization systems are those endo-symbiotic–host interactions that could be exploited for producing second-generation lignocellulosic ethanol and other biotechnologies. Many endo-symbiotic termite gut microbes are obligate in nature, or unable to survive independently from the host organism. Therefore, cultivation methods to study insect gut microbes are limited, with the termite endo-symbionts remaining largely unknown for nearly a century. Only in recent decades have molecular tools emerged that are capable of investigating termite gut microbiota; these tools include PCR, DGGE (denaturing gradient gel electrophoresis), and FISH (fluorescent *in situ* hybridization). Using these approaches, it was determined that insect gut microbes carried out at least two primary symbiotic functions: (1) the provisional supply of nutrients to the host from the gut microbiota; and (2) the lignocellulosic degradation of biomass by microbes to fuel their own energy demands.[21,22] Until recently however, data supporting the function in biomass deconstruction of endo-symbionts was limited and thus poorly understood in insects such as termites—although it was widely hypothesized that gut symbiotic microbes were integral for lignocelluloses degradation.

13.2.4 Biological Pretreatment

In addition to enzymatic saccharification, natural biomass-utilization systems have also been exploited for pretreatment. In particular, white-rot fungus has been widely used for biological pretreatment. Biological pretreatment is the process of disrupting the lignin–carbohydrate complexes, to reduce the recalcitrance, by enzymes and small biomolecules produced by microorganisms.[23] Biological pretreatment occurs under mild reaction conditions to degrade biomass and increase the saccharification efficiency. Additionally, the process is environmentally friendly, requires minimum energy input, and no additional fermentation inhibitors are produced. For these reasons, biological pretreatment has been touted as one of the most viable pretreatment technologies.[24,25] The most promising and well-used microorganism for biological pretreatment is the white-rot fungus, which belongs to the class *Basidiomycetes*.[26] In 1983, Hatakka analyzed the pretreatment effects of 19 different white-rot fungi on wheat straw enzymatic saccharification.[27] The study found that pretreatment with these white-rot fungi could increase the cellulose enzymatic saccharification to glucose and the efficiency could be comparable with that of alkali treatment. This led to an explosion of scientific discussion and research. Scientists from around the world began working on the discovery of new white-rot fungi strains and developing biological pretreatment technologies with white-rot fungi. In addition to white-rot fungi, brown-rot fungi and soft-rot fungi were also developed for use in biological pretreatment. Hundreds of fungi strains have been explored for use in biological pretreatments, including: *Pleurotus ostreatus*, *Ceriporiopsis subvermispora*, *Coriolus versicolor*, *Dichomitus squalens*, *Pycnoporus cinnabarinus*, *Ischnoderma benzoinum*, *Phanerochaete chrysosporium*, *Trametes versicolor*, *Ceripora lacerata*, *Stereum hisutum*, *Polyporus brumalis*, *Irpex lacteus*, *Merulius tremellosus* and others. Of all the investigated fungal species, only a select few have high lignin degradation selectivity that could greatly increase cellulose hydrolysis and saccharification efficiency. In addition to fungi, bacteria can also be used for biological pretreatment using both anaerobic and aerobic systems. Anaerobic degradation utilizes mainly mesophilic, rumen-derived bacteria.[28,29] Compared to anaerobic systems, aerobic bacterial pretreatment has a higher efficiency for degrading high-lignin biomass.[30,31] Although biological pretreatment has been successfully demonstrated to increase saccharification efficiency in the laboratory, the process time is long and the saccharification efficiency is relatively low compared with chemical/physical pretreatment methods. Despite the low saccharification efficiency and long processing time associated with fungal biological pretreatment methods, recent biological pretreatment with termite conversion systems has been shown to take 4–5 h; from modification of lignin components in the foregut/midgut to the hindgut for biomass hydrolysis and processing.[32] For future biomass-conversion technologies, reducing the pretreatment duration with a concomitant increase in efficiency are the key technical barriers for realizing large-scale biomass-conversion applications. However, the molecular mechanisms in white-rot fungus can be explored and the components can be used to reverse-design relevant conversion processes.

13.2.5 The Importance of Next-Generation -Omics Technologies for the Reverse-Design of Natural Biomass-Utilization Systems

The emergence of next-generation -omics technologies—genomics, meta-genomics, proteomics and transcriptomics—has clarified the symbiotic biocatalyst systems present in insects such as termites.[16,18] The traditional view of biomass deconstruction by termites was predicated by the assumption that 'higher' termites lacked cellulolytic hindgut flagellate protozoa, and therefore symbiotic microbes were solely responsible for biomass conversion.[33,34] When a cellulase gene of termite origin was discovered in 1998, the role for termite gut symbionts shifted away from biomass degradation and instead towards nitrogen fixation and anaerobic respiration.[35] However, meta-genomic and proteomic efforts to study the 'higher' wood-feeding termite hindgut revealed a community of cellulose-degrading microbes—even discovering a bacterial gene encoding a novel endo-xylanase—yet no evidence of lignin degradation was shown.[18] This work corroborated previous genome-level studies indicating the microbial roles for biomass deconstruction and further provided extensive gene-centered details of 'higher' termite hindgut microbes.[18] Termite host tissues, *via* recent transcriptomic efforts, have been shown to produce a litany of cellulases and hemicellulases, as well as lignases, and contribute to lignocellulose degradation.[36–38]

Despite the controversy over which plays a more significant role, the symbionts or host, synergistic digestive mechanisms for biomass conversion are relevant for the basic and applied reverse-design of natural biomass-utilization systems. From this perspective, -omics technologies—coupled with data annotation and high-throughput functional screening—will continue to aid novel gene discovery and elucidate important biocatalyst systems.[19,39] Systems biology approaches that collectively integrate various -omics-derived data have made possible the holistic study of natural biomass-conversion processes.[16]

13.3 Biomass Processing Optimization Based on Natural Biomass-Utilization Systems

Multiple enabling elements are required to improve current biomass-conversion technologies: first, biocatalysts with effective inhibitor-removal capacities will be needed to increase the conversion efficiency; second, microbial strains capable of consolidating the hydrolysis and fermentation steps to reduce enzyme load will reduce costs; and third, down-stream compatible pretreatment methods must be made available to further reduce costs. Achieving a comprehensive understanding of natural biomass-utilization systems will guide the reverse-design of biorefineries for lignocellulosic biomass conversion. Next-generation -omics technologies are already paving the way to defining various biocatalyst co-regulatory networks, integrating systems with the molecular-level understanding of naturally occurring biomass-degradation systems.

13.3.1 The Reverse-Design of Natural Biomass-Utilization Systems as an Enabling Tool for Consolidated Biomass Processing

The reverse-design of natural biomass-conversion processes, enabled by next-generation -omics technologies and systems biology, could give rise to a powerful energy-conversion process called "consolidated biomass processing" (CBP). Many of the studies relevant to biomass-conversion technologies have focused on the discovery and production of cellulases and other lignocelluloses hydrolytic enzymes. Although important for biotechnology applications, the hydrolysis of PCW components only represents one of the technological barriers when considering a cost-competitive model for second-generation lignocellulose-derived bioethanol. Consolidated biomass processing represents an overall schema for a cost-competitive model of the lignocellulosic production of bioethanol for natural biomass-utilization system reverse-designed biorefineries. The discovery of cellulases and other hydrolytic enzymes is only one aspect of biomass conversion, and alone cannot offer a cost-competitive solution to second-generation lignocellulosic biofuels.

However, integration of all the biomass processing elements could enable CBP as the ultimate cost-effective conversion process. These processes include pretreatment, enzymatic hydrolysis, and the fermentation of released C5 and C6 sugars into ethanol. CBP could lower the technology costs associated with these processes through several efforts: first, enzyme–microbe synergy derived from natural biomass-utilization systems by high-throughput -omics technologies *via* systems biology; second, the use of robust thermophilic organisms capable of cellulose hydrolysis; third, simultaneous saccharification and fermentation of cellulose to ethanol using synthetic biology tactics; and fourth, decreasing the effect of hydrolysis inhibition on microbes inherent in the ethanol fermentative processes (Figure 13.1).

13.3.2 Consolidated Biomass Processing Enabled Progress

The reverse-design of biorefineries by CBP research initiatives has made considerable progress, primarily in two complementary synthetic biology strategies. The first approach features engineered cellulytic microbes capable of increased hydrolysis, thus improving overall product yields. The second approach utilizes engineered, non-cellulase-producing microbes, for increased fermentation yields of the released sugar monomers. Therefore, from a

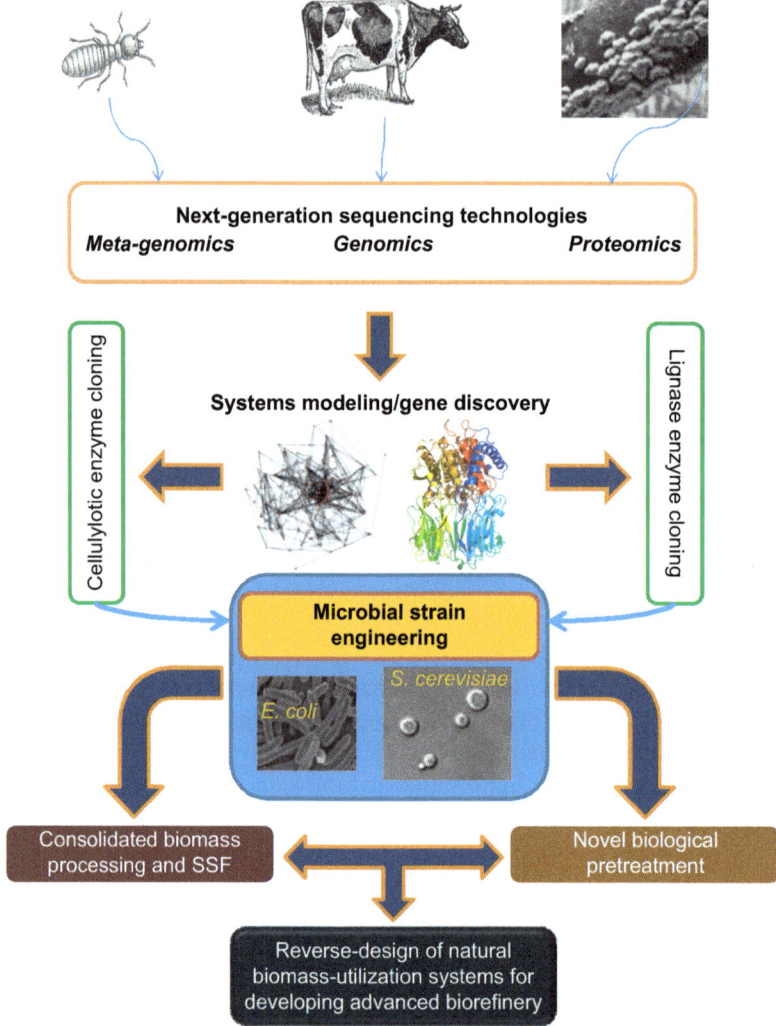

Figure 13.1 Flowchart showing natural biomasss-utilization systems for the reverse-design of biorefinery. Next-generation -omics technology will enable novel gene discovery and systems biology modeling of complex regulatory elements in cellulose-utilizing organisms like termites, cattle, and white-rot fungi. Novel or efficient cellulolytic (cellulose/hemicellulose-degrading) enzymes as well as lignases are cloned based on mined sequence data. Cloned genes are introduced into candidate microbial strains capable of consolidating biomass processing and/or providing an effective biological treatment strategy. SSF = simultaneous saccharification and fermentation.

synthetic biology perspective, recent studies concerning fundamental biological conversion processes have shown promise for developing CBP as feasible method for second-generation lignocellulosic biofuels.

Naturally occurring cellulytic-capable microorganisms will play a fundamental role in developing synthetic organisms capable of producing sugar yields that meet requirements for industrial use. Metabolic engineering was used to improve the cellulytic properties of *Clostridium cellulolyticum*, a mesophilic anaerobe prevalent in insect gut communities.[18,39] Inefficient regulation of the carbon flux from pyruvate leads to an accumulation of intracellular growth inhibitory compounds.[40,41] Conceptually, a decrease in cellulose-degrading microbial populations should impede hydrolysis efficiency in biomass-conversion technologies. It was shown that a heterologous expression of pyruvate dehydrogenase and alcohol dehydrogenase from *Zymomonas mobilis* in recombinant *C. cellulolyticum* could increase cellulose consumption and the concentration of ethanol by 150 and 43%, respectively.[42] Essentially, a metabolic engineering approach transformed the naturally weak cellulytic activity of *C. cellulolyticum* into a highly fermentative anaerobic bacterium.[42] The report thus highlighted the feasibility of metabolic engineering to increase overall cellulose hydrolysis in natively weak cellulytic organisms by eliminating pyruvate accumulation, a known bacterial growth inhibitory compound.

Development of recombinant non-cellulytic microbes is another essential component to addressing the sustainability of CBP technologies. Currently, high-temperature steam explosion stands as one of the most promising pretreatment options for lignocellulosic biomass. As a result, the harsh conditions directly impact the selection of microbes for use in a CBP biorefinery, obviously implying the importance of thermophilic organisms capable of cellulose degradation. Equally important, though, is the ethanol tolerance for microbes used in CBP, as high ethanol concentrations will inhibit cellulose hydrolysis and negatively impact the overall biomass-conversion efficiency.[43] For this reason, highly ethanol-tolerant thermophilic bacterial strains have received considerable attention.

Metabolic engineering of a thermophilic organisms was first achieved in *Thermoanaerobacterium saccharolyticum*, and showed that gene-transfer tools were not only possible for thermophiles, but also that ethanol production could be increased.[41,44] Eliminating lactic acid production *via* gene knockout in *T. saccharolyticum* was reportedly accompanied by a proportional increase in acetic acid and ethanol yields.

Overall, the systems biology analysis of natural biomass-utilization systems will provide the key components including microbes and enzymes for reverse-design. The recently developed synthetic biology approaches will allow us to reverse-engineer microbes that can be used for a more consolidated processing strategy based on natural biomass-utilization systems.

References

1. B. D. Solomon, J. R. Barnes and K. E. Halvorsen, *Biomass Bioenergy*, 2007, **31**, 416.
2. M. Balat, *Energy Sources, Part A*, 2009, **31**, 1242.

3. S. Kumar, S. P. Singh, I. M. Mishra and D. K. Adhikari, *Chem. Eng. Technol.*, 2009, **32**, 517.

4. P. Kumar, D. M. Barrett, M. J. Delwiche and P. Stroeve, *Ind. Eng. Chem. Res.*, 2009, **48**, 3713.

5. M. Carrier, A. Loppinet-Serani, D. Denux, J.-M. Lasnier, F. D. R. Ham-Pichavant, F. O. Cansell and C. Aymonier, *Biomass Bioenergy*, **35**, 298.

6. E. M. Rubin, *Nature*, 2008, **454**, 841.

7. L. R. Lynd, M. S. Laser, D. Bransby, B. E. Dale, B. Davison, R. Hamilton, M. Himmel, M. Keller, J. D. McMillan, J. Sheehan and C. E. Wyman, *Nat. Biotechol.*, 2008, **26**, 169.

8. S. P. S. Chundawat, G. T. Beckham, M. E. Himmel and B. E. Dale, *Annu. Rev. Chem. Biomol. Eng.*, 2011, **2**, 121.

9. B. Yang and C. E. Wyman, *Biofuels Bioprod. Biorefin.*, 2008, **2**, 26.

10. M. E. Himmel, S.-Y. Ding, D. K. Johnson, W. S. Adney, M. R. Nimlos, J. W. Brady and T. D. Foust, *Science*, 2007, **315**, 804.

11. J. Perez, J. Munoz-Dorado, T. de la Rubia and J. Martinez, *Int. Microbiol.*, 2002, **5**, 53.

12. C. F. Delwiche, L. E. Graham and N. Thomson, *Science*, 1989, **245**, 399.

13. F. Chen and R. A. Dixon, *Nat. Biotechnol.*, 2007, **25**, 759.

14. T. L. Erwin, *Coleopterists Bull.*, 1982, **36**, 74.

15. A. D. Chapman, ed. *Numbers of Living Species in Australia and the World*, Report for the Australian Biological Resources Study, Canberra, Australia, 2006.

16. W. Shi, R. Syrenne, J.-Z. Sun and J. S. Yuan, *Insect Sci.*, 2010, **17**, 199.

17. J. A. Breznak and A. Brune, *Annu. Rev. Entomol.*, 1994, **39**, 453.

18. F. Warnecke, P. Luginbuhl, N. Ivanova, M. Ghassemian, T. H. Richardson, J. T. Stege, M. Cayouette, A. C. McHardy, G. Djordjevic, N. Aboushadi, R. Sorek, S. G. Tringe, M. Podar, H. G. Martin, V. Kunin, D. Dalevi, J. Madejska, E. Kirton, D. Platt, E. Szeto, A. Salamov, K. Barry, N. Mikhailova, N. C. Kyrpides, E. G. Matson, E. A. Ottesen, X. Zhang, M. Hernandez, C. Murillo, L. G. Acosta, I. Rigoutsos, G. Tamayo, B. D. Green, C. Chang, E. M. Rubin, E. J. Mathur, D. E. Robertson, P. Hugenholtz and J. R. Leadbetter, *Nature*, 2007, **450**, 560.

19. A. Brune, *Trends Biotechnol.*, 1998, **16**, 16.

20. J. S. Yuan, Y. Qu, S. Li and C. N. Stewart, *Biofuels*, **2**, 487.

21. A. E. Douglas, *Annu. Rev. Entomol.*, 1998, **43**, 17.

22. G. Tokuda and H. Watanabe, *Biol. Lett.*, 2007, **3**, 336.

23. Y. Sun and J. Cheng, *Bioresour. Technol.*, 2002, **83**, 1.

24. F. A. Keller, J. E. Hamilton and Q. A. Nguyen, *Appl. Biochem. Biotechnol.*, 2003, **108**, 27.

25. J. Lee, *J. Biotechnol.*, 1997, **56**, 1.

26. D.-L. Huang, G.-M. Zeng, C.-L. Feng, S. Hu, C. Lai, M.-H. Zhao, F.-F. Su, L. Tang and H.-L. Liu, *Bioresour. Technol.*, 2010, **101**, 4062.

27. A. I. Hatakka, *Appl. Microbiol. Biotechnol.*, 1983, **18**, 350.

28. Z.-H. Hu and H.-Q. Yu, *Process Biochem.*, 2005, **40**, 2371.

29. Z.-H. Hu, S.-Y. Liu, Z.-B. Yue, L.-F. Yan, M.-T. Yang and H.-Q. Yu, *Environ. Sci. Technol.*, 2007, **42**, 276.
30. A. Arora, L. Nain and J. K. Gupta, *World J. Microbiol. Biotechnol.*, 2005, **21**, 303.
31. A. Mshandete, L. Björnsson, A. K. Kivaisi, S. T. Rubindamayugi and B. Mattiasson, *Water Res.*, 2005, **39**, 1569.
32. J. Ke, D. Laskar, D. Gao and S. Chen, *Biotechnol. Biofuels*, 2012, **5**, 11.
33. L. R. Cleveland, *Proc. Natl. Acad. Sci. U. S. A.*, 1923, **9**, 424.
34. X. Zhou, J. A. Smith, F. M. Oi, P. G. Koehler, G. W. Bennett and M. E. Scharf, *Gene*, 2007, **395**, 29.
35. H. Watanabe, H. Noda, G. Tokuda and N. Lo, *Nature*, 1998, **394**, 330.
36. M. E. Scharf, Z. J. Karl, A. Sethi and D. G. Boucias, *PLoS One*, 2011, **6**, e21709.
37. A. Tartar, M. Wheeler, X. Zhou, M. Coy, D. Boucias and M. Scharf, *Biotechnol. Biofuels*, 2009, **2**, 25.
38. M. R. Coy, T. Z. Salem, J. S. Denton, E. S. Kovaleva, Z. Liu, D. S. Barber, J. H. Campbell, D. C. Davis, G. W. Buchman, D. G. Boucias and M. E. Scharf, *Insect Biochem. Mol. Biol.*, 2010, **40**, 723.
39. W. Shi, S.-Y. Ding and J. Yuan, *BioEnergy Res.*, 2011, **4**, 1.
40. E. Guedon, M. L. Desvaux and H. Petitdemange, *J. Bacteriol.*, 2000, **182**, 2010.
41. L. R. Lynd, W. H. V. Zyl, J. E. McBride and M. Laser, *Curr. Opin. Biotechnol.*, 2005, **16**, 577.
42. E. Guedon, M. L. Desvaux and H. Petitdemange, *Appl. Environ. Microbiol.*, 2002, **68**, 53.
43. T. Williams, J. Combs, B. Lynn and H. Strobel, *Appl. Microbiol. Biotechnol.*, 2007, **74**, 422.
44. S. G. Desai, M. L. Guerinot and L. R. Lynd, *Appl. Microbiol. Biotechnol.*, 2004, **65**, 600.

CHAPTER 14

The Ruminant Animal as a Natural Biomass-Conversion Platform and a Source of Bioconversion Agents

PAUL J. WEIMER[a,b]

[a] Agricultural Research Service, US Department of Agriculture, USA and US Dairy Forage Research Center, Madison, WI 53706, USA; [b] Department of Bacteriology, University of Wisconsin, Madison, WI 53706, USA
Email: paul.weimer@ars.usda.gov

14.1 Introduction

Plant biomass is, on a tonnage basis, the most abundant organic material on the surface of the Earth, and it is a major component of the Earth's carbon cycle. Because of its abundance, its accessibility, and its biodegradability, plant biomass is also the foundation of most schemes for the sustainable production of biofuels. The majority of the biomass in most plants resides in the cell wall, which consists primarily of three types of polymer—cellulose, hemicelluloses and lignin. As lignin is only slowly biodegradable and hemicelluloses are readily hydrolyzed to sugars by acidic pretreatment,[1] most bioconversion research has been aimed at improving the degradation of cellulose, the most abundant of the three polymers.

RSC Energy and Environment Series No. 10
Biological Conversion of Biomass for Fuels and Chemicals: Explorations from Natural Utilization Systems
Edited by Jianzhong Sun, Shi-You Ding and Joy Doran-Peterson
© The Royal Society of Chemistry 2014
Published by the Royal Society of Chemistry, www.rsc.org

Most current and proposed biomass-based (non-photosynthetic) biofuel production systems seek to utilize bioconversion agents that exhibit high cellulose biodegradation capacity, and in general these systems rely on some combination of cellulose-degrading enzymes and sugar-fermenting microorganisms, with ethanol as the biofuel product of interest.[2] The bioconversion platform that has received the most attention is the simultaneous saccharification and fermentation (SSF) platform employing cellulase enzymes and hexose-fermenting yeasts such as *Saccharomyces cerevisiae*. Though advances have been made in developing fungal strains secreting large amounts of cellulase, and in improved yeast strains that produce high titers of ethanol, SSF processes are currently limited by the high cost of enzyme production, due in part to the requirement of a separate reactor dedicated to aerobic cultivation of the cellulase-producing fungus. Historically, a second limitation of the SSF process has been the general inability of industrial yeast strains to ferment pentose sugars. Most engineered pentose-fermenting strains either utilize glucose prior to utilizing xylose[3,4] and/or utilize xylose more slowly than glucose.[5]

An alternative platform for cellulosic biomass conversion is consolidated bioprocessing (CBP), which employs anaerobic microorganisms (usually bacteria) that produce their own cellulase enzymes and ferment the hydrolytic products (typically soluble cellulose oligomers), with ethanol as the desired product. The most well-studied CBP bacteria are the thermophilic *Clostridium thermocellum*[6,7] and the mesophilic *Clostridium phytofermentans*.[8] CBP is theoretically a simpler and more elegant process than SSF, as the entire bioconversion process can be carried out in a single bioreactor. However, CBP organisms generally produce additional end-products (acetic, lactic, and formic acids, along with H_2 and CO_2), thus lowering the ethanol yield, and are sensitive to product inhibition, resulting in ethanol titers too low for cost-efficient recovery.

The conventionally described forms of both SSF and CBP ethanol-production systems share the fact that they are both artificial constructs. While the microbial agents in both platforms were originally isolated from natural environments, they are not particularly abundant there, and getting these agents to perform well in a managed industrial habitat has required intensive effort. While these pure culture systems have been developed to maximize productivity, the advances have come at a cost. Successive improvements in productivity have narrowed the adaptable range of the microbial agents, to the point where they have lost the opportunity, or perhaps even the capacity, to interact with other microbial species with whom they have co-evolved. For these monocultures, the bioreactor has become, in essence, a life-support system whose residents have lost the ability to adapt to upsets in the physical/chemical/biological conditions that continuously occur in nature. Moreover, as will be discussed below, the rates of biomass conversion in these managed systems do not necessarily outpace those in more natural systems.

With these facts in mind, it is instructive to examine natural systems of biomass degradation to identify opportunities for further improving biomass conversion in more managed systems. This chapter will focus on a single natural bioconversion system, the ruminant animal, particularly the grazing cow. Over 60 years ago, Hungate famously developed the analogy of the ruminant as a self-feeding, self-replicating mobile cellulose-degrading bioreactor.[9] This brilliant and prescient observation by the father of rumen microbiology will serve as an outline for our analysis. In essence, the grazing ruminant animal further consolidates CBP to handle all aspects of biomass conversion, from harvesting to mechanical pretreatment, to fermentation, to the recovery of fermentation products.[9,10] In examining this system, some intriguing questions arise: How does biomass conversion in the ruminant compare quantitatively with that in managed systems? What novel process features have been developed by the ruminant animal? Can these features be incorporated into managed biomass-to-fuels systems, or can they replace less desirable process steps? And finally, can the ruminal fermentation, or some of its microbes and enzymes, serve as a basis for the production of fuels in managed bioreactors?

14.2   The Ruminant Animal

Ruminant animals represent one of the most widespread and abundant plant bioconversion systems on Earth. Ruminants are classified within the suborder Ruminantia of the order Artiodactyla, class Mammalia. Of the 155 recognized species, 121 are classified within the family Bovidae, which includes not only domestic livestock species (cattle, sheep, goats, water buffalo, yak) but also many familiar wild species that dominate grasslands worldwide. The bovids vary substantially in size and in feeding behavior.[11] Many species are concentrate selectors or browsers that preferentially consume the most digestible parts of the plant. Others, particularly cattle (*Bos taurus* and *Bos indicus*), have relatively large muzzles, are thus are less selective in their feeding behavior and tend to consume diets high in roughage. Thus cattle are of particular interest for processing recalcitrant plant cell walls, the major component of most grassland plants and bioenergy crops. In fact, from a processing standpoint, the ruminant animal in general, and grazing cattle in particular, represents perhaps the pinnacle of the natural bioconversion of cellulosic biomass. The anatomical and physiological adaptations of ruminants to their diet are superbly detailed in the monograph of Van Soest,[11] and are briefly summarized here.

Ruminants are popularly considered to have four stomachs, but in fact have a single stomach separated into four specialized chambers. All arise from the same embryological tissue, and they are ordered (in terms of passage of materials) as the rumen, reticulum, omasum and abomasum. The rumen and reticulum are usually considered as a single unit (the reticulorumen), and in this analysis will be referred to together as the "rumen". The rumen evolved in grazing or browsing animals as a means of slowing the rate of passage of fibrous feeds to permit an anaerobic fermentation by slowly growing fibrolytic microbes. The chief fermentation products are volatile fatty acids (VFAs, C2–C6 monocarboxylic acids) that are absorbed through the rumen wall, which is highly papillated to increase absorption into the bloodstream of the host animal. Once absorbed, the acids are used by the host tissues as their primary source of energy and anabolic precursors.

Digesta exit the (reticulo)rumen through the reticuloomasal orifice, which is lined with folded tissue (leaves) that capture larger feed particles for muscular propulsion back into the reticulum, where the fermentation process continues. Digesta escaping into the omasum are subjected to partial dewatering and salt removal prior to passage into the fourth chamber, the abomasum. This chamber, a homolog to the gastric stomach in non-ruminants, carries out an acid hydrolysis of proteins (including both those present in the feed and those arising from microbial protein synthesis in the rumen), for subsequent absorption in the small intestine. Further passage brings the digesta to the colon, and some of these digesta enters a cecum where additional fermentation occurs.

14.2.1   Feedstock Pretreatment, Bovine Style

Plant cell walls, the bulk of the bovine diet in nature and in domestic pastoral systems, are naturally recalcitrant to rapid biodegradation, and engineered bioconversion systems almost invariably include physical or chemical pretreatment of the feedstock to enhance the ability of enzymes or microbes to access the component plant polysaccharides from tissues having various degrees of lignification.[12] The ruminant has no option for chemical pretreatment, but instead has developed an unusually effective physical pretreatment process based on frequent rechewing of ingested feed, a process known as "rumination". Feed is first ingested by rapid eating (approximately one bite per second),[13,14] during which the feed is chewed but a little during its translocation to the rumen. Once there, the feed is hydrated and then returned by regurgitation (over the course of hours) to the buccal cavity for chewing. The physical process of chewing elegantly solves the problem of how to increase the surface area of the feed particles and expose their contents to the ruminal microflora, without making the particles too small to compromise the feed particles' ability to stimulate additional rumination and salivation. By crushing the feed particles between the stationary upper teeth and the bank of lower teeth that are moved by jaw motion in a roughly circular pattern, feed particles are stripped or delaminated along their length, in a manner similar to tearing sheets off a tablet of paper (as opposed to the more difficult task of cutting the tablet into blocks). This motion uniquely provides a dramatic increase in surface area without greatly reducing the length of the fiber particles, all at a minimal energy cost (estimated at approximately 1 per cent of the gross energy of the feed particle).[10]

Compared to eating, rumination involves substantially more chews over a substantially longer time (Table 14.1), and the total number of bites and chews (over 5×10^4 per day on high-forage diets) is impressive. Rumination time varies among animals and is primarily determined by a combination of feeding strategy, body size, and feed composition. In general, the rumination time is greater in grazers than in browsers, and increases with both body mass (which affects level of feed intake) and roughage content of the diet (usually quantified

Table 14.1   Mean values for eating, drinking and ruminating activity, and for the composition of rumen contents, determined from 12 cows fed total mixed rations that were high or low in neutral detergent fiber (NDF). Data from ref. 13.

Cow activity/per day	High fiber (35% NDF)	Low fiber (25% NDF)
Dry matter intake/kg	18.7^b	22.8^a
Water intake		
Free water/L	74.4^b	82.2^a
Total water/L^c	96.9^b	99.0^a
Eating time/min	349^a	294^b
Ruminating time/min	496^a	383^b
Eating bites	$22\ 783^a$	$18\ 618^b$
Ruminating chews	$31\ 232^a$	$23\ 867^b$
Total bites	$54\ 015^a$	$42\ 485^b$
Rumen contents		
Digesta volume/L	98.8^a	89.8^b
Digesta + headspace volume/L	111.5	108.0
Wet digesta/kg	86.4^a	75.7^b
Dry matter/kg	11.3	10.8
Dry matter/%^{d}	13.0^b	14.3^a
Density/kg L^{-1e}	0.88^a	0.84^b

^{a,b}Averages having different superscripts within rows differ ($P<0.05$).
^{c}Includes water in feed.
^{d}kg of dry matter per kg wet digesta.
^{e}kg wet digesta per liter of wet digesta.

as the neutral detergent fiber, or NDF, *i.e.*, the fraction of dry weight that is insoluble in boiling sodium dodecyl sulfate solution at pH 7).[15] There appears to be an upper limit of rumination time of *ca.* 10–11 hours per day, so that at high levels of intake there may be insufficient time for proper rumination, resulting in reduced ruminal digestibility during a shorter ruminal residence time.[16]

14.2.2   The Ruminal Fermentation

The basic conversions of biomass components to fermentation products are depicted in Figure 14.1. Most of the organic components of biomass (lignin and long-chain fatty acids being the main exceptions) are subjected to conversion in the rumen to a mixture of VFAs, the gases methane and CO_2, and microbial cell mass. Each of the various conversions is carried out by a number of different species, some with overlapping catabolic capabilities.

14.2.2.1   Carbohydrates

Cellulose, the most abundant carbohydrate in whole plant material, is fermented with first-order kinetics[14–19] at rate constants that can approximate $0.1\,h^{-1}$. Storage polysaccharides such as cereal grain starches[20–22] and grass fructans[23]

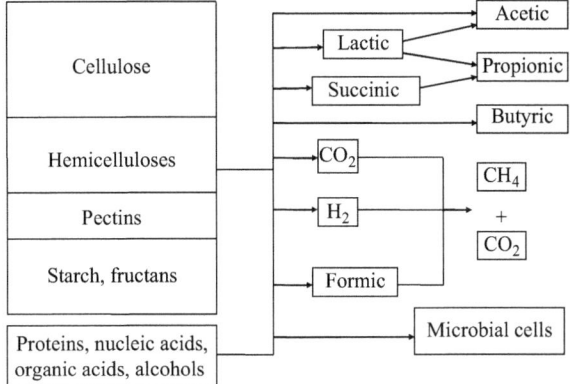

Figure 14.1 Overview of the ruminal conversion of biomass components to fermentation products and microbial cells. Almost all of the components of the biomass, except for lignin and some lipids, can potentially be converted during the anaerobic fermentation process.
Adapted from ref. 10.

are fermented more rapidly and to near completion, and soluble sugars are fermented more rapidly yet. The products of these fermentations are a mixture of acetic, propionic and butyric acids, along with methane and CO_2. Hungate[24] described the following general carbohydrate fermentation stoichiometry:

$$56\,\text{Hexose} \rightarrow 62\,\text{Acetic} + 22\,\text{Propionic} + 16\,\text{Butyric} + 60.5\,CO_2 + 33.5\,CH_4 + 27\,H_2O$$

Hemicelluloses and their major components, the pentose sugars, are all readily fermented and result in the formation of similar products, although in different ratios.[21] Pectins from both forage[25] and citrus[26] are degraded more rapidly than any other cell-wall component ($k_d \sim 0.3\,\text{h}^{-1}$), with an unusually high ratio of acetic acid to other products. Di- and tricarboxylic organic acids, which may comprise several percent of the dry weight of grasses, are fermented either mostly to acetic acid (for citric, aconitic, and quinic acids) or to acetic/propionic mixtures (malic and malonic acids).[27]

14.2.2.2 Proteins

Proteins are hydrolyzed to peptides and amino acids, which are then fermented to a variety of VFAs, including both C2–C6 straight-chain and C4–C5 branched-chain acids, along with NH_3 and CO_2 [ref. 24]. The rate and extent of protein fermentation is largely a function of the solubility of the proteins, and the degree to which they are protected from enzymatic attack *via* sequestration within plant cell walls, or by covalent linkage to tannins or to carbohydrates (the latter usually being an intentional result of feed processing to promote passage of the protein undegraded to the abomasum).[28]

Feed proteins are hydrolyzed to peptides and amino acids by extracellular proteases and peptidase enzymes elaborated by many microbial (primarily bacterial) species. Once formed, these hydrolytic products are readily fermented by a variety of ruminal bacteria. There appears to be two functional classes of peptide and amino acid fermenters. The first are nutritionally versatile species such as *Prevotella ruminicola*; many of these species can also ferment hemi-celluloses and various soluble sugars.[29,30] The second are the obligate amino acid fermenters originally identified as *Clostridium aminophilum*, *C. sticklandii* and *Peptostreptococcus anaerobius*,[31] and now known to include a wide variety of other taxa.[32] These specialist species ferment amino acids at about 10 times the specific activity of the first group, which has led to their being called "hyperammonia producing bacteria (HAB)". The HAB appear to be present in only low numbers in the rumen (probably due to their sensitivity to bacteriocins and to certain feed additives such as monensin), but rapidly take over trypticase-containing enrichment cultures *in vitro*.[31]

14.2.2.3   Nucleic Acids

Nucleic acids are rapidly degraded in the rumen by nucleases.[33,34] The resulting ribose and deoxyribose are readily fermented by many ruminal bacteria. Nucleic acid bases are fermented to varying extents, primarily to C1–C4 VFAs.[22,35]

14.2.2.4   Lipids

Lipid utilization in the rumen is relatively limited. Triglycerides are hydrolyzed by several bacterial species (*e.g.*, the specialist bacterium *Anaerovibrio lipolytica*), and the resulting glycerol is readily fermented (probably by many species) to a mixture of propionic and acetic acids, and CO_2 [ref. 36]. Long-chain fatty acids are typically carried through the rumen with little modification,[37] although some unsaturated fatty acids are subjected to isomerization or, if electron donors are available, are reduced to the corresponding saturated fatty acids. These latter "biohydrogenation" reactions may serve as a detoxification mechanism, as saturated fatty acids are considerably less toxic to ruminal bacteria than unsaturated fatty acid precursors.[38] Because of the toxicity of their fatty acid moieties and their tendency to coat feed particles, lipids present in forages or added exogenously as dietary fats can suppress certain processes such as fiber digestion[39] or methanogenesis.[40]

14.2.2.5   Aromatic Compounds

Lignin interferes with the biodegradation of plant cell walls in most microbial habitats, and the rumen is no exception. First-order rate constants of fiber in mature forages (0.04 to 0.07 h^{-1}) are much lower than those of the same forages harvested at a more immature and less lignified growth stage (0.08 to 0.3 h^{-1}).[41] Both the rate and extent of NDF digestion display strong negative correlations with measured lignin content across and within forage species.[42] Although it appears that certain low-molecular-weight lignin model compounds can be slowly degraded by mixed ruminal microflora *in vitro*,[43] polymer lignin is regarded as indigestible under anaerobic conditions,[44] and in fact is widely used as a marker for indigestible residue in the measurement of the rate of passage through the digestive tract.[11] By contrast, ruminal microbes display some ability to degrade tannins, particularly the low-molecular-weight (non-condensed) tannins.[45]

The ruminal fermentation has some similarities to the well-known process of anaerobic digestion that occurs in natural environments (such as aquatic sediments and anoxic soils) and in domestic wastewater treatment (Figure 14.2). Both processes rely on a diverse community of microbes to convert complex organic matter to simpler products under anaerobic conditions. In anaerobic digestion, VFAs of ≥ 3 carbons are oxidized by proton-reducing acetogenic bacteria to yield acetate, H_2 and CO_2. The acetate is cleaved to methane and CO_2 by a specialized group of archaea, the aceticlastic methanogens; this latter conversion accounts for approximately 70% of the methane produced in anaerobic digesters.[46] The ruminal fermentation may be regarded as an incomplete anaerobic digestion process, in which VFAs accumulate and are used by the host for energy and biosynthetic reactions, rather than being converted to methane and CO_2. VFA accumulation results from a lack of proton-reducing acetogens and aceticlastic methanogens, both of which grow too slowly to prevent washout from the rumen.

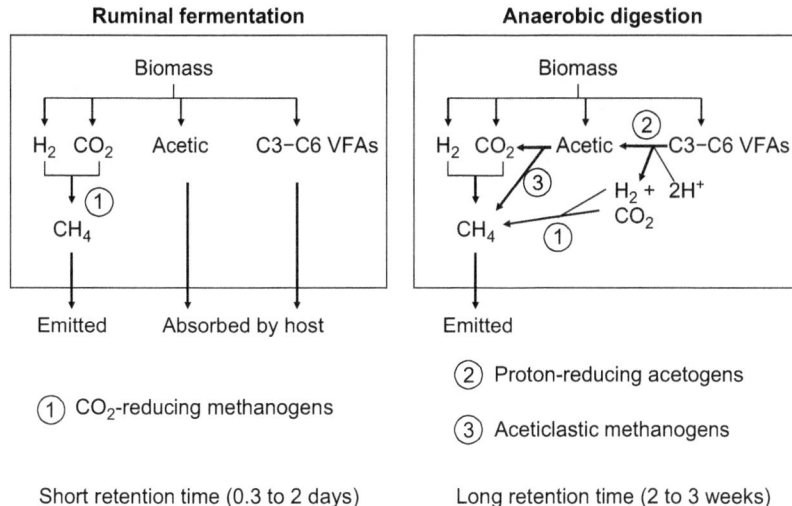

Figure 14.2   Comparison of ruminal fermentation with anaerobic digestion. In the rumen, fermentation is incomplete owing to the absence of proton-reducing acetogens and aceticlastic methanogens, which cannot grow at rates sufficient to keep up with the rapid dilution rate of the ruminal contents. This results in the accumulation of VFAs, which are used by the host animal for energy and biosynthetic precursors. Both processes also release substantial amounts of CO_2.

14.2.3 The Rumen as a Bioreactor

As the central organ in a complex multi-compartment digestive system, the rumen functions as a fermentation vat whose properties resemble many of the desirable features of an industrial bioreactor.[9,10] Temperature is controlled, in this case near 39 °C, the growth optimum of most ruminal microbes, by a combination of fermentation heat and body heat, with the drinking of water serving as an effective check on over-heating. pH is regulated by a combination of buffering (by saliva, produced in amounts of up to 100 L per day, and by the feeds themselves), rapid VFA absorption by the rumen epithelial cells, and passage of a portion of the VFAs from the rumen.

The rumen in a dairy cow may approach 100 L in size, and contains approximately two-thirds of the total volume of digesta in the animal.[47] As indicated in Table 14.1, the rumen features a high solids loading, near the 20% target established by the US Department of Energy for engineered cellulosic bioconversion processes.[48] Because microbes (mostly bacteria) are present in high numeric density ($>10^{10}$ cells per g rumen contents)[49] both on the degrading feed particle surface and in the rumen liquid phase, there is little need for mixing of the rumen contents to achieve contact of the feed with the microbes. Mixing is accomplished by gentle muscular contractions that enhance the contact of VFA products with the rumen epithelial cells (for VFA absorption), and that facilitate the flow of the liquid phase down-stream to the omasum, and the bidirectional flow of solids (down-stream to the omasum, and up through the esophagus to the mouth for rumination).

The passage of materials through the rumen resembles that of a plug–flow reactor, whose loading rates are determined by the meal pattern of the animal, but attenuated by the backward flow of solids for rumination. The large size of the rumen has developed as a means of slowing the rate of passage of solid feeds, to accommodate the growth rates of fiber-degrading microbes, which are slower than those of starch- and sugar-degrading bacteria. As a result, the net rate of passage of the solid phase of digesta (0.01 to 0.05 h^{-1}) is substantially below that of the liquid phase (*ca.* 0.15–0.2 h^{-1}). Selective retention of solids is achieved by a combination of settling; gas-assisted buoyancy of particles into a relatively immobile mat atop the rumen liquor, and laminae (leaves) at the ruminal/omasal orifice that collect large particles for backflushing into the rumen. The mat is particularly important as it traps particles, increasing their overall retention time, and it provides a separate habitat that may allow fungi a means of prolonging their residence in the rumen. Protozoal numbers tend to increase dramatically near the end of a diurnal feeding cycle, and it has been demonstrated that they are able to sequester in the ventral portion of the rumen after feeding.[50] This settling has been proposed to be facilitated by intra-cellular storage of relatively dense starch granules engulfed by the protozoa soon after the cow eats, the net effect of which would be to reduce the washout rate of these relatively slow-growing microbes from the rumen.

Rates of conversion of cellulose in the rumen appear to exceed those of other natural microbial habitats. For a grazing cow consuming 20 kg of forage dry matter (DM) per day, and a ruminal cellulose digestibility of 60%, the approximate rate of conversion can be calculated:

$$(20 \, \text{kg forage DM per day}) \times (0.35 \, \text{kg cellulose} \, [\text{kg forage DM}]^{-1}) \times 0.6$$

$$= 4.2 \, \text{kg cellulose digested per day}.$$

At a ruminal volume of 80 L, the volumetric rate of cellulose degradation is approximately 52.5 g L^{-1} per day, or 2.2 g L^{-1} per hour.

Plant cell wall material leaving the rumen can potentially be degraded in the lower tract, but the extent of this degradation is relatively minor[51] owing perhaps to the rapid rate of passage through the tract and the relatively recalcitrant nature of the partially degraded digesta. However, at high levels of intake and unfavorable ruminal environmental conditions (*e.g.*, low pH), up to 30% or more of the total tract fiber digestion has been reported to occur post-ruminally.[52]

14.3 The Ruminal Microbiota and their Metabolic Capabilities

The rumen contains representatives of all four microbial groups (bacteria, archaea, protists and fungi), each with different roles in the fermentation (Table 14.2). On a cell number basis, bacteria far exceed the populations of any

Table 14.2 Major microbial inhabitants of the rumen, their abundance and activities.

Microbial group	*Microbial domain*	*Per cent of microbial mass*	*Major activities[a]*
Bacteria	Bacteria	50–90	Degradation of fiber, starch and other carbohydrates Degradation of proteins, nucleic acids, organic acids Lipid biohydrogenation Net production of microbial cells (protein for host)
Protists (Protozoa)	Eucarya	10–40	Predation of bacteria Engulfment and slow fermentation of starch Produce H_2 and host endosymbiotic methanogens
Fungi	Eucarya	1–8	Disruption (primarily physical) of plant tissues Fiber degradation
Archaea	Archaea	<1–3	Thermodynamic displacement of H_2 production

[a]Bacteria, protozoa, and fungi all produce VFAs utilized by the host animal.

of the other groups, though on a total cell mass basis, protists (protozoa) can approach that of the bacteria,[24] and the fungi and archaea account for much smaller proportions.[11,53,54] Depending on one's definition of what constitutes a microbial species, the rumen contains hundreds to thousands of species,[55] and thus represents a tremendous amount of genetic diversity for new or improved bioconversion capabilities. Most of our knowledge of the conversion of feed components by the microflora or by specific groups of microbes has been obtained from *in vitro* studies.

14.3.1   *In vitro* Studies with Mixed Ruminal Microorganisms

Early studies on rumen microbiology were conducted with the unfractionated microbial community of the rumen, hereafter referred to as "mixed ruminal microflora" (MRM). Even after the isolation of pure cultures of ruminal microbes and the laboratory reconstitution of defined mixed cultures, MRM have remained important subjects of study for determining fermentation stoichiometries, and for evaluating forage quality and the effects of feed additives.[15] More recently, the methods of molecular microbial ecology have been used to characterize the composition of MRM (reviewed by Deng *et al.*[56] and McSweeney *et al.*[57]).

For both *in vitro* studies and analysis of the composition of the ruminal community, it is necessary to obtain representative samples of ruminal contents. Ruminal samples are sometimes collected from a tube inserted directly through the animal's mouth and into the rumen ("stomach tubing") or from a large-diameter needle inserted into the rumen though the left side of the animal ("ruminocentesis"). However, these two sampling methods result in microbial communities of inferior quality; the former method recovers large amounts of saliva, and both methods recover only liquid and some small feed particles. Because the bulk of the rumen bacterial and fungal populations (particularly the fibrolytic bacteria most desired for biofuels applications) are associated with feed particles across a wide range of particle sizes,[58,59] it is desirable to obtain a more representative sample of ruminal contents. This can be achieved by manual collection of both liquid-phase ("rumen liquor") and solid-phase ("digesta") from the rumen of animals at slaughter, but is more commonly collected from animals surgically fitted with a permanent ruminal fistula (sometimes referred to as a "cannula", Figure 14.3). Use of live fistulated animals is preferred over slaughtered animals in that the former permit frequent access to the rumen on-demand and over an extended period of time. In addition, the use of live animals allows the researcher to control the diet of the animal (for example, to adapt the animal to a particular feed component or ingredient to facilitate adaptation of the MRM) and to collect samples at known times after feed presentation. Ruminal sampling of fistulated animals is accomplished by removal of the soft rubber or elastomer plug and insertion of the hand (covered with an arm-length polythene glove) directly into the rumen, allowing the removal of both liquid and solid fractions. The sampled materials are typically placed in prewarmed thermos bottles, which are then tightly covered for transport to the laboratory, to retain heat and minimize exposure of the largely anaerobic population to oxygen.

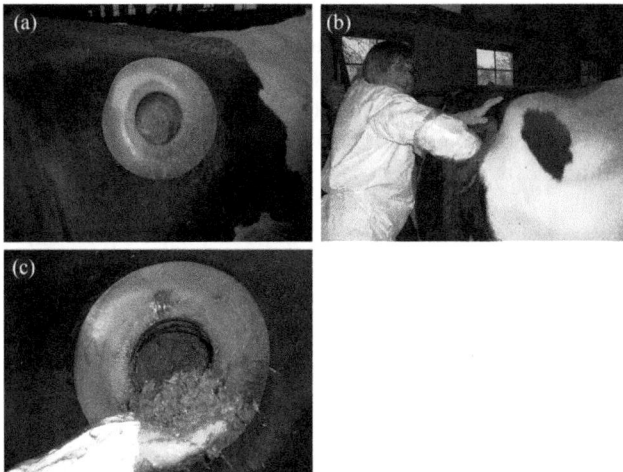

Figure 14.3 Sampling for ruminal contents through a rumen fistula. (a) A surgically implanted fistula. The initial surgery involves implanting a smaller fistula, which is replaced several weeks later by a fistula of this size ($\sim 10\,cm$ inside diameter). (b) A researcher sampling whole rumen contents (solid and liquid). (c) Solid digesta upon removal from the rumen. The solids are enriched with fiber-degrading bacteria, which require contact with the substrate for effective degradation.

Once collected, MRM are most commonly used for *in vitro* digestibility studies. To assure representative inocula into relatively small fermentation vessels, specific amounts of liquid are recovered by squeezing through four or more layers of cheesecloth into a graduated cylinder while gassing with CO_2. Compared to whole ruminal contents, filtration through cheesecloth reduces the rate of hemicellulose fermentation by 30–40% and the rate of pectin fermentation by 21–50%, but does not affect the rate of starch fermentation.[60] It is thus useful to further process the ruminal sample to enhance recovery of solid-adherent microbes. For this a fraction of the solid digesta retained by the cloth is weighed and placed in a blender jar with a volume of anaerobic buffer (typically those of Goering and Van Soest[15] or of McDougall[61]) equal to that of the squeezed liquid. This suspension is then blended briefly ($\sim 1\,min$) under CO_2, and then filtered through cheesecloth into the previous liquid filtrate. The use of a blended inoculum increases the recovery of the solid-adherent bacteria into the liquid phase, resulting in a more active community for plant-cell-wall degradation. *In vitro* assays are usually conducted with a 30 to 40% by volume amount of this diluted inoculum, which translates to a 15 to 20% inoculum of ruminal fluid. This concentration typically contains an excess of cellulolytic bacteria (*i.e.*, no decrease in the rate of cellulose degradation is observed in diluted cultures until the original inoculum concentration is reduced by a factor of at least six).[62]

In vitro incubations have been shown to substantially alter the relative proportions of specific microbial groups. Protozoa fare especially poorly under *in vitro* conditions: although they constitute nearly half the microbial biomass in the rumen itself under some feeding conditions,[11] they typically disappear within 24 h of *in vitro* incubation. Maintenance of ruminal protozoa in culture is also notoriously difficult and appears to require nearly daily transfer.[63,64] *In vitro* culture stability is but one issue that has dogged the study of the ruminal protozoa. These microbes contain both endo-symbiotic (intra-cellular) and epi-symbiotic (attached to the protozoal exterior) bacteria, and co-culturing these bacterial populations makes it difficult to exclude non-symbiotic bacteria from the culture, thus complicating interpretations of physiological experiments.

The bacterial community is considerably more durable than the protozoal community, although *in vitro* incubation has been shown to decrease the relative proportion of *Prevotella* (the most abundant genus in the rumen) and to shift the relative proportions of the cellulolytic populations (*viz.*, *Ruminococcus albus* at the expense of *R. flavefaciens*).[65] Shifts in the proportions of individual bacterial species were consistent despite differing *in vitro* treatments, suggesting that population shifts are due primarily to 'bottle effects' inherent in the *in vitro* methodology. Regardless of these shifts in the relative population sizes of individual taxa, it is important to note that the essential features of *in vitro* fermentations, in terms of such metrics as substrate utilization rates and fermentation product ratios, are generally similar to those in the rumen itself.[66,67] This suggests that there is considerable functional redundancy within the ruminal microbial community.

14.3.2   *In vitro* Studies with Defined Cultures

While studies with MRM have produced a wealth of information on the quantitative aspects of the ruminal fermentation (rate and extent of digestion, fermentation product distribution, and microbial protein yield), understanding the contributions of specific microbial groups or of interactions among microbial species has required the study of defined cultures. These can either contain a single strain of a single species ("pure culture") or a known combination of several strains representing the same or different species ("defined mixed culture"). For research applicable to biofuel production, the focus has been on bacteria that degrade cellulose and/or hemicelluloses. Cellulolytic species that have been isolated from the rumen are listed in Table 14.3. Three predominant species (*Fibrobacter succinogenes*, *Ruminococcus flavefaciens* and *Ruminococcus albus*) have been quantitatively characterized with respect to hydrolysis kinetics, fermentation products formed, growth yields and maintenance co-efficients and growth rates and affinity constants for soluble cello-dextrins.[68–72] Many of these variables have been measured as a function of growth rate and pH. Genomic analysis of these species is well advanced relative to most other ruminal bacteria that have been isolated in pure culture, which has stimulated an examination of the polysaccharide hydrolases produced by these bacteria.[73–75] In addition to the three predominant ruminal cellulolytic bacteria, there is a host of other species whose isolation has been reported from the rumen (Table 14.3).[24,76] For the most part, these species have been isolated

Table 14.3 Described species of ruminal cellulolytic bacteria.

Species	Strain[a]	Gram[b]	Reported fermentation products	Genome status[c]
Predominant species:				
Fibrobacter succinogenes	S85	−	Succinic, acetic, formic acids, CO_2	Completed
Ruminococcus albus	7	+	Acetic acids, ethanol, H_2, CO_2	Completed
Ruminococcus albus	8	+	Acetic acid, ethanol, H_2, CO_2	Draft
Ruminococcus flavefaciens	FD-1	+	Acetic, succinic, formic acids, CO_2	Draft
Fibrobacter succinogenes	S85	−	Succinic, acetic, formic acids, CO_2	Completed
Minor species:				
Butyrivibrio fibrisolvens[c]	Various	+	Formic, butyric, lactic acids, ethanol, H_2, CO_2	—[e]
Cillobacterium cellulosolvens	—[d]	+	Lactic acid	—
Cellulomonas fimi	—[d]	+	Formic, acetic, lactic acids	—
Clostridium lochheadii	—[d]	+	Acetic, formic, butyric acids, ethanol, H_2, CO_2	—
Clostridium longisporum	B6405	+	Formic, acetic, butyric, acids, ethanol, H_2, CO_2	—
Micromonospora ruminantium	ATCC 27728	+	Lactic, acetic, propionic, butyric acids	—

[a]Strain listed is one studied in the most detail.
[b]Cell wall type (Gram-positive or Gram-negative).
[c]Not all strains are cellulolytic.
[d]No strains, including original culture described by Hungate[24], are currently available.
[e]Genome sequences of strain 16/4 from a human source is available.

from only one or a few rumen samples, are thought to be present in only minor quantities, and/or are represented by strains that have either been lost from culture, or are not widely available for more detailed study.

As a group, the three predominant species of ruminal cellulolytic bacteria (RCB) are among the most active of all mesophilic cellulolytic microbes, being able to sustain a maximal growth rate constant on crystalline cellulose of nearly $0.1\,h^{-1}$.[77] The three species share several other characteristics: they are strictly anaerobic and do not produce spores or other structural adaptations to resist environmental insult; fiber degradation is dependent upon direct adherence to fiber particles; the temperature range for growth is narrow, with an optimum near 39 °C; and the pH range for growth is also narrow (between about 6.0 and 7.0),[78] although adherent cells continue to digest fiber during periods of non-growth.[62] All three species use ammonia as a nitrogen source, with little capability for incorporating amino acids or peptides into cellular nitrogen. Branched-chain VFAs (isobutyric, isovaleric, and 2-methylbutyric) are required for growth, as are certain vitamins (particularly biotin).[24] The ruminococci require other growth factors present in ruminal fluid or yeast extract. Some growth factors remain undefined, although it is known that *R. albus* requires 3-phenylpropionic acid for growth on cellulose (but not on cellobiose);

this requirement appears to be related to a need to synthesize a capsule or glycocalyx to facilitate adherence of cells to the fiber. In the ruminal environment, most organic micronutrients are provided by other microbial species.[79] For industrial application of pure cultures of the cellulolytic bacteria, it would be necessary to supply these micronutrients, though some (e.g., vitamins and microminerals) might be provided by the plant feedstock.

All of the described species of ruminal cellulolytic bacteria carry out mixed acid fermentations to yield a variety of products. It is interesting that, despite their many similarities in their capacity to degrade cellulose and hydrolyze hemicelluloses, and similarities in physiological characteristics, the predominant RCB differ in the organization of their fibrolytic systems. Some ruminococci produce cell-bound enzyme complexes organized into cellulosomes, with dockerin domains on catalytic enzymes bound to cohesin domains on a cell-bound scaffoldin protein.[73,75] By contrast, *F. succinogenes* lacks these structures, and instead organizes its fibrolytic enzymes on the cell surface in a manner that awaits characterization.[74] The organization of proteins on the surface appears to be unique, in that this species possess the unusual capacity to orient its cells along the crystallographic axis of the cellulose fiber.[80,81] *F. succinogenes* S85 and *R. albus* 7 appear to be able to hydrolyze a similar

Table 14.4 Polysaccharide degradation patterns in the predominant ruminal cellulolytic bacteria.

Polysaccharide	Linkage[a]	*F. succinogenes S85* Hydrolyzed	Utilized	*R. albus 7* Hydrolyzed	Utilized
Cellulose	β-1,4-Glc	+	+	+	+
Lichenan	Mixed β-1,3- and β-1,4 Glc			+	+
Homoxylan	β-1,4-Xyl	+	−	+	+
Xylan	β-1,4-Xyl (with 4-*O*-MeGlcA substituents)	+	−	+	+
Xyloglucan	β-1,4-Glc (with β-1,6-Xyl substituents)	+	−	+	Nt[b]
Glucomannan	Mixed β-1,4-Glc and -Man	+	−	+	+
Inulin	β-2,1-Fru	+	−	+	Nt
Citrus pectin	α-1,4-GalA (with substantial methylation)	+	−	+	Nt
Phlein (fructan)	β-2,6-Fru	+	−	+	Nt
Arabinogalactan	β 1,3-Gal with some β-1,6-Gal side-chains capped with various monosaccharides	−	−	−	−
Chitosan	β-1,4-GlcNH₂ and *N*-AcGlcNH₂	−	−	Nt	Nt
Curdlan	β-1,3-Glc	−	−	−	−
Laminarin	β-1,3-Glc and β-1,6-Glc (mixed linkage, 3 : 1)	−	−	−	−
Starch	α-1,4-Glc (with α-1,6 branches)	−	−	−	−

[a]Fru = fructose, Gal = galactose, GalA = galacturonic acid, Glc = glucose, GlcA = glucuronic acid, GlcNH₂ = glucosamine, Man = mannose, NAcGlcNH₂ = *N*-acetylglucosamine, 4-*O*-MeGlcA = 4-*O*-methylglucuronic acid.
[b]Nt = not tested.

range of polysaccharides (Table 14.4), although the latter species appears far better at utilizing the hydrolytic products than does the former, which is virtually a cellulose specialist.

Hemicellulolytic capability is more widely spread among the ruminal microbiota than is cellulolytic activity.[82] Some cellulolytic bacteria can hydrolyze various hemicelluloses, including the most abundant xylan, though they vary in their ability to utilize the hydrolytic products. A number of non-cellulolytic species can actively degrade hemicelluloses as well.[83,84] Many members of the genus *Prevotella* are hemicellulolytic, as revealed by examination of the genomes of *P. rumincola* 23 and *P. bryantii* B_14,[85] and specific studies focused on xylan degradation by *P. bryantii* B_14.[86] Once hydrolyzed, hemicellulose oligomers appear to be readily fermented by many different ruminal species.[87]

14.4 Enzymes from Ruminal Bacteria

Most ruminant animals consume large amounts of a plant-based diet that consists of a variety of plant parts and plant species, and are thus a logical source in the search for novel plant cell wall-degrading enzymes. Fibrolytic enzyme production is encoded within the genomes of many different microbial species. With the sequencing of the genomes of individual fibrolytic bacterial species,[73–75] as well as several community meta-genomes,[88,89] a large array of polysaccharide hydrolases and related enzymes has been revealed (see Table 14.5).

The three predominant RCB species share a requirement for attachment to plant cell walls for effective degradation. Unlike the cellulases from RCB, enzymes for hemicellulose hydrolysis are likely distributed both on the cell surface and secreted into the extra-cellular medium.[86] Most of these enzymes contain carbohydrate-binding modules (CBMs) and several of the enzymes represent unique classes in the CAZy database of carbohydrate-active enzymes.[90]

When considering the potential use of enzymes from ruminal microbes in the industrial production of fuels from biomass, two important facts must be kept in mind. The first is that enzyme titers produced by anaerobic fibrolytic bacteria are much lower than those in, for example, industrially used aerobic fungi or bacteria. This lower titer is due in part to the fact that production has not been optimized through strain selection, metabolic engineering, or recombinant means. However, it is also due to the fact that enzymes from anaerobic bacteria tend to have high specific activities, which is in large measure a reflection of the fact that they are normally presented primarily on a cellular surface in an either complexed or uncomplexed form, and work together with the producing organism to deconstruct plant biomass, a phenomenon termed "cell-enzyme synergy" by Lynd *et al.*[2] Many of these enzymes have very unremarkable activities when separated from the cells that produce them. This fact, combined with the likely difficulty of producing the enzymes in sufficient quantity, will probably limit their industrial utility in their native form.

Table 14.5 Fibrolytic genes within the completed genomes of ruminal cellulolytic bacteria.

	F. succinogenes S85[a]	*R. albus 7*[b]
Genome size (bp)	3 842 635	4 482 082
Chromosome	3 842 635	3 685 404
Plasmids	–	420 706
		352 645
		15 907
		7420
Total predicted open reading frames	3085	3872
Cellulase organization	Cell surface, organization unknown	Cell surface, non-cellulosomal
CAZy Families/sequences[c]		
Glycosyl hydrolases	27+/100	35 + /97
Glycosyl transferases	11+/ 53	6/24
Polysaccharide lyases	5/12	4/7
Carbohydrate esterases	6/17	7/18
Carbohydrate-binding modules	9 + /63	10/128

[a]Suen *et al.*[74]
[b]Suen *et al.*[75]
[c]From the CAZy database.[90] Numbers of families that include " + " indicate the presence of sequences not yet classified into families.

14.5 The Potential Use of Ruminal Microbes for Fuel and Chemical Production

14.5.1 Pure Cultures

As noted above, the predominant RCB are among the most active cellulose degraders in nature, and thus would seem likely candidates for use in industrial biomass fermentations to produce biofuels or bulk chemicals. However, several traits militate against such use. First, these species are relatively demanding in pure culture. They are extremely sensitive to oxidizing conditions, and usually require prereduced media for growth. They require various growth factors (vitamins, branched-chain VFAs, and/or 3-phenylpropionic acid) that are normally provided through cross-feeding by other microbes but which must be added exogenously to cultivate these bacteria in pure culture. The two pure culture fermentations that would appear to be of most interest to an industrial microbiologist are those of *F. succinogenes*, which produces succinate as its major product, and *R. albus*, which produces ethanol and H_2. The yield of succinate (up to 1.2 mol succinate per mol anhydroglucose) by *F. succinogenes* exceeds that of most other succinate producers, but the maximal succinate accumulation in bioreactors after 40–72 h of batch culture is only about 2 g L^{-1}.[91,92] Ethanol production by *R. albus* appears to be limited both by ethanol toxicity and sensitivity to low pH resulting from the production of acetate as a co-product. Despite decades of effort from scientists worldwide, no recombinant genetic system has been developed for any of the RCB species. This currently places these bacteria off-limits for the entire field of genetic and metabolic engineering.

A further complicating factor in industrial fermentations is the readiness with which these species participate in symbiotic associations with other bacterial species. For example, *F. succinogenes* readily grows with succinate utilizers such as *Selenomonas ruminantium* (resulting in this case in decarboxylation of succinate to propionate) and *R. albus* readily carries out inter-species H_2 transfer to, among others, fumarate reducers,[93] CO_2-reducing acetogens[94] or methanogenic archaea,[95] resulting in the production of alternative reduced products and virtual elimination of ethanol and H_2 accumulation (Figure 14.4).

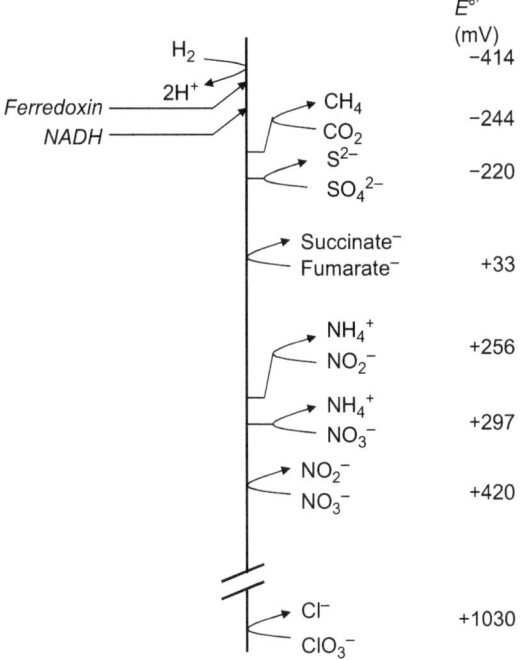

Figure 14.4 Standard reduction potentials at pH 7 ($E^{\circ\prime}$) for redox transformations that influence the disposal of reducing equivalents during ruminal oxidation of organic matter and H_2. Electron-accepting reactions having numerically greater $E^{\circ\prime}$ values (*e.g.*, nitrate reduction) are thermodynamically more favorable than those having numerically lower $E^{\circ\prime}$ values (*e.g.*, CO_2 reduction to methane), and are more easily oxidized by common electron carriers such as NADH or ferredoxin. Reductions using these exogenously fed electron acceptors can decrease methane production and result in the redistribution of the relative proportions of fermentation end-products. However, the products of these reductions are typically of little or no use to the animal, and are not useful as fuels when produced in an extra-ruminal fermentation.

These tendencies raise the prospect that culture contamination, which would drastically decrease the amounts of desired end-products, could be a concern under industrial conditions.

Certain non-fibrolytic ruminal bacteria have unusual metabolic capabilities that may merit their use in industrial fermentations to produce bulk chemicals. *Selenomonas ruminantium* has been shown to produce up to 23 g of propionate and 10 g of acetate per liter from lactate in fed-batch reactors after 72 h incubation.[96] A strain of *Clostridium kluyveri* isolated from the rumen has been reported to produce up to 12.8 g caproic acid per liter from ethanol and acetate.[97] Interestingly, this species is likely a transient in the rumen, as its population is much higher in silages than in the rumen of silage-fed cows.

14.5.2 Mixed Ruminal Microflora

The many shortcomings of individual ruminal bacterial species for industrial biofuels and chemical production seem to evaporate when one examines the bioconversion capabilities of the MRM as a whole. The species-diverse microbial community of the rumen is well-adapted for rapidly processing a wide variety of biomass materials with little external management.[22,24] Facultative anaerobes consume oxygen that would otherwise inhibit the strict anaerobes.[49] Several fibrolytic species participate in the degradation of the various plant-cell-wall polysaccharides. Individual microbial species and physiological types cross-feed nutrients and remove intermediates (Figure 14.5). The net effect is the nearly complete degradation of the physically accessible, non-lignin components of the biomass feedstock, with the production of substantial concentrations of VFA products.

14.6 Extra-Ruminal Fermentation of Biomass by Mixed Ruminal Bacteria

The elegant and comprehensive conversion of plant biomass by the ruminant provides both an inspiration and a roadmap for replicating the process outside of the animal ("extra-ruminally") in a bioreactor. In fact, the rumen has been likened to a bioreactor having numerous characteristics desired for effective conversion of plant biomass. As noted above, these characteristics include: (1) high solids loading (typically around 15% w/v); (2) temperature control within a few degrees of the optimum temperature for the microflora (*viz.*, 39 °C); (3) pH control; (4) gentle mixing requiring very little power input; (5) regulation of solid and liquid passage through the reactor, with preferential retention of the solid fraction containing the degrading biomass; and (6) continuous absorptive removal of the primary energetic products (*viz.*, VFAs). Added to this are important process features or capabilities such as non-aseptic operation and a wide substrate range that are unmatched by conventional CBP or SSF biofuel systems. Several questions also arise regarding the practicality of producing

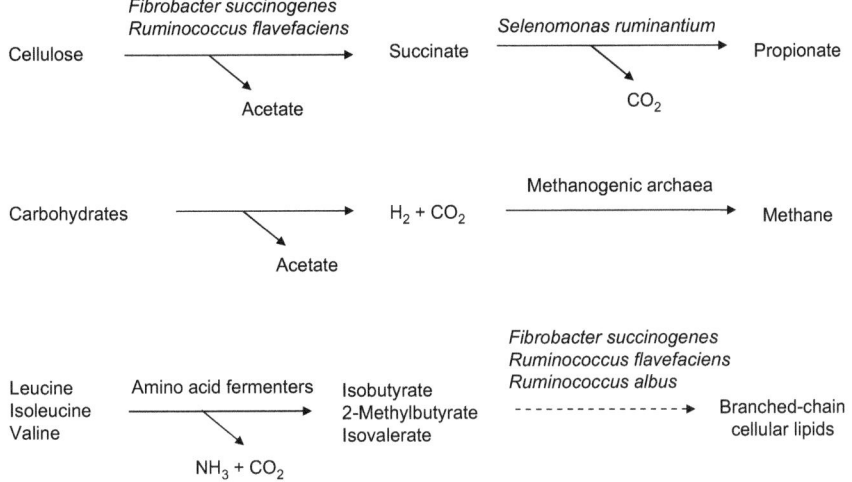

Figure 14.5 Examples of interactions among ruminal microbes. Solid lines indicate catabolic reactions. Dashed lines indicate anabolic reactions.

VFA extra-ruminally from biomass, in particular culture stability, scalability and the integration of process steps to mimic those of the ruminant animal.

14.6.1 Non-Aseptic Operation

A striking feature of the ruminal habitat is its ability to retain a stable mixed culture of microbes despite the continuous input of allochothonous (non-native) microbes during feed and water ingestion. This ability to operate under non-aseptic conditions is one of the chief advantages of an extra-ruminal fermentation employing mixed ruminal microflora. This non-aseptic operation thus resembles that of an anaerobic digester and starkly contrasts with monoculture-based biofuel systems, whose potential for contamination is rarely addressed in the biofuels literature. The vulnerability of conventional monoculture-based fermentation processes to contamination is a major concern to the corn ethanol industry. Despite steam-sterilization of reactors and subsequent loading with a corn mash subjected to temperatures of 110 °C to facilitate gelatinization and enzymatic hydrolysis of the starch, contamination of the yeast culture by lactic acid bacteria is a widespread and expensive issue.[98] Liberal use of antibiotics (the corn ethanol industry is now a major consumer of antibiotics in the USA) is sometimes necessary to limit losses of fermentable sugars to about 3%. Cellulosic ethanol processes may not fare much better. While the severity of some physical and chemical pretreatment processes will essentially sterilize most biomass materials, consideration still needs to be given to the fermentation medium itself, as well as the bioreactor and its associated plumbing. Formulae for calculating energy requirements for sterilization have been developed by Deindorfer and Humphrey,[99] and these requirements turn out to be substantial. These authors provide an example in which sterilization of a 1.2×10^4 gallon (4.6×10^4 L) reactor with medium using 50 psig steam at 91 kg min^{-1} required 62 min to heat from 55 to 120 °C, and 11 min of hold time at 120 °C; this calculates to $\sim6.63\times10^3$ kg steam. At a conservative cost of $3.00/million BTU natural gas and a combustion efficiency of 81.7%, this translates to $\sim$$8.50 to sterilize this relatively small reactor. Additional energy is required to bring the initial temperature to 55 °C, though this can be captured in cooling water used to bring the reactor temperature back to 30 °C.

14.6.2   Wide Substrate Range

The ability of biofuel production systems to use both the cellulose and hemicellulose components of biomass is regarded as essential for practical applications of cellulosic ethanol. Acidic pretreatments of cellulosic biomass result in the hydrolysis of hemicelluloses to produce a mixture of sugars, most of which are pentoses. An ongoing debate with respect to ethanol production by SSF is whether *Saccharomyces cerevisiae* yeast can be developed that ferment pentoses at the same rate as, and simultaneously with, hexoses. By contrast, engineered CBP processes envision either using strains of cellulolytic bacteria that can use pentoses (*e.g.*, *Clostridium phytofermentans*) or conducting hemicellulose fermentations in a separate reactor (*e.g.*, using *Thermoanaerobacterium saccharolyticum*). These debates ignore the important fact that plant biomass is not composed of only hexoses and pentoses, but also contains substantial amounts of protein, lipids, nucleic acids, organic acids and secondary metabolites. Conventional biofuel production scenarios consider these components to be unusable substrates that will pass into the waste stream along with the lignin fraction. By contrast, as noted in Section 14.2.2 above, extra-ruminal biomass fermentations retain the capacity to degrade most of these non-carbohydrate organic materials (excluding lignin and long-chain fatty acids). Proteins, some nucleic acid bases, and the glycerol moiety of triglycerides, are readily fermented by MRM, as are most di- and tricarboxylic acids that can make up several percent of forage dry matter.

This broad range of utilizable substrates is also shared with conventional anaerobic digestion systems. Interestingly, Lopez *et al.* proposed using anaerobic digestion for treating the 1.5×10^5 Mg ruminal contents and 3×10^4 Mg blood produced in Uruguayan slaughterhouses annually.[100] It would seem worthwhile to consider instead the use of an extra-ruminal fermentation for this and other complex but degradable waste streams.

14.6.3   Culture Regenerability and Stability

Eastern gamagrass-fed batch cultures of MRM have been carried through 123 successive sequential transfers at 2 to 3 day intervals with little effect on the concentrations of VFAs produced,[24] and MRM have been carried in continuous culture over a period of months in specialized systems that feed ground forages periodically while feeding liquid continuously.[101–103] *In vitro* ruminal fermentations result in changes relative to the proportions of individual bacterial species, without substantial modification of the output of the fermentations. As noted above, *in vitro* cultivation typically results in the loss of the protozoal population;[11] it is likely that this would increase the overall efficiency of substrate conversion, as *in vivo* protozoal defaunation has been shown to have no discernible effect on fiber utilization and to increase microbial protein yield in the rumen.[104]

14.6.4 Scalability

The large number of domestic ruminants worldwide (1.4×10^9 cattle and water buffalo plus 1.84×10^9 sheep and goats)[105] constitute approximately 2×10^{11} L of fermentation capacity, making this the world's largest commercial fermentation. As the maximal rumen size is approximately 100 L, a question might be raised regarding the scalability of ruminal fermentation to the scale equivalent to modern ethanol plants. In principle, it would seem that ruminal fermentation is eminently scalable. The heat produced by the fermentation is minimal and even at a large scale would likely require only modest heating or cooling capability in an engineered system. Mixing requirements are likely to be minimal, as in the rumen itself, except perhaps during a discharge cycle to remove undegraded residues.

14.6.5 Observed Operational Characteristics

Several reports have evaluated the potential of using extra-ruminal fermentations for biomass conversion, generally within the context of a waste-treatment process. In keeping with this approach and the known capabilities of the ruminal microflora to process unmodified feeds, the substrates in these studies were subjected to physical pretreatment (grinding), but not chemical pretreatment. Early work indicated that domestic refuse was more extensively degraded *in vitro* by a mixed ruminal inoculum than by an inoculum from an anaerobic digester,[106] and addition of mixed ruminal inoculum to anaerobic digesters improved substrate conversion.[107] More recently, there has been an intensive effort in China to use ruminal microbes to digest harvested cattails (*Typha latifolia* Linn.), which are grown in constructed wetlands used for phytoremediation of various organic and inorganic wastes. Hu *et al.* carried out fermentations at a 2 L scale and reported that operation for 125 h at an initial feedstock concentration of 12.2 g volatile solids (*i.e.*, organic material) per liter resulted in 66% conversion and a VFA yield of 0.56 g VFA per g of volatile solid consumed.[108] By contrast, operation at pH 5.8 resulted in only 29% conversion of feedstock, mostly restricted to the soluble or non-cellulosic fraction. Reduced conversion at pH 5.8 is in accord with the well-known growth inhibition of ruminal cellulolytic bacteria at pH values below 6.0.[77,78] Application of a central composite experimental design and response surface methodology to a set of 10 fermentations revealed that, at all feedstock concentrations, the pH optimum was 6.9.[109] At an initial feedstock volatile solids concentration of 12.2 g L^{-1} and a 120 h run time, 65% conversion of feedstock yielded 0.41 g VFAs and 0.11 g microbial cell mass per g of volatile solids. These results suggest that extra-ruminal fermentations can provide high yields and high conversion rates of VFAs.

14.6.6 Modeling the Ruminal Fermentation

The benchmark model for the ruminal fermentation is the rumen sub-model[110,111] of the Cornell Net Carbohydrate and Proteins System, a commercial software package that predicts cattle nutrient requirements.[112] Modifications to the model[113] have improved its popularity, although, the model still basically is built around a "1985 rumen" (J. B. Russell, personal communication), i.e., before additional research yielded some important principles of rumen microbiology. A simple model describing cellulose degradation by ruminal bacteria that incorporates some of the more recent advances has been developed in the user-friendly STELLA® modeling language[114] but has not yet been expanded to describe the utilization of other feed components.

Recently, Zhao et al.[115] modeled the extra-ruminal fermentation of plant biomass (specifically, cattail herbage). The model is based on principles of anaerobic digestion and includes 11 separate bioprocesses carried out by a total of 4 different microbial groups: cattail degraders (i.e., microbes that carry out polysaccharide hydrolysis), sugar fermenters, VFA utilizers, and H_2 utilizers. Only the carbohydrate fraction of the biomass is considered as the substrate, and (as in the Cornell model) this is divided into slowly-digesting and rapidly-digesting pools (essentially corresponding to the structural and non-structural carbohydrates). Degradation rates of the slowly-digesting pool are based on surface limitation, long known as the primary factor limiting cellulose and NDF limitation by mixed ruminal microbes.[19,116] The model displays good prediction of substrate consumption and product (VFAs, methane, CO_2, microbial cells), despite the simplifying factors and the uncertain physiological capabilities of the "VFA utilizers". Overall, the most important factor determining cattail consumption and product formation in the fermentation was the rate of hydrolysis of the slowly-digested fraction of the biomass.

14.6.7 Challenges in Integrating Process Features

It is significant to note that the current literature and modeling efforts have used biomass loading rates of around one-tenth that of the rumen itself, and have either not controlled the rumen pH, or have artificially controlled the pH by exogenous addition of alkali without removing VFAs during the fermentation. Neither condition adequately simulates the rumen under in vivo conditions. As Van Soest has rightly noted, "No presently available device can simulate the absorption of VFAs, rumination, and the complex differential passage that occurs in the living [ruminant] animal".[11] There is little doubt that successful extra-ruminal production of VFAs at the industrial scale, and economic feasibility will require the integration of several process steps that are carried out seamlessly in the animal. Attaining this goal for some process steps will likely be simple, while for others it will be challenging. For example, mimicking the effective physical pretreatment resulting from rumination would require some clever engineering. Moreover, the ability of the animal to sustain high loading rates of substrate requires rapid recovery of the VFAs from the rumen liquor, which is achieved *in vivo* through absorption by rumen epithelial cells. Without such absorption, accumulation of the acidic VFA products—even with exogenous pH control—is likely to limit the rate and extent of fermentation, particularly fiber digestion. These challenges provide some interesting opportunities for further research.

14.7 Volatile Fatty Acid Conversion: A Practical Route to Fuels?

The preceding discussion emphasizes the many attractive features of the ruminal fermentation which, by and large, are transferable to extra-ruminal operation in dedicated bioreactors. The main challenge for the biofuels researcher is that the primary fermentation end-products, VFAs, while retaining almost three-quarters of the energy content of the original feedstocks, are not directly usable as fuels for transportation, industrial, or domestic use. VFAs are corrosive, malodorous, and—despite their name—are not sufficiently volatile for use in internal combustion engines. It is thus necessary to convert the VFAs to more practical fuel compounds.[10,114] Several potential routes will be discussed in turn. Most involve some form of chemical (non-biological) conversion, although some additional biochemical steps may be involved.

14.7.1 Rumen-Derived Anaerobic Digestion

The rumen-derived anaerobic digestion (RuDAD) process differs from other processes discussed below in that it is a purely biological process, and its sole fuel product is methane. In RuDAD, a conventional anaerobic digestion process is separated into two stages: an acidogenic fermentation stage in which biomass is converted to VFAs, and a methanogenic stage in which the VFAs are converted to methane and CO_2.[118] Splitting the process into two stages affords the advantage of separate process control for each stage: the ruminal fermentation proceeds optimally at pH values between 6 and 7, while the methanogenic stage proceeds optimally in the pH range of 7 to 8. By employing this system under continuous culture conditions, loading rates of up to 25.8 g of volatile solids per liter per day were achieved, along with methane yields of 0.438 L methane per g of volatile solids, the latter value is equivalent to 98% of the theoretical value. The RuDAD process has been operated over a period of several months, during which time digester upsets were easily corrected by re-establishment of initial conditions and re-inoculation.[119]

14.7.2 Lipid-Accumulating Yeasts

The biosynthesis of long-chain fatty acids from acetate and esterification to produce triglycerides is a common feature of both bacterial and eukaryotic metabolism. Most organisms produce long-chain fatty acids and triglycerides only in the amounts needed for the synthesis of cell membranes. However, some species of yeast (*e.g.*, *Cryptococcus curvatus*, *Yarrowia lipolytica*) over-produce triglycerides and store them intra-cellularly as lipid vesicles. These lipids can serve as feedstocks for industrial biodiesel production using conventional transesterification chemistry. *C. curvatus* is of particular interest because it can use VFAs to produce these lipids.[120] Although the process was demonstrated using effluent from a VFA-accumulating anaerobic digestion process as a growth substrate for the yeast, it would be worthwhile to use an effluent from an extra-ruminal biomass fermentation as a potential feedstock for comparison with the anaerobic-digestion-based fermentation.

14.7.3 MixAlco

Currently, the best-developed process for converting VFAs to fuels is the MixAlco process pioneered by Holtzapple[121,122] and depicted schematically in Figure 14.6. MixAlco is a hybrid process that combines a modified anaerobic digestion of biomass to VFAs, followed by a chemical conversion of VFA anions to C3–C8 alcohols, with 2-propanol being the most abundant product. The fermentation step employs a variety of inocula (typically, sewage digester sludge) and the addition of chemical agents (*e.g.*, iodoform, CHI_3) to inhibit methanogenesis, prevent VFA oxidation, and facilitate the formation of longer-chain VFAs over an incubation period of 2 to 3 weeks. These VFAs are then subjected to a series of non-biological unit operations, which ultimately convert dried CaVFA salts to a mixture of ketones, which are subsequently hydrogenated to alcohols. Individual process steps differ considerably in their

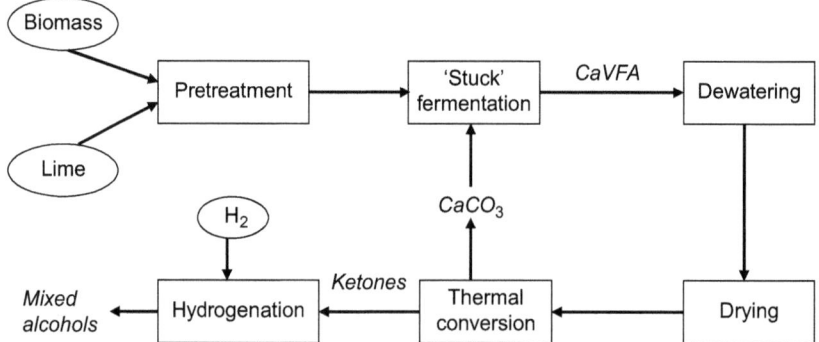

Figure 14.6 Schematic of the MixAlco process for the production of mixed alcohols. The blocks represent process steps and the ovals represent inputs. The primary outputs of the individual steps are italicized. Adapted from ref. 122.

complexity. The fermentation is relatively straightforward, while the recovery of VFAs (which involves countercurrent extraction with an amine solvent) is more complex. On the whole, the proposed engineering details suggest high capital costs but low operating costs. Within the context of this chapter, the key feature of the MixAlco process is that it might be readily adapted to substitute an unmodified extra-ruminal fermentation for the 'stuck' anaerobic-digestion step. Such a substitution could substantially shorten fermentation times from weeks to days, though the effect of the ruminal mixture of C2–C4 VFAs on the overall feasibility of the system warrants investigation.

14.7.4 Electrolysis to Hydrocarbons and Related Compounds

VFAs can be electrolytically decarboxylated at an anode surface to form alkyl radicals.[123] Once formed, these radicals can undergo various reactions, including: (1) "tail-to-tail" condensation to produce alkanes (Kolbe synthesis); (2) loss of an additional electron to form alkenes; (3) hydroxylation to form alcohols (Hofer–Moest reaction); or (4) reaction with alcohols to form esters (Figure 14.7). At the opposite electrode (cathode) surface, protons are reduced to H_2,

$$2H^+ + 2e^- \rightarrow H_2$$

which can be recovered for use to power the electrolysis or to reduce alkenes to alkanes. Electrolysis requires a potential of <3 V, in order to prevent anodic decomposition of water and evolution of O_2.

Levy *et al.*[124] characterized the effect of the electrolysis conditions on the formation of the different anodic products, using pure or mixed VFA species,

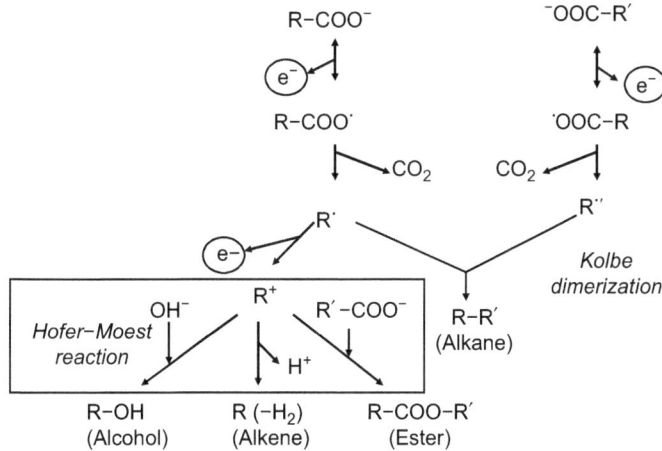

Figure 14.7 Electrochemical reactions for the conversion of VFAs to hydrocarbons and related compounds by one-electron or two-electron routes. In an electrochemical cell under aqueous conditions, these reactions occur at the anode, while at the cathode, protons are reduced to H_2.

and proposed the use of a 'stuck' anaerobic digestion to generate VFAs for electrolysis (though they did not perform electrolysis on fermentation broths). In principle, these electrolytic reactions could be carried out using broths from extra-ruminal biomass fermentations conducted with MRM inocula. At a typical molar ratio of ruminal VFAs of 6 acetic : 2 propionic : 1 butyric, the primary product of Kolbe electrolysis would be propane, a widely used fuel for fleet vehicles and for rural domestic heating and cooking. Producing longer chain hydrocarbons would require a shift in the VFA end-product ratio to increase in the average chain length. The ruminal fermentation has been shown to yield a range of VFA product ratios both *in vivo* and *in vitro*, depending on the feed and on the presence of certain modulators, although acetate is almost always the major product. Propionic acid formation tends to be enhanced in feeds higher in starch[21] or glycerol,[36] or when methanogenesis is artificially suppressed by the feeding of monensin[125,126] or electron sinks such as nitrate.[127] Butyric acid production appears to be enhanced by feeding corn distiller's grains.[128] Addition of species such as the caproic-acid-producing *Megasphaera elsdenii*[129] or *Clostridium kluyveri*[130] may also be a means of improving the VFA product ratios.

14.8   Co-Product Formation

The economics of fuel and commodity chemical manufacture by fermentation is complicated by the fact that the feedstocks represent a high percentage of the total production cost. Economically viable production systems will thus probably require co-products to add value to the overall process. Ethanol production from corn benefits from the sale of two co-products, CO_2 and distiller's grains, each of which is produced in mass amounts approximately equivalent to the ethanol itself. CO_2 is purified and compressed for use in beverage carbonation, dry ice production and other applications. Distiller's grains, usually sold as "distiller's dried grains and solubles" (DDGS) are composed primarily of yeast cells and undegraded corn fiber, and are used extensively as a component of livestock feeds; approximately 85% of the DDGS sold in the USA are fed to cattle.[131] The advantages of DDGS include a high protein and energy content, and relatively low cost. Disadvantages include variable product quality (though this is improving), and a relative deficiency in several essential amino acids (particularly methionine). The high fat and phosphorus content of DDGS can cause problems in some applications, particularly when feeding dairy cows: the metabolism of some fats can alter mammary lipogenesis and result in milk fat depression, and elevated P content of manure limits the amount of manure that can be applied to some lands (owing to their potential for run-off and subsequent eutrophication of fresh waters).

For cellulosic ethanol, the issue of co-products has received a surprising lack of attention, given the marginal economics of these processes. For yeast-based SSF, co-products would be expected to include CO_2 and a modified type of DDG whose energy and protein contents would be diluted by relatively indigestible residual biomass components (lignin and any poorly accessible polysaccharides). For engineered CBP, one would expect lower total yields of a DDG residue (owing to the generally low cell yields of CBP organisms such as *Clostridium thermocellum*), although the residue might have a more favorable amino acid profile (owing to the generally higher methionine content of bacterial protein).

In the rumen, the co-products of VFA production include methane and CO_2 (lost by eructation by the animal), indigestible residues of feed (mostly highly lignified plant fiber) and microbial cells. The latter, upon leaving the rumen, are degraded in the acidic abomasum to amino acids, which are absorbed in the small intestine to provide a major fraction of the host's amino acid requirements. The amino acid profile of rumen microbial cell protein is regarded as superior to that of virtually any commercial protein feed, particularly with regard to the content of methionine and lysine, the first two limiting nutrients for milk production.[132,133] The methionine content of ruminal bacterial cells has been shown to be substantially higher than that of the protozoal fraction.[134] It thus appears that microbial cell proteins could likely be a high-value co-product of an industrial extra-ruminal fermentation, and would be produced at yields somewhat above that of other bacterial CBP systems (owing to the relatively high yield of rumen bacterial cells).[135]

A second potential use of the residue from an industrial extra-ruminal fermentation would be as an wood adhesive. Some rumen bacterial cells produce an abundant extra-cellular glycocalyx that assists in binding the cells to the degrading feed particle.[136] The combination of this glycocalyx and the bacterial cells themselves renders a stickiness to the fermentation residue. The glycocalyx and the adherent bacterial cells cannot be easily removed from the residual substrate. However, studies with pure cultures of RCB have shown that, if the residue is of sufficiently small particle size, it can be effectively used as a co-adhesive with the commonly used wood adhesive, phenol–formaldehyde (PF). Under certain conditions, up to 73% of the weight of PF could be replaced with a fermentation residue without loss of shear strength, or wood failure values that define the utility of an adhesive.[137,138] However, handling these solid residues during plywood manufacturing is currently impractical, and more research is needed to overcome these handling problems.

14.9 Concluding Remarks

In the quest for an integrated and efficient system for the bioconversion of cellulosic biomass, the ruminant animal has a head start over humanity of at least 10 million years. Ruminants, in the process of their evolution, developed a mutualistic symbiosis with a complex microbial community for the fermentation of recalcitrant plant biomass to a high-energy fuel, the VFAs. The cow and her bioreactor, the rumen, presaged many of our own industrial engineering innovations: effective physical pretreatment; the continuous stirred tank reactor; the separation of solid from liquid phases; and the continuous extraction of fermentation products. The ease with which the ruminal fermentation may be run in bioreactors provides a strong incentive for the more complete integration of these innovations into an industrial VFA platform, and for developing novel or improved chemistries for converting energetic VFA products to hydrocarbon fuels.

References

1. A. Aden and T. Foust, *Cellulose*, 2009, **16**, 535.
2. L. R. Lynd, P. J. Weimer, W. H. Van Zyl and I. S. Pretorius, *Microbiol. Mol. Biol. Rev.*, 2002, **66**, 506.
3. N. W. Y. Ho, Z. Chen and A. P. Brainard, *Appl. Environ. Microbiol.*, 1998, **64**, 1852.
4. T. W. Jeffries and Y. S. Jin, *Appl. Microbiol. Biotechnol.*, 2004, **63**, 495.
5. S. J. Ha, J. M. Galazka, S. R. Kimm, J. H. Chio, X. Yang, J. H. Seo, N. L. Glass, J. H. D. Cate and Y. S. Jin, *Proc. Natl. Acad. Sci. U. S. A.*, 2011, **108**, 504.
6. L. R. Lynd, W. H. Van Zyl, J. E. McBride and M. Laser, *Curr. Opin. Biotechnol.*, 2005, **16**, 577.
7. X. Shao, M. Jin, A. Guseva, C. Liu, V. Balan, D. Hogsett, B. E. Dale and L. R. Lynd, *Bioresour. Technol.*, 2011, **102**, 8040.
8. M. Jin, V. Balan, C. Gunawan and B. E. Dale, *Biotechnol. Bioeng.*, 2011, **108**, 1290.
9. R. E. Hungate, *Bacteriol. Rev.*, 1950, **14**, 1.
10. P. J. Weimer, J. B. Russell and R. E. Muck, *Bioresour. Technol.*, 2009, **100**, 5323.
11. P. J. Van Soest, *Nutritional Ecology of the Ruminant*, Cornell University Press, Ithaca, 2nd edn, 1994.
12. N. Mosier, C. Wyman, B. Dale, R. Elander, Y. Y. Lee, M. Holtzapple and M. Ladisch, *Bioresour. Technol.*, 2005, **96**, 673.
13. R. G. Dado and M. S. Allen, *J. Dairy Sci.*, 1995, **78**, 118.
14. F. Bargo, F. L. D. Muller, J. E. Delahoy and T. W. Cassidy, *J. Dairy Sci.*, 2002, **85**, 1477.
15. H. K. Goering and P. J. Van Soest, *Forage Fiber Analysis (Apparatus, Reagents and Some Applications)*. Agriculture Handbook No. 379, Agricultural Research Service, US Department of Agriculture, Washington DC, 1975.
16. J. G. Welch, *J. Anim. Sci.*, 1982, **54**, 885.
17. D. R. Waldo and L. W. Smith, *J. Dairy Sci.*, 1972, **55**, 125.
18. P. J. Van Soest, *Fed. Proc.*, 1973, **32**, 1804.
19. P. J. Weimer, J. M. Lopez-Guisa and A. D. French, *Appl. Environ. Microbiol.*, 1990, **56**, 2421.
20. D. E. Bauman, C. L. Davis and H. F. Bucholtz, *J. Dairy Sci.*, 1971, **54**, 1282.
21. M. Murphy, R. L. Baldwin and L. J. Koong, *J. Anim. Sci.*, 1982, **55**, 411.
22. P. J. Weimer, *Bioresour. Technol.*, 2011, **102**, 3254.

23. A. Ziolecki, W. Guczynska and M. Wojciechowicz, *Lett. Appl. Microbiol.*, 1992, **15**, 244.
24. R. E. Hungate, *The Rumen and its Microbes*, Academic Press, New York, 1966.
25. R. D. Hatfield and P. J. Weimer, *J. Sci. Food Agric.*, 1995, **69**, 185.
26. C. M. Gradel and B. A. Dehority, *Appl. Environ. Microbiol.*, 1972, **23**, 332.
27. J. B. Russell and P. J. Van Soest, *Appl. Environ. Microbiol.*, 1984, **47**, 155.
28. G. A. Broderick, R. J. Wallace and E. R. Ørskov, in *Physiological Aspects of Digestion and Metabolism in Ruminants*, ed. T. Tsuda, Y. Sasaki and R. Kawashima, Academic Press, Orlando, 1991, p. 541.
29. R. J. Wallace and M. L. Brammall, *Microbiology*, 1985, **131**, 821.
30. R. J. Wallace, C. Atasoglu and C. J. Newbold, *J. Anim. Sci.*, 1999, **12**, 139.
31. J. B. Russell, H. J. Strobel and G. J. Chen, *Appl. Environ. Microbiol.*, 1988, **54**, 872.
32. S. C. P. Eschenlauer, N. McKain, N. D. Walker, N. R. McEwan, C. J. Newbold and R. J. Wallace, *Appl. Environ. Microbiol.*, 2002, **68**, 4925.
33. R. H. Smith and A. B. McAllan, *Br. J. Nutr.*, 1970, **24**, 545.
34. A. B. McAllan and R. H. Smith, *Br. J. Nutr.*, 1973, **29**, 331.
35. M. A. Cotta, *Appl. Environ. Microbiol.*, 1995, **56**, 3867.
36. N. A. Krueger, R. C. Anderson, L. O. Tedeschi, T. R. Callaway, T. S. Edington and D. J. Nisbet, *Bioresour. Technol.*, 2010, **101**, 8469.
37. R. I. Mackie, B. A. White and M. P. Bryant, *Crit. Rev. Microbiol.*, 1991, **14**, 449.
38. A. E. Maczulak, B. A. Dehority and D. L. Palmquist, *Appl. Environ. Microbiol.*, 1981, **42**, 856.
39. T. Hino and Y. Nagatake, *Anim. Sci. Technol.*, 1993, **64**, 121.
40. F. Dohme, A. Machmüller, A. Wasserfallen and M. Kreuzer, *Lett. Appl. Microbiol.*, 2001, **32**, 47.
41. L. W. Smith, H. K. Goering and C. H. Gordon, *J. Dairy Sci.*, 1972, **55**, 1140.
42. H. G. Jung and K. P. Vogel, *J. Anim. Sci.*, 1986, **57**, 206.
43. H. Kajikawa, H. Kudo, T. Kondo, K. Jodai, Y. Honda and M. Kawuhara, *FEMS Microbiol. Lett.*, 2000, **187**, 15.
44. T. K. Kirk and R. L. Farrell, *Annu. Rev. Microbiol.*, 1987, **41**, 465.
45. A. H. Smith, E. Zoetendal and R. I. Mackie, *Microb. Ecol.*, 2005, **50**, 197.
46. D. L. Klass, *Science*, 1984, **223**, 4640.
47. R. Parra, in *The Ecology of Arboreal Folivores*, ed. G. G. Montgomery, Smithsonian Institution Press, Washington DC, 1978, p. 205.
48. J. D. McMillan, E. W. Jennings, A. Mohagheghi and M. Zuccarello, *Biotechnol. Biofuels*, 2011, **4**, 29.
49. J. B. Russell, *Rumen Microbiology and its Role in Ruminant Nutrition*, James B. Russell Publishing, Ithaca, 2002.
50. B. A. Dehority and P. A. Tirabasso, *J. Gen. Microbiol.*, 1989, **135**, 539.

51. D. R. Mertens and L. O. Ely, *J. Anim. Sci.*, 1979, **49**, 1085.
52. M. J. Ulyatt, D. W. Dellow, C. S. W. Reid and T. Bauchop, in *Digestion and Metabolism in the Ruminant*, ed. I. W. McDonald and A. C. I. Warner, University of New England Publishing, Armidale, 1974, p. 119.
53. P. H. Janssen and M. Kirs, *Appl. Environ. Microbiol.*, 2008, **74**, 3619.
54. P. J. Weimer, D. M. Stevenson, D. R. Mertens and E. E. Thomas, *Appl. Microbiol. Biotechnol.*, 2008, **80**, 135.
55. M. Kim, M. A. Morrison and Z. Yu, *FEMS Microbiol. Ecol.*, 2011, **76**, 1.
56. W. Deng, D. Xi, H. Mao and M. Wanapat, *Mol. Biol. Rep.*, 2007, **35**, 265.
57. C. McSweeney, S. Kang, E. Gagen, C. Davis, M. Morrison and S. Denman, *Rev. Bras. Zootec.*, 2009, **38**, 341.
58. W. M. Craig, G. A. Broderick and D. B. Ricker, *J. Nutr.*, 1987, **114**, 56.
59. C. W. Forsberg and K. Lam, *Appl. Environ. Microbiol.*, 1977, **33**, 528.
60. M. Marounek, S. Bartos and P. Brezina, *Z. Tierphysiol. Tierernahrg. Futtermittelkde*, 1985, **53**, 50.
61. E. I. McDougall, *Biochem. J.*, 1948, **43**, 99.
62. F. Mouriño, R. Akkarawongsa and P. J. Weimer, *J. Dairy Sci.*, 2001, **48**, 848.
63. G. S. Coleman, in *Symbiotic Associations*, ed. R. Dubos and A. Kessler, Society for General Microbiology, Reading, 1963, p. 298.
64. G. S. Coleman and D. C. Sandford, *J. Gen. Microbiol.*, 1978, **107**, 359.
65. P. Weimer, D. M. Stevenson, M. B. Hall and D. R. Mertens, *Anim. Feed Sci. Technol.*, 2011, **169**, 68.
66. D. Kaiser and J. H. Weniger, *Arch. Tierzucht.*, 1994, **37**, 535.
67. R. A. Prins, W. C. Cline-Thiel and A. T. Van't Klooster, *Agric. Environ.*, 1981, **6**, 183.
68. S. G. Pavlostathis, T. L. Miller and M. J. Wolin, *Appl. Environ. Microbiol.*, 1988, **54**, 2655.
69. S. G. Pavlostathis, T. L. Miller and M. J. Wolin, *Appl. Environ. Microbiol.*, 1988, **54**, 2660.
70. Y. Shi and P. J. Weimer, *Appl. Environ. Microbiol.*, 1992, **58**, 2583.
71. P. J. Weimer, *Arch. Microbiol.*, 1993, **160**, 288.
72. Y. Shi and P. J. Weimer, *Appl. Environ. Microbiol.*, 1996, **62**, 1084.
73. M. E. Berg Miller, D. A. Antonopoulos, M. T. Rincon, M Band, A. Bari, T. Akraiko, A. Hernandez, J. Thimmapuram, B. Henrissat, P. M. Coutinho, I. Borovok, S. Jindou, R. Lamed, H. J. Flint, E. A. Bayer and B. A. White, *PLoS One*, 2009, **4**, e6650.
74. G. Suen, P. J. Weimer, D. M. Stevenson, F. O. Aylward, J. Boyum, J. Deneke, C. Drinkwater, N. N. Ivanova, N. Mikhailova, O. Chertkov, L. A. Goodwin, C. R. Currie, D. Mead and P. J. Brumm, *PLoS One*, 2011, **6**, e18814.
75. G. Suen, D. M. Stevenson, D. Bruce, O. Chertkov, A. Copeland, J. F. Cheng, C. Detter, J. C. Detter, L. A. Goodwin, C. S. Han, L. J. Hauser, M. L. Land, A. Lapidus, S. Lucas, L. Meincke, S. Pitluck, R. Tapia, T. Woyke, J. Boyum, D. Mead and P. J. Weimer, *J. Bacteriol.*, 2011, **193**, 5574.

76. M. P. Bryant, N. Small, C. Bouma and I. M. Robinson, *J. Bacteriol.*, 1958, **76**, 529.
77. P. J. Weimer, *J. Dairy Sci.*, 1996, **79**, 1496.
78. J. B. Russell and R. L. Baldwin, *Appl. Environ. Microbiol.*, 1980, **39**, 604.
79. M. Morrison, R. I. Mackie and A. Kistner, *Appl. Environ. Microbiol.*, 1990, **56**, 3220.
80. H. Kudo, K.J. Cheng and J. W. Costerton, *Can. J. Microbiol.*, 1987, **33**, 267.
81. P. J. Weimer and C. L. Odt, *Am. Chem. Soc. Symp. Ser.*, 1995, **618**, 291.
82. B. A. Dehority and H. W. Scott, *J. Dairy Sci.*, 1967, **50**, 1136.
83. B. A. Dehority, *J. Bacteriol.*, 1965, **89**, 1515.
84. B. A. Dehority, *Appl. Microbiol.*, 1967, **15**, 987.
85. J. Purushe, D. E. Fouts, M. Morrison, B. A. White, R. I. Mackie, The North American Consortium for Rumen Bacteria, P. H. Coutinho, B. Henrissat and K. E. Nelson, *Microb. Ecol.*, 2010, **60**, 721.
86. D. Dodd, Y. H. Moon, K. Swaminathan, R. I. Mackie and I. K. McCann, *J. Biol. Chem.*, 2010, **285**, 30261.
87. M. A. Cotta, *Appl. Environ. Microbiol.*, 1993, **59**, 3557.
88. J. M. Brulc, D. A. Antonopoulos, M. E. Berg Miller, M. K. Wilson, A. C. Yannarell, E. A. Dinsdale, R. E. Edwards, E. D. Frank, J. B. Emerson, P. Wacklin, P. M. Coutinho, B. Henrissat, K. E. Nelson and B. A. White, *Proc. Natl. Acad. Sci. U. S. A.*, 2009, **16**, 1948.
89. M. Hess, A. Sczyrba, R. Egan, T.-W. Kim, H. Chokhawala, G. Schroth, S. Luo, D. S. Clark, F. Chen, T. Zhang, R. I. Mackie, L. A. Pennacchio, S. G. Tringe, A. Visel, T. Woyke, Z. Wang and E. M. Rubin, *Science*, 2011, **331**, 463.
90. B. L. Cantarel, P. M. Coutinho, C. Rancurel, T. Bernard, V. Lombard and B. Henrissat, *Nucleic Acids Res.*, 2009, **37**, D233.
91. R. Gokarn, M. A. Eitman, S. A. Martin and K. E. Eriksson, *Appl. Biochem. Biotechnol.*, 1997, **68**, 69.
92. Q. Li, J. A. Siles and I. P. Thompson, *Appl. Microbiol. Biotechnol.*, 2010, **88**, 671.
93. E. L. Iannotti, D. Kafkewitz, M. J. Wolin and M. P. Bryant, *J. Bacteriol.*, 1973, **114**, 1231.
94. T. L. Miller, E. Currenti and M. J. Wolin, *Appl. Microbiol. Biotechnol.*, 2000, **54**, 494.
95. T. L. Miller and M. J. Wolin, *Appl. Environ. Microbiol.*, 1995, **61**, 3832.
96. D. C. Eaton and A. Gabelman, *US Pat.*, 5 137 736, 1992.
97. P. J. Weimer and D. M. Stevenson, *Appl. Microbiol. Biotechnol.*, 2012, **94**, 461.
98. A. Muthiyan and S. C. Ricke, *Bioresour. Technol.*, 2009, **100**, 5033.
99. F. H. Deindorfer and A. E. Humphrey, *Appl. Microbiol.*, 1958, **7**, 256.
100. I. Lopez, M. Passeggi and L. Borzacconi, *Water Sci. Technol.*, 2006, **54**, 231.
101. J. W. Czerkawski and G. Breckenridge, *Br. J. Nutr.*, 1977, **38**, 371.

102. W. H. Hoover, B. A. Crooker and C. J. Sniffen, *J. Anim. Sci.*, 1976, **43**, 528.

103. R. M. Teather and F. D. Sauer, *J. Dairy Sci.*, 1987, **73**, 666.

104. J. P. Jouany, D. I. Demeyer and J. Grain, *Anim. Feed Sci. Technol.*, 1988, **21**, 229.

105. United Nations Food and Agricultural Organization, *Statistical Year-book*, United Nations, New York, 2005.

106. H. J. M. Camp, G. J. M. Verkley, H. J. Gizjen and G. D. Vogels, *Biol. Waste*, 2006, **30**, 309.

107. S. B. Barnes and J. Keller, *Water Sci. Technol.*, 2003, **28**, 155.

108. Z. H. Hu and H. Q. Yu, *Waste Manage.*, 2006, **26**, 1222.

109. Z. H. Hu, H. Q. Yu and J. C. Zheng, *Bioresour. Technol.*, 2006, **97**, 2103.

110. J. B. Russell, J. D. O'Connor, D. G. Fox, P. J. Van Soest and C. J. Sniffen, *J. Anim. Sci.*, 1992, **70**, 3551.

111. C. J. Sniffen, J. D. O'Connor, P. J. Van Soest, D. G. Fox and J. B. Russell, *J. Anim. Sci.*, 1992, **70**, 3562.

112. D. G. Fox, L. O. Tedeschi, T. P. Tylutki, J. B. Russell, M. E. Van Amburgh, L. E. Chase, A. N. Pell and T. R. Overton, *Anim. Feed Sci. Technol.*, 2004, **112**, 29.

113. M. E. Van Amburgh, L. E. Chase T. R. Overton, D. A. Ross, E. B. Recktenwald, R. J. Higgs and T. P. Tylutki, *72nd Cornell Nutrition Conference*, Cornell University, 2010.

114. J. B. Russell, R. E. Muck and P. J. Weimer, *FEMS Microbiol. Ecol.*, 2009, **67**, 183.

115. S. H. Zhao, Z. B. Yue, B. J. Ni, Y. Mu, H. Q. Yu and H. Harada, *Water Res.*, 2009, **43**, 2047.

116. S. G. Pavlostathis and E. Giraldo-Gomez, *Crit. Rev. Environ. Control*, 1991, **21**, 411.

117. H. N. Chang, N. J. Kim, J. Kang and C. M. Jeong, *Biotechnol. Bioproc. Eng.*, 2010, **15**, 1.

118. H. J. Gijzen, K. B. Zwart, F. J. M. Verhagen and G. P. Vogels, *Biotechnol. Bioeng.*, 1988, **31**, 418.

119. K. B. Zwart, H. J. Gijzen, P. Cox and G. D. Vogels, *Biotechnol. Bioeng.*, 1988, **32**, 719.

120. Z. Chi, Y. Zheng, J. Ma and S. Chen, *Int. J. Hydrogen Energy*, 2011, **26**, 9542.

121. M. T. Holtzapple and C. B. Granda, *Appl. Biochem. Biotechnol.*, 2008, **156**, 95.

122. M. Holtzapple, N. Ross, N. Chang, V. Chang, S. Adelson and C. Brazel, *Appl. Biochem. Biotechnol.*, 1999, **79**, 609.

123. S. Torii and H. Tanaka, in *Organic Electrochemistry*, ed. H. Lund and O. Hammerich, Marcel Dekker, New York, 4th edn, 2000, p. 499.

124. P. F. Levy, J. E. Sanderson, E. Ashare and S. R. deRiel, in *Liquid Fuel Developments*, ed. D. L. Wise, CRC Press, Boca Raton, 1983, p. 159.

125. D. I. Demeyer and C. J. Van Nevel, in *Digestion and Metabolism in the Ruminant*, ed. I. W. McDonald and A. C. I. Warner, University of New England Publishing, Armidale, 1975, p. 366.

126. C. J. Nevel and D. I. Demeyer, *Environ. Monit. Assess.*, 1996, **42**, 73.
127. S. M. van Zijderveld, W. J. J. Gerrits, J. A. Apajalahti, J. R. Newbold, J. Dijkstra, R. A. Leng and H. B. Perdok, *J. Dairy Sci.*, 2010, **93**, 5856.
128. D. H. Kleinschmit, D. J. Schingoethe, K. F. Kalscheur and A. R. Hippen, *J. Dairy Sci.*, 2006, **89**, 4784.
129. J. Guitterez, R. E. Davis, L. L. Lindahl and E. J. Warwick, *Appl. Microbiol.*, 1959, **7**, 16.
130. W. R. Kenealy, Y. Cao and P. J. Weimer, *Appl. Microbiol. Biotechnol.*, 1995, **44**, 507.
131. http://www.ethanol.org/index.php?id = 38&parentid = 8#Distillers%20Grain.
132. G. A. Broderick, in *Forage Quality, Evaluation and Utilization*, ed. G. C. Fahey Jr., M. D. Collins, D. R. Mertens and L. E. Moser, American Society for Agronomy, Madison, 1994, p. 200.
133. C. G. Schwab, *Anim. Feed Sci. Tech.*, 1996, **59**, 87.
134. W. G. Bergen, D. B. Purser and J. H. Cline, *J. Anim. Sci.*, 1968, **27**, 1497.
135. J. B. Russell and R. L. Baldwin, *Appl. Environ. Microbiol.*, 1979, **37**, 537.
136. J. W. Costerton, K. J. Cheng, G. G. Geesey, T. I. Ladd, J. C. Nickel, M. Dasgupta and T. J. Marrie, *Annu. Rev. Microbiol.*, 1987, **41**, 435.
137. P. J. Weimer, A. H. Conner and L. F. Lorenz, *Appl. Microbiol. Biotechnol.*, 2003, **63**, 29.
138. P. J. Weimer, R. G. Koegel, L. F. Lorenz, C. R. Frihart and W. R. Kenealy, *Appl. Microbiol. Biotechnol.*, 2005, **66**, 635.

CHAPTER 15

Tipula abdominalis: a Natural Biorefinery with Novel Microbial Enzymes Useful for Pectin-Rich Biomass Deconstruction

DANA M. SCHNEIDER,[a] EMILY D. HENRIKSEN,[b] WHITNEY E. BOLAND,[b] MEREDITH C. EDWARDS[b] AND JOY DORAN-PETERSON*[b]

[a] School of Chemistry and Biochemistry, Georgia Institute of Technology, NAI Center for Ribosomal Origins and Evolution, Atlanta, GA 30332, USA; [b] Department of Microbiology, University of Georgia, Athens, GA 30602, USA
*Email: jpeterso@uga.edu

15.1 Introduction

Insects are the largest taxonomic group of animals on Earth and insects that consume lignocellulose and host gut microbial consortia have become of special interest for the application of microbial conversion of lignocellulose to biofuels, including ethanol, butanol, and hydrogen. One of the major challenges in lignocellulose conversion is the need for robust and inexpensive enzymes to deconstruct lignocellulose into fermentable sugars.[1] The gut microbial community of lignocellulose-degrading insects may be mined for these enzymes, as well as for

RSC Energy and Environment Series No. 10
Biological Conversion of Biomass for Fuels and Chemicals: Explorations from Natural Utilization Systems
Edited by Jianzhong Sun, Shi-You Ding and Joy Doran-Peterson
© The Royal Society of Chemistry 2014
Published by the Royal Society of Chemistry, www.rsc.org

novel microorganisms themselves for the application of improved biofuel production technology.

As plants have evolved recalcitrant structures to resist predation, so have their consumers evolved mechanisms to overcome that resistance. For microorganisms, these mechanisms include enzymes that deconstruct plant polymers to sugar moieties. Many animals, including insects, host a gut microbial community that facilitates digestion of a recalcitrant lignocellulolytic diet. Lignocellulose–ecosystem–insect–microbial; interactions may be viewed as a whole process, a natural biorefinery.

This chapter describes the natural biorefinery concept from ecosystem to biotechnology using *Tipula abdominalis* as the model insect, with the application of a novel enzyme obtained from this system in biofuel production.

15.2 Overall Ecosystem Biorefinery

Detritus (dead organic matter) in the form of lignocellulolytic leaf litter provides the major contribution of organic material for carbon and energy in small streams in forested watersheds.[2,3] Ecosystem refinement (decomposition) of detritus reduces organic matter size and releases nutrients for re-entry into food webs.[4] Few animals synthesize all of the lignocellulolytic enzymes required to digest plant litter, thus animal nutrition requires the activity of microbes, either in free-living inter-relationships or associated in the animal gut.

Detritus refinement occurs in three main phases: leaching, conditioning, and fragmentation (Figure 15.1).[4-7] Leaching is the rapid (within 24 h) loss of soluble organic matter (or dissolved organic matter, DOM) from coarse

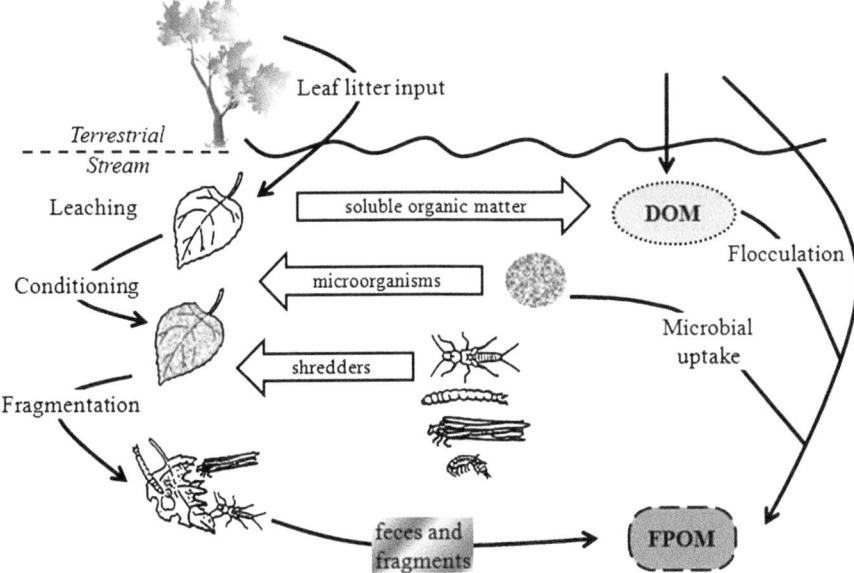

Figure 15.1 Leaf and detritus processing in riparian streams.

particulate organic matter (CPOM) shortly after immersion.[4] "Conditioning" refers to microbial colonization and the chemical–structural modification of leaf material, which increases palatability for detritivorous macroinvertebrates, or shredders.[4] Fragmentation results in the reduction of detritus leaf litter from coarse to fine particulate organic matter (FPOM).[4]

15.2.1 Macroinvertebrate Shredders: a Consortium of Natural Biorefineries

Contributing to the fragmentation process of detritus refinement are shredders: a functional feeding group of insects that consume CPOM, primarily, conditioned leaf litter detritus. In small riparian stream ecosystems, leaf litter comprises the majority of carbon and energy input;[8] however, many organisms are unable to degrade this lignocellulosic material. By refining leaf litter into a form that other organisms can use, shredders influence the bioavailability of carbon and energy within the ecosystem.

Shredders, and detritivores in general, do not seem to produce all the necessary lignocellulolytic enzymes themselves to digest the abundant plant polymers of their diets.[9–11] Some shredders host gut microbial communities that facilitate the digestion of lignocellulose.[12,13] The necessity of microbially mediated hydrolysis and fermentation for the digestion and assimilation of plant polymers has been demonstrated in three different genera of shredders, including the aquatic cranefly, *Tipula abdominalis*.[13]

15.2.2 *Tipula abdominalis*: a Natural Biorefinery Hosting a Gut Microbial Community

Tipula abdominalis is an aquatic crane fly, which is found in riparian stream ecosystems (Figure 15.2(a)). *T. abdominalis* larvae (Figure 15.2(b)) consume conditioned leaf litter throughout the fall, winter, and spring. The gut morphology of *T. abdominalis* larvae consists of two main compartments: the midgut and hindgut (Figure 15.2(c)). The midgut is highly alkaline, at pH 11, while the hindgut is neutral, at pH 7.[14] Studies suggest that proteolysis occurs in the alkaline conditions in the midgut, dissociating protein complexes from plant polymers, which are then more accessible for saccharification and microbial fermentation in the pH-neutral hindgut.[13,15–18]

Figure 15.2   *Tipula abdominalis* (a) adult (photo: Jill Stuckey), (b) larva (photo: Theresa Rogers), and (c) diagram of the gut.

The *T. abdominalis* larval hindgut hosts a dense and diverse microbial community.[12,19] Because the greatest number of bacteria are associated with the hindgut wall, and attachment to epithelial cells is likely a mechanism for prolonged interactions,[20] it is hypothesized that much of the resident bacteria are found in the wall-associated subpopulation.

The *T. abdominalis* larval gut is a model natural biorefinery (Figure 15.3).[21] In this model, the biomass substrate (conditioned leaf litter) is ingested by larvae. Biomass is pretreated during ingestion, in which maceration decreases the particle size and increases the surface-area-to-volume ratio. The biomass is further pretreated in the alkaline midgut, in which proteolysis degrades

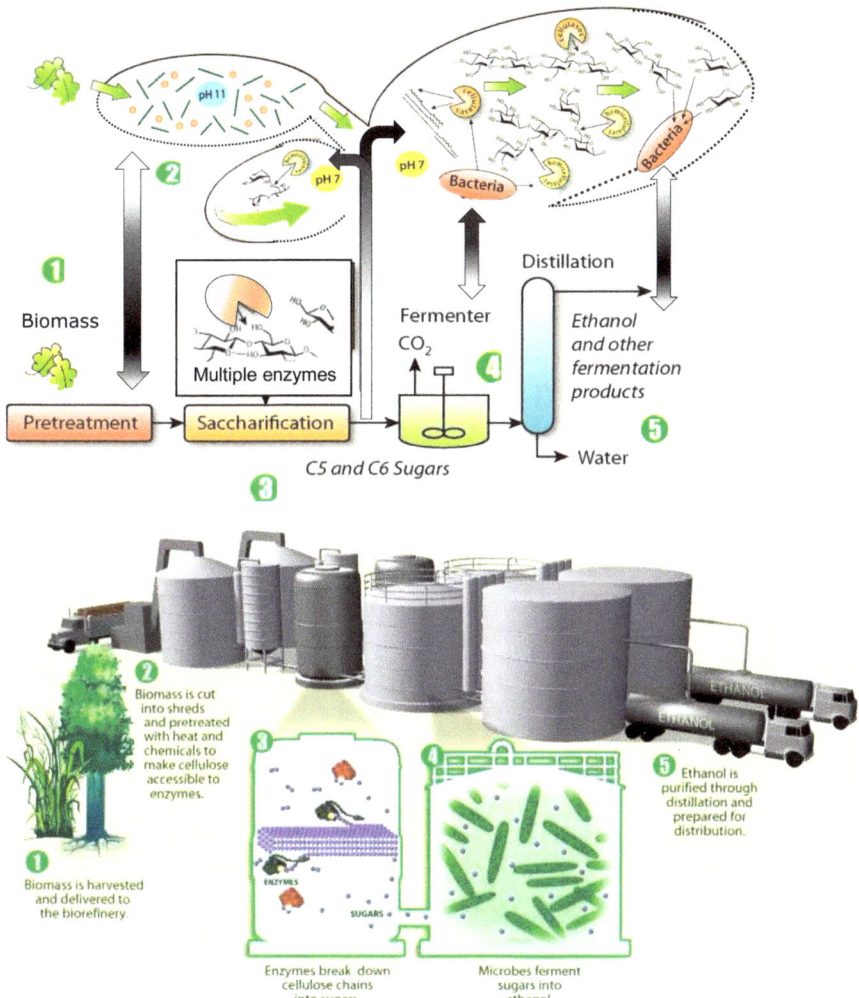

Figure 15.3 Model of the *T. abdominalis* larval gut as a natural biorefinery: ① biomass particle size reduction; ② pretreatment; ③ saccharification *via* enzymes; ④ fermentation; and ⑤ production of useful products.

complexed proteins making polysaccharide polymers more accessible for further processing. In the neutral pH hindgut, bacterial enzymes refine plant polymers into monomeric units. Released sugars are then consumed by bacteria and converted to acetate and other fermentation products, which can be transported across the gut to the hemolymph to support larval energy and growth requirements.[18] Lastly, waste and by-products are excreted and are valuable to other organisms in the ecosystem. This model of the *T. abdominalis* larva as a natural biorefinery can be applied to developments in technology for industrial biomass refinery processes. In particular, bacteria, or their enzymatic capabilities, can be harnessed for biotechnological applications, such as fuel production from biomass.

15.3 *T. abdominalis* Hindgut Bacterial Community Phylogenetic Profile

Three main bacterial phyla are associated with the *T. abdominalis* larvae: *Bacteroidetes*, *Firmicutes*, and *Proteobacteria* (Figure 15.4). *Betaproteobacteria* dominate the bacterial communities associated with the leaf litter diet and fecal cast, and *Clostridia* dominate the hindgut bacterial communities. The communities associated with the ingested leaf litter diet and fecal casts are quite different from the communities associated with the hindgut. Bacterial communities from the hindgut had greater diversity than from the leaf diet and fecal casts, as indicated by Chao1 estimations and the Shannon diversity index (Table 15.1).[22]

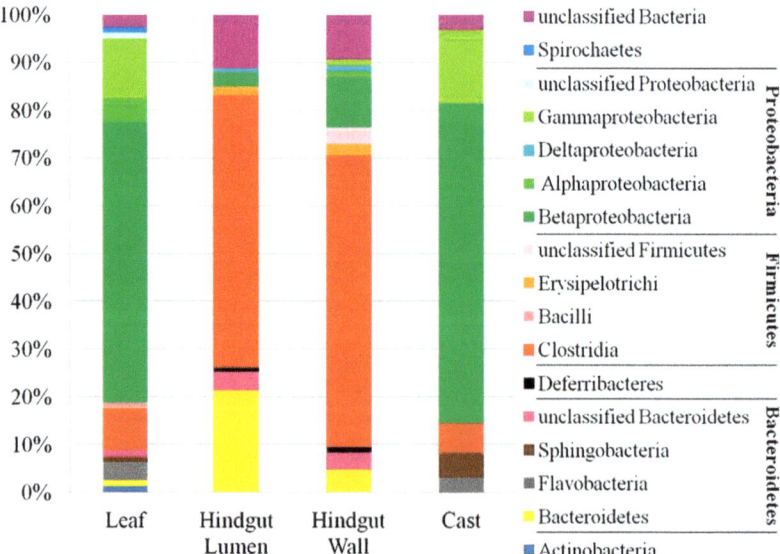

Figure 15.4 Bacterial community composition associated with the *T. abdominalis* leaf litter diet, hindgut lumen, hindgut wall, and fecal casts. Phylogenetic distribution of 16S rRNA clones by class, confidence threshold 95%.[22]

Table 15.1 Community richness and diversity calculations for 16S rRNA gene clone libraries at Operational Taxonomic Unit (OTU)$_{0.97}$ [ref. 22].

Library	N^a	S^b	$Chao1^c$	H^d	$H/H_{max}{}^e$	$Coverage^f$
Leaf	80	39	64.3	3.258	0.89	0.71
Hindgut lumen	148	75	126.75	3.924	0.91	0.69
Hindgut wall	85	48	106.67	3.602	0.93	0.61
Cast	97	34	61.2	3.110	0.88	0.82

^{a}Total number of sequences.
^{b}Number of OTUs at 97% sequence similarity.
^{c}Estimated number of OTUs at 97% sequence similarity, $Chao1 = S + (n_1)^2/(2n_2)$, where n_1 is the number of singletons and n_2 is the number of doubletons.
^{d}Shannon diversity index.
$^e H_{max} = \ln(S)$. ^{f}Calculated from Good's equation: $Coverage = 1 - (n_1/N)$.

15.4 Firmicutes

The *T. abdominalis* hindgut bacterial community is dominated by obligate anaerobes – *Clostridia*. However, facultative anaerobes, particularly the *Firmicutes*, are more readily isolated from the hindgut and also have desirable enzymatic activity to degrade plant polymers (Table 15.2).[19] Five isolates including *Paenibacillus amylolyticus* 27C64, and four other microorganisms with identical ribotypes can degrade numerous model plant polymer substrates (highlighted in Table 15.2).

15.5 Paenibacillus

Paenibacillus amylolyticus 27C64, isolated from the hindgut of *T. abdominalis*[19] is a Gram-positive, spore-forming, facultative anaerobe with wide-ranging enzymatic capabilities.[23,24] *P. amylolyticus* is believed to play a significant role in the *T. abdominalis* hindgut community due to its extensive enzyme repertoire, its resistance to multiple classes of anti-microbial compounds, and the ability of this organism to inhibit the growth of other microorganisms by producing its own anti-microbial compounds.[19,21] This organism has been isolated from larvae in three states (Michigan, North Carolina, and Georgia), both aerobically and anaerobically. *P. amylolyticus* 27C64 does not produce one major fermentation product such as ethanol nor butanol, and is thus not a good native candidate as the fermentative organism. However, as it has exo-enzymes with activities desired for a wide range of biomass conversions to fermentable products, *P. amylolyticus* 27C64 is an attractive candidate for biotechnology applications as a source of novel enzymes to engineer into ethanologens (or other fermentation biocatalysts) for consolidated bioprocessing.

15.5.1 Expression Library

Genomic DNA from *P. amylolyticus* 27C64 was prepared using Qiagen kits (Valencia, CA) and after partial digestion with Sau3AI and agarose gel extraction of 2–5 kb fragments, *P. amylolyticus* 27C64 genomic fragments were ligated into BamHI digested pUC19, transformed into *E. coli* DH5α or *E. coli* LY40A by heat–shock, and grown on LB agar with 50 mg L^{-1}

Table 15.2 *T. abdominalis* hindgut bacteria: *Firmicutes* isolates and their enzymatic activity on model plant polymers: carboxymethylcellulose (CMC)[42]; starch (Difco 272100); xylan[43]; polygalacturonate (PGA)[26]; and methylumbelliferyl conjugates of cellobiopyranoside (MUC), arabinofuranoside (MUA), glucoside (MUG), mannopyranoside (MUM), and xyloside (MUX).[44]

Isolate	% Similarity	Cultured strain match	CMC	Starch	Xylan	PGA	MUA	MUX	MUC	MUG	MUM
10C22	100	Bacillus circulans	+	−	−	−	−	−	−	−	−
11C46	100		−	+	−	−	+	+	−	+	−
12C70	100	Bacillus cohnii	−	−	−	−	−	−	−	+	−
13C100	99.8	Bacillus firmus	+w	+	−	−	−	−	+	+	−
14C111			+	−	−	−	−	−	+w	−	−
15C59*2	99.8	Bacillus fusiformis	−	−	−	−	−	−	−	−	−
20C75	96.6	Bacillus sphaericus	−	−	−	−	−	−	−	−	−
17C67	100	Bacillus silvestris	−	−	−	−	−	−	−	−	−
18C68			−	−	+	−	−	−	−	−	−
21C27	100	Bacillus megaterium	−	−	−	−	−	+	+	+	−
22C25	100	Bacillus weihenstephanensis	−	−	−	−	−	−	−	−	−
23C26	99.8	Bacillus pumilus	−	−	−	−	+	+w	+	+	−
24C57	99.2	Bacillus thuringiensis	+	+	−	−	−	+	+	−	−
27C64*5	99.1	Paenibacillus amylolyticus	+	+	+	+	+	+	+	+	+
29C58s	97.6		−	−	−	−	−	−	−	−	−
31C28	96.2	Paenibacillus agaridevorans	−	+	+	−	+	+	+	+	+
32C23	95.0		+	+w	+	−	+	+	+	+	+
33C53L*2	98.0	Paenibacillus glycanilyticus	−	−	−	−	−	−	−	−	−
35C82	100	Staphylococcus aureus	−	−	−	−	−	−	−	−	−
36C7*3	99.0	Staphylococcus saprophyticus	−	−	−	−	−	−	−	−	−
39C61W*2	100	Staphylococcus warneri	−	−	−	−	+	+	+	−	−
41C21	100	Bacillus muralis	−	−	−	−	−	−	+	−	−

*n = n isolates with identical partial 16S rRNA gene sequences and identical enzymatic profiles for the substrates tested in this study. + represents hydrolysis of the substrate; − represents no observable hydrolysis of the substrate; +w represents weak hydrolysis of the substrate.

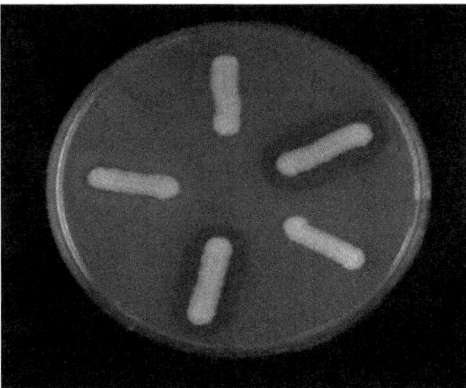

Figure 15.5 Polygalacturonic acid plate at pH 7.0 with clockwise from the top: LY40A, LY40A + PelB, LY40A, LY40A + PelB, LY40A + PelA. Note the large zones of clearing around the PelB containing cultures. A visible, but much smaller zone of clearing is seen with LY40A + PelA under these conditions.

ampicillin, 1 mg L^{-1} X-Gal, and 2.5 mg L^{-1} isopropyl β-D-1-thiogalactopyr-anoside (IPTG).[25] Insert-containing transformants of LY40A were screened for pectinase activity on polygalacturonase medium.[26] After growth on the polygalacturonase medium, plates were flooded with 2 N HCl and pectinase-producing colonies were identified by the appearance of clearing or halos surrounding the colony growth (Figure 15.5).

15.5.2 Pectinase Activity

The pectin backbone can consist of a homopolymer of α-1,4-D-galacturonic acid (homogalacturonan) or repeats of the disaccharide α-1,2-L-rhamnose-α-1,4-D-galacturonic acid (rhamnogalacturonan-I), and, typically, 30–80% of galacturonic acid residues are methylated. Homogalacturonan can be substituted with xylose or apiose, while rhamnogalacturonan-I is often substituted with galactose, arabinose, or galactan.[27,28]

Degradation of pectin requires methylesterases and depolymerases. Pectin methylesterases are responsible for the hydrolysis of methylester linkages at random from the polygalacturonic acid backbone.[29] Pectin depolymerases act upon polygalacturonate backbones and belong to one of two families: lyases or polygalacturonases. Lyases cleave by β-elimination giving a Δ-4,5-unsaturated product, while polygalacturonases are responsible for hydrolytic cleavage of the polygalacturonate chain.[30,31] There are two types of lyases: pectate lyases, which cleave unesterified polygalacturonate (or pectate), and pectin lyases, which cleave methyl esterified pectin. Pectate lyases are classified into families based on amino acid similarity.[32]

Pectate lyase assays were performed essentially as described[33,34] with transformant cell extracts prepared by sonication or by using the supernatant from the culture to measure the activity in secreted protein. The enzyme assay mixture contained 0.2% (w/v) polygalacturonic acid (PGA, Sigma) or pectin with varying degrees of methylesterification (MP Biomedicals). The formation of Δ-4,5-unsaturated products at 235 nm for 1–3 min was observed. One unit of enzyme activity was defined as the amount of enzyme that produces 1 μmol Δ-4,5-unsaturated product per minute under the assay conditions. Specific activity was reported as U mg^{-1} protein and the Bradford method was used to determine protein concentration.[35]

15.6   PelA and PelB

Two pectinase-positive clones were identified after screening approximately 6000 clones on polygalacturonase medium and were subcloned for further study.[36] The first clone had an open reading frame (ORF) of 669 bp (*pelA*, GenBank accession no. GU289919) and the second contained a single ORF of 1176 bp (*pelB*, GenBank accession no. GU289920). The first gene, *pelA*, encodes a 222 amino acid protein and the protein demonstrated its highest activity on polygalacturonic acid, but retained 60 and 56% of maximum activity on 8.5 and 90% methylated pectin, respectively. CaCl$_2$ was required for activity, and the optimal conditions were pH 10.5, 45 °C, and 1.5 mM CaCl$_2$ (Figure 15.6). PelA has a high similarity (95%) to PelA from *P. barcinonensis*,

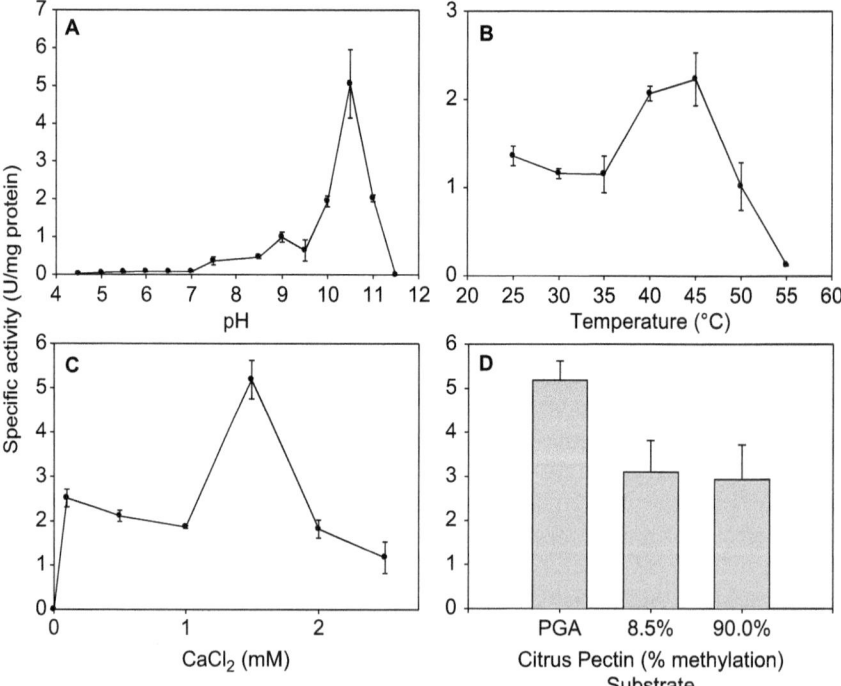

Figure 15.6   PelA optima for pH (a), temperature (b), CaCl$_2$ (c), and activity on different pectic substrates (d).
Adapted from ref. 36.

and is a subclass of the pectate lyase family 3 from saprophytic, non-pathogenic bacteria. PelB is a 302 amino acid protein demonstrating its highest activity on methylated pectin (20–34%), followed by polygalacturonic acid (Figure 15.7). In contrast to PelA, PelB had a broad temperature optimum of 30–45 °C, with a pH optimum of 9.5. PelB showed low (≤28%) homology to family 1 pectate lyases. *P. amylolyticus* 27C64 PelA and PelB were the first pectate lyases described in *P. amylolyticus* and show an unusual combination of pectin and pectate lyase activities by degrading both polygalacturonic acid and highly methylated pectin.[36]

15.6.1 Use of PelB to Improve Fermentation Parameters

Pectinases are important for the degradation of lignocellulose, where pectin can comprise a significant portion of the plant structure in many sources of biomass, including sugarbeet pulp, citrus peel, and apple pomace.[37,38] Sugarbeet pulp is composed of 24% cellulose, 25% hemicellulose, and 24% pectin, with low concentrations of lignin (<8%) and protein (<8%) on a dry weight basis.[37] As an agricultural residue, its fermentation to ethanol for fuel is renewable and sustainable. The most costly components of these fermentations are the enzymes needed to degrade the substrate into fermentable carbohydrates. If the fermenting organism could also produce at least some of the enzymes needed for plant-cell-wall deconstruction, this could potentially reduce the enzyme cost. To investigate this possibility using the novel pectate lyase from the insect-associated bacterium, the gene encoding PelB was transformed into an ethanol-producing bacterium. Further studies were conducted with PelB because of its higher specific activity and the higher concentrations of secreted recombinant protein when compared to PelA (Figures 15.6 and 15.7).

15.6.1.1 Construction of JBP25

Ethanologenic *Escherichia coli* LY40A was generated by addition of the cellobiose phosphotransferase system from *Klebsiella oxytoca* to enable strain LY40A to transport and metabolize the dimeric form of glucose, namely cellobiose.[39,40] The gene encoding PelB was cloned into a pUC-based plasmid and transformed into strain LY40A to generate strain JBP25 using standard molecular techniques (Figure 15.8).[39,40] Polygalacturonase activity was confirmed prior to fermentation by clearing around the colony growth on polygalacturonase medium as previously described.

15.6.1.2 Sugarbeet Pulp Fermentations and Analysis of Ethanol Production

Fermentations were performed essentially as described previously,[37] except as indicated. Sugarbeet pulp moisture content was calculated using a Denver Instrument IR 35 Moisture Analyzer and enough solids for a final

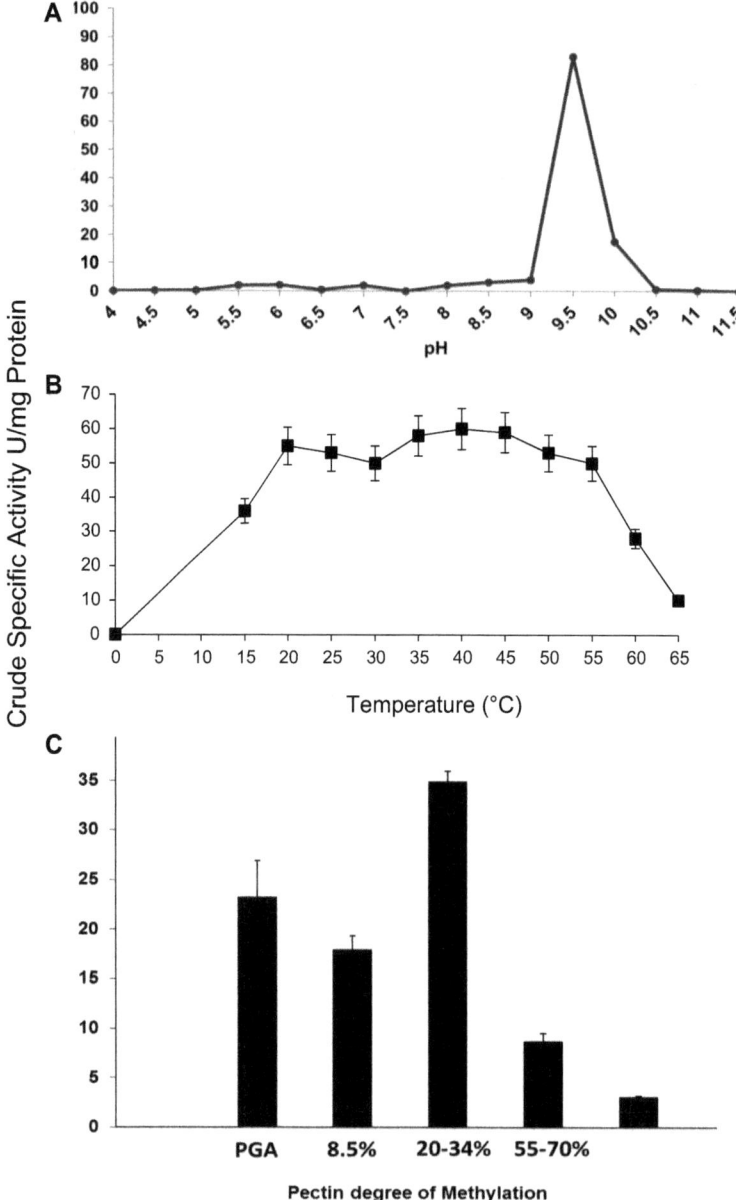

Figure 15.7 PelB optima for pH (a), temperature (b), and activity on different pectic substrates (c).
Adapted from ref. 36.

concentration of 10% w/v was added to 2X Luria broth. Fermenters were auto-claved, placed in a water bath at 45 °C, and mixed with magnetic stirrers. The pH was adjusted to 9.5 and maintained using a Jenco 3671 pH controller and

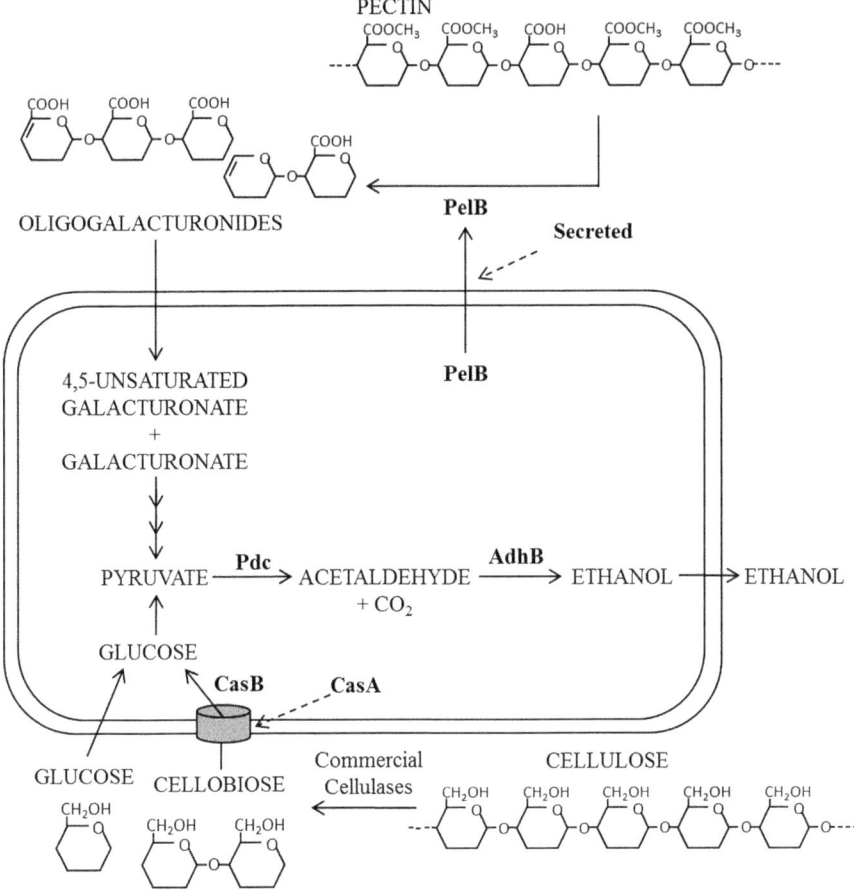

Figure 15.8 Depiction of the capabilities of JBP25. The ethanol production pathway
was engineered into *E. coli* using the pyruvate decarboxylase (Pdc) and
alcohol dehydrogenase (AdhB) genes from *Zymomonis mobilis*.[39] The
cellobiose phosphotransferase system (CasA CasB) was added for uptake
and metabolism of cellobiose.[40] JBP25 resulted when the gene encoding
PelB from *Paenibacillus amylolyticus* was added to the bacterium containing the *pdc*, *adhB*, and *casAB* genes.

KOH (6 and 10 M) in order to optimize the enzyme activity of PelB. Alkali
treatment of pectin-rich materials will impact the structure of the pectin,
however, all fermenters were treated in the same fashion and a discussion of
alkali pretreatment in its own right is beyond the scope of this paper. Super-
natants from overnight cultures of LY40A and JBP25 were harvested, added to
the fermenters, and the mixtures were incubated at 45 °C for 8 h. Visible reductions in viscosity were observed in the fermenters containing the supernatant from JBP25 (Figure 15.9, inset). After 8 h, the pH was then dropped to
4.5 using H_2SO_4 to obtain the optimum pH for fungal cellulases and pectinases

that were added to facilitate further plant-cell-wall degradation to fermentable carbohydrates. Celluclast and pectinase from *Aspergillus niger* (Novozymes) were added to the vessels at concentrations of 10 filter paper units $(FPU)\,g^{-1}$ dry wt and 10 polygalacturonase units $(PGU)\,g^{-1}$ dry wt, respectively. After 16 h, the pH was increased to 5.5 with KOH and the temperature was decreased to 35 °C, optimal fermentation conditions for the ethanologen, and maintained at these values throughout the fermentation. Each fermenter was inoculated to an OD_{550} of 1.0 with *E. coli* strains LY40A or JBP25.

Fermentations were conducted in triplicate with samples collected every 24 h until completion.

To quantify ethanol production, gas chromatography (GC) was performed as previously described;[41] fermentation supernatant samples were filtered with a 0.22 μm filter prior to analysis. Ethanol concentration in the fermenters after inoculation (Time 0) reached 11.1 g L^{-1} for the LY40A fermentations and over $15\,g\,L^{-1}$ for the JBP25 fermentations (Figure 15.9). Presumably neither fermentation reached the maximum theoretical yield of $20\,g\,L^{-1}$ due to the very low total enzyme loading. Adding enzymes to excess (additional cellulase and pectinase enzymes) resulted in both LY40A and JBP25 achieving ethanol titers close to the theoretical maximum.

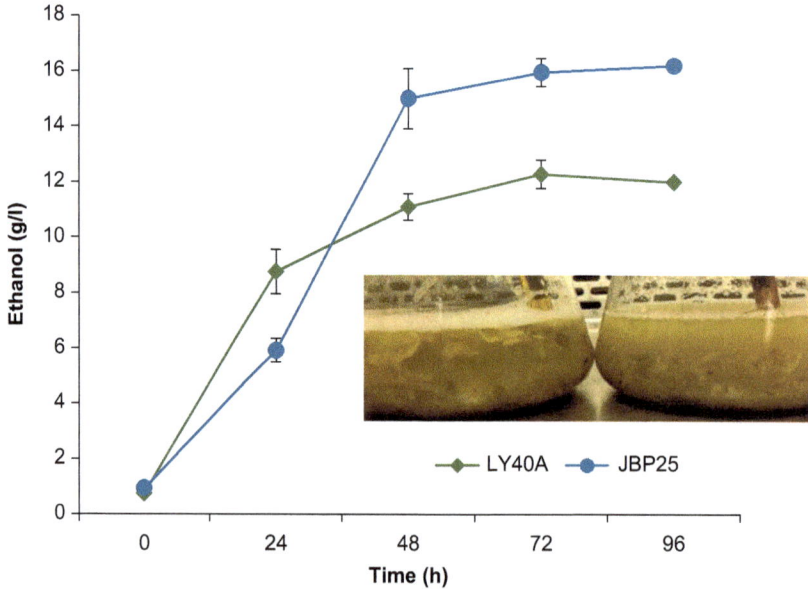

Figure 15.9 Ethanol production from fermentations using LY40A or JBP25 as biocatalysts. "Time 0" indicates inoculation with either organism at a starting OD_{550} of 1.0. The pH was maintained at 5.5 and the temperature was 35 °C during the fermentation. The inset is a photograph of the fermentations 24 h after inoculation.

15.7 Summary

Lignocellulose-feeding insects such as detritivores play a crucial role in carbon cycling in small headwater streams. One of these detritivores, *Tipula abdominalis*, was examined for the community structure of its associated microbiota. The hindgut-associated bacteria are thought to play a role in decomposition of the leaf litter diet of the insect. Microorganisms were isolated from the hindgut using aerobic and anaerobic cultivation techniques. *Paenibacillus amylolyticus* isolates were recovered using both techniques from animals captured in three states in the USA. A genomic library from *P. amylolyticus* 27C64 was screened for various enzyme activities useful in biomass conversions for fuel production. A novel pectate lyase, PelB, was discovered with unusual activity on both polygalacturonate and highly methylated pectin. The addition of PelB to pectin-rich fermentations with very low commercial enzyme loadings increased ethanol concentrations by 26%, even without optimization. These experiments indicate the potential use of insect-associated microbial enzymes for enhancing fermentation yields from biomass. A suite of enzyme activities is needed for complete degradation of plant cell walls and the addition of low levels of commercial enzymes was still essential to complement the activity of the novel pectate lyase. The addition of PelB with dual activities increased ethanol yields and reduced the amount of commercial enzyme required, and adding additional synergistic enzyme activities obtained from the *P. amylolyticus* library screening may further decrease the amount of commercial enzyme needed for biomass conversion to fuels. An alternative approach may be to engineer the *P. amylolyticus* bacterium itself, as it is able to produce multiple enzyme activities. However, it produces more than one fermentation product in its native state. Studies are underway to optimize products from lignocellulosic biomass using this approach, however they are beyond the scope of the current study. Whether using insect-associated microorganisms or their enzymes, lessons learned from the natural biorefinery of *Tipula abdominalis* show promise for biotechnology applications, especially in bioconversions for fuel production. In addition to harvesting other enzymes from bacteria associated with the insect gut, further exploration into the mechanisms of the pretreatment of biomass inside the insect, especially the alkaline environment of the midgut, could elucidate other processes that could be mimicked to further enhance the production of fuels and chemicals from recalcitrant biomass substrates.

Acknowledgements

Partial support for this work was provided by the US Department of Energy No. DOE DE-EE-0000410 and the US Department of Agriculture No. 01495.

References

1. S. Octave and D. Thomas, *Biochimie*, 2009, **91**, 659.
2. S. Fisher and G. Likens, *Ecol. Monogr.*, 1973, **43**, 421.

3. J. B. Wallace, S. L. Eggert, J. L. Meyer and J. R. Webster, *Science*, 1997, **277**, 102.
4. K. W. Cummins, *BioSci.*, 1974, **24**, 631.
5. J. R. Webster and E. F. Benfield, *Annu. Rev. Ecol. Syst.*, 1986, **17**, 567–594.
6. L. Maltby, in *The Rivers Handbook*, ed. P. Calow and G. E. Petts, Blackwell Scientific Publications, Oxford, 1992, vol. 1, p. 331.
7. M. O. Gessner, E. Chauvet and M. Dobson, *Oikos*, 1999, **85**, 377.
8. R. Vannote, G. Minshall, K. W. Cummins, J. Sedell and C. Cushing, *Can. J. Fish. Aquat. Sci.*, 1980, **37**, 130.
9. F. Barlocher and B. Kendrick, *Oikos*, 1974, **26**, 55.
10. F. Barlocher and C. W. Porter, *J. N. Am. Benthol. Soc.*, 1986, **5**, 58.
11. K. H. Walters and L. A. Smock, *Hydrobiologia*, 1991, **220**, 29.
12. M. J. Klug and S. Kotarski, *Appl. Environ. Microbiol.*, 1980, **34**, 408.
13. R. L. Sinsabaugh, A. E. Linkins and E. F. Benfield, *Ecology*, 1985, **66**, 1464.
14. M. M. Martin, *Invertebrate–Microbial Interactions: Ingested Fungal Enzymes in Arthropod Biology*, Cornell University Press, Ithaca, 1987.
15. C. Canhoto and M. A. S. Garca, *Can. J. Zool.*, 2006, **84**, 1087.
16. T. M. Clark, *J. Chem. Ecol.*, 1999, **25**, 1945.
17. M. A. S. Garca and F. Barlocher, *Aquat. Insects*, 1998, **21**, 11.
18. D. L. Lawson and M. J. Klug, *J. N. Am. Benthol. Soc.*, 1989, **8**, 85.
19. D. M. Cook, E. D. Henriksen, R. Upchurch and J. B. D. Peterson, *Appl. Environ. Microbiol.*, 2007, **73**, 5683.
20. R. J. Dillon and V. M. Dillon, *Annu. Rev. Entomol.*, 2004, **49**, 71.
21. D. M. Cook and J. B. Doran-Peterson, *Insect Sci.*, 2010, **17**, 303.
22. D. M. Cook, PhD thesis, *University of Georgia*, 2010.
23. O. Shida, H. Takagi, K. Kadowaki, L. K. Nakamura and K. Komagata, *Int. J. Syst. Bacteriol.*, 1997, **47**, 299.
24. C. Ash, F. G. Priest and M. D. Collins, *Antonie Van Leeuwenhoek*, 1993, **64**, 253.
25. J. Sambrook, E. F. Fritsch and T. Maniatis, *Molecular Cloning: A Laboratory Manual*, Cold Spring Harbor Laboratory, New York, 2nd edn, 1989.
26. M. P. Starr, A. K. Chatterjee, P. B. Starr and G. E. Buchanan, *J. Clin. Microbiol.*, 1977, **6**, 379.
27. W. G. T. Willats, L. McCartney, W. Mackie and J. P. Knox, *Plant Mol. Biol.*, 2001, **47**, 9.
28. B. L. Ridley, M. A. O'Neill and D. Mohnen, *Phytochemistry*, 2001, **57**, 929.
29. J. R. Whitaker, *Enzyme Microb. Technol.*, 1984, **6**, 341.
30. R. S. Jayani, S. Saxena and R. Gupta, *Process Biochem.*, 2005, **40**, 2931.
31. T. Sakai, T. Sakamoto, J. Hallaert and E. J. Vandamme, *Adv. Appl. Microbiol.*, 1993, **39**, 231.
32. P. M. Coutinho and B. Henrissat, in *Recent Advances in Carbohydrate Bioengineering*, ed. H. J. Gilbert, G. Davies, B. Henrissat and B. Svensson, RSC Publishing, Cambridge, 1999.
33. A. Collmer, J. L. Ried and M. S. Mount, in *Methods in Enzymology*, ed. W. A. Wood and S. T. Kellogg, Academic Press, San Diego, 1988.

34. M. Soriano, A. Blanco, P. Diaz and F. I. Javier Pastor, *Microbiology*, 2000, **146**, 89.
35. M. Bradford, *Anal. Biochem.*, 1976, **72**, 248.
36. W. E. Boland, E. DeCrescenzo Henriksen and J. B. Doran-Peterson, *Appl. Environ. Microbiol.*, 2010, **76**, 6006.
37. J. B. Doran, J. Cripe, M. Sutton and B. Foster, *Appl. Biochem. Biotechnol.*, 2000, **84–86**, 141.
38. J. B. Doran Peterson, *Ethanol Producer Magazine*, 2003, **9**, 26.
39. M. C. Edwards, E. DeCrescenzo Henriksen, L. Sharma, L. P. Yomano, B. C. Gardner, L. O. Ingram and J. B. Doran-Peterson, *Appl. Environ. Microbiol.*, 2011, **77**, 5184.
40. L. O. Ingram, T. Conway, D. P. Clark, G. W. Sewell and J. F. Preston, *Appl. Environ. Microbiol.*, 1987, **53**, 2420.
41. J. B. Doran Peterson, S. Brandon, A. Jangid, E. DeCrescenzo-Henriksen, B. Dien and L. O. Ingram, in *Biofuels: Methods and Protocols, Methods in Molecular Biology*, ed. J. R. Mielenz, 2009, vol. 581, p. 263.
42. W. A. Wood, and S. T. Kellogg, *Biomass Part A, Cellulose and Hemicellulose*, Academic Press, San Diego, CA, 1988, 160.
43. F. Mondou, F. Shareck, R. Morosoli and D. Kluepfel, *Gene*, 1986, **49**, 323–330.
44. K. R. Sharrock, *J. Biochem. Biophys. Methods*, 1988, **17**, 81–106.

CHAPTER 16

Cloning, Mutation and Over-Expression of Lignocellulase Genes

WEILAN SHAO

Biofuels Institute, School of Environment, Jiangsu University, Zhengjiang, Jiangsu 212013, P. R. China
Email: weilanshao@gmail.com

16.1 The Mechanisms of Lignocellulose Degradation

Lignocellulose is the largest renewable carbohydrate source on Earth, and fermentable sugars can be derived from two of its main components, cellulose and hemicellulose. The three main components of lignocellulose are cellulose, hemicellulose and lignin. The proportions of these components in a lignocellulose fiber depends on the source of the fiber, the age of the fiber, and the extraction conditions used to obtain the fibers.[1] Cellulose, the main structural component providing strength and stability to the plant cell walls, is a long β-1,4-linked chain of glucose molecules, and hydrogen bonds between different layers of the polysaccharides contribute to the resistance of crystalline cellulose to degradation. Hemicellulose, the second most copious constituent of lignocellulose, is not a chemically well defined compound but rather a family of heteropolysaccharides, composed of different 5- and 6-carbon monosaccharide units. These units link cellulose fibers into microfibrils, cross-link with lignin, and consist of sugars including glucose, xylose, galactose, arabinose and mannose, creating a complex network of bonds that provide structural

RSC Energy and Environment Series No. 10
Biological Conversion of Biomass for Fuels and Chemicals: Explorations from Natural Utilization Systems
Edited by Jianzhong Sun, Shi-You Ding and Joy Doran-Peterson
© The Royal Society of Chemistry 2014
Published by the Royal Society of Chemistry, www.rsc.org

strength. Lignin is a three-dimensional polymer of phenylpropanoid units composed of three major phenolic components, *p*-coumaryl alcohol, coniferyl alcohol and sinapyl alcohol. Lignin can be considered as the cellular glue providing the plant tissue and the individual fibers with compressive strength and the cell wall with stiffness, in addition to providing resistance to chemical and physical damage. The lignin content of the fibers influences the structure, properties, morphology, flexibility and rate of hydrolysis.[2] These structures are naturally resistant to enzyme degradation. However, some methods for biomass pretreatment have been developed to increase the reactivity and digestibility of lignocellulosic materials, and these make the enzymatic hydrolysis of lignocellulosic biomass possible.[3–5]

Enzymatic degradation of lignocellulose is recognized as the hydrolysis method with the greatest potential for improvement and cost reduction.[6] The hydrolysis of lignocellulose requires a number of different cellulases, hemicellulases and ligninases. A sequential breakdown of the linear cellulose chains to glucose requires the activities of endo-glucanase, exo-gluconase, and glucosidase. These enzymes are distributed widely in animal guts, soil, rivers, hot springs and many other environments where microorganisms can grow. The rate of natural cellulase degradation of crystalline cellulose, however, is one of the bottlenecks for either thermochemical pretreatment, which can only destroy hemicellulose and lignin, or enzymatic hydrolysis of lignocellulose. Furthermore, lignin is a tough barrier protecting cellulose from enzyme attachment in lignocellulosic biomass; the degradation of lignin does not release any fermentable sugar, but the presence of ligninases including laccase, lignin peroxidase, Mn peroxidase, and ferulic acid esterase has been shown to enhance the hydrolysis of cellulose and hemicelluloses. Hemicelluloses form another barrier for the resistance of the cell wall to cellulase degradation. The difficulty for enzymatic degradation of hemicelluloses lies in the complex glycoside keys involved in the heteropolysaccharides including glucuronoarabinoxylan, galactomannan, xyloglucan and so on.[7,8] Hemicellulases must be capable of hydrolyzing branched chains containing different sugars and functional groups; these enzymes include xylanase, xylosidase, arabinosidase, mannosidase, acetylesterase, glucuronidase and galactosidase.

Although there are many enzymes involved in the complete degradation of lignocellulose, it is not necessary to apply all these enzymes to the degradation of a specific type of lignocellulose. Different lignocellulosic biomasses contain not only different ratios of cellulose, hemicelluloses and lignin, but also different ratios of xylan- and mannan-type hemicelluloses (Table 16.1).[7,9] For example, mannan constitutes a major type of hemicellulases in softwood, while very little mannan exists in the lignocellulosic biomass of agricultural crops belonging to grasses (*Poaceae*). Thus mannan-degradation enzymes, including mannanase, mannosidase and galactosidase, can be ignored when the feedstock is a grass type of agricultural crop. Furthermore, the compositions and structures of xylan-type hemicelluloses vary among the major groups of lignocellulosic biomass (Table 16.2).[10,11] Theoretically, the debranching enzyme arabinosidase, in addition to glucuronidase, is required for grass xylan, whereas

Table 16.1 Chemical constituents of different types of lignocellulosic biomass.[a]

Material	Cellulose/%	Major hemicelluloses		Lignin/%	Ash/%[c]
		Xylan/%	Mannan/%		
Hardwood[b]					
Beech	45	20.8	0.9	22	0.5
Poplar	48	17.4	4.7	21	ND
Softwood[b]					
Spruce	41	8.6	16.3	27	0.5
Pine	41	8.8	11.7	28	—
Grasses (*Poaceae*)[b]					
Wheat straw	38	21.2	0.3	21	9.4
Rice straw	34	24.5	—	13	14.4
Bagasse[a]	40	21.1	0.3	25	ND

[a]All values are given as a percentage of the dry weight of the material; bagasse is the residue after the juice is extracted from sugar cane.
[b]"Hardwood", "softwood" and "grasses" are popular terms. Corresponding to plant taxonomy, softwoods are gymnosperm plants; grasses and hardwoods are angiosperm plants, belonging to monocots and dicots, respectively.
[c]These data were adapted from refs. 7 and 13.

Table 16.2 The major types of xylans in lignocellulosic biomass.

Material	Xylan type	Amount/%[a]	Composition/ molar ratio[b]	DP_n[b]	Reference
Hardwoods	Acetyl glucuronoxylan	10–3	Xyl : Ac : GlcUA = 10 : 7 : 1	200	10
Softwoods	Arabino-glucuronoxylan	7–15	Xyl : Ara : GlcUA = 10 : 1.3 : 2	100	10
Grasses	Arabino-glucuronoxylan	20–35	Xyl : Ara : GlcUA = 8 : 1 : 1	70	11

[a] Values are given as a percentage of the dry weight of the materials.
[b]Abbreviations: Xyl = xylose; Ac = acetate; GlcUA = glucuronic acid; Ara = arabinose; and DP_n = degree of polymerization.

acetylesterase, instead of arabinosidase, is required for hardwood xylan degradation, *e.g.*, Lucena *et al.* found that wood-feeding termites, in contrast to litter-feeding termites, might not be the best source of enzymes that degrade bagasse.[12] Therefore, the design of an optimum enzyme 'cocktail' for each type of lignocellulose greatly simplifies enzyme systems.

16.2 Cloning and Identification of Lignocellulase Genes from Animal Guts

It is an indisputable fact that cellulases exist widely in microorganisms and plants. Many fungi and bacteria have complex cellulase systems which can hydrolyze lignocellulose to produce glucose and xylose. In plants, cellulase plays the role of hydrolyzing cell walls in different stages of plant development, for example, in fruit ripening and in stalk abscission. In 1924, Cleveland proved that some arthropods and herbivores can use plants as a food source because their bodies contain a large number of bacteria that participate in the hydrolysis of cellulose.[13] In 1963, Stradine and Whitaker detected cellulase and chitinase activity in the sterile hepatopancreas of snail (*Helix pomatia*), and demonstrated that cellulase and chitinase activities were not correlative with bacterial content in snail digestive fluid.[14] In 1998, Smant obtained four cDNA from endo-β-1,4-glucan enzymes (EGs) from two different genera of parasitic nematodes on plants using molecular biological methods. In that same year, Watanabe screened the cDNA library of the termites *Reticulitermes speratus* using its endo-β-1,4-glucanase anti-serum, and acquired its endo-β-1,4-glucanase cDNA using rapid amplification of cDNA ends (RACE), further proving that endogenous cellulases exist in animals.[15–16]

Some complex lignocellulase systems of fungi and bacteria, *e.g.*, *Trichoderma reesei* and *Clostridium thermocellum*, have been extensively investigated, but no single enzyme system can completely degrade lignocellulosic biomass at the reasonable rate required for industrial processing. Insects can efficiently digest lignocellulose as their main food source probably because of the synergistic functions of the lignocellulase systems of symbionts in their guts. The ability of insects to feed on wood, straw and foliage has recently prompted extensive investigations into the mechanisms for the efficient digestion of lignocellulose, *e.g.*, wood-feeding termites have been found to accomplish lignocellulose digestion using specialized gut physiology, endogenous enzymes, and enzymes produced by gut symbionts. The symbiotic systems in insects can not be simply amplified because most of the symbiotic microorganisms are not able to grow under artificial culture conditions. However, the guts of these lignocellulose-feeding insects can be important resources of novel genes, from which lignocellulases of high efficiency can be over-produced by recombinant DNA techniques.[9]

Extensive effort has been put into identifying, cloning and expressing lignocellulases from insects and their microbiota, and all kinds of glycoside hydrolases and carbohydrate-binding modules have been found in symbiotic systems in insect guts.[12,17,18] Strategies for the discovery and characterization of cellulolytic enzymes from insects have been reviewed by Willis *et al.*[19] Purification of enzymes from insect guts was the major challenge in early studies because of insufficient starting material, yet many endogenous and symbiont-produced cellulases were purified and some of them were subsequently cloned from insect guts.[20–25] Microbial communities in lignocellulose-consuming insects represent a rich resource of novel lignocellulose-degrading enzymes; these enzymes and corresponding genes can be obtained *via* enrichment, isolation and identification of gut microbes. Doran-Peterson's research group has cultured bacteria from various areas of the *Tipula abdominalis* alimentary tract with various media under both aerobic and anaerobic conditions. They successfully isolated and identified the cellularlytic bacteria strain 27C64 and a novel strain T202[T] from the *T. abdominalis* larval gut.[26,27]

The construction and screening of DNA and cDNA libraries has proven to be an efficient strategy to find cellulases including endo-glucanases, cellobiohydrolases and glucosidases from insect guts.[19] In recent years, molecular approaches such as meta-genomics, functional meta-genomics, meta-transcriptomics, and meta-protoeomics have produced large amounts of data to enable the quick investigation of gene findings.[18,28,29]

In order to discover novel lignocellulases or auxiliary proteins of high efficiency, the genes in databases need to be cloned and heterologously expressed, so that the properties such as specific activity, stability and substrate specificity of the enzymes can be determined by using purified recombinant enzymes. Many genes from either insects or their microbiota have been successfully expressed in *Escherichia coli* or eukaryotic systems such as *Spodoptera* SF9 cell cultures.[19] However, the capacity to evaluate the biochemical activities encoded by the genes in databases is still the major limiting factor, and efficient gene expression is required for finding novel enzymes with potential applications from insect guts.

16.3 Gene-Expression Systems for the Production of Recombinant Lignocellulases

16.3.1 Criteria for the Selection of an Expression Platform

In molecular biotechnology and bioengineering, the high-yield production of recombinant proteins is achieved by cloning the target gene into an expression vector, and introducing the recombinant vector into a corresponding host such as bacterium, yeast, plant or animal cells, where the target gene is expressed. Gellissen *et al.* have demonstrated the use of gene-expression platforms ranging from Gram-negative and Gram-positive prokaryotes, to several yeasts and filamentous fungi to mammalian and plant cells. All of these expression platforms have special favorable characteristics, but also limitations and drawbacks.[30] Therefore, to select an expression platform to give an efficient and economical expression of a gene product, considerations must include the intended use, timeframe, the availability of resources, and the characteristics of the recombinant enzymes.

There are special criteria for the selection of an expression platform for the production of technical enzymes. Many industrial processes such as lignocellulose degradation, pulp bleaching or waste water treatment produce low-value products with little or no commercial value. Practical use of biocatalysts is currently not affordable for most of these industries but demand for a robust, economically viable process is increasing as companies seek to improve their 'green footprint'. Production of these industrial enzymes should be very fast with extremely low total cost, which is a challenge to current gene-expression techniques.

The diversity of gene-expression systems, gene resources, enzyme types, and enzyme properties must be considered for the heterologous gene expression of lignocellulases. When preparing samples for characterizing a gene product, one can try any type of expression platform as long as enough enzyme can be expressed and purified for the determination of its properties, *e.g.*, insect β-1,4-endo-glucanase was expressed in insect cells of *Spodoptera* SF9 for high fidelity.[19]

Besides the activity and stability of industrial enzymes such as lignocellulases, the potential of an enzyme application largely depend on the rate, the amount, and the cost of enzyme production. Microbial expression platforms are suitable for this purpose. There have been few side-by-side comparisons of bacterial and fungal hosts; sometimes several different hosts need to be tested for increasing expression levels of a single gene. Hundreds (if not thousands) of genes have been successfully over-expressed in *E. coli*, while only a few proteins can be heterologously expressed in filamentous fungi at the high levels and rates found in *E. coli*. For ultimate applications, in contrast to medicinal protein production which can tolerate high production costs and the need for proper post-translational modification, lignocellulases need to be produced in large quantities, with high activity, and at low cost.

16.3.2 Gene-Expression Vector Systems for *E. coli*

The bacterial host *E. coli* is often the first choice for the expression of recombinant enzymes, because it is easy, fast and inexpensive to cultivate, and many vector systems have been developed for this host. To reach a high level of expression in *E. coli*, the foreign gene is usually under the control of a regulatory promoter, which plays important roles in reducing the adverse effects of the recombinant protein on host cells, decreasing the degradation of target proteins by cellular protease from the host cells, and increasing the production of active recombinant protein. Using operator genes and their repressors from different sources, many *E. coli* expression vector systems have been developed in the last 20 years. The best known vectors are those containing the *lac* promoter and its hybrids, the bacteriophage p_L promoter and the T7 promoter, which are respectively identified as the "*lac/tac/trc* system", the "p_L system" and the "T7 system".[32] Recently, a heat–shock promoter regulated by alternative sigma factor, Sigma 32 of *E. coli* has been successfully employed to control foreign gene expression in a series of expression vectors designated as the "Hsh system".[33] Knowledge of the regulation mechanism and properties for these expression systems is helpful to consider when choosing a system to express an enzyme.

In the *lac/tac/trc* expression system, vectors carry the *lac* promoter, or its hybrid *tac* or *trc* promoter. Under the control of one of these promoters, the transcription initiation of the target gene is repressed by the repressor of the *lac* operon in the absence of lactose or one of its analogs such as isopropyl-β-D-thiogalactopyranoside (IPTG). In both experimental and commercial production settings, the expression of target genes is induced by adding IPTG or lactose as the inducing agent into the culture of *E. coli* harboring vectors of the *lac/tac/trc* system, wherein the inducing agent releases the repressor and allows transcription to initiate. However, the high cost and toxicity of IPTG limits its wide use for the production of proteins for medical and many industrial applications. Lactose is cheaper than IPTG as an inducing agent, but is not as effective as the latter.

In the p_L system bacteriophage λ, the early transcription promoters p_L and p_R are regulated by a repressor encoded by the *cI* gene. The p_L has been used in expression vectors to control the expression of target genes *via* the gene product of *cIts857*, which is a temperature-sensitive mutant of *cI*. In cells harboring these vectors, the repressor binds to p_L and represses the transcription of the target gene at low temperatures but not at elevated temperatures, and thus gene expression is induced by raising the temperature of a culture.[34] However, application of the p_L system to large-scale cultures is limited because a method is lacking for rapidly raising the temperature from about 30 to 40 °C in industrial settings to promote effective induction.[35]

In the T7 system, the bacteriophage T7 promoter is used in vectors to control the expression of the target gene, and the transcription is specifically performed by T7 RNA polymerase. The gene encoding T7 RNA polymerase has been integrated into the chromosome of host cells under the control of the *lac* or p_L promoter, and its expression is induced by IPTG or temperature shift. The bacteriophage T7 promoter is the strongest among all the promoters used in *E. coli* expression systems, but growth inhibition and/or inclusion body formation are sometimes associated with high expression levels.[36] Meanwhile, the T7 system faces the same problems as other systems with inducing agents or raising temperatures.

16.3.3   Properties of Vectors of the Hsh System for *E. coli*

The typical vector for the Hsh system is pHsh which contains a synthesized promoter recognized and regulated by the sigma factor (σ^{32}, encoded by the *rpoH* gene) for heat–shock reactions of *E. coli* (Figure 16.1). When *E. coli* is subjected to a quick rise of temperature, σ^{32} recognizes the heat–shock promoters of a group of heat–shock protein-encoding genes, resulting in the expression of heat–shock proteins. The DNA sequences of heat–shock promoters are known, and their consensus sequences are different from those of the general promoters recognized by σ^{70} [refs. 37 and 38]. The differences between the two consensus sequences are shown below.

General promoters: - - - - - - - - - - TTGACA-*16* ~ ~ ~ *18 N* - TATAAT

Heat–shock promoters: - - -C - C - CTTGAA-*13* ~ *15 N*-CCCCAT-T

Although the physiological functions and regulatory mechanisms of the heat–shock system in *E. coli* are well understood, heat–shock promoters were not used effectively as promoters to regulate the expression of foreign genes in plasmid vectors until recently. This may be due to the fact that the heat–shock reaction lasts only 20 minutes after *E. coli* is subjected to an increase in

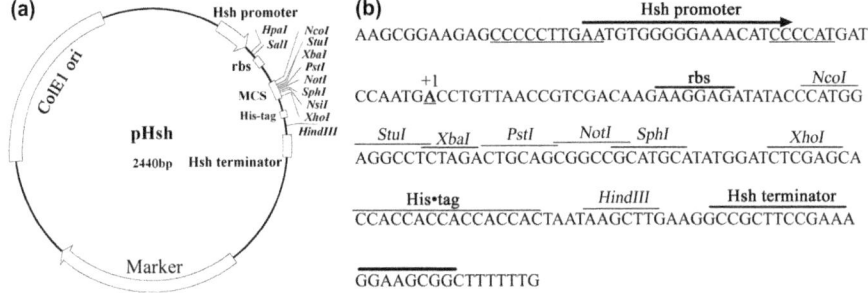

Figure 16.1 A map of the expression vector pHsh. (a) The main elements of the plasmid, including ColE1 replication from pUC19, the ribosome-binding site (RBS), the multiple-cloning site (MCS), a His-tag sequence, and the artificial Hsh promoter and Hsh terminator which initiate and terminate the transcription controlled by σ^{32}; "Marker" can be a chloramphenicol, an ampicillin or a kanamycin resistance gene. (b) The DNA sequence of the Hsh promoter, MCS, RBS, and Hsh terminator.

temperature. Commercial application of this system may have also been discouraged by the apparent difficulty in quickly raising the temperature of a large volume of culture medium, as in the case of the using p_L system. However, it has been shown that gene expression under the control of the Hsh promoter in the vector pHsh can be continued for up to 10 h after a heat–shock induction.[39–41] Meanwhile, a procedure named "flow-in-heat" has been designed and successfully used to perform temperature up-shift in fermenters so that heat–shock induction of gene expression is not more difficult in large-scale applications.[33] By using pHsh vectors, many recombinant lignocellulases have been produced in fermenters at high levels and at low cost, *e.g.*, the laccase of *Thermus thermophilus*, and the xylanase and β-glucosidase of *Thermotoga maritime* are produced in 25 L fermenters at 3.7×10^4, 2×10^6 and 7.3×10^4 U L^{-1} of culture, respectively.

Gene expression in the plasmid vectors of the Hsh system employs the regulation mechanism of the heat–shock system of *E. coli*, while other expression systems are directly or indirectly regulated by repressors such as gene products of *lac*I and *cI(ts)*857. Although its transcription may not be as strong as bacteriophage promoters, the Hsh promoter allows its plasmids to employ a replicon having a very high copy number. The Hsh expression system, along with the induction methods of flow-in-heat have the following advantages: (1) the expression vectors of the Hsh system achieve high recombinant protein yield (U L^{-1} or mg L^{-1}); (2) gene expression is induced by a temperature shift instead of a costly chemical inducing agent such as IPTG; (3) the small molecular size of Hsh vectors allows the modification of the sequences of the target gene *in situ*; and (4) there is no special host strain requirement due to the existence of heat–shock systems in all strains of *E. coli*.

In practice, the existence and expression of a foreign gene can stress living cells; expression plasmids are continuously subjected to mutation or modification in cells and the subsequent result is that the recombinant cells harboring

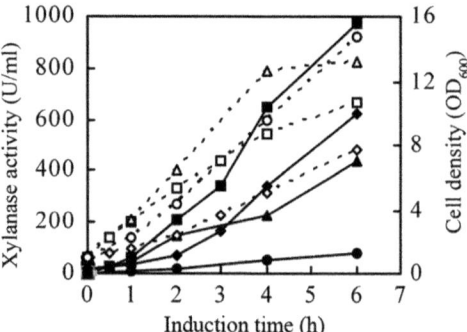

Figure 16.2   Gene expression in Hsh vectors in *E. coli* grown in Terrific Broth.
The cell density (empty) and xylanase activity (filled) in the culture of
pHsh-xynIII transformed *E. coli* grown at a constant temperature of
30 °C (-○-, -●-), 37 °C (-△-, -▲-), or 42 °C (-◇-, -◆-), or during the
temperature shift from 30 to 42 °C (-□-, -■-).

modified plasmids suffer less stress and grow faster with less activity or less
expression of a target product. Therefore, to obtain the maximal expression of
a recombinant enzyme, the expression plasmids of the correct sequence must be
purified from a fresh culture and dissolved in TE (Tris-HCl and EDTA) buffer
for long-term usage and storage. Expression plasmids should not be stored in
transformed cells because old transformants may contain modified expression
plasmids. Furthermore, there is a special requirement for the application
of expression vectors regulated by temperature up-shift, in which gene
expression is under the control of σ^{32} or temperature-sensitive repressors such
as $cI(ts)857$: recombinant cells should not be cultivated at temperatures above
30 °C during the procedures before gene expression is induced, which includes
cultivation during molecular cloning, sequence improving, gene transformation, inoculum enlargement and so on. Although a sharp up-shift instead of a
slow rise of temperature can increase gene-expression levels by about 50%, the
basal expression of a target gene is increasing as the growth temperature is
raised (Figure 16.2).[33] Therefore, 30°C and lower temperatures are favorable
for reducing basal expression, maintaining healthy growth of recombinant cells
and avoiding mutations of expression plasmids, this is especially true when the
target protein is toxic to the host cell.

16.4   Improving the Expression of Recombinant Enzymes

16.4.1   Increasing Gene-Expression Levels by Site-Directed
Mutagenesis

Expression platforms and even different expression systems within the same
platform have sequence specificity, and few natural gene sequences can be expressed to maximal levels without modification to fit a particular expression
system. In *E. coli* expression systems, strong and regulated transcription has to
be provided by the promoter in each vector, *e.g.*, the T7 expression system uses

the promoter from bacteriophage T7 to control the transcription of target genes, which is probably the strongest promoter known. The Hsh system uses heat–shock promoters from *E. coli* to control the transcription of target genes, which allows host cells to grow to a higher cell density in the presence of recombinant plasmids of very high copy numbers. While the promoters in expression vectors offer the basal advantages of strong and regulated transcription, the gene-expression levels still differ greatly from gene to gene, indicating that post-transcriptional regulation, in addition to the toxicity of the gene product to the host, is very important for improving gene-expression levels. Post-transcriptional regulation mechanisms have been studied in *E. coli* for 20 years; some known elements that limit expression levels include the Shine–Dalgarno (SD) sequence, the space between the SD region and the start codon (ATG), the mRNA secondary structure, and codon usage.[42,43] Occasionally, some genes are difficult to over-express in *E. coli* for unknown reasons.

To express an industrial enzyme for large-scale production, the overall strategy should follow these procedures: (1) investigate the properties of a target enzyme to make sure it is the best for your application purpose; (2) clone the gene into an expression vector, and analyze the possible limitations in gene over-expression, which include the SD/ATG space, the mRNA secondary structure, the GC contents, and codon usage of the target gene; and (3) design protocols for site-directed mutagenesis, the goal in this stage is to improve gene-expression levels and therefore no mutation should cause changes to amino acid residues in the gene product. After modifying the sequences to fit an efficient vector and a new host, many genes can be expressed to high levels, with a few exceptions where the recombinant protein is unstable in host cells or toxic to cell growth. Many lignocellulase genes have been modified by site-directed mutagenesis to achieve over-production in *E. coli*. These include xylanase A, xylanase B, xylosidase, arabinosidase, glucuronidase, and cellulase (endo-glucanase).[33,39–41,44,45] The *xynA* gene of *Thermomyces lanuginosus* can be used as an example to describe the methods for improving gene-expression levels and further improving the solubility of the recombinant protein without changing the amino acid residues of the enzyme.

The *xynA* gene of the thermophilic fungus *T. lanuginosus* encodes a relatively stable xylanase (XynA), with catalytic activity over a broad range of pH values.[46] The xylanase is free of cellulase activity, and the hydrolysis products consist of xylooligosaccharides together with a small amount of xylose. The mature xylanase of *T. lanuginosus* has 196 amino acids with 31 amino acids (16%) encoded by rarely used codons (fraction <12%) in *E. coli* according to the codon usage table (http://www.kazusa.or.jp/codon/).[40] Four rarely used codons (coding for T2, P4, S6 and H10) lie in the first 10 amino acids. Furthermore, R58 is encoded by the rarest codon AGA, which is demonstrated to have a detrimental effect on protein expression. L105 and R117 are encoded by CUA and CGA, respectively, which are the next rarest codon pairs in *E. coli*. To eliminate the negative effect of rare codons on protein expression, a mutant gene was designed using preferred codons to replace the rare codons. The codon-replaced gene "xynA1" was synthesized through primer splicing and nested-PCR, and cloned into an expression vector, a 2442-bp plasmid pHsh, to construct the expression plasmid pHsh-xynA1. The gene-expression level at this stage was only 1.3% of the total cell protein, and therefore, further improvement of the gene sequence was desired.[40]

The mRNA secondary structure of pHsh-xynA1 in the translational initiation region was predicted using the MFOLD algorithm available online (http://www.bioinfo.rpi.edu/applications/mfold/rna/form1-2.3.cgi); the message was arbitrarily defined from the transcription start site ACC to 70 nt down-stream; the MFOLD analysis was conducted under default conditions, except that the folding temperature was changed from 37 to 42 °C. A hairpin structure was found in the SD region, and the SD sequence and AUG were located at a spacing of 15 nt, and were subjected to modification to generate an optimized plasmid pHsh-xynA2. After site-directed mutagenesis by using inverse-PCR, the original hairpin structure was completely destroyed, and the SD sequence was released to a large loop enclosing the start codon; the free energy of the 70 nt was changed from –6.56 to –4.96 cal mol^{-1} (Figure 16.3), meanwhile, the spacing between AUG and the SD sequence was reduced to 8 nt in pHsh-xynA2. The expression level was increased from 1.3 to 13% of the total cell protein after this stage of gene modifications.[40] Gene over-expression was achieved after the xylanase gene in expression vector pHsh was optimized in SD/ATG space, codon usage, and potential mRNA secondary structure, however, the majority of the recombinant xylanase was insoluble and in the inactive inclusion body form.[40]

16.4.2 Soluble Expression of Aggregation-Prone Protein

Recombinant protein aggregation, *e.g.*, inclusion body formation, upon gene over-expression in *E. coli* is often a problem for *E. coli* as a gene-expression platform for the industrial production of enzymes. Inclusion bodies are densely packed particles of aggregated protein found in the cytoplasmic or periplasmic spaces of *E. coli* during high-level expression of heterologous protein.[47] Practically, most people would not examine whether there are dense electron-refractive particles, and simply believe that all recombinant proteins in precipitants are in inclusion bodies, which can include unstable proteins denatured during cell growth or after cell disruption. However, when considering how inclusion bodies are formed and how to reduce or eliminate them, we have to indicate that low stability is a dominant property of many enzymes, and thus the precipitation of denatured proteins should not be counted in inclusion body formation. Formation of inclusion bodies is determined by several factors: insolubility of the product at the concentrations being produced, inability to fold correctly in the bacterial environment, and/or lack of appropriate bacterial chaperone proteins. Accordingly, some strategies are commonly employed to reduce inclusion bodies: (1) controlling the rate of protein synthesis; (2) enabling secretion into the periplasm; (3) fusing the target protein with a soluble polypeptide; and (4) co-expression of chaperone genes.

Fungal xylanase A of *T. lanuginosus* is relatively stable with a small molecular mass; it forms typical inclusion bodies in both the cytoplasmic and

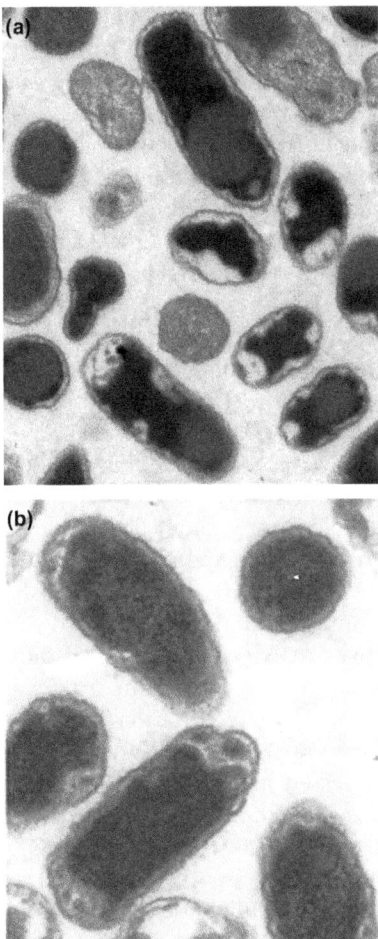

Figure 16.3 Inclusion bodies of *T. lanuginosus* XynA formed in cytoplasm and periplasm of *E. coli*. Electronic micrographs of inclusion bodies in *E. coli* cells transformed by plasmid pHsh-xynA2 for intracellular expression (a), or pET-xynA2 for secretive expression of XynA (b).

periplasmic spaces of *E. coli* during high level expression under normal conditions (Figure 16.3). Soluble recombinant xylanase A was not increased by controlling the rate of protein synthesis *via* growing the recombinant cells at temperatures as low as 20 °C. A second strategy fused the *xynA* gene into a gene encoding a soluble xylosidase/arabinosidase, XarB. Unfortunately, XarB failed to rescue XynA from inclusion body formation, but was captured into the precipitant by XynA when an intra-cellular protein was expressed from the fused gene *xarB-xynA* in the pHsh vector.[44]

Secretion of recombinant proteins into the periplasm of *E. coli* has been reported to benefit the production of certain recombinant proteins resulting in higher protein solubility, correct folding, and facilitated down-stream processing, although inclusion bodies may still form in the periplasm. Furthermore, increasing the concentration of chaperones in a heat–shock system regulated by σ^{32} in *E. coli* has been shown to assist in the correct folding of the target protein, and in some instances, co-expression of DnaK-DnaJ can greatly increase the soluble proportion of recombinant proteins in the cytoplasm.

Two methods were employed for the secretion of recombinant XynA of *T. lanuginosus*: (1) cloning the gene *xynA2* into the expression vector pET-20b(+) with the fusion of the signal peptide sequence to obtain the expression plasmid pET-xynA2; and (2) adding a signal peptide encoding sequence at the 5′-end of *xynA* to generate the expression plasmid pHsh-ex-xynA2 from pHsh-xynA2, so that pHsh-ex, a vector for secretive expression, was constructed for the Hsh system (GenBank accession no. FJ715939). Interestingly, only a small proportion of XynA was soluble in the cell-free extracts, and most of the enzyme was in inclusion bodies when the XynA was expressed from pET-xynA2. However, XynA was produced by cells harboring pHsh-ex-xynA2 in a soluble form without detectable inclusion bodies.[44]

The secretion of XynA does not reduce the inclusion body formation in a pET vector, but gives a soluble expression in a pHsh vector where xylanase activity is increased from 47 (in pHsh-xynA2) to 221 U mL^{-1} (in pHsh-ex-xynA2).[40,44] This improved activity is the result of the increased expression of chaperons and the secretion of recombinant proteins into the periplasm of *E. coli*. Skelly *et al.* reported the correlation between σ^{32} levels and *in vitro* expression of *E. coli* heat–shock genes.[48] When the heat–shock transcription factor σ^{32} was over-expressed in *E. coli*, the enzyme activity of preS2-S9-β-galactosidase was increased.[49] An increased concentration of σ^{32} was observed in the cells carrying an expression plasmid with >200 copies of the Hsh promoter either at constant temperature of 30 °C or after a temperature up-shift.[44] The findings not only explain how high-level expression is achieved in the Hsh expression system, but also indicate that σ^{32} of *E. coli* can play an important role in the soluble expression of secreted recombinant proteins. While the mechanism for the pHsh-ex mediated soluble expression of an aggregation-prone protein is not completely understood, a novel strategy is offered to prevent or reduce inclusion body formation during the heterologous expression of genes from eukaryotes in prokaryotic systems, provided the expression level is high enough.

No growth inhibition of recombinant cells is found when a xylanase is over-expressed as intra-cellular protein in *E. coli*.[39,41] However, the recombinant cells stop growing within about 2–3 h after a periplasmic xylanase is induced to over-express in the expression vector pHsh-ex. This phenomenon can result from overload of the cell membrane, which is supported by the fact that the mutation from xynA2 to xynA-m1 caused an increase of gene expression along with a significant decrease in the final cell density. Therefore, alternative induction protocols must be designed to increase the number of starting cells. The culture and induction techniques are applicable to the performance of pHsh-ex-mediated expression of other proteins to be exported into the periplasm.[44]

16.5   A New Technique for Improving the Properties of Recombinant Enzymes

The total cost of an industrial enzyme depends on its properties such as specific activity and stability, in addition to the efficiencies in industrial-scale enzyme production.[19,50] For example, many enzymes have a half-life of about 30 min under reaction conditions, therefore, if the enzyme's half-life is 30 min longer, only half the amount of enzyme will be consumed to complete the same task. However, natural enzymes have different sequences, often for identical functions, and there is little knowledge available regarding the relationship between an amino acid sequence and an enzyme property, and in most cases, it is not possible to predict which mutation of sequences will improve an enzyme property. Therefore, directed-evolution methods are increasingly being used on industrial enzymes to improve their properties including substrate specificity, specific activity, thermostability *etc.*[51–54]

The success of directed-evolution experiments hinges on the efficiency of the methods used to create random mutagenesis libraries and screen libraries for mutants with properties of interest.[55] As library diversity represents a crucial parameter in the directed evolution of a target gene for improved functionality, various protocols have been established for creating random mutagenesis libraries.[56] Random mutagenesis, along with genetic selection or high-throughput screening, constitutes an important approach to identifying critical regions of proteins, studying structure–function relations and developing novel proteins with desirable properties.[55]

Many methods for the directed evolution of enzymes were reviewed by Lutz and Patrick.[57] Some of the newer *in vitro* DNA mutagenesis approaches were detailed in both PCR and non-PCR categories.[58] One of the most commonly used random mutagenesis methods is error-prone PCR, which introduces random mutations during PCR by reducing the fidelity of the DNA polymerase.[59] The natural error rate of the polymerase can be altered and enhanced by modifying standard PCR methods.[60] This technique has the advantage of developing new enzymatic properties without a structural understanding of the targeted enzyme, and often yields unique mutations that could not be predicted.[60]

The early techniques of error-prone PCR generally involved the following steps: (1) amplifying the target gene as the PCR template under error-prone conditions, to generate amplified target sequences that contain random mutations; (2) treating the terminal ends of the amplified target sequence using restriction endo-nucleases; (3) ligating the treated target sequence into a suitable expression vector using a DNA ligase; and (4) transforming the expression vector containing the target sequences into a suitable host cell, to obtain a population of cells, or a mutagenesis library, which contains the various target sequences. This process is similar to a cloning or subcloning process of the target sequence, except that in conventional cloning or subcloning, only a small number of transformants need to be obtained, while in the construction of a mutagenesis library, generally tens of thousands of transformants are needed in order to realistically obtain a suitable target mutant.

In order to improve efficiency, a mega-primer-based PCR method for whole plasmids, referred to as "MEGAWHOP" was established and its use was followed by a few improvements in generating random mutagenesis libraries.[61–66] Mega-primer-based PCR uses randomly mutagenesized target sequences as PCR primers and an expression vector as the template to yield double-stranded plasmids by using high-fidelity DNA polymerase. Amplification products are treated with Dpn I, which degrades the template expression vector, but leaves intact the amplified product. This process effectively modifies steps (2) and (3) of the above process, and yields libraries that are virtually free of plasmids containing no or multiple inserts. Nevertheless, the method of mutagenizing the mega-primers also involves a restriction enzyme digestion of the PCR product and of the expression vector, making the process complicated and low in efficiency, and the amplification products using the mega-primers are linear sequences, which are not amenable to forming a circular plasmid having two nicks, resulting in a relatively low transformation efficiency. Furthermore, for the process to be satisfactory, the size of the mega-primers should be in the range of 500 and 1000 bases, thereby limiting the size of the target sequence. Therefore, there is a need for a faster, simpler and more universally applicable method for generating random mutagenesis libraries.

A newly published method for directed evolution is a one-step construction of a mutagenesis library by error-prone PCR of the target gene in an expression plasmid and is called "*in situ* error-prone PCR". The unique features of this method include error-prone PCR performed with a pair of primers amplified from the whole vector with a different selection marker, and a thermostable DNA ligase is involved in the PCR mixture so that a ligation occurs in every thermal cycle of error-prone PCR.[67] In this process random mutations are introduced into a target gene while keeping the expression plasmid intact. The experimental materials for *in situ* error-prone PCR are: (1) a pair of expression vectors of the same series which have the same sequences except those of different selection marker genes, *e.g.*, pHsh-amp and pHsh-kana (GenBank accession nos. FJ571619, FJ571621); (2) a pair of short primers for inverse PCR, which are complementary to the regions flanked on both sides of the multiple-cloning sites in the expression vectors; and (3) a thermostable DNA ligase from the hyperthermophilic bacterium *Thermotoga maritima*.[68]

The principal procedures in the *in situ* error-prone PCR method are illustrated in Figure 16.4. These include: (1) a target gene is cloned into the vector carrying the first selection marker resulting in expression plasmid 1; (2) most of the vector carrying the second selection marker is amplified by using a pair of short primers (filled, short arrows) and DNA polymerase of high fidelity to produce linear DNA fragments named "vector-primers"; and (3) PCR amplification is performed under error-prone conditions by using vector primers with plasmid 1 as the template in the presence of a thermostable DNA ligase. In the thermocycles, the single strands of denatured vector primer and template DNA anneal to a complementary strand. Taq DNA polymerase then synthesizes a new strand of the target gene with a random mutation, and the thermostable DNA ligase repairs the nick to produce new circular strands. The new circular

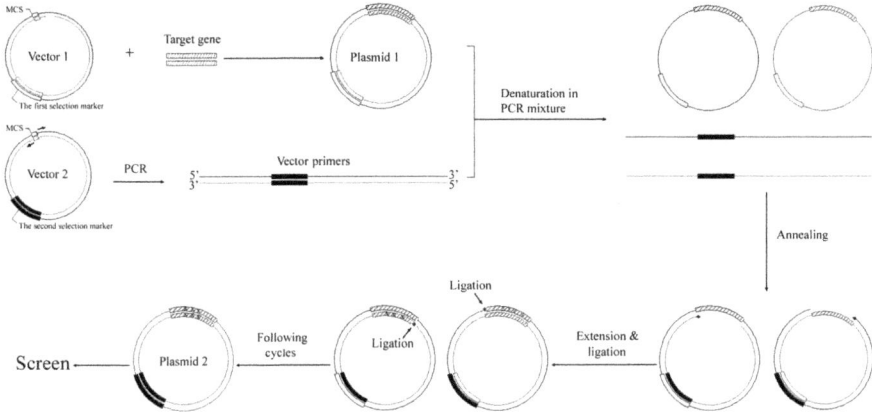

Figure 16.4 Outline of the *in situ* error-prone PCR method for creating random mutation libraries.

strands can be templates in subsequent PCR cycles, and a library of completely circular, nick-free plasmids that comprises both the second selection marker and a mutated target region is generated. These nick-free plasmids will be transformed into suitable host cells, which are selected based on the second selection marker. The mutated target sequences are then expressed, followed by the selection of the mutants with the desired mutations to achieve directed evolution of the target gene.

A mutant with desirable mutations obtained in accordance with the steps above can be subjected to a new round of *in situ* error-prone PCR. In this application, the expression plasmids, comprising a mutated target gene with one or more desired mutations is used as a template, and further mutation, selection and directed evolution are performed. A pair of vector primers comprising another selection marker is prepared and employed to support a new selection.

In conclusion, thermostable DNA ligase mediated *in situ* error-prone PCR conveniently and effectively creates random mutagenesis libraries, and has been successfully employed to improve xylanase and cellulase gene products.[67] The advantages of *in situ* error-prone PCR include: (1) random mutagenesis libraries can be created by one-step error-prone PCR followed by a direct transformation of competent *E. coli* cells without restriction enzyme digestion and ligation; (2) variable-length DNA fragments can be used as a mutation target, and there is no length limitation on the mutation target gene; (3) the transformants harboring error-prone PCR products can be directly screened using an alternative to antibiotics without the interference of starting plasmids; and (4) the accumulation of positive mutations can be obtained by multiple rounds of *in situ* error-prone PCR without any subcloning.

16.6 Summary

Complete degradation of lignocellulose requires the synergistic action of three groups of enzymes: cellulases, hemicellulases and ligninases. The current limitations of enzymatic degradation of lignocellulose lie in the low activities and high costs of these enzymes. Many fungi and bacteria possess full enzyme systems for the complete degradation of lignocellulose, however, the hydrolysis efficiency of these microbial systems is far too low to meet the requirements of industrial processes. On the other hand, some insects such as termites can efficiently digest lignocellulose as their main food source. To find novel enzymes for the efficient degradation of lignocellulose, many cellulolytic enzymes and their encoding genes have been identified from insect guts and gut microbiota.[19,29] Techniques for meta-omics including meta-genomics, functional meta-genomics, meta-transcriptomics, and meta-protoeomics produce large amounts of data to enable the quick progression of new findings.[29] With thousands of new genes, gene expression has become a bottleneck for identification, characterization, evaluation, and over-production of each enzyme. Both currently and for the foreseeable future, the major limiting factor is the capacity to evaluate the biochemical activities of the gene products,[31] and significant improvements in yield and cost reductions are expected to make large-scale fermentation of lignocellulosic substrates possible.[8] Therefore, molecular biotechnology is now a crucial technique for finding novel enzymes, improving enzyme properties, and increasing enzyme yields. Tools and techniques have been developed in recent years for the improvement of recombinant enzyme production and the enzyme properties desired for cost-efficient industrial processes. These new developments include the Hsh expression system for *E. coli*, heat–shock induction of gene expression in industrial fermenters, soluble expression of aggregation-prone protein, *in situ* error-prone PCR, to name a few. The extensive combination of enzymology, entomology, molecular biotechnology and bioengineering is enhancing the prospects for the enzymatic degradation of lignocellulosic materials.

Acknowledgements

This author wishes to thank Dr Juergen Wiegel for critical review of the manuscript. The author also thanks Dr Ruoxue Liu for preparing some data. The work related to development of the Hsh system was supported by the National Natural Science foundation of China (grant no. 30970062) and the Nanjing Program of Science and Technology (grant no. 201101094).

References

1. N. Reddy and Y. Q. Yang, *Trends Biotechnol.*, 2005, **23**, 22.
2. J. C. Del Rio, G. Marques, J. Rencoret, A. T. Martinez and A. Gutierrez, *J. Agric. Food Chem.*, 2007, **55**, 5461.
3. B. E. Dale, *US Pat.*, 4 600 590, 1986.
4. M. Kurakake, W. Kisaka, K. Ouchi and T. Komaki, *Appl. Biochem. Biotechnol.*, 2001, **90**, 251.
5. J. G. Gardner, L. A. Zeitler, W. J. S. Wigstrom, K. Engel and D. H. Keating, *Biotechnol. Lett.*, 2012, **34**, 81.

6. C. E. Wyman, *Trends Biotechnol.*, 2007, **25**, 153.
7. M. Pauly and K. Keegstra, *Plant. J.*, 2008, **54**, 559.
8. H. Jorgensen, J. B. Kristensen and C. Felby, *Biofuels, Bioprod. Biorefin.*, 2007, **1**, 119.
9. J. Sun, and X. J. Zhou, in *Recent Advances of Entomological Research: from Molecular Biology to Pest Management*, ed. T. X. Liu and L. Kang, Springer, Berlin, 2010, p. 441.
10. R. L. Whistler and C.-C. Chen, in *Wood Structure and Composition*, ed. M. Lewin and I. S. Goldstein, Marcel Dekker, New York, 1991, p. 287.
11. N. S. Thompson, in *Wood and Agricultural Residues*, ed. J. Soltes, Academic Press, New York, 1983, p. 101.
12. S. A. Lucena, L. S. Lima, L. S. A. Cordeiro, C. Sant'Anna, R. Constantino, P. Azambuja, W. D. Souza, E. S. Garcia and F. A. Genta, *Biotechnol. Biofuels*, 2001, **4**, 51.
13. L. R. Cleveland, *Biol. Bull.*, 1924, **46**, 117.
14. G. A. Stradine and D. R. Whitaker, *Can. J. Biochem. Physiol.*, 1963, **41**, 1621.
15. G. Smant, J. P. Stokkermans, Y. T. Yan, J. M. de Boer, T. J. Baum, X. H. Wang, R. S. Husseyi, F. J. Gommers, B. Henrissat, E. L. Davis, J. Helder, A. Schots and J. Bakker, *Proc. Natl. Acad. Sci. U. S. A.*, 1998, **95**, 4906.
16. H. Watanabe, H. Noda, G. Tokuda and N. Lo, *Nature*, 1998, **394**, 330.
17. S. J. Lee, K. S. Lee, S. R. Kim, Z. Z. Gui, Y. S. Kim, H. J. Yoon, I. Kim, P. D. Kang, H. D. Sohn and B. R. Jin, *Comp. Biochem. Physiol., Part B: Biochem. Mol. Biol.*, 2005, **140**, 551.
18. F. Warnecke, P. Luginbühl, N. Ivanova, M. Ghassemian, T. H. Richardson, J. T. Stege, M. Cayouette, A. C. McHardy, G. Djordjevic, N. Aboushadi, R. Sorek, S. G. Tringe, M. Podar, H. G. Martin, V. Kunin, D. Dalevi, J. Madejska, E. Kirton, D. Platt, E. Szeto, A. Salamov, K. Barry, N. Mikhailova, N. C. Kyrpides, E. G. Matson, E. A. Ottesen, X. Zhang, M. Hernández, C. Murillo, L. G. Acosta, I. Rigoutsos, G. Tamayo, B. D. Green, C. Chang, E. M. Rubin, E. J. Mathur, D. E. Robertson, P. Hugenholtz and J. R. Leadbetter, *Nature*, 2007, **450**, 560.
19. J. D. Willis, C. Oppert and J. L. Jurat-Fuentes, *Insect Sci.*, 2010, **17**, 184.
20. S. R. Marata, W. R. Terra and C. Ferreira, *Insect Biochem. Mol. Biol.*, 1995, **25**, 835.
21. S. R. Marata, E. H. P. Andrade, C. Ferreira and W. R. Terra, *Eur. J. Biochem.*, 2004, **271**, 4169.
22. F. A. Genta, W. R. Terra and C. Ferreira, *Insect Biochem. Mol. Biol.*, 2003, **33**, 1085.
23. L. Li, J. Frohlich, P. Pfeiffer and H. Konig, *Eukaryotic Cell*, 2003, **2**, 1091.
24. A. M. Scrivener and M. Slaytor, *Insect. Biochem. Mol. Biol.*, 1994, **24**, 223.
25. M. Sugimura, H. Watanabe, N. Lo and H. Saito, *Eur. J. Biochem.*, 2003, **270**, 3455.
26. D. M. Cook and J. Doran-Peterson, *Insect Sci.*, 2010, **17**, 303.
27. T. E. Rogers and J. Doran-Peterson, *Insect Sci.*, 2010, **17**, 291.

28. Y. Hongoh, L. Ekpornprasit, T. Inoue, S. Moriya, S. Trakulnaleamsai, M. Ohkuma, N. Noparatnaraporn and T. Kudo, *Mol. Ecol.*, 2006, **15**, 505.
29. W. B. Shi, R. Syrenne, J. Z. Sun and J. S. Yuan, *Insect Sci.*, 2010, **17**, 184.
30. G. Gellissen, A. W. M. Strasser and M. Suckow, in *Production of Recombinant Proteins*, ed. G. Gellisson, Wiley, Weinheim, 2005, p. 1.
31. G. Banerjee, J. S. Scott-Craig and J. D. Walton, *Bioenerg. Res.*, 2010, **3**, 82.
32. J. Sambrook and D. W. Russell, *Molecular Cloning: a Laboratory Manual*, Cold Spring Harbor Laboratory Press, New York, 3rd edn, 2001.
33. H. Wu, J. Pei, Y. Jiang, X. Song and W. Shao, *Biotechnol. Lett.*, 2010, **32**, 795.
34. C. M. Elvin, P. R. Thompson, M. E. Argall, P. Hendry, N. P. J. Stamford, P. E. Lilley and N. E. Dixon, *Gene*, 1990, **87**, 123.
35. A. N. Glazer and H. Nikaido, *Microbial Biotechnology*, Freeman and Company, New York, 1995.
36. D. Russell, in *Gene Expression Systems*, ed. J. M. Fernandez and J. P. Hoeffler, Academic Press, London, 1999, p. 9.
37. J. H. Miller, *A Short Course in Bacterial Genetics*, Cold Spring Harbor Laboratory Press, New York, 1992, p. 17.1.
38. P. C. T. Turner, A. G. Mclennan, A. D. Bates and M. R. H. White, *Instant Notes in Molecular Biology*, BIOS Scientific Publishers, Abingdon, 1997.
39. H. Wu, J. Pei, G. Wu and W. Shao, *Enzyme Microb. Technol.*, 2008, **42**, 230.
40. E. Yin, Y. Le, J. Pei, W. Shao and Q. Yang, *World J. Microbiol. Biotechnol.*, 2008, **24**, 275.
41. W. Qu and W. Shao, *Biotechnol. Lett.*, 20 11, **33**, 1407.
42. L. Gold, *Annu. Rev. Biochem.*, 1988, **57**, 199.
43. K. E. Rudd and T. D. Schneider, in *A Short Course in Bacterial Genetics,* Cold Spring Harbor Laboratory Press, New York, 1992, p. 17.19.
44. Y. Le, J. Peng, H. Wu, J. Sun and W. Shao, *PLoS One*, 2011, **6**, e18489.
45. W. Shao, Y. Xue, A. Wu, I. Kataeva, J. Pei, H. Wu and J. Wiegel, *Appl. Environ. Microbiol.*, 2011, **77**, 719.
46. A. Schlacher, K. Holzmann, M. Hayn, W. Steiner and H. Schwab, *J. Biotechnol.*, 1996, **49**, 211.
47. G. A. Bowden, A. M. Paredes and G. Georgiou, *Bio/Technology*, 1991, **9**, 725.
48. S. Skelly, T. Coleman, C. Fu, N. Brot and H. Weissbach, *Proc. Natl. Acad. Sci. U. S. A.*, 1987, **84**, 8365.
49. J. G. Thomas and F. Baneyx, *Mol. Microbiol.*, 1996, **21**, 1185.
50. L. R. Lynd, M. S. Laser, D. Bransby, B. E. Dale, B. Davison, R. Hamilton, M. Himmel, M. Keller, J. D. McMillan, J. Sheehan and C. E. Wyman, *Nat. Biotechnol.*, 2008, **26**, 169.
51. J. R. Cherry and A. L. Fidantsef, *Curr. Opin. Biotechnol.*, 2003, **14**, 438.
52. R. R. Chirumamilla, R. Muralidhar, R. Marchant and P. Nigam, *Mol. Cell Biochem.*, 2001, **224**, 159.
53. V. G. H. Eijsink, S. Gaseidnes, T. V. Borchert and B. van den, *Biomol. Eng.*, 2005, **22**, 21.

54. L. Yuan, I. Kurek, J. English and R. Keenan, *Mol. Biol. Rev.*, 2005, **69**, 373.
55. M. Olsen, B. Iverson and G. Georgiou, *Curr. Opin. Biotechnol.*, 2000, **11**, 331.
56. L. G. Otten and W. J. Quax, *Biomol. Eng.*, 2005, **22**, 1.
57. S. Lutz and W. M. Patrick, *Curr. Opin. Biotechnol.*, 2004, **15**, 291.
58. M. M. Ling and B. H. Robinson, *Anal. Biochem.*, 1997, **254**, 157.
59. R. C. Cadwell and G. F. Joyce, *PCR Methods Appl.*, 1992, **2**, 28.
60. D. W. Leung, E. Chen and D. V. Goeddel, *Technique*, 1989, **1**, 11.
61. J. Brons-Poulsen, N. E. Petersen, M. Horder and K. Kristiansen, *Mol. Cell. Probes*, 1998, **12**, 345.
62. M. Kammann, J. Laufs, J. Schell and B. Gronenborn, *Nucleic Acids Res.*, 1989, **17**, 5404.
63. K. Miyazaki and M. Takenouchi, *Biotechniques*, 2002, **33**, 1033.
64. G. Sarkar and S. S. Sommer, *Biotechniques*, 1990, **8**, 404.
65. R. Tyagi, R. Lai and R. G. Duggleby, *BMC Biotechnol.*, 2004, **4**, 2.
66. D. Wei, M. Li, X. Zhang and L. Xing, *Anal. Biochem.*, 2004, **331**, 401.
67. W. Shao, Y. Le and J. Pei, *US Pat.*, 1012-0004142-A1, 2012.
68. Y. Le, J. Peng, J. Pei, H. Li, Z. Duan and W. Shao, *Enzyme Microb. Technol.*, 2010, **46**, 113.

CHAPTER 17

Cellulose-Dissolving Systems and their Effects on Enzymatic Hydrolysis

KUN WANG,[a] FENG XU[a] AND RUN-CANG SUN*[a,b]

[a] Beijing Key Laboratory of Lignocellulosic Chemistry, Beijing Forestry University, Beijing 100083, P. R. China; [b] State Key Laboratory of Pulp and Paper Engineering, South China University of Technology, Guangzhou 510640, P. R. China
*Email: rcsun3@bjfu.edu.cn

17.1 Introduction

The transition from a fossil-fuel-based economy to a renewable carbohydrate economy will inevitably take place in the foreseeable future. It will be driven by the need for secure and sustainable energy supplies and the desire to diminish the greenhouse effect.[1–3] First-generation bioethanol is mainly produced from sugar- or starch-based agricultural crops. Although this type of bioethanol is produced at a competitive cost, the raw material supply will not be sufficient to meet the increasing demand for food, posing risks to national food security. One of the most promising options to meet this challenge is the production of bioethanol from various lignocellulosic materials, such as agricultural residues (*e.g.*, wheat straw, sugarcane bagasse, and corn stover) and forest residues (*e.g.*, sawdust and thinning rests), as well as dedicated crops (*e.g.*, salix and switchgrass) using second-generation technologies.[4] The production of lignocellulosic bioethanol

RSC Energy and Environment Series No. 10
Biological Conversion of Biomass for Fuels and Chemicals: Explorations from Natural Utilization Systems
Edited by Jianzhong Sun, Shi-You Ding and Joy Doran-Peterson
© The Royal Society of Chemistry 2014
Published by the Royal Society of Chemistry, www.rsc.org

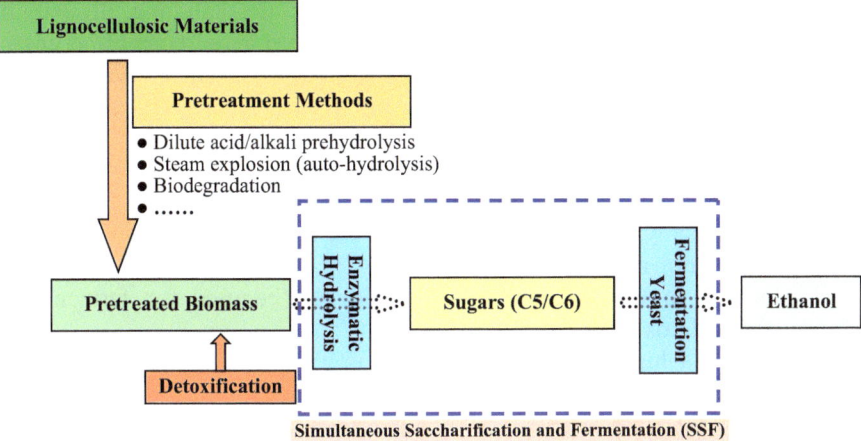

Figure 17.1 Process flow-sheet of bioethanol production from lignocellulosic materials. Reproduced from ref. 5 with permission. © Elsevier, 2010.

via biological conversion generally comprises the following main steps: pretreatment; detoxification; enzymatic hydrolysis; yeast fermentation; and recovery and concentration of ethanol. A simplified process flow-sheet for bioethanol production from lignocellulosic materials is shown in Figure 17.1.[5]

Compared with starchy biomass, lignocellulosic biomass is often described as "recalcitrant" due to its complex structural and chemical mechanisms for resisting assault on its structural sugars from the microbial and animal kingdoms.[6,7] The cost-effective release of soluble fermentable sugars from lignocellulose is among the most costly steps for emerging biorefineries. However, there are many reasons behind the poor hydrolysis of cellulose. In addition to the differences in intrinsic reactivity between β-1,4-glycosidic bonded cellulose and α-1,4-linked starch, the individual cellulose chains are joined by a network of inter- and intra-molecular hydrogen bonding and van der Waals forces, which cause cellulose to be packed into a much more highly ordered crystalline structure than starch. In addition, cellulose has fewer chain ends than starch due to its higher degree of polymerization (DP = 300–2000 *vs.* 17–26). Furthermore, cellulose produces insoluble cellodextrins during hydrolysis while starch produces soluble maltoolignosaccharides. To complicate the scenario further, cellulose is tangled with hemicelluloses and lignin in nature.[8] An enzymatic process is regarded as the most attractive way to degrade cellulose to glucose. The enzyme-catalyzed hydrolysis of cellulose is controlled by two crucial stages: the adsorption of enzymes on cellulosic particles, and the formation of enzyme–substrate complexes. The associated substrate-related factors include the presence of hemicelluloses and lignin, cellulose crystallinity and degree of polymerization, as well as the accessible external and internal surfaces of cellulose. In the bioconversion route, pretreatment is essential to open up the biomass structure and make the process more efficient.[9] It has a great impact on all the other steps in the process, *e.g.*, enzymatic hydrolysis, fermentation, down-stream processing and wastewater handling, digestibility of the cellulose, fermentation toxicity, stirring power requirement and energy demand in the down-stream processes and wastewater treatment demands. An effective pretreatment should have a number of features. It has to:[4]

- result in the high recovery of all carbohydrates;
- result in the high digestibility of cellulose in the subsequent enzymatic hydrolysis;
- produce no, or a very limited amount of, sugar or lignin degradation products so that the pretreated liquid can be used for fermentation without detoxification;
- result in a high solids concentration as well as high concentrations of liberated sugars in the liquid fraction;
- have a low energy demand or be performed in such a way that the energy can be reused in other process steps as secondary heat; and
- have low capital and operational costs.

A number of pretreatment methods have been investigated over the years, including dilute acid, steam explosion, hot water, lime, and organic solvents, all of which have drawbacks for large-scale applications.

Recently, a series of cellulose-soluble solvents have been employed as pretreatment reagents to disrupt the tight packing arrangement of cellulose fibrils in the crystalline domains, such as *N*-methylmorpholine-*N*-oxide (NMMO), ionic liquids (ILs), concentrated phosphoric acid, a LiOH/urea system, and a LiCl/*N*,*N*-dimethylacetamide (LiCl/DMAc) system, *etc.* As a low-energy and environmentally friendly process, dissolution of cellulose disrupts the hydrogen-bonding networks in cellulose by forming new H-bonds between the solvents and cellulose. The cellulose is regenerated by rapid precipitation of the dissolved cellulose doped with an anti-solvent such as water, methanol, ethanol, or acetone, *etc.* The original architecture of the cellulose fibers is predominantly lost due to the crystal structure transformation. Consequently, a rapid decomposition of cellulose can be achieved because of the lower packing density and inflated structure.[10] Compared with the conventional pretreatment methods, the unique advantage of its mild operation conditions (temperature <130 °C and ambient pressure) makes it a potentially important pretreatment method for enhancing the enzymatic hydrolysis of cellulose.

In this chapter, we provide a summary and research update for cellulose-dissolving systems including our own results obtained so far. The dissolving mechanisms of cellulose in different solvents are provided, and the discussion is mainly focused on the ramifications of cellulose solvents for subsequent enzymatic hydrolysis of the regenerated cellulose.

17.2 Pretreatment with Cellulose Solvents

For the past 150 years, intensive efforts have been devoted to developing new solvent systems for dissolving cellulose, which is the most abundant polymer in nature. The distinguished discovery of the structural features of macro-molecules was honored with the award of a Nobel Prize to Hermann Staudinger in 1953, presumably the first Nobel Prize for a polymer chemist.[11] There is still no consensus on the process necessary to disrupt the hydrogen-bond system of cellulose to initiate solvation of the cellulose chains. Based on the nature of the interactions between cellulose and solvents, cellulose solvents can be classified into four categories:[12]

- cellulose acts as a base, the solvent acts as an acid, *e.g.*, H$_2$SO$_4$ and tri-fluoroacetic acid;
- cellulose acts as an acid, the solvent acts as a base, *e.g.*, KOH and Triton;
- cellulose acts as a ligand, the solvent acts as a complexing agent, *e.g.*, Cuam and Cadoxen; and
- cellulose acts as a reactive compound, and is converted to a soluble transient derivative or intermediate, *e.g.*, xanthate, trifluoroactate.

There is a huge amount of published data in the field of cellulose solvents. In particular, the development of cellulose-dissolving pretreatment for improving the yield and rate of enzymatic hydrolysis has become popular in recent years. Our discussion will focus on the relevant achievements, and this approach will be illustrated by dividing it into five widely used cellulose solvents: amine oxides, Li-salt-containing polar/aprotic solvents, NaOH/urea aqueous solution, concentrated phosphoric acid, and ILs.

17.2.1 Amine Oxides

The first patent on a procedure to dissolve cellulose in different tertiary amine oxides was issued to Graenacher and Sallmann in 1939.[13] About 30 years later, Johnson developed a series of cyclic mono NMMO compounds and achieved better results.[14] NMMO belongs to a family of cyclic, aliphatic, tertiary amine oxides, and its most prominent feature is its highly polar N–O group with a dipole moment of 4.38 mD.[15] The nature of the N–O group is best described as a coordinate covalent bond. The bond is either presented as ionic, with a positive charge on the nitrogen and a negative charge on the oxygen, as do-native, with an arrow pointing at the oxygen, or as a simple single bond, with no indication of its special character (Figure 17.2).[16] It is commonly believed that the solvation power of NMMOs originates in their ability to disrupt the hydrogen-bond network of cellulose and to form solvent complexes by

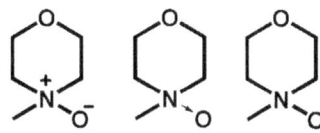

Figure 17.2 Three possible formulae for *N*-methylmorpholine-*N*-oxide (NMMO). Reproduced from ref. 16 with permission. © Elsevier, 2001.

establishing new hydrogen bonds between the macromolecule and the solvent (Figure 17.3).[17,18] However, details of the physicochemical process of cellulose dissolution are still far from being understood. In 1979, McCorsley and Varga developed a method to produce highly concentrated cellulose solutions of up to 23% by treating cellulose with NMMO and water, and subsequently removing water with an applied vacuum.[19] Hydration with one water molecule per NMMO molecule significantly decreased the melting point from 170 °C for pure NMMO to about 74 °C and improved the dissolution strength for cellulose. However, water and cellulose exhibit competitive behavior with regard to hydrogen-bond formation with NMMO, with water being evidently preferred. NMMO with a water content of up to 17% (w/w), corresponding to a "1.2 hydrate", is able to dissolve cellulose. At higher water contents, cellulose is precipitated.[16] The ternary phase diagram for the basic cellulose–NMMO–water system is shown in Figure 17.4.[20] Lyocell technology based on

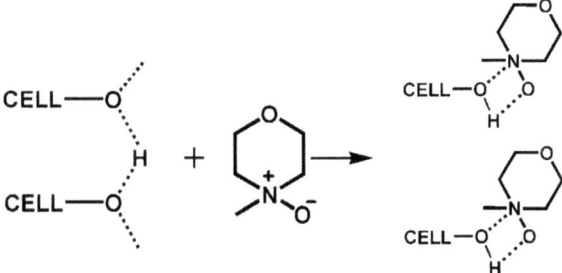

Figure 17.3 Proposed mechanism for the dissolution of cellulose in NMMO.

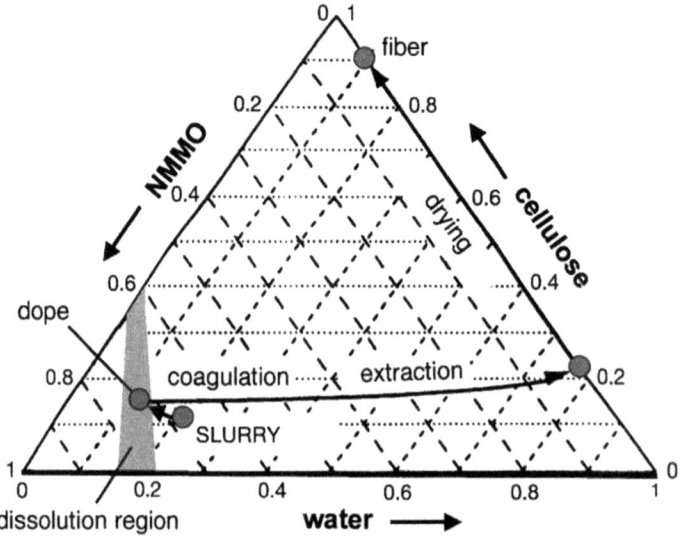

Figure 17.4 Ternary phase diagram for the cellulose–NMMO–water.
Reproduced from ref. 20 with permission. © Elsevier, 2001.

dissolving cellulose in NMMO has been established and used to manufacture diverse cellulosic products.

Recently, this technology has been applied as a pretreatment method for second-generation bioethanol production, encouraged by global energy shortages and environmental problems. Avicel, extensively used microcrystalline cellulose, has been employed to study the effectiveness of NMMO pretreatment on enhancing cellulose hydrolysis. Untreated Avicel released only 2.5 g L^{-1} reducing sugars after 12 h hydrolysis, and leveled off at 2.8 g L^{-1} after 24 h hydrolysis. The yield of cellulose-to-glucose was approximately 23% for the untreated Avicel after 72 h enzymatic hydrolysis. In comparison, at least a two-fold enhancement of the hydrolysis rate was observed in the first 12 h of hydrolysis from regenerated cellulose. Kuo and Lee attributed this phenomenon to the higher fraction of β-glucosidic bonds accessible to cellulase by disrupting the crystalline domains and increasing the surface area, resulting from the rapid precipitation of dissolved Avicel.[21] Cellulose I, the typical spatial conformation in native cellulose, has a monoclinic unit cell with most likely anti-parallel cellobiose chain segments running in opposite directions along the fiber axis. After precipitation from solution using an aqueous medium, the separation of the sheets in any particular domain or crystallite during the dissolution process would facilitate their movement into the lower free-energy position, cellulose II.[22] Cellulose II also represents a monoclinic unit cell, consisting of two anti-parallel-running cellulose chain segments, known as the most stable structure of technical relevance.[23] However, the lattice dimensions of cellulose I and II are different, especially the lattice angle. In addition, the hydrogen-bonding system of cellulose II appears to be more complicated than that of cellulose I. El-Wakil and Hassan used the other commercial microcrystalline cellulose (MCC) as the starting material and compared the structural changes of regenerated cellulose dissolved in the FeTNa, NaOH/thiourea or NMMO systems.[24] They reported that the regenerated cellulose prepared from NMMO showed the lowest crystallinity and broadening of the cellulose II X-ray pattern as well as the smallest crystallite size. In addition, it had the lowest degree of polymerization and a porous structure with many fibrils on the surface (Figure 17.5). It was surprising to find that a smaller amount of reducing sugars was released from the enzymatic hydrolysis of regenerated cellulose dissolved in NMMO at 130 °C than released at 100 °C.[21] This is probably due to the occurence of cellulose degradation at higher temperature during Lyocell treatment. In the NMMO–cellulose system, homolytic reactions could introduce 2-keto structures into cellulose or cause random chain scission. Meanwhile, heterolytic reactions could oxidize the aldehyde groups to the corresponding carboxylic acids as well as convert the hydroxyl groups into carbonyls to a minor extent.[16] Cellulose degradation during NMMO pretreatment not only reduces the hydrolysis yield of regenerated cellulose, but also inhibits the cellulase activity and results in a slower hydrolysis rate.[21]

Having demonstrated the significant effect of NMMO on enhancing enzymatic hydrolysis, cotton,[25,26] sugarcane bagasse,[21] hybrid poplar,[27] spruce (*Piceaabies*) and oak (*Quercusrobur*)[28] were then taken as model lignocellulosic materials for pretreatment. After regeneration from NMMO, the crystal

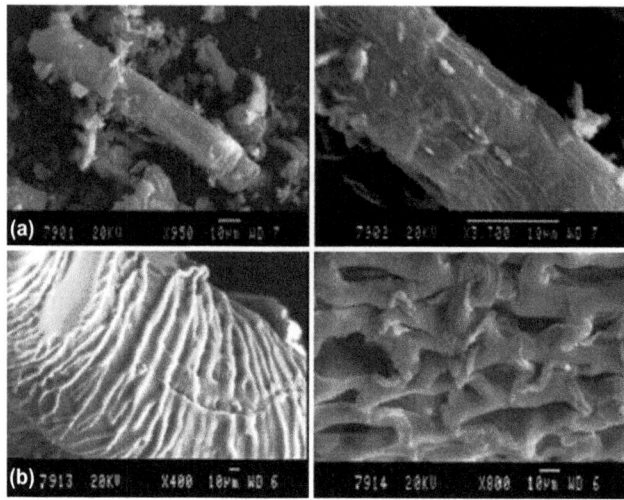

Figure 17.5   Scanning electron micrographs of (a) MCC and (b) NMMO-regenerated cellulose.
Reproduced from ref. 24 with permission. © Wiley & Sons, 2008.

structure of cotton was transformed from cellulose I to cellulose II or to an amorphous structure, and the degree of polymerization was lower than half that of the untreated cotton. Consequently, more than three-fold and approximately four-fold enhancements of the initial hydrolysis rate and saccharification conversion were achieved, respectively.[26] Jeihanipour *et al.* compared the influence of three modes of action of NMMO (dissolution 85%, ballooning 79%, and swelling 73%) on subsequent enzymatic hydrolysis.[25] Although the cellulose swelling and ballooning processes significantly changed the physical properties and increased the volume of cellulose, the gross structure of cellulose maintained. The dissolution mode was found to be the most effective pretreatment for bioethanol production, as it introduces new hydrogen bonds into the reorganized state after regeneration. For sugarcane bagasse, the morphology analysis indicated a much thinner fiber bundle and a rough, irregular and corrugated surface on the regenerated sample after NMMO pretreatment (Figure 17.6). Kuo and Lee further studied the effects of different parameters (temperature, time and substrate concentration) in detail, and concluded that 130 °C and 1 h were the optimal conditions and different sugarcane bagasse/NMMO concentrations (from 5 to 20 wt%) did not affect enzymatic hydrolysis of the regenerated bagasse.[21] No negative effect on ethanol fermentation was observed and about 0.15 g ethanol per g of bagasse was produced by simultaneous saccharification and fermentation (SSF) with cellulase and *Z. molibis*. Spruce and oak, as the typical trees for softwood and hardwood, were dissolved in 85% NMMO and then regenerated in boiling distilled water.[28] Under the best conditions (130 °C, 2 h), a 72.2% glucose yield (% of the theoretical yield) from regenerated oak was obtained after 24 h enzymatic hydrolysis, which is 4.8-fold higher than that obtained from untreated

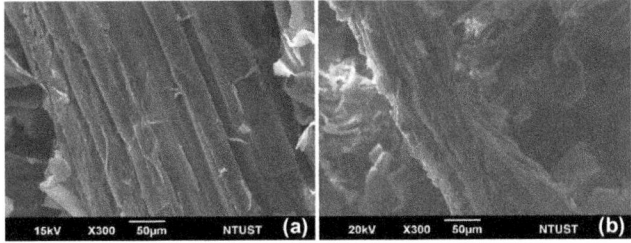

Figure 17.6   Scanning electron micrographs of (a) untreated bagasse and (b) bagasse regenerated from 130 °C, 1 h pretreatment with NMMO (b).
Reproduced from ref. 21 with permission. © Elsevier, 2009.

oak (14.9%). In comparison, NMMO pretreatment had more drastic effects on the softwood spruce. The enzymatic hydrolysis of untreated spruce for 24 h only resulted in a 7.1% glucose yield, and an 83.5% yield was obtained from pretreated spruce at 130 °C for 3 h. Similar results were observed for an SSF experiment. The ethanol yields were 85.4 and 89.0% of the theoretical yield from regenerated oak and spruce, which were 4.6 and 13.1 times those from untreated oak and spruce, respectively. Meanwhile, no significant inhibitory effect of NMMO on baker's yeast was observed during glucose fermenting in the presence of 1% NMMO. These results clearly indicated that the cellulose II structure is much more reactive (accessible) than the cellulose I structure.[8] The hydrophobic forces between the molecular chains could act as a major factor in the resistance to enzymatic hydrolysis by cellulase.[29] The parallel chains in cellulose I only stack through hydrophobic interactions such as van der Waal force along the 010 direction. However, the anti-parallel chains in cellulose II are stacked through hydrophobic interactions and further stabilized with inter-molecular hydrogen bonds.[23] Therefore, the much weaker hydrophobic interactions in cellulose II than in cellulose I could be one of the most important reasons for the obvious promotion of enzymatic hydrolysis.

Employing NMMO technology, various cellulosic products have been developed and manufactured on a large scale. Based on the above reported results, it seems that NMMO pretreatment could be a practicable method for bioethanol production from lignocellulosic materials. Spontaneous decomposition and side-reactions could cause the consumption of this expensive solvent and safety issues need to be thoroughly investigated to avoid any risk in the industrial pretreatment process.

17.2.2   Lithium-Salt-Containing Polar/Aprotic Solvents

Lithium chloride–*N*,*N*-dimethylacetamide (LiCl/DMAc) was first discovered to dissolve polyamides and chitin in 1972.[30,31] Then, its application quickly spread to dissolving cellulose,[32,33] and it was extensively employed for cellulose derivatization under homogeneous reaction conditions.[34–37] DMAc can be substituted in the solvent mixture with *N*,*N*-dimethylformamide (DMF),

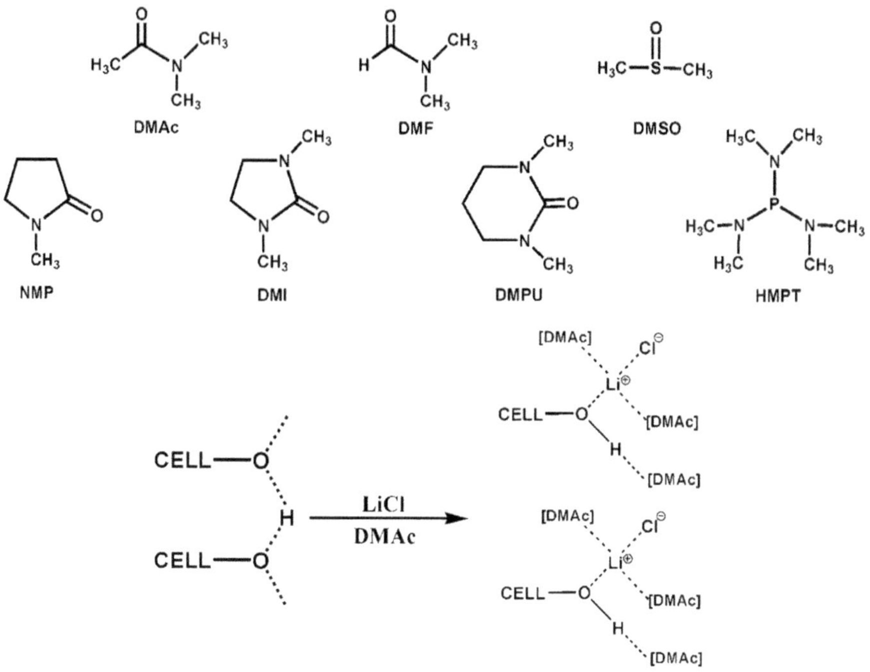

Figure 17.7 The structures of several polar/aprotic organic liquids and a proposed mechanism for the dissolution of cellulose in LiCl/DMAc.

dimethyl sulfoxide (DMSO), *N*-methyl-2-pyrrolidone (NMP), 1,3-dimethyl-2-imidazolidinon (DMI), *N,N*-dimethylpropylene urea (DMPU), and hexamethylphosphorictriamide (HMPT) (Figure 17.7). Taking DMAc as an example, extensive investigation and discussion devoted to the mechanism for cellulose dissolution in LiCl-containing solvents has been reported previously,[38,39] and these observations have suggested that the unique ability of the LiCl/DMAc system to dissolve cellulose can be attributed to the interaction of the lithium moieties of solvated, undissociated ion pairs in the LiCl molecule with the hydroxyl group oxygen atoms of cellulose (Figure 17.7). This solvent system can dissolve cellulose with a molecular weight of more than 10^6 g/mol under ambient conditions without severe degradation or other undesirable reactions. Therefore it can be utilized analytically to investigate the molecular properties of cellulose, such as chain dimensions and flexibility in the dissolved state.[40,41]

However, application of this solvent system for the pretreatment of cellulose has been seldom reported. Recently, our research group compared the influence of the LiCl/DMAc system and other cellulose-dissolving systems on subsequent enzymatic hydrolysis.[27] Once regenerated from the LiCl/DMAc system, the X-ray diffraction (XRD) pattern of cellulose exhibited similar peaks to untreated cellulose; the crystallinity index also slightly decreased (Figure 17.8). This indicates that solvent penetration during the dissolution process is limited to the amorphous regions. Decrystallization of the crystalline domains results

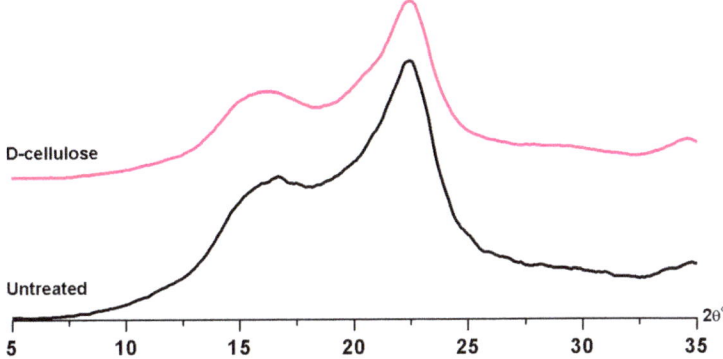

Figure 17.8 X-ray diffraction spectra of untreated cellulose, and cellulosic samples regenerated from LiCl/DMAc.

in the peeling of thin layers from the cellulose structure, which probably retains some of the molecular ordering of the original structure. As the LiCl and DMAc are washed away, the peeled layers regenerate into a paracrystalline matrix, which is slightly more disordered than the crystalline cellulose, but sufficiently distinct from amorphous cellulose.[42] Furthermore, decrystallization was not obvious under the given conditions (130 °C, 30 min) according to kinetic analysis.[43] The multi-stage process of cellulose dissolution includes the disintegration of the crystalline regions (ΔH_{fusion}), transition to the amorphous regions ($\Delta H_{\text{transition}}$), solvation of the polymer macromolecules ($\Delta H_{\text{interaction}}$), and mixing of solvated polymer molecules with solvent (ΔH_{mixing}). The total enthalpy of cellulose dissolution is given by:

$$\Delta H_{\text{solution}} = \Delta H_{\text{fusion}} + \Delta H_{\text{transition}} + \Delta H_{\text{interaction}} + \Delta H_{\text{mixing}}$$

All terms are exothermic except ΔH_{fusion}, which is associated with breaking the H-bonds in crystalline regions. Therefore, the overall process is exothermic and is favored by lower temperatures. Ramos *et al.* reported that the I_c (index of crystallinity) values of all celluloses investigated were constant at 150 °C and slightly decreased as a function of decreasing temperature to 130 °C.[43] This small change in the crystal structure of cellulose led to the release of only $0.02\,\text{mg}\,\text{mL}^{-1}$ more glucose than the untreated sample in the first 8 h of enzymatic hydrolysis, and improved the finally cellulose conversion by 15% (from 33 to 48%). These results indicate that the LiCl/DMAc system is not an efficient pretreatment for bioethanol production, although it has been successfully adopted as a reaction medium for the chemical modification of cellulose. In addition, incomplete recycling of this expensive solvent system still prevents its large-scale application.

17.2.3 NaOH/Urea Aqueous Solutions

In recent years, a series of publications has appeared on the dissolution and modification of cellulose in mixtures of an aqueous base with urea at cold

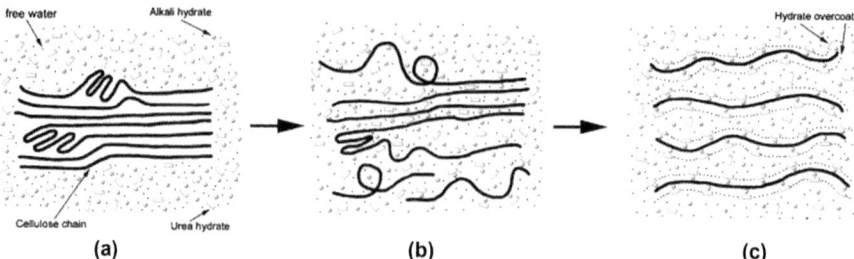

Figure 17.9   Schematic of the dissolution process of cellulose in LiOH/urea and
NaOH/urea aqueous solutions pre-cooled to −10 °C: (a) cellulose
bundles in the solvent; (b) swollen cellulose in the solution; and (c) a
transparent cellulose solution.
Reproduced from ref. 46 with permission. © John Wiley & Sons, 2005.

temperatures.[44,45] Cellulose with a molecular weight below 1.2×10^5 g/mol can
be rapidly dissolved in 7 wt% NaOH/12 wt% urea aqueous solution precooled
to −10 °C within 2 min or even within several seconds.[46] The rapid dissolution
of cellulose suggests a dynamic self-assembly process resulting in the formation
of a large inclusion complex (IC) with NaOH–urea–water clusters through a
new hydrogen-bonded network structure, which is in a highly stable state at low
temperatures.[47] The chemical shifts of C=O indicated that the interaction
between NaOH and urea in the solution plays an important role in the
solvation of cellulose, effectively preventing the self-association of cellulose
macromolecules and improving the stability of the cellulose solution.[48] The
NaOH 'hydrates' were bound to cellulose chains, while the urea hydrates served
as hydrogen-bonding donors and receptors preventing the approach of cellulose molecules towards each other in solution. This led to a good dispersion of
the cellulose, resulting in an actual solution (Figure 17.9).[49] The dissolution
powers of other similar systems are in the order: LiOH/urea > NaOH/urea ≫
KOH/urea. The solvent molecules containing Li^+ and Na^+ ions, having
relatively small ionic radii, and high charge densities could easily penetrate
cellulose and expand it into the swollen and dissolved state; whereas the relatively
large K^+ ions could be imbibed by the cellulose chains.[46] When the cellulose was
recovered from the solution, a crystal transition from cellulose I into II was
observed.[27,46] In this case, the corner O_6 atoms of cellulose rotated from '*gt*'
(*gauche–trans*) to '*tg*' (*trans–gauche*) positions, forming the 'O_2–O_6' inter-chain
hydrogen bonds dominant in cellulose II (Figure 17.10). The work of
Laszkiewicz,[50] Isogai and Atalla,[51] and Zhou et al.[48,52] suggests that there exists
an upper DP limit of cellulose for each solution system, and cellulose with a
higher DP value (beyond the corresponding upper limit) cannot be dissolved in
that solution. Wang et al. utilized cellulase to tailor cellulose molecules to the
desired molecular weight to improve their dissolution in a NaOH/urea solution.[53]
They found that the reduction of cellulose's molecular weight by enzymatic
treatment was the most important factor controlling the solubility of cellulose in
cold NaOH/urea solutions, much more important, in fact, than crystallinity.

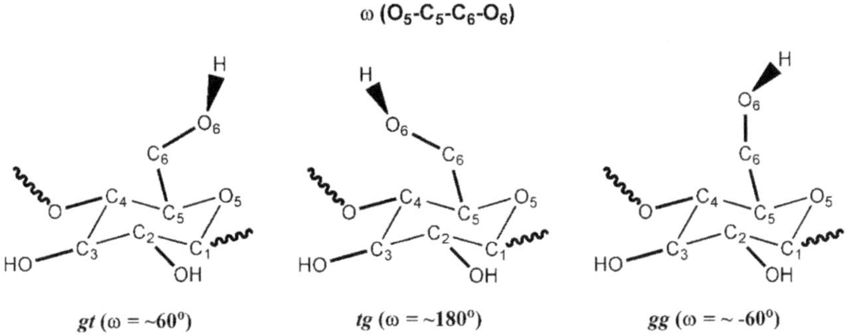

$$\omega \; (O_5\text{-}C_5\text{-}C_6\text{-}O_6)$$

$gt \; (\omega = \sim\text{-}60°)$ $tg \; (\omega = \sim\text{-}180°)$ $gg \; (\omega = \sim \text{-}60°)$

Figure 17.10 The glucose rotamer conformations (ω is an angle).

As a non-polluting and non-toxic solvent system for cellulose, the use of NaOH/urea aqueous solutions is a potential pretreatment process for bio-ethanol production. After a 1 h enzymatic hydrolysis, $6.2\,\text{mg}\,\text{mL}^{-1}$ of reducing sugars was released from regenerated cotton; however, only $2.3\,\text{mg}\,\text{mL}^{-1}$ of reducing sugars was obtained from untreated cotton. Final saccharification conversions of around 85% were achieved by employing this pretreatment for both cotton[26] and hybrid poplar.[27] For spruce, a loose structure of fiber bundles was created by this pretreatment method as compared to the untreated fiber bundles. Moreover, parts of the fiber bundles were broken down to small pieces by the chemical treatment (Figure 17.11). Zhao *et al.* found that alkaline treatment using NaOH without urea remarkably improved the bioconversion of cellulose to glucose.[54] The concentration of 2% NaOH was critical because the bioconversion rate and efficiency were dramatically increased over the original 1%. The cold temperature significantly increased the enzymatic hydrolysis efficiency, and the addition of urea further increased the hydrolysis rate and glucose yield. Similar results were obtained on releasing xylose and mannose from hemicelluloses. Although it is difficult to lower the operation temperature in a large-scale industrial operation, it can be realized by processing in winter in Northern areas where extra energy to cool the pretreatment system is not required.

17.2.4 Concentrated Phosphoric Acid

Walseth first developed a procedure for producing high-reactivity cellulose suitable for cellulase activity studies by swelling air-dried cellulose in 85% phosphoric acid.[55] However, the same system can act either as a swelling agent or as a dissolution agent, depending on the properties of the cellulose and the operating conditions.[56] Zhang *et al.* observed a phase transition from cellulose swelling to cellulose dissolution when the phosphoric acid concentration was greater than a critical value (~ 80.5% for Avicel).[57] The modification of the original cellulose by adding a small amount of water to form a slurry is the most important step for rapid cellulose dissolution, because it efficiently avoids the formation of highly viscous cellulose-dissolved gels outside the dry cellulose

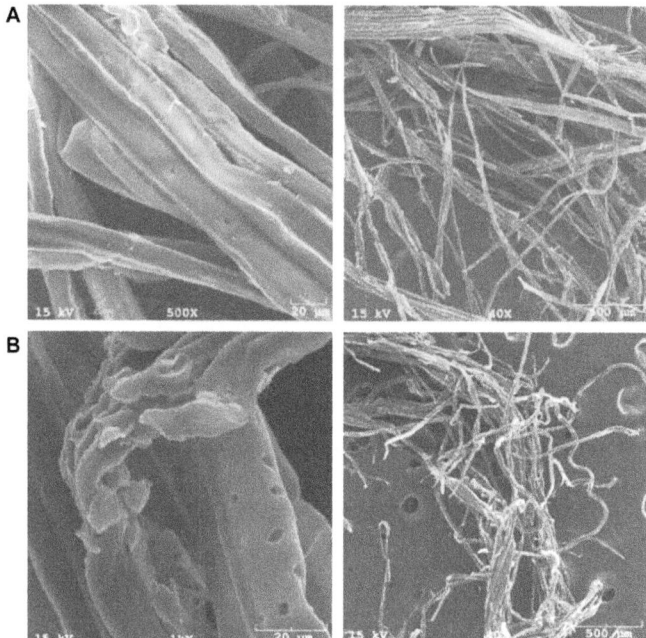

Figure 17.11   Scanning electron micrographs of wood fiber bundles: (a) before; and
(b) after NaOH/urea treatment at cold temperatures.
Reproduced from ref. 54 with permission. © John Wiley & Sons, 2008.

particles. The phosphoric acid concentration required for the phase transition
of cellulose lies within a narrow range. Thereafter, the further addition of a
small volume of concentrated phosphoric acid caused the acid concentration to
increase to a level high enough to dissolve cellulose. Two main processes are
involved in the formation of clear cellulose solutions: (1) an esterification reaction between the alcoholic hydroxyl groups of cellulose and phosphoric acid
to form cellulose phosphate (H_3PO_4 + cellulose ↔ cellulose-o-PO_3H_2); and (2)
competitive hydrogen-bond formation between the hydroxyl groups of the
cellulose chains and water molecules or hydrogen ions.[56] Meanwhile, another
by-reaction is the acid hydrolysis of β-glucosidic bonds in cellulose, but this
acid hydrolysis can be minimized by decreasing the dissolution temperature.[58,59] Swelling Avicel (treated with 77% phosphoric acid) resulted in smaller
particle sizes, and many small holes on the surfaces of the residual cellulose
particles. In contrast, regenerated Avicel from the clear solution did not show
any new supramolecular structure in its cellulosic particles or fibers after precipitation (Figure 17.12). The crystal configuration of cellulose was gradually
transformed from cellulose I to amorphous cellulose with increasing reaction
time, and cellulose II was determined to be the transition state.[60] The regenerated cellulose had a low activation energy and a high adsorption isotherm
constant, which led to rapid increase in the initial enzymatic hydrolysis
rate.[59,60] In addition, final conversions of 95 and 92% were achieved after 72 h

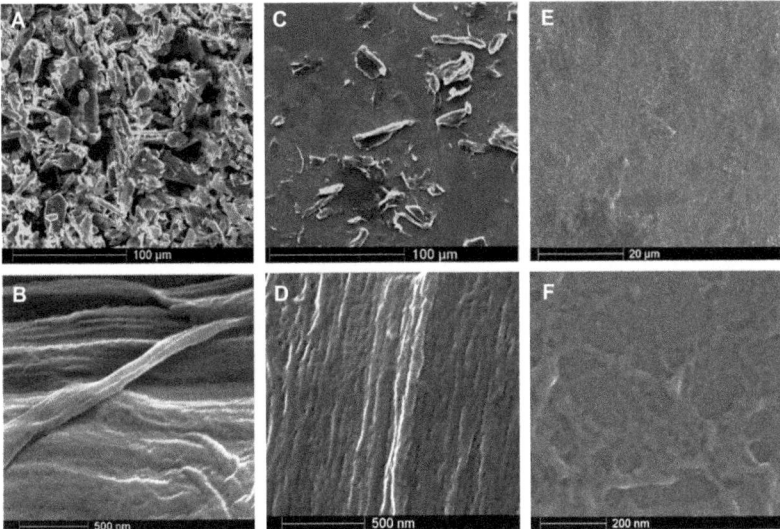

Figure 17.12 Scanning electron micrographs of: (a) and (b) intact Avicel; (c) and (d) cellulose treated with 77% phosphoric acid; and (e) and (f) cellulose treated with 83% phosphoric acid at two different magnifications. Reproduced from ref. 57 with permission. © American Chemical Society, 2006.

enzymatic hydrolysis for regenerated cotton[26] and poplar,[27] respectively, although a mixed structure of cellulose I and II, and a cellulose II structure alone were determined by FT-IR and XRD analysis, respectively. Our research data established a strong correlation between the crystal structure transformation of cellulose and its hydrolysis kinetics. The XRD spectra of cellulose clearly illustrated that the peak of plane 020 ($2\theta = 21.9°$) was almost totally covered by the adjacent peak around $2\theta = 20.2°$ (plane 110) after regeneration from concentrated phosphoric acid. Blackwell *et al.* assumed that the inter-molecular hydrogen bonds between adjacent cellulose molecules were located entirely in the same lattice plane (the 020 plane).[61] This indicates that the concentrated phosphoric acid prevents the formation of inter-molecular hydrogen bonds during the regeneration process. The hydrogen ion from phosphoric acid is relatively small and could enter the space between adjacent cellulose chains. Consequently, it is predominant in the competitive hydrogen-bond formation between cellulose chains, and between cellulose chains and other hydrogen ions such as those from water. The complete transformation from cellulose I to cellulose II was the most likely reason for the highest yield of glucose and conversion of regenerated cellulose.[27]

Based on these initial findings, Zhang *et al.* developed cellulose solvent- and organic solvent-based lignocelluloses fractionation (CSLF or COSLIF) methods to separate lignocellulosic materials under modest reaction conditions (*e.g.*, $\sim 50\,°C$ and atmospheric pressure) by using phosphoric acid, acetone and water.[62–65] Figure 17.13 shows the conceptual processes involved in this

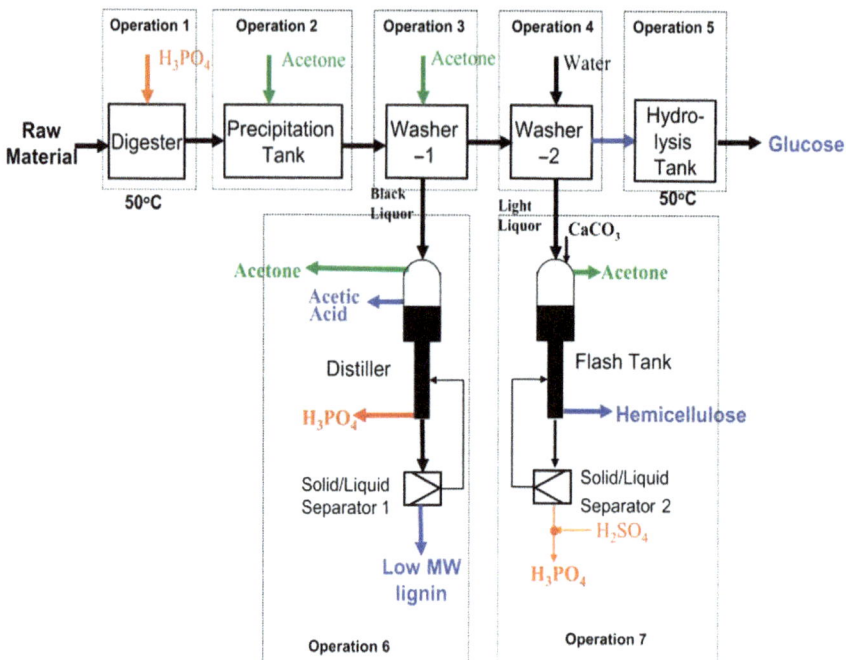

Figure 17.13 Conceptual flowchart of the COSLIF process with recycling of concentrated phosphoric acid and acetone.
Reproduced from ref. 62 with permission. © John Wiley & Sons, 2007.

lignocelluloses fractionation technology. The key ideas of COSLIF are: (1) removal of partial lignin and hemicelluloses (eliminating the major obstacles to cellulose hydrolysis and allowing cellulase to access the substrate more efficiently); (2) decrystallization of cellulose fibers by a cellulose solvent (providing better cellulose accessibility to cellulase); and (3) modest reaction conditions (causing a decrease in sugar degradation, less inhibitor formation, lower utility consumption, and less capital investment).[60] The pretreated Avicel and α-cellulose were completely converted to soluble sugars within 3 h. For herbaceous cellulose (corn stover and switchgrass) and hardwood (hybrid poplar) lignocelluloses, the pretreated cellulosic samples were ∼94% hydrolyzed after 12 h and *ca.* 96–97% hydrolyzed after 24 h. This technology resulted in increase in digestibility by a factor of at least 11 for softwood (Douglas fir) lignocelluloses, although the final conversion was only ∼75% because of inefficient lignin removal.[62] Next, the effects of COSLIF pretreatment conditions (acid concentration, reaction time, and temperature) on enzymatic cellulose digestibility were further investigated.[63] Pretreated industrial hemp hurbs had a very high hydrolysis rate and digestibility when the phosphoric acid concentration exceeded 81%. Their results indicated that 30 min was not long enough to efficiently dissolve biomass using concentrated phosphoric acid, and no difference in glucan digestibility between the samples treated for 60 and 120 min, respectively, was observed. With regard to reaction temperature, there was some improvement in enzymatic glucan digestibility when temperatures increased from 40 to 50 °C but no further improvement at higher temperatures (60 °C). On the basis of a maximum sugar release from biomass, the best pretreatment condition for hemp hurds was finally identified as 84.0% H_3PO_4, 50 °C and 60 min. Moreover, Zhang's group have produced comprehensive research work on this pretreatment process, and have discussed the supramolecular structure, substrate accessibility and reaction parameters in detail.[60,64,65] For intact switchgrass, the structural features of cellulose I were clearly revealed by XRD (significant peaks corresponding to the 101 and 002 lattice planes were identified) and CP/MAS ^{13}C NMR (signals around 65 ppm were assigned to the highly ordered interiors). COSLIF pretreatment led to a significant reduction in the crystalline fraction and Crystallinity Index (CrI) value.[64] The accessibility of cellulose to cellulase (CAC) is one of the most important (rate-limiting) factors influencing enzymatic hydrolysis rates.[58] After being pretreated using COSLIF, switchgrass exhibited a drastic increase in CAC by 16.3-fold, leading to high digestibilities and hydrolysis rates even at low enzyme loadings.[64] This technology has also been extensively applied to bamboo,[66] Bermuda grass, reed and rapeseed stover,[67] and achieved good performance for the bioconversion of cellulose to glucose. The COSLIF process produces more highly digestible materials than other pretreatment methods. Dilute acid (DA) pretreatment disrupts the lignocellulosic structure mainly by dissolving hemicelluloses, leaving the original microfibrous cellulose structures intact (Figure 17.14).[65] In contrast, COSLIF-pretreated corn stover had a nearly two-fold higher hemicellulose composition. However, it had no clear fibrous structure and almost two-fold higher CAC levels. For enzymatic hydrolysis, COSLIF-pretreated biomass had much higher digestibility and hydrolysis rates than DA-pretreated biomass. This result was attributed to the fact that efficient structure destruction and high substrate accessibility are the primary determinants of the rate or the extent of glucan enzymatic digestibility, rather than the efficiency of hemicellulose removal. Soaking in aqueous ammonia (SAA) is another way to remove a large amount of lignin and dissolve hemicellulose.[68] Under optimal conditions (10% w/w ammonia, 140 °C, 1 : 20 solid-to-liquid ratio, 14 h), SAA resulted in a substrate with greatly reduced acid-insoluble lignin (74% removal), and increased CAC by 1.4-fold compared to the untreated switchgrass. The final conversion yield for SAA-pretreated biomass with low cellulase loading (3 filter paper units (FPUs) g^{-1} glucan) was 42% without bovine serum albumin (BSA) blocking, and 58% with BSA blocking. In contrast to SAA, COSLIF process introduced 34% acid-insoluble lignin removal, but dramatically increased CAC by 16-fold. A final digestibility of 85% was achieved under the same enzymatic conditions, demonstrating that increasing cellulose accessibility is a more important pretreatment consideration than delignification for effectively releasing sugars from recalcitrant lignocellulosic materials.[69]

COSLIF is a new lignocelluloses pretreatment technology. It has several advantages, including high glucan digestibility, high hydrolysis rate, low cellulase use, modest reaction conditions, and higher revenues from co-products (acetic acid, lignin, and hemicelluloses). However, this technology still faces

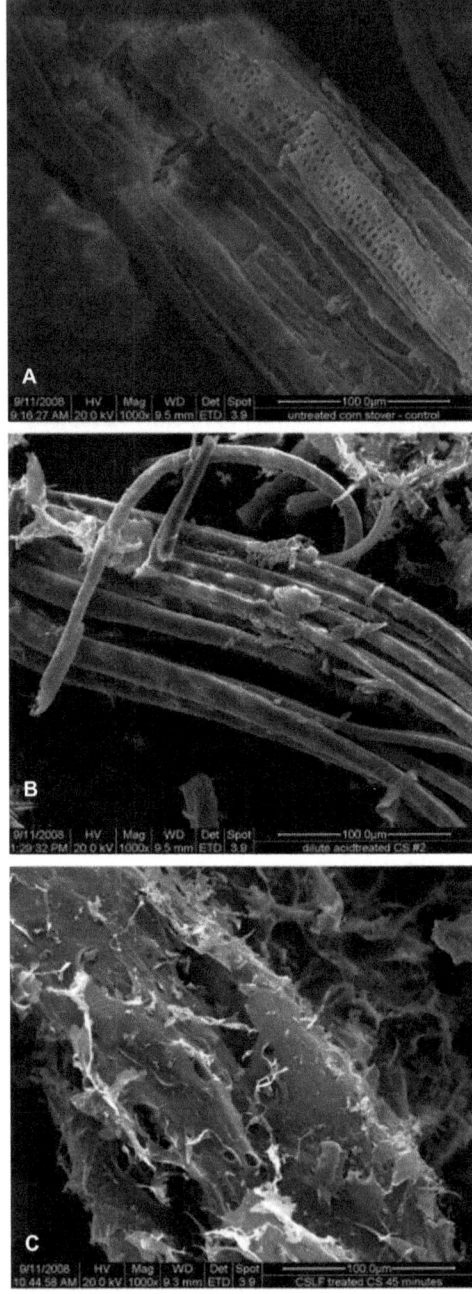

Figure 17.14   Scanning electron micrographs of corn stover: (a) before pretreatment; (b) pretreated with dilute acid; and (c) with COSLIF for 45 min. Reproduced from ref. 65 with permission. © John Wiley & Sons, 2009.

with some problems, such as high processing costs for the efficient recycling of both solvents. In addition, this technology is not yet commercially proven and its economic viability for use in large-scale biorefining remains to be demonstrated. In summary, COSLIF technology remains at an early stage of development and a more detailed economic analysis based on rigorous Aspen-plus models is needed to understand its potential for practical application.[65]

17.2.5 Ionic Liquids

Recently, the use of ILs to treat lignocellulosic materials has become a hot research topic. ILs are a group of salts that exist as liquids at relatively low temperatures (<100 °C). Interest in developing and investigating the properties and applications of ILs in many fields of science and technology has intensified as a result of the introduction of the principles of "green" chemistry.[70] Specifically, due to their high thermal stability and nearly complete non-volatility, ILs are becoming attractive alternatives to volatile and unstable organic solvents. In 2002, Swatloski *et al.* first reported the use of an IL as a solvent for cellulose, both for the regeneration of cellulose and for the chemical modification of the polysaccharide.[71] This result opened up a new class of solvents for the cellulose research community.

The most common examples of these ILs include the salts of organic cations, such as alkylimidazolium $[R_1R_2IM]^+$, alkylpyridinium $[RPy]^+$, tetraalkylammonium $[NR_4]^+$, or tetraalkylphosphonium $[PR_4]^+$, and anions, such as hexafluorophosphate $[PF_6]^-$, tetrafluoroborate $[BF_4]^-$, nitrate $[NO_3]^-$, methanesulfonate (mesylate) $[CH_3SO_3]^-$, trifluoromethanesulfonate (triflate) $[CF_3SO_3]^-$, and bis-(trifluoromethanesulfonyl)amide $[Tf_2N]^-$, as well as several low-melting-point chloride, bromide, and iodide salts (Figure 17.15).[72] Despite the increasing number of ILs developed for dissolving cellulose, the mechanism of this dissolution process is not well understood. Figure 17.16 shows the proposed mechanism for the solvation of cellulose in ILs.[73] The oxygen and hydrogen atoms of the cellulose form electron donor–electron acceptor (EDA) complexes with the charged species of the IL. The solvation of cellulose by ILs involves the formation of stoichiometric H-bonding between the hydroxyl protons of the solutes and the anions of the ILs,[74] the strength of which is three

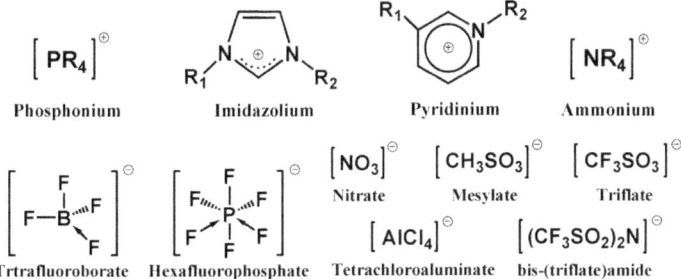

Figure 17.15 Commonly used cations and anions in ionic liquids.

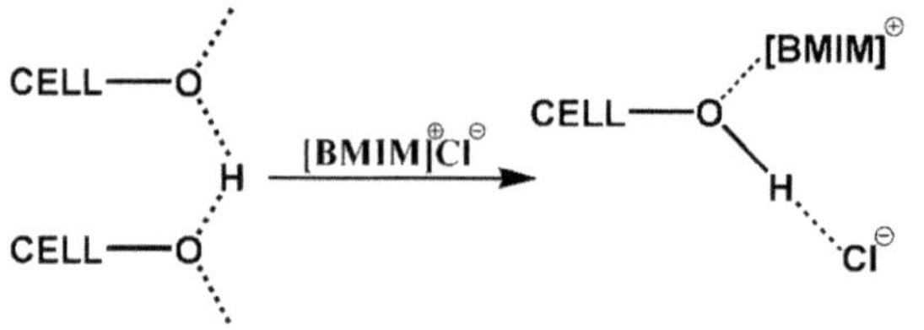

Figure 17.16   Proposed dissolution mechanism for cellulose in [BMIM]Cl.
Reproduced from ref. 73 with permission. © Elsevier, 2008.

times stronger than the H-bonding energy from water and methanol.[75] In addition, based on molecular dynamics simulations, the interactions between imidazolium cations and glucose play an important role in forming hydrophobic interactions. The simulation showed that the cellulose oligomer conformation is similar to the conformation of crystalline cellulose in water. However, [C$_2$MIM][OAc]-solvated oligomers show a totally different backbone conformational distribution of inter-molecular rotamers. The rotatable bond (ω-angle) has three stable staggered rotamers: gauche–trans (*gt*), trans–gauche (*tg*), and gauche–gauche (*gg*) (Figure 17.10). When dissolved in [C$_2$MIM][OAc], the *gt* is still the dominant rotamer; however, the population of *tg* rotamers increased, while *gg* rotamer conformations decreased (Figure 17.17).[75] The most successful cations for cellulose dissolution are based on the methylimidazolium and methylpyridinium (nitrogen-containing) cores with different side-chains, although the mechanisms have not been probably explained. The maximum dissolution power is achieved with the C4 side-chain, and incorporated hydroxyl atoms seem to enhance cellulose solubility. This could be due to the additional polarity of the heteroatomic substituents on the imidazolium ring.[76] The addition of an excess of a polar solvent like water, acetone, dichloromethane, acetonitrile, or mixture of them could precipitate cellulose from the homogeneous solution. XRD spectra clearly indicated that the crystal transformation from cellulose I to cellulose II or amorphous cellulose occurred during dissolution and regeneration from ILs, varying among the different starting materials. ILs broke inter-molecular and intra-molecular hydrogen bonds and changed the conformations, consequently destroyed the original crystalline form of cellulose. Moreover, the coagulation process was too transitory to favor cellulose recrystallization.[77]

Initial reports on the use of ILs to pretreat cellulose to enhance its susceptibility to enzymatic hydrolysis employed either 1-butyl-3-methylimidazolium chloride ([BMIM]Cl) or 1-allyl-3-methylimidazolium chloride ([AMIM]Cl), using Avicel as a model crystalline material.[78,79] The initial enzymatic hydrolysis rate was approximately 50-fold higher for regenerated Avicel as compared to untreated cellulose, measured by soluble reducing sugar assay. After about 2 h of enzymatic hydrolysis, the regenerated cellulose reaction mixtures appeared transparent whereas untreated cellulose mixtures were opaque. Increasing the incubation time (from 10 min to 3 h) and the temperature (from 130 to 150 °C) had little effect on the rate of glucose

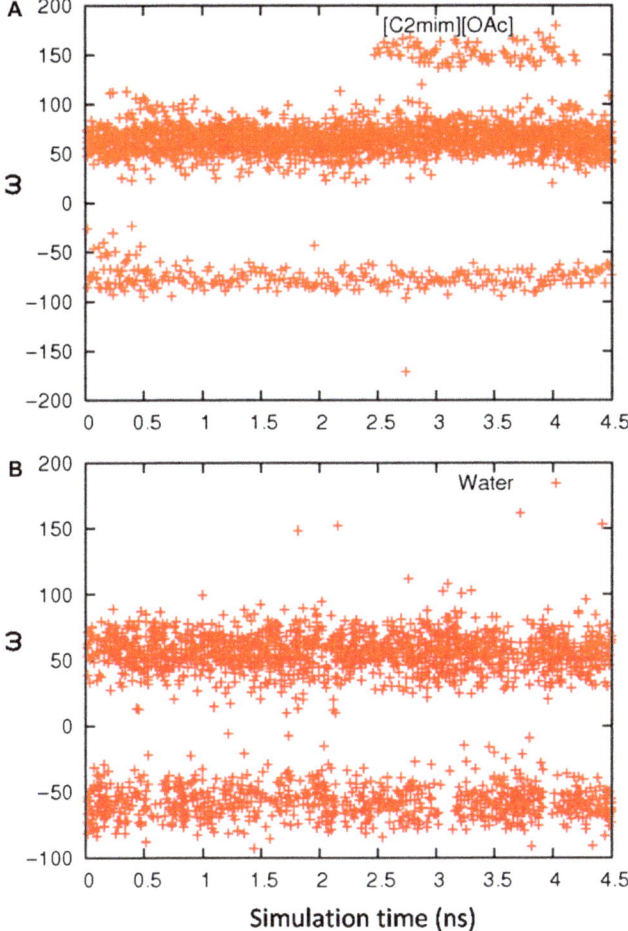

Figure 17.17 Molecular dynamics trajectories of the glucose rotamer conformation (A: dissolving in [C2mim][OAc]; B: dissolving in Water) (*gt*, $\omega \approx 60°$; *tg*, $\omega \approx 180°$; *gt*, $\omega \approx -60°$).[75]

formation and the final conversion. The risk of partial degradation of the cellulose increased with increasing temperature.[80] Meanwhile, regenerated Avicel exhibited much lower crystallinity (a 60–75% reduction) and higher cellulase adsorption capacity (about 2-fold). After 6 h, >95% conversion of cellulose to reducing sugars was achieved with an enzyme/regenerated Avicel ratio of 3:20 (w/w) at 50 °C. It is worth noting that IL residues should be carefully removed during the regeneration process, since their presence had detrimental effects on cellulase activity, depending on the amount of IL remaining.[8] Microwave irradiation further increased the dissolution of cellulose in ILs and significantly reduced the degree of polymerization of the regenerated cellulose.[81] Similar bioconversion improvements were observed when applying ILs pretreatment to real lignocellulosic materials.[82–84] Li *et al.*

screened different kinds of ILs (1-ethyl-3-methylimidazole acetate [EMI-M][OAc], 1-ethyl-3-methylimidazolium diethyl phosphate [EMIM]DEP, 1-butyl-3-methylimidazolium chloride [BMIM]Cl, 1-ethyl-3-methylbutylpyridinium diethyl phosphate [EMBy]DEP, and 1-ethyl-3-methylimidazolium dibutyl phosphate [EMIM]DBP), and selected [EMIM]DEP as the candidate pretreatment solvent in view of its low viscosity and its potential to accelerate enzymatic hydrolysis.[83] They also found that increasing the pretreatment temperature (from 130 to 150 °C) and time (from 10 min to 2 h) and the variety of anti-solvent (methanol, ethanol or water) did not appear to affect the bioconversion of regenerated wheat straw, consistent with the results of Dadi *et al.*[78,79] However, over a wider range of temperatures (from 70 to 150 °C), more reducing sugars were released by increasing the temperature from 70 to 130 °C. After 26 h of fermentation of hydrolyzates, following enzymatic hydrolysis, the ethanol production was 0.43 g per of glucose from the regenerated wheat straw. The presence of lignin in the lignocellulosic materials obviously decreased the solubility of cellulose in ILs.[85,86] The maximum solubility of lignin is achieved when the Hildebrand solubility parameter (δ_H) values of the polymer and solvent are identical.[87] [EMIM][OAc] was selected as the optimal IL for achieving the compromise between high lignin solubility and low maple wood flour solubility. When 40% of the lignin was removed, the cellulose crystallinity index dropped below 45%, resulting in >90% of the cellulose present in maple flour being hydrolyzed by *Trichodermaviride* cellulase.[82] In contrast, [EMIM][OAc] completely solubilized both the cellulose and lignin in switchgrass. Confocal fluorescence images showed that the lignin-rich sclerenchyma and middle lamella in plant tissues disintegrated immediately and simultaneously as primary and secondary cell walls started to separate from the middle lamella. Intact stem sections formed a very viscous solution during the first 30–50 min, and complete solubilization was accomplished after 2.5–3 h (Figure 17.18). After addition of water to this cellulose solution, fibrous structures were formed, and the regenerated switchgrass rejected lignin during this process. Finally, 72.5% of IL-pretreated switchgrass was converted to sugar after 24 h enzymatic hydrolysis, as compared to 16.5% for the untreated biomass.[88]

The use of ILs has provided a new platform for improving the saccharification process during the conversion of carbohydrates in lignocellulosic biomass into fermentable sugars. Due to their present high cost, the recovery and recycling of ILs is required for commercial use in biomass pretreatment. In addition, the efficient fractionation of lignin and hemicelluloses dissolved in ILs is critical to prolonging the lifetime of the recycled ILs and decreasing production costs by co-converting the whole biomass into higher value products. The dissolved lignin can be recovered with acetone and subsequent evaporation.[85] However, the recovery of hemicelluloses from ILs has never been reported.

17.3 Perspectives

There is no doubt that the use of cellulose-dissolving systems as a pretreatment process for second-generation bioethanol production is a relatively new

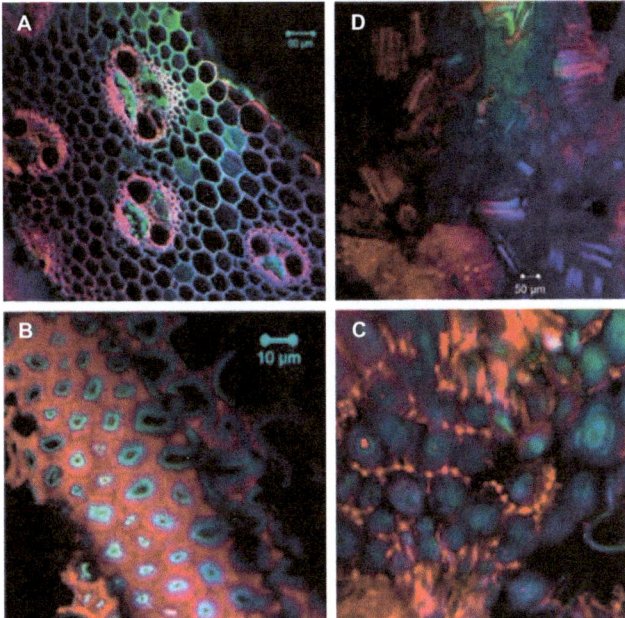

Figure 17.18 *In situ* dynamic study of switchgrass dissolution in [EMIM][OAc]. Confocal fluorescence images of a switchgrass stem section: (a) before pretreatment; (b) after 20 min of pretreatment; and (c) after 50 min of pretreatment. (d) The complete breakdown of the organized plant cell wall structure was observed after 2 h.
Reproduced from ref. 88 with permission. © John Wiley & Sons, 2009.

technology, still in the initial stages. In contrast to traditional pretreatment methods, cellulose solvents have unique advantages including milder operation conditions, fewer degradation by-products and higher cellulose accessibility to cellulase. With the further exploitation of new solvent systems for improving the stability and solubility of cellulose, a new path to realizing the commercial production of lignocellulosic bioethanol will open up. Based on the concept of biorefinery, cellulose solvents provide the possibility of clearly fractionating all the major components of lignocellulosic biomass, and producing value-added co-products to decrease the processing costs. The co-production of a variety of products from multi-component feedstock is a marker of mature industries. In addition, the development of fractionation processes, such as simultaneously recycling solvents and isolating pollutants, may further prolong the lifetime of these relatively expensive solvents and accelerate the large-scale industrialized production of 'green' fuels.

Acknowledgements

This work was supported by the State Forestry Administration (grant no. 201204803), National Science Foundation of China (grant no. 30930073), and Ministry of Science and Technology (grant no. 973-2010CB732204).

References

1. B. Hahn-Hagerdal, M. Galbe, M. F. Gorwa-Grauslund, G. Liden and G. Zacchi, *Trends Biotechnol.*, 2006, **24**, 549.
2. Y. Lin and S. Tanaka, *Appl. Microbiol. Biotechnol.*, 2006, **69**, 627.
3. L. R. Lynd, J. H. Cushman, R. J. Nichols and C. E. Wyman, *Science*, 1991, **251**, 1318.
4. M. Galbe and G. Zacchi, *Adv. Biochem. Eng. Biotechnol.*, 2007, **108**, 41.
5. K. Wang and R. C. Sun, in *Cereal Straw as a Resource for Sustainable Biomaterials and Biofuels: Chemistry, Extractives, Lignins, Hemicelluloses and Cellulose*, ed. R. C. Sun, Elsevier, Oxford, 2010, p. 267.
6. S. Ding and M. E. Himmel, *J. Agric. Food Chem.*, 2006, **54**, 597.
7. M. E. Himmel, S. Y. Ding, D. K. Johnson, W. S. Adney, M. R. Nimlos, J. W. Brady and T. D. Foust, *Science*, 2007, **315**, 804.
8. H. Zhao, C. L. Jones, G. A. Baker, S. Xia, O. Olubajo and V. N. Person, *J. Biotechnol.*, 2009, **139**, 47.
9. R. P. Chandra, R. Bura, W. E. Mabee, A. Berlin, X. Pan and J. N. Saddler, *Adv. Biochem. Eng. Biotechnol.*, 2007, **108**, 67.
10. K. Igarashi, M. Wada and M. Samejima, *FEBS J.*, 2007, **274**, 1785.
11. H. Ringsdorf, *Angew. Chem. Int. Ed.*, 2004, **43**, 1064.
12. T. Liebert, in *Cellulose Solvents: for Analysis, Shaping and Chemical Modification*, ed. T. F. Liebert, T. J. Heinze and K. J. Edgar, American Chemical Society, Washington DC, 2010, p. 1.
13. B. Graenacher and R. Sallmann, *US Pat.*, 2179181, 1939.
14. L. Johnson, *US Pat.*, 3447939, 1969.
15. P. Linton, *J. Am. Chem. Soc.*, 1940, **62**, 1945.
16. T. Rosenau, A. Potthast, H. Sixta and P. Kosma, *Prog. Polym. Sci.*, 2001, **26**, 1763.
17. K. M. Harmon, A. C. Akin, P. K. Keefer and B. L. Snider, *J. Mol. Struct.*, 1992, **269**, 109.
18. M. M. Ioleva, A. S. Goikhman, S. I. Banduryan and S. P. Papkov, *Vysokomol. Soedin., Ser. B.*, 1983, **25**, 803.
19. C. C. McCorsley and J. K. Varga, *US Pat.*, 4142913, 1979.
20. H. P. Fink, P. Weigel, H. J. Purz and J. Ganster, *Prog. Polym. Sci.*, 2001, **21**, 1473.
21. H. Kuo and C. K. Lee, *Bioresour. Technol.*, 2009, **100**, 866.
22. B. G. Ranby, *Acta. Chem. Scand.*, 1952, **6**, 128.
23. P. Langan, Y. Nishiyama and H. Chanzy, *Biomacromolecules*, 2001, **2**, 410.
24. N. A. El-Wakil and M. L. Hassan, *J. Appl. Polym. Sci.*, 2008, **109**, 2862.
25. A. Jeihanipour, K. Karimi and M. J. Taherzadeh, *Biotechnol. Bioeng.*, 2010, **105**, 469.
26. H. Kuo and C. K. Lee, *Carbohydr. Polym.*, 2009, **77**, 41.
27. K. Wang, H. Y. Yang, F. Xu and R. C. Sun, *Bioresour. Technol.*, 2011, **102**, 4524.
28. M. Shafiei, K. Karimi and M. J. Taherzadeh, *Bioresour. Technol.*, 2010, **101**, 4914.

29. M. Wada, M. Ike and K. Tokuyasu, *Polym. Degrad. Stab.*, 2010, **95**, 543.
30. M. Huglin, *Light Scattering from Polymer Solutions*, Academic Press, New York, 1972.
31. S. L. Kwolek, P. W. Morgan, J. R. Schaefgen and L. W. Gulrich, *Macromolecules*, 1977, **10**, 1390.
32. C. L. McCormick, *US Pat.*, 4278790, 1981.
33. A. F. Turbak, *US Pat.*, 4302252, 1981.
34. G. T. Ciacco, T. F. Liebert, E. Frollini and T. Heinze, *Cellulose*, 2003, **10**, 125.
35. A. El-Seoud, G. A. Marson, G. T. Ciacco and E. Frollini, *Macromol. Chem. Phys.*, 2000, **201**, 882.
36. B. Morgenstern and H. W. Kammer, *Polymer*, 1999, **40**, 1299.
37. H. Pionteck, W. Berger, B. Morgenstern and D. Fengel, *Cellulose*, 1996, **3**, 127.
38. C. L. McCormick, P. A. Callais and B. H. Hutchinson, *Macromolecules*, 1985, **18**, 2394.
39. S. Stryuk, J. Eckelt and B. A. Wolf, *Cellulose*, 2005, **12**, 145.
40. H. Aono, D. Tatsumi and T. Matsumono, *Biomacromolecules*, 2006, **7**, 1311.
41. M. Yanagisawa and A. Isogai, *Biomacromolecules*, 2005, **6**, 1258.
42. J. C. Z. Duchemin, R. H. Newman and M. P. Staiger, *Cellulose*, 2007, **14**, 311.
43. L. A. Ramos, J. M. Assaf, O. A. El-Seoud and E. Frollini, *Biomacromolecules*, 2005, **6**, 2638.
44. L. Zhang and J. Zhou, *Chin. Pat.*, 00114486.3, 2000.
45. J. Zhou and L. Zhang, *Polym. J.*, 2000, **32**, 866.
46. J. Cai and L. Zhang, *Macromol. Biosci.*, 2005, **5**, 539.
47. J. Cai, L. Zhang, J. Zhou, H. Qi, H. Chen, T. Kondo, X. Chen and B. Chu, *Adv. Mater.*, 2007, **19**, 821.
48. J. Zhou, L. Zhang and J. Cai, *J. Polym. Sci. Part B: Polym. Phys.*, 2004, **42**, 347.
49. J. Cai, L. Zhang, S. Liu, Y. Liu, X. Xu, X. Chen, B. Chu, X. Guo, J. Xu, H. Cheng, C. C. Han and S. Kuga, *Macromolecules*, 2008, **41**, 9345.
50. B. Laszkiewicz, *J. Appl. Polym. Sci.*, 1998, **67**, 1871.
51. A. Isogai and R. H. Atalla, *Cellulose*, 1998, **5**, 309.
52. J. Zhou, Y. Qin, S. L. Liu and L. Zhang, *Macromol. Biosci.*, 2006, **6**, 84.
53. Y. Wang, Y. Zhao and Y. Deng, *Carbohydr. Polym.*, 2008, **72**, 178.
54. Y. Zhao, Y. Wang, J. Y. Zhu, A. Ragauskas and Y. Deng, *Biotechnol. Bioeng.*, 2008, **99**, 1320.
55. S. Walseth, *Tappi J.*, 1952, **35**, 228.
56. D. Klemm, B. Philipp, T. Heinze, U. Heinze and W. Wagenknecht, in *Comprehensive Cellulose Chemistry*, Wiley-VCH, Weinheim, 1998, vol. 1.
57. Y. H. P. Zhang, J. Cui, L. R. Lynd and L. R. Kuang, *Biomacromolecules*, 2006, **7**, 644.
58. Y. H. P. Zhang and L. R. Lynd, *Biomacromolecules*, 2005, **6**, 1510.
59. Y. H. P. Zhang and L. R. Lynd, *Biotechnol. Bioeng.*, 2006, **94**, 888.

60. Y. H. P. Zhang, Z. Zhu, J. Rollin and N. Sathitsukasanoh, in *Cellulose Solvents: for Analysis, Shaping and Chemical Modification*, ed. T. F. Liebert, T. J. Heinze and K. J. Edgar, American Chemical Society, Washington DC, 2010, p. 365.
61. J. Blackwell, in *Cellulose Chemistry and Technology*, ed. J. C. Arthur Jr, American Chemical Society, Washington DC, 1977, p. 206.
62. Y. H. P. Zhang, S. Y. Ding, J. R. Mielenz, J. B. Cui, R. T. Elander, M. Laser, M. E. Himmel, J. R. McMillan and L. R. Lynd, *Biotechnol. Bioeng.*, 2007, **97**, 214.
63. G. Moxley, Z. Zhu and Y. H. P. Zhang, *J. Agric. Food Chem.*, 2008, **56**, 7885.
64. N. Sathistukasnoh, Z. Zhu, S. Wi and Y. H. P. Zhang, *Biotechnol. Bioeng.*, 2011, **108**, 521.
65. Z. Zhu, N. Sathistukasnoh, T. Vinzant, D. J. Schell, J. D. McMillan and Y. H. P. Zhang, *Biotechnol. Bioeng.*, 2009, **103**, 715.
66. N. Sathitsuksanoh, Z. Zhu, T. J. Ho, M. D. Bai and Y. H. P. Zhang, *Bioresour. Technol.*, 2010, **101**, 4926.
67. H. Li, N. J. Kim, M. Jiang, J. W. Kang and H. N. Chang, *Bioresour. Technol.*, 2009, **100**, 3245.
68. T. H. Kim and Y. Y. Lee, *Bioresour. Technol.*, 2005, **96**, 2007.
69. A. Rollin, Z. Zhu, N. Sathitsuksanoh and Y. H. P. Zhang, *Biotechnol. Bioeng.*, 2011, **108**, 22.
70. A. El-Seoud, A. Koschella, L. C. Fidale, S. Dorn and T. Heinze, *Biomacromolecules*, 2007, **8**, 2629.
71. R. P. Swatloski, S. K. Spear, J. D. Holbrey and R. D. Rogers, *J. Am. Chem. Soc.*, 2002, **124**, 4974.
72. A. Pinkert, K. N. Marsh, S. Pang and M. P. Staiger, *Chem. Rev.*, 2009, **109**, 6712.
73. L. Feng and Z. L. Chen, *J. Mol. Liq.*, 2008, **142**, 1.
74. R. C. Remsing, R. P. Swatloski, R. D. Rogers and G. Moyna, *Chem. Commun.*, 2006, **28**, 1271.
75. H. B. Liu, K. L. Sale, B. M. Holmes, B. A. Simmons and S. Singh, *J. Phys. Chem. B*, 2010, **114**, 4293.
76. T. Heinze, K. Schwikal and S. Barthel, *Macromol. Biosci.*, 2005, **5**, 520.
77. H. Zhang, J. Wu, J. Zhang and J. He, *Macromolecules*, 2005, **38**, 8272.
78. A. P. Dadi, S. Varanasi and C. A. Schall, *Biotechnol. Bioeng.*, 2006, **95**, 904.
79. A. P. Dadi, C. A. Schall and S. Varanasi, *Appl. Biochem. Biotechnol.*, 2007, **137–140**, 407.
80. S. Barthel and T. Heinze, *Green Chem.*, 2006, **8**, 301.
81. S. H. Ha, N. L. Mai, G. An and Y. M. Koo, *Bioresour. Technol.*, 2011, **102**, 1214.
82. S. H. Lee, T. V. Doherty, R. J. Lindardt and J. S. Dordick, *Biotechnol. Bioeng.*, 2009, **102**, 1368.
83. Q. Li, Y. C. He, M. Xian, G. Jun, X. Xu, J. M. Yang and L. Z. Li, *Bioresour. Technol.*, 2009, **100**, 3570.

84. H. Zhao, G. A. Baker and J. V. Cowins, *Biotechnol. Prog.*, 2010, **26**, 127.
85. N. Sun, M. Rahman, Y. Qin, M. L. Maxim, H. Rodriguez and R. D. Rogers, *Green Chem.*, 2009, **11**, 646.
86. M. Zavrel, D. Bross, M. Funke, J. Büchs and A. C. Spiess, *Bioresour. Technol.*, 2009, **100**, 2580.
87. S. H. Lee and S. B. Lee, *Chem. Commun.*, 2005, **27**, 3469.
88. S. Singh, B. A. Simmons and K. P. Vogel, *Biotechnol. Bioeng.*, 2009, **104**, 68.

CHAPTER 18

What We Can Learn From Natural Biomass-Utilization Systems for Developing Novel Bioreactors

YINHUA WAN,*[a] BENKUN QI,[a] JIANMIN XING,[a] QIANG LIAO,[b] JIANZHONG SUN[c] AND SHULIN CHEN[d]

[a] National Key Laboratory of Biochemical Engineering, Institute of Process Engineering, Chinese Academy of Sciences, Beijing 100190, P. R. China; [b] The State Key Lab of Mechanical Transmission, College of Power Engineering, Chongqing University, Chongqing 40030, P. R. China; [c] Biofuels Institute, Jiangsu University, Zhenjiang 212013, P. R. China; [d] Department of Biological Systems Engineering, Washington State University, Pullman, WA 99164, USA
*Email: yhwan@home.ipe.ac.cn

18.1 Biomimetics: Nature-Inspired Technology Innovation

Throughout 3.4 billion years of evolutionary history, life on Earth has produced a variety of living structures which have experimented with successful solutions to adapt to environmental conditions. This evolutionary information was coded into living organisms' genes and passed from one generation to another. It is useful to investigate the solutions and strategies resulting from long-term nature-inspired evolution, and then possibly apply them to technology to solve the engineering problems that humans encounter. Biological

RSC Energy and Environment Series No. 10
Biological Conversion of Biomass for Fuels and Chemicals: Explorations from Natural Utilization Systems
Edited by Jianzhong Sun, Shi-You Ding and Joy Doran-Peterson
© The Royal Society of Chemistry 2014
Published by the Royal Society of Chemistry, www.rsc.org

systems have superior capabilities and the imitation and adaption of their features and characteristics could significantly improve our technology.[1]

Biomimetics is the study of the structure and function of biological systems as models for the design and engineering of processes, substances, devices, or systems. It covers a highly inter-disciplinary field, involving the understanding of biological functions, structures, and the principles of various objects found in nature, and the design and fabrication of various materials and devices of commercial interest by engineers, material scientists, chemists and others.[2] Therefore, biomimetics can be applied to a large number of technological fields. It is an innovation method and thought to facilitate technological development and foster scientific basics.[3]

18.2 Natural Biomass-Utilization Systems

With the increasing growth in population, and the development of economies across the world, a conflict between our demand for fossil fuels (coal, oil and gas) and their supply has emerged. Burning of fossil fuels produces large amounts of carbon dioxide, which contributes to global warming, and the production of biofuels from renewable feedstock is the only foreseeable source of energy that can replace non-renewable fossil fuels in the long term.

Lignocellulosic plant biomass is composed of 25–55% cellulose, 25–50% hemicellulose and 10–40% lignin, it represents the largest reservoir of potentially fermentable sugars on Earth.[4] The main technological bottleneck that impedes the large-scale bioconversion of lignocellulosic feedstock for the production of fuels and other commodity products is the lack of economical and efficient technologies to break down the recalcitrance of lignocellulose for the efficient production of fermentable sugars. Pretreatment is a prerequisite step during the bioconversion process, and is required because of the complexity of the structure and components of cellulosic materials. The aim of pretreatment is to disrupt the shield formed by lignin and hemicellulose, reduce the degrees of crystallinity and polymerization of cellulose, and to make cellulose more susceptible towards enzymatic attack.[5] Pretreatment has been estimated as the most expensive step during the lignocellulose-to-ethanol bioconversion processes, accounting for about 20% of the total cost.[6] Over the years, a large number of pretreatment methods have been developed and can be classified into four types: biological; chemical; physical; and physicochemical.[7] The current technology in lignocellulose pretreatment, however, still has some severe limitations that need to be overcome, including: the requirement for high-temperature and high-pressure reaction conditions; the loss of fermentable sugars; the need for the addition of one or more chemicals; and the production of inhibitory compounds, *etc.*[8–10] Enzymatic hydrolysis is another important step during lignocellulose bioconversion. During this step, cellulose and hemicellulose are converted into their sugar constituents, which can be fermented by microbial biocatalysts to produce a wide variety of biofuels and chemicals.[11] Nevertheless, the cost and hydrolyzing efficiency of cellulase is quite far removed from large-scale commercial digestion of lignocellulosic feedstock. Even if significant progress has been made in the reduction of the cost of cellulase, these catalytic proteins are still expensive.[12] Summing up the above, it can be concluded that energy-efficient and cost-effective biochemical utilization of lignocellulosic biomass still requires important breakthroughs in several processes, in particular, pretreatment technology, efficient biocatalysts, and efficient and continuous bioreactor design for the digestion of a variety of lignocellulosic biomass types.[13]

In nature, there is a large number of existing good examples of the efficient utilization of plant biomass. Of these examples, herbivorous mammals and lignocellulose-feeding insects are the most worth mentioning.[13,14] Ruminants and termites, as representatives of herbivorous mammals and lignocellulose-feeding insects, respectively, possess incredible lignocellulose-degradation capabilities.[15–17] Ruminants, subsisting on a diet of herbaceous plants, can be considered as having the world's largest commercial fermenter in their rumen, with a net volume of about 2×10^{11} L.[18] Termites are the most efficient cellulose digester reported to date. Cellulose digestion is in the range of 74–99% for lower termites and 91–97% for higher termites.[4] It has been reported that the volumetric rate of lignocellulose degradation measured from the VFA (volatile fatty acid) production rate in the rumen is around 18 g chemical oxygen demand (COD)-based VFA (VFA-COD)/$L_{rumen} \cdot d$ (Lrumen is the volume of rumen, L; d is the time of lignocellulosic degradation, day), while the estimated value for termites is 225 g VFA-COD/$L_{tract} \cdot d$ (Ltract is the volume of termites' tract, L), more than 12 times that of the rumen.[19]

Lignocellulose digestion by both herbivorous mammals and lignocellulose-feeding insects has been extensively investigated in the past half a century. The mechanisms of breakdown of cellulosic substrates are being gradually uncovered. No doubt, the results from this research will inevitably shed light on overcoming the current technology bottlenecks in lignocellulose bioconversion. This chapter mainly address the novel bioreactors developed from processing bionics by studying natural biomass-utilization systems, with an emphasis on the simulation of mammalian herbivores because they have been extensively investigated in this field. This chapter aims to provide an overview of the present development status of the biomimetic reactors used for the digestion of lignocellulosic biomass.

18.2.1 Plant-Based Feed Digestion by Mammalian Herbivores

Mammalian herbivores cluster into two groups, depending on the difference in the gut position where fiber digestion takes place, *i.e.*, foregut fermenters and hindgut fermenters.[20] With ruminants such as sheep, kangaroos, giraffes, and cattle as representatives, foregut fermenters chew their cud and digest the plant-based feed mainly in the rumen. While hindgut fermenters, such as elephants, horses, rhinoceri, mole rats, gorillas and so on, do not chew their cud. Ingested fiber passes through the stomach and small intestine, and fiber digestion takes place in the large intestine.[21] Regular contraction of the caecum and colon mixes the digesta with microbial inocula for digestion. It also promotes the adsorption of fermentation products such as VFAs through the epithelial surface. Plant-based feed digestion by mammalian herbivores is characterized by the fact that the animals themselves do not produce endogenous cellulolytic enzymes, fiber digestion is accomplished by the concerted action of extraneous enzymes secreted by symbiotic microorganisms in their gastrointestinal tracts.[22,23]

Due to the extensive information published on fiber digestion by foregut fermenters in recent years, rumen metabolism will be only summarized in the following. After the plant-based materials are ingested, they first undergo pretreatment before entering the rumen. The pretreatment of biomass is accomplished by mechanical grinding-chewing (mastication), in strong contrast to current lignocellulose bioconversion processes.[18] Upon entering the rumen, the intermittent and powerful motion of the reticulorumen (reticulum and rumen) forces the digesta to be inoculated with microorganisms and thoroughly mixed, which drives the hydrolysis and fermentation of plant-based biomass. During digestion, particle sizes of the digesta are reduced and their densities are increased. The fine and dense substrate particles settle at the bottom of the rumen, while the large particles circulate in the upper portion of rumen. It is during this period that the digesta in the upper reticulorumen likely undergo regurgitation and rechewing.[19] In the rumen, ingested feed stays as long as 72 h due to the retention of particles by the omasum laminae, which is similar to the retention of particles in a continuous fermentor, with the purpose of increasing the opportunities for biocatalytic breakdown of the cellulosic substrate.[24]

It is worth mentioning that the most important feature of rumen digestion is rumination, which involves regurgitating the semi-digested feed from the rumen and chewing it again.[25–27] Rumination has two purposes, one is to increase the digestible surfaces for enzymatic and microbial attack, the other is to release the trapped enzymes and microbes in digested particles, facilitating their adhesion to new digestible surfaces.[19] Microbial fermentation in the rumen produces VFAs, which show inhibitory effects on cellulolytic microorganisms. The rumenant, however, has evolved a problem-solving strategy, that is on-site continuous adsorption of VFAs by the rumen epithelium.[28,29] This process can be seen as an analogy for the reaction–separation coupling process existing in the fields of chemical and biochemical engineering.

18.2.2 Digestion of Lignocellulose by Termites

Compared to plant-based biomass digestion by mammalian herbivores, digestion of lignocellulose by termites is more complex. The ingested wood in the mouth is ground to particles through the action of physical structures, especially the mandibles. Then the wood particles go from salivary glands, through the foregut and the midgut, finally reaching the hindgut where end-product VFAs, mainly acetate, are produced.[30] During the entire digestion process, various lignocellulolytic enzymes, secreted by the termite itself and by symbiotic organisms, are produced in sequence and act synergistically. The polysaccharide fraction of the lignocellulosic substrate is almost totally digested with the help of combined physical, chemical and biochemical actions.[13,31] Watanabe and Tokuda figuratively divide the wood digestion in the termite gut into three stages: mechanical grinding in the mandibles and the proventriculus, hydrolysis of the polysaccharide fraction coupled with product recovery in the midgut (higher termites) and fermentation in the hindgut paunch.[30] For higher termites, the midgut is the main site of cellulose digestion, and the resulting products (mainly glucose and cellobiose) are quickly assimilated through the midgut wall.[32] While for lower termites, enzymatic hydrolysis of cellulose occurs in the hindgut due to the inhabitation of protozoa.[33] The pH of the hindgut ileum is alkaline, from 10 to 12,[34] thus, alkaline treatment of the substrate takes place at this site. The paunch region of the hindgut is not completely anaerobic due to an existing oxygen gradient. This feature is likely related to lignin degradation that requires oxygen.[35,36]

18.3 Bioreactor Design learned from Natural Biomass-Utilization Systems

18.3.1 Anaerobic Bioreactor Design learned from the Rumen

The rumen can be seen as an efficient bioreactor, in which lignocellulosic materials are digested, producing VFAs as the product. In recent years, researchers have developed a variety of bioreactors for the degradation of lignocellulosic substrates with a rumen microbial consortium. These bioreactors range from very simple batch reactors[37–40] to more complex ones which can accomplish continuous operation and on-site removal of VFAs.[41–45] According to differences in application, the developed reactors can be divided into two groups, one is used to simulate rumen fermentation to estimate the amount of fermentation products as nutrients for the host animal; the other is used to improve biogas production from lignocellulosic waste, as used in rumen-derived anaerobic digestion systems.

18.3.1.1 Rumen Simulation Bioreactors

In vivo experimentation is the best method by which to study the interactions between rumen microbes and their hosts. It has some disadvantages, however, such as a long experimental period, difficultly in controlling the experimental conditions and high operational costs. Therefore, there have many attempts to develop simple, stable, and economical fermenter systems that would mimic rumen function. Batch bioreactors are used in the *in vitro* digestion of forage crops,[46,47] but in batch bioreactors, the cultivation time is short because of the decrease in microbial activity, and changes in microflora and end-product inhibition, resulting from the build-up of products and the decrease in pH. Batch reactors do not resemble the rumen at all due to the fact that rumen fermentation is a continuous cultivation system with continuous input of substrates and buffer, and continuous output of chyme.[48] Thus, a continuous-culture apparatus better resembles real rumen reactors. In a continuous-culture bioreactor, substrate and buffer are continuously or semi-continuously fed in, while at the same time accompanied by removal of the fermentation products. Continuous-culture systems can be subdivided into single-flow systems with the same turnover rates for liquids in the chyme, and dual-flow systems with different turnover rates.[49]

As early as the 1960s scientists designed a number of single-flow devices for maintaining the rumen microbial population in continuous culture. In 1963, Rufener Jr et al. described the design, operation, and an apparatus for the continuous culture of mixed rumen microbiota.[50] Using the apparatus, long-term cultivation in vitro was achieved. On the basis of the device reported by Rufener Jr *et al.*, Slyter *et al.* designed an improved and simplified apparatus, whose components were commercially available. It was also simple to construct and operate.[51] Abe and Kumeno invented a permeable continuous-culture apparatus by combining a dialysis system with the continuous culture, which could simulate both the removal of the end-products of fermentation and the flow of ingesta.[52] The biggest problem with single-flow systems is that the same turnover rates for solids and liquids causes a partial loss of protozoal components. The presence of this phenomenon also deviates too much from real rumen metabolism.

During real rumen metabolism, the retention of solids adsorbing microorganisms by omasum laminae guarantees the complete digestion of ingested feed, which in turn leads to different turnover rates for solids and liquids. It has been reported that the turnover rate for liquids is in the range of 4% per h to 10% per h, higher than that for solids, which varies between 2% per h and 7% per h.[53] In a continuous-culture system, a high dilution rate is needed in order to efficiently remove end-products, but this operation leads to the washout of substantial amounts of protozoa. The solids, therefore, need to have a lower turnover rate than liquids, as is the case in the rumen.

It has been reported so far that there are two ways to accomplish the different turnover rates for solids and liquids in dual-flow systems, either by artificially compartmentalizing the rumen by placing the feed in a permeable bag, or by filtering the outflow to retain solids in the fermenter. In the former way, the solid inoculum from the rumen was placed in one nylon or terylene bag, while the feed to be digested was placed in another bag, then both bags were placed in a vessel filled with artificial saliva. Solid digesta residues were manually replaced by a new bag of substrate at various intervals.[54–56] As may be imagined, the introduction of new bags requires opening the apparatus, which exposes the anaerobic rumen microorganisms to oxygen, causing detrimental effects on microbial activity. In view of this limitation, Gizzi *et al.* set up an improved rumen simulation system by using a modified plastic syringe and a tap. The modified apparatus could maintain anaerobic conditions during the whole fermentation period. The apparatus could also automatically correct the acidity produced during fermentation.[57] In addition to the problem of exposure of the fermenter content to oxygen, immobilization of the feed in the bag produces other problems. One is that the solids turnover time is fixed by the replacement interval of the bags. Solids with different digestibilities and densities are all removed, together with important microbial populations that are attached to the solid particles.

In the case of the dual-flow system, Hoover *et al.* conducted pioneering work. They described an apparatus with a dual effluent-removal system, which was designed to simulate the different turnover rate for liquids and solids. In their apparatus, in addition to one overflow port that allowed liquid and solid fermentation media to pass through, each fermentation jar was equipped with a filtered output for the removal of primarily liquid media. This design permits the removal of liquids and solids at different rates. However, the filter may become clogged over a long period of operation because of the heterogeneity of the fermentation media, thus requiring replacement at 48 h intervals.[58] In the rumen, different turnover rates for liquids and solids are achieved by the natural stratification of the fermentation contents. On the basis of above mechanism, Teather and Sauer developed an artificial rumen design that could function in the same manner in the rumen.[59] The critical component was a custom-fabricated 'T' glass fitted in the bottom third of the fermenter as the overflow. Level with the overflow, there was a helical blade that gently mixed the feed material by slicing through it. The bioreactor allowed the maintenance of the natural stratification of solids into floating, suspended and sedimented components according to their density. Solids subject to thorough digestion have greatly reduced particle sizes, thus tend to sink to the bottom of the reactor. The solid particles left the fermenter according to their gravity, similar to the manner that feed particles pass through the rumen. The advantage of this apparatus over previous designs is that it succeeds in maintaining rumen microbial populations and fermentation parameters at normal levels. Nevertheless, the fermenter outflow is susceptible to clogging, especially when fiber-rich substrates are used for fermentation. Based on the apparatus designed by Teather and Sauer, Muetzel *et al.* described a modified continuous-culture system, in which program-controlled stirring speeds allowed the formation of a raft mat where protozoa could survive, similar to that found in the rumen. The fermenter featured an automatically controlled feed-dispensing system and a non-clogged outflow.[60] It seemed to resemble rumen fermentation quite well in terms of end-product composition and some enzyme activities, but a change in the microbial community composition was reported by the authors.

Generally, the widely used continuous-culture bioreactor consists of eight units, *i.e.*, the fermentation unit; the temperature control unit; the pH control unit; the stirring unit; the buffer transportation unit; the gas transportation unit; the effluent collection unit; and the gas collection unit. Different systems vary in the working volume of the fermenter, the stirring manner, the compositions of buffer, the type of gas input to the fermenter, the outflow of solids and liquids, the presence of a dialysis system or not, *etc.* Table 18.1 lists the key parameters of some rumen simulation bioreactors reported in the literature. Researchers designed these apparatus to investigate the metabolism of ingested feed in the rumen. It has been found that eukaryotic organisms, especially protozoa and fungi, play an important role in the digestion of lignocellulosic materials using these reactors.[52,55,58]

Table 18.1 Key parameters of some rumen simulation bioreactors reported in the literature.

Stirring manner	Gas type	Heating manner	Outflow of liquids and solids	Composition of buffer	Dialysis system	Reference
Magnetic stirring	—	Water bath	Single flow	60% Artificial saliva, 40% tap Water	Yes	51
Mechanical stirring	N_2	Warm water spray	Dual flow	60% Artificial saliva, 40% tap Water	No	58
Mechanical stirring	95 N_2; 5% CO_2	Water bath	Single flow	Artificial saliva	Yes	52
Mechanical stirring	—	Water bath	Dual flow	Artificial saliva	No	54
Mechanical stirring	N_2	Water bath	Dual flow	Artificial saliva	No	59
Mechanical stirring	CO_2	Water bath	Single flow	Artificial saliva	No	61
Mechanical stirring	CO_2	Electric heating	Dual flow	Phosphate buffer	Yes	62
Mechanical stirring	N_2	Water bath	Dual flow	Artificial saliva	No	57
Mechanical stirring	CO_2	Water jacket	Dual flow	Bicarbonate–phosphate buffer	No	60
No stirring	No gas	Water bath	Single flow	60% Artificial saliva, 40% tap water	Yes	50
Mechanical stirring	95 N_2; 5% CO_2	Water bath	Dual flow	Bicarbonate–phosphate buffer	No	56

18.3.1.2 *The Rumen-Derived Anaerobic Digestion System*

As mentioned previously, the rumen fermentation system is one of the most efficient ecosystems in terms of the degradation of lignocellulosic materials in nature. Its high efficiency results from the synergistic action of a variety of rumen microorganisms, including bacteria, fungi, protozoa, archeaea, *etc.*[63,64] For this reason, the application of rumen populations as inocula for the digestion of agro-industrial cellulosic residues in industrial-scale digesters has been the research target of investigators since the 1980s. However, to realize this target, the following two requirements must be met.[65] One is that the growth and performance of microbial biomass in *in vitro* fermentation should be similar to that *in vivo* when replacing natural substrates like grass with lignocellulosic waste; the other is that the digester should perform well over long periods of operation.

The rumen-derived anaerobic digestion (RUDAD) system is the most well-known type of anaerobic digestion system using rumen inocula, apart from rumen simulation technology. The RUDAD reactor configuration is a hydraulic flush reactor, with solid and hydraulic retention times uncoupled by means of a filter. It should be noted that unlike rumen simulation technology, the RUDAD systems developed to date do not implement the key parameters of rumen digestion in addition to using rumen microorganisms as inocula, and they are mainly used for the production of biogas, a renewable energy.

To the best of our knowledge, biogas production from a one-stage RUDAD system was first reported by Barnes and Keller.[66] In 2004, they described a laboratory-scale anaerobic sequencing batch reactor (ASBR) operating with a rumen-based microbial inoculum. They found that the VFA production rate from cellulose in the ASBR was comparable with that obtained in a continuous system with solid/liquid separation.[66] Fantozzi and Buratti also reported biogas production from the digestion of animal and vegetal biomass with rumen fluid in a single-stage continuously stirred tank reactor (CSTR).[67]

The most commonly used RUDAD is a two-stage system developed by Gijzen *et al.*[68] In one reactor, the waste cellulose was hydrolyzed and converted to VFAs using rumen-based inocula, then the VFA-rich liquid effluent was shifted to another reactor for the production of biogas. In comparison to the one-stage system, two-stage anaerobic digestion shows improved substrate conversion and process stability, protects methanogen from shock loads, removes compounds toxic to methanogens, and enriches biogas.[69] Multiple studies have been performed on the application of two-stage RUDAD systems and have shown promising results. Gijzen *et al.* developed a novel high-rate reactor, which can be used during the first stage of the two-stage system. The reactor showed a high conversion rate of cereal straws into VFAs when employing rumen microorganisms.[70] When the high-rate reactor was combined with a high-rate upflow anaerobic sludge blanket (UASB) methanogenic reactor, the highly efficient conversion of filter paper cellulose into biogas was achieved.[68] With the cellulosic fraction of domestic refuse as a substrate, results from Zwart *et al.* demonstrated the stability of two-stage RUDAD reactors over extended periods of time.[71] Recently, Nair *et al.* developed a novel two-stage anaerobic digestion system which can accommodate high substrate loading rates, superior to conventional systems.[44] From the available references, it seems that the two-stage RUDAD progress is a promising prospect for industrial-scale applications. However, problems are likely to be encountered when scaling up this process because the required area of filter, used for filtration of the liquid effluent from the first stage, would be greater than that of the reactor itself, due to the low flux limitations of the filter.[72]

18.3.2 Biomimetic Design learned from Lignocellulose-Feeding Termites

To the best of our knowledge, the present research regarding termites has focused on lignocellulolytic enzymes and the microorganisms present in the digestive tract of termites. There is little literature available on biomimetic design derived from lignocellulose-feeding termites. Therefore, in this section, we provide the assumption on biomimetic design deduced from reported references regarding digestion of cellulosic materials by termites.

Pretreatment of lignocellulosic feedstock is a prerequisite for efficient enzymatic hydrolysis and the result of pretreatment has a direct effect on the cost of down-stream unit operations.[73] It has long been recognized that biological pretreatment is less energy intensive and more economical compared to other pretreatments. Fungal pretreatment usually requires several weeks, much longer than thermochemical pretreatments. Lignocellulose degradation by termites, however, occurs in hours.[74] The high efficiency demonstrated by lignocellulose-feeding termites shows great potential for future applications. The pretreatment strategy used by termites is to modify the lignin structure, instead of degrading it, through the collaborative action of host and symbiont-derived enzymes in the alimentary canal.[15] Recently, Ke *et al.* reported the combined process of the biological modification of lignin and physical chewing during the chewing process of the lower termite *Coptotermes formosanus*.[75] This finding opens up novel viewpoints for developing effective biomass pretreatment technologies. With the mechanisms of lignin unlocking being gradually revealed, termite-based pretreatment strategies could be mimicked in the near future. The development of termite-inspired low-cost, high-efficiency and environmentally friendly pretreatment technologies will most likely become a research hotspot in the next decade.

18.3.3 The Application of Chemical Reactor Theory to Natural Biomass-Utilization Systems

In order to discover how variations in gut configuration and digestive tactics maximize an animal's net rate of energy conversion, researchers modeled animal guts with chemical reactor components and then used the principles of reactor design to identify the variables that characterize digestive strategies.[76] Three kinds of ideal reactors recognized by chemical reactor theory are most frequently modeled. They are batch reactors (BRs), plug flow reactors (PFRs) and CSTRs.[76,77] In a BR, materials are added to the reaction vessel and mixed thoroughly. At the end of the reaction, the resulting products and any unreacted raw materials are all removed. In a PFR, materials enter and flow continuously in a tuber reaction vessel and leave the vessel in the same order that they entered. Materials are perfectly mixed in the radial direction, but mixing or diffusion along the flow path is negligible. In a CSTR, materials flow through the reaction vessel steadily and are mixed completely. The three ideal reactors can serve, separately or in combination, as the model of portions of digestive tracts of particular animals. Investigators use mass balance laws and chemical kinetics to predict the performance of gut reaction models with respect to digestive reactions.[76] Penry and Jumars proposed that the ruminant gut be modeled as two reactors in series, a CSTR followed by a PFR, while the gut of hindgut fermenters be modeled as a PFR followed in series by a CSTR.[76] Alexander modeled the gut structure of higher vertebrate herbivores as CSTR–PFR–CSTR in series. Fermentation occurred in the CSTRs, and digestion but no fermentation occurred in the PFR.[78] By comparing the passage of fluid and particulate materials through the hindgut of mammals to chemical reactors, Hume concluded that the small intestine resembled a PFR; movement of digesta through diverticula such as the cecum best fit a CSTR model; and flow of digesta in the proximal colon was similar to a modified PFR because the incoming reactants were mixed both radially and axially.[79] Jumars introduced axial variation in a tubular gut by modeling it as a sequence of discrete CSTRs connected in series.[80] Logan developed a time-dependent PFR model that could fit the performance of tubular animal guts.[81]

18.4 Summary

At present, our society is facing various unprecedented challenges, of which, energy shortages, urgently need to be addressed. For years, researchers across the world have been focusing their effort on developing highly efficient bio-conversion systems for the production of renewable biofuels from lignocellulosic biomass. However, to date, the developed systems are far behind natural biomass-utilization systems. Rumen and guts of termites can be viewed as the most efficient bioreactors for lignocellulose bioconversion in nature. Their high efficiency results from not only the favorable reaction conditions in these natural bioreactors, but also from the special configurations of the organs involved and the concerted action of the various microorganisms inhabiting these organs. The former can be easily modeled as a bioreactor, but as for the latter, there is still huge amounts of work to be done. On the one hand, we need to further deepen our understanding of lignocellulose digestion mechanisms in natural biomass-utilization systems through intensive research. On the other hand, we need to simulate the natural-biomass utilization systems as closely as possible through novel biomimetic design. With the concerted effort of scientists and engineers, efficient and cost-effective bioconversion of lignocellulosic biomass will be realized in the near future.

References

1. Y. Bar-Cohen, *Bioinspiration Biomimetics*, 2006, **1**, P1.
2. B. Bhushan, *Philos. Trans. R. Soc., A*, 2009, **367**, 1445.
3. H. Stachelberger, P. Gruber and I. C. Gebeshuber, in *Biomimetics–Materials, Structures and Processes: Examples, Ideas and Case Studies*, ed. P. Gruber, D. Bruckner, C. Hellmich, H.-B. Schmiedmayer, H. Stachelberger and I. C. Gebeshuber, Springer-Verlag, Berlin, 2011.
4. Y. Sun and J. Cheng, *Bioresour. Technol.*, 2002, **83**, 1.
5. P. Kumar, D. M. Barrett, M. J. Delwiche and P. Stroeve, *Ind. Eng. Chem. Res.*, 2009, **48**, 3713.
6. B. Yang and C. E. Wyman, *Biofuels Bioprod. Biorefin.*, 2008, **2**, 26.

7. P. Alvira, E. Tomas-Pejo, M. Ballesteros and M. J. Negro, *Bioresour. Technol.*, 2010, **101**, 4851.
8. V. B. Agbor, N. Cicek, R. Sparling, A. Berlin and D. B. Levin, *Biotechnol. Adv.*, 2011, **29**, 675.
9. A. T. W. M. Hendriks and G. Zeeman, *Bioresour. Technol.*, 2009, **100**, 10.
10. L. da Costa Sousa, S. P. S. Chundawat, V. Balan and B. E. Dale, *Curr. Opin. Biotechnol.*, 2009, **20**, 339.
11. B. Qi, X. Chen and Y. Wan, *Bioresour. Technol.*, 2010, **101**, 4875.
12. J. D. Stephen, W. E. Mabee and J. N. Saddler, *Biofuels Bioprod. Biorefin.*, 2012, **6**, 159.
13. J. Sun and X. J. Zhou, in *Recent Advances in Entomological Research: From Molecular Biology to Pest Management*, ed. T.-X Liu and L. Kang, Higher Education Press, Beijing, 2010, p. 261.
14. M. Morrison, P. B. Pope, S. E. Denman and C. S. McSweeney, *Curr. Opin. Biotechnol.*, 2009, **20**, 358.
15. M. E. Scharf and D. G. Boucias, *Insect Sci.*, 2010, **17**, 166.
16. M. Ohkuma, *Appl. Microbiol. Biotechnol.*, 2003, **61**, 1.
17. J. B. Russell, R. E. Muck and P. J. Weimer, *FEMS Microbiol. Ecol.*, 2009, **67**, 183.
18. P. J. Weimer, J. B. Russell and R. E. Muck, *Bioresour. Technol.*, 2009, **100**, 5323.
19. A. Bayane and S. R. Guiot, *Rev. Environ. Sci. Bio/Technol.*, 2011, **10**, 43.
20. R. E. Ley, M. Hamady, C. Lozupone, P. J. Turnbaugh, R. R. Ramey, J. S. Bircher, M. L. Schlegel, T. A. Tucker, M. D. Schrenzel and R. Knight, *Science*, 2008, **320**, 1647.
21. C. E. Stevens and I. D. Hume, *Comparative Physiology of the Vertebrate Digestive Systems*, 2nd edn, 2004, Cambridge University Press, Cambridge.
22. G. Iason, *Proc. Nutr. Soc.*, 2005, **64**, 123.
23. G. R. Iason and S. E. Van Wieren, *Digestive and Ingestive Adaptations of Mammalian Herbivores to Low-Quality Forage*, Blackwell Publishing, Oxford, 1999, p. 337.
24. R. I. Mackie, *Integr. Comp. Biol.*, 2002, **42**, 319.
25. J. G. Welch, *J. Anim. Sci.*, 1982, **54**, 885.
26. K. Chap, L. P. Milligan and P. M. Kennedy, *Can. J. Anim. Sci.*, 1984, **64**, 339.
27. D. H. Bae, J. G. Welch and B. E. Gilman, *J. Dairy Sci.*, 1983, **66**, 2137.
28. A. C. Storm, N. B. Kristensen and M. D. Hanigan, *J. Dairy Sci.*, 2012, **95**, 2919.
29. C. E. Stevens and B. K. Stettler, *Am. J. Physiol.*, 1966, **210**, 365.
30. H. Watanabe and G. Tokuda, *Annu. Rev. Entomol.*, 2010, **55**, 609.
31. A. Brune, *Trends Biotechnol.*, 1998, **16**, 16.
32. A. Fujita, M. Hojo, T. Aoyagi, Y. Hayashi, G. Arakawa, G. Tokuda and H. Watanabe, *J. Wood Sci.*, 2010, **56**, 222.
33. T. Yoshimura, *Wood Res. (Kyoto, Jpn.)*, 1995, **82**, 68.
34. J. F. Harrison, *Annu. Rev. Entomol.*, 2001, **46**, 221.

35. T. Kuhnigk, E. M. Borst, A. Ritter, P. Kämpfer, A. Graf, H. Hertel and H. König, *Syst. Appl. Microbiol.*, 1994, **17**, 76.
36. A. Brune, E. Miambi and J. A. Breznak, *Appl. Environ. Microbiol.*, 1995, **61**, 2688.
37. Z. H. Hu, H. Q. Yu, Z. B. Yue, H. Harada and Y. Y. Li, *Biochem. Eng. J.*, 2007, **37**, 219.
38. Z. H. Hu, G. Wang and H. Q. Yu, *Biochem. Eng. J.*, 2004, **21**, 59.
39. C. O'Sullivan, P. C. Burrell, W. P. Clarke and L. L. Blackall, *Bioresour. Technol.*, 2008, **99**, 4723.
40. A. Bernalier, G. Fonty and P. Gouet, *Anim. Feed Sci. Technol.*, 1991, **32**, 131.
41. L. P. Broudiscou, Y. Papon and A. F. Broudiscou, *Reprod., Nutr., Dev.*, 1999, **39**, 255.
42. G. Colon and J. C. Sager, *Life Support Biosphere Sci.*, 2001, **7**, 291.
43. R. Dalhoff, A. Rababah, V. Sonakya, N. Raizada and P. A. Wilderer, *Water Sci. Technol.*, 2003, **48**, 163.
44. S. Nair, Y. Kuang and P. Pullammanappallil, *Environ. Technol.*, 2005, **26**, 1003.
45. R. C. Araujo, S. Calsamiglia, M. Rodriguez-Prado, S. Cavini and A. Ferret, *J. Dairy Sci.*, 2010, **93**, 706.
46. J. M. A. Tilley and R. A. Terry, *Grass Forage Sci.*, 1963, **18**, 104.
47. K. H Menke, L. Raab, A. Salewski, H. Steingass, D. Fritz and W Schneider, *J. Agric. Sci.*, 1979, **93**, 217.
48. R. E. Hungate, *The Rumen and its Microbes*, Academic Press, New York, 1966, p. 533.
49. Y.-H. Jiang, W.-J. Shen, D.-P. Bu, H.-J. Yang and J. Shen, *Chin. Anim. Husb. Vet. Med.*, 2011, **38**, 212.
50. W. H. Rufener Jr, W. O. Nelson and M. J. Wolin, *Appl. Microbiol.*, 1963, **11**, 196.
51. L. L. Slyter, W. O. Nelson and M. J. Wolin, *Appl. Microbiol.*, 1964, **12**, 374.
52. M. Abe and F. Kumeno, *J. Anim. Sci.*, 1973, **36**, 941.
53. D. C. Church, *The Ruminant Animal: Digestive Physiology and Nutrition*, Prentice Hall, New Jersey, 1988, p. 145.
54. J. W. Czerkawski and G. Breckenridge, *Br. J. Nutr.*, 1977, **38**, 371.
55. R. A. Weller and A. F. Pilgrim, *Br. J. Nutr.*, 1974, **32**, 341.
56. J. H. Aafjes and J. K. Nijhof, *Br. Vet. J.*, 1967, **123**, 436.
57. G. Gizzi, R. Zanchi and F. Sciaraffia, *Anim. Feed Sci. Technol.*, 1998, **73**, 291.
58. W. H. Hoover, B. A. Crooker and C. J. Sniffen, *J. Anim. Sci.*, 1976, **43**, 528.
59. R. M. Teather and F. D. Sauer, *J. Dairy Sci.*, 1988, **71**, 666.
60. S. Muetzel, P. Lawrence, E. M. Hoffmann and K. Becker, *Anim. Feed Sci. Technol.*, 2009, **151**, 32.
61. M. Fuchigami, T. Senshu and M. Horiguchi, *J. Dairy Sci.*, 1989, **72**, 3070.
62. T. Hino, M. Sugiyama and K. Okumura, *J. Gen. Appl. Microbiol.*, 1993, **39**, 35.

63. P. N. Hobson, in *The Rumen Microbial Ecosystem*, ed. P. N. Hobson and C. S. Stewart, Chapman & Hall, London, 2nd edn, 1997, p. 1.
64. C. G. Orpin, *Anim. Feed Sci. Technol.*, 1984, **10**, 121.
65. H. J. Gijzen, K. B. Zwart, P. T. Gelder and G. D. Vogels, *Appl. Microbiol. Biotechnol.*, 1986, **25**, 155.
66. S. P. Barnes, J. Keller, S. J. Hall and C. Fux, *Anaerobic Rumen SBR for Degradation of Cellulosic Material*, IWA Publishing, London, 2004, p. 305.
67. F. Fantozzi and C. Buratti, *Bioresour. Technol.*, 2009, **100**, 5783.
68. H. J. Gijzen, K. B. Zwart, F. J. M. Verhagen and G. P. Vogels, *Biotechnol. Bioeng.*, 1988, **31**, 418.
69. A. K. Kivaisi and M. Mtila, *World J. Microbiol. Biotechnol.*, 1997, **14**, 125.
70. H. J. Gijzen, H. J. Lubberding, F. J. Verhagen, K. B. Zwart and G. D. Vogels, *Biol. Wastes*, 1987, **22**, 81.
71. K. B. Zwart, H. J. Gijzen, P. Cox and G. D. Vogels, *Biotechnol. Bioeng.*, 1988, **32**, 719.
72. M. Walker, Ph. D thesis, University of Southampton, 2008.
73. B. D. Solomon, J. R. Barnes and K. E. Halvorsen, *Biomass Bioenergy*, 2007, **31**, 416.
74. S. Chen, X. Zhang, D. Singh, H. Yu and X. Yang, *Biofuels*, 2010, **1**, 177.
75. J. Ke, D. D. Laskar, D. Gao and S. Chen, *Biotechnol. Biofuels*, 2012, **5**, 11.
76. D. L. Penry and P. A. Jumars, *BioSci.*, 1986, **36**, 310.
77. I. D. Hume, *Acta Zool. Sin.*, 2002, **48**, 1.
78. R. M. N. Alexander, L. De Bruyn, R. N. Hughes, P. W. Skelton and C. G. Jones, *Philos. Trans. R. Soc., B*, 1991, **333**, 249.
79. I. D. Hume, in *Gastrointestinal Microbiology*, ed. R. I. Mackie and B. A. White, Chapman & Hall, New York, 1997, p. 84.
80. P. A. Jumars, *Am. Nat.*, 2000, **155**, 544.
81. D. J. Logan, A. Joern and W. Wolesensky, *J. Theor. Biol.*, 2002, **216**, 5.

CHAPTER 19

Techno-Economic Analysis and Life-Cycle Assessment of Lignocellulosic Biomass to Sugars Using Various Pretreatment Technologies

LING TAO,* ERIC C. D. TAN, ANDY ADEN AND
RICHARD T. ELANDER

National Renewable Energy Laboratory, Golden, CO 80401, USA
*Email: Ling.Tao@nrel.gov

19.1 Introduction

Lignocellulosic biomass has been considered as a renewable source of sugars for biofuels and bio-products. This biomass resource could be increased from the current 473 million dry tons (annually) to nearly 1.1 billion dry tons by 2030 under a conservative set of assumptions about future increases in crop yields.[1,2] It is necessary to overcome the chemical and structural properties that have evolved in biomass to prevent its disassembly,[3] economically and sustainably. Therefore, how to sustainably and cost effectively break down lignocellulosic biomass into usable sugars is the key to sugar upgrading to potentially more value-added renewable biofuels (such as bioethanol and biohydrocarbons) and bio-products (such as biopolymers). To overcome the natural resistance of

RSC Energy and Environment Series No. 10
Biological Conversion of Biomass for Fuels and Chemicals: Explorations from Natural Utilization Systems
Edited by Jianzhong Sun, Shi-You Ding and Joy Doran-Peterson
© The Royal Society of Chemistry 2014
Published by the Royal Society of Chemistry, www.rsc.org

plant cell walls to microbial and enzymatic deconstruction, known as "biomass recalcitrance",[3] a synergistic effort between pretreatment and enzymatic hydrolysis has been studied extensively for the last decade.

Pretreatment of biomass for biofuel production is one of the most crucial steps as it has a significant impact on not only enzymatic hydrolysis, fermentation and down-stream processing,[4] but also on the overall process economics and sustainability perspectives. Typically, hydrolysis yields in the absence of pretreatment are <20% of theoretical yields, whereas yields after pretreatment often exceed 90% of theoretical values.[5] The objective of biomass pretreatment is to make the cellulose and hemicellulose accessible to the enzymes without the unnecessary degradation of sugars to unusable compounds. As the first step in the biochemical conversion process, pretreatment plays a critical role in preparing biomass for enzymatic conversion to C5 and C6 sugars and, in some processes, directly hydrolyzing a portion of structural carbohydrates to oligomeric and monomeric sugars.

The Consortium for Applied Fundamentals and Innovation (CAFI) was formed in 2000[6] to collaboratively develop and publish data on leading biomass pretreatment options. The pretreatment step has been projected to be one of the most expensive capital investments in the biochemical conversion process, as well as having a significant impact on the down-stream conversion steps, such as enzymatic hydrolysis and fermentation. Six leading pretreatment technologies were studied in the course of CAFI's efforts. They were ammonia fiber explosion (AFEX), dilute acid using H_2SO_4 (DA), liquid hot water (LHW), Lime, soaking aqueous ammonia (SAA) or ammonia recycle percolation (ARP), and sulfur dioxide impregnated steam explosion (SO_2). Techno-economic analysis (TEA) has been used as an important tool for R&D facilitation within collaborative projects conducted by the CAFI team. The first CAFI project (CAFI 1) used a consistent source of corn stover to determine glucose and xylose yields from various pretreatment processes upon pretreatment and subsequent enzymatic hydrolysis. TEA for each pretreatment process was conducted using these reported sugar yields.[7] The CAFI project team has recently extended this approach to switchgrass in the CAFI 3 project.[8] However, TEA has not been applied to CAFI 2 projects using hybrid poplar as the feedstock.

Aggressive renewable fuel policies will necessitate unprecedented biomass feedstock production[1] and the installation of fuel conversion and distribution infrastructure. The US cellulosic biofuels industry is currently in its nascent stages. Conversion technologies have been actively developed and process improvements to these technologies have been investigated with the goal of cost optimization, thus minimizing risk and expediting industry learning. Expansion of the cellulosic biofuels industry at the scale needed to meet the Renewable Fuel Standard Goals[9] is expected to come with some environmental impact (see, for example, ref. 10). Hence, in addition to the economic feasibility, overarching concerns such as the environmental sustainability also need to be addressed for sustainable biofuel production. In this chapter, life-cycle assessment (LCA) is conducted to assess the life-cycle impact and environmental sustainability associated with the production of sugar intermediates *via* the evaluated pretreatment technologies. LCA results provide a better understanding of the processes from an environmental aspect, and consequently a more informed assessment and comparison of the technologies can be made.

As the biofuel industry develops, TEA coupled with LCA will play a key role in process development and the targeting of technical and economic barriers for these new fuels and feedstocks.[11] The production cost of sugar from lignocellulosic biomass is important to the biofuels industry. The biomass sugar after pretreatment and enzymatic hydrolysis is often a mixture of different fermentable sugars, degraded sugar products, salts, and other chemicals. The performance of microorganisms in using this sugar mixture has been studied extensively for fermentation to ethanol,[12] and also in other microorganisms for fermenting to products other than ethanol, such as butanol or higher hydrocarbons. It has been demonstrated that not only can glucose and xylose be effectively utilized; minor sugars of arabinose, mannose and galactose could also be utilized.[13] In addition to microbial engineering,[14–16] sugar intermediates can be catalytically upgraded to biofuel or bio-products. For instance, glucose in aqueous solution can be hydrogenated into sorbitol using ruthenium catalysts at 100 °C.[17] Biomass-derived carbohydrates were also converted into liquid alkanes by dehydration/hydrogenation over metal catalysts.[18] Approaches using aqueous solutions of sugars derived from lignocellulosic biomass to produce liquid hydrocarbons have been demonstrated by Virent Inc.,[19] Dumesic and his co-workers,[18,20,21] and a few other researchers.[22–24] Production of other chemicals from the biomass-derived sugars has also been demonstrated.[25] The purpose of this analysis is to estimate the sugar production costs for biofuels and bio-products from lignocellulosic biomass. Down-stream biofuels and bio-product conversion and upgrading processes are not included in the analysis.

In this chapter, the process economic analysis and LCA of the six pretreatment processes studied in the CAFI 2 project are reported using hybrid poplar as the biomass feedstock. The study includes monomer and oligomer sugar yields (glucose, xylose, arabinose, mannose and galactose), total installed capital costs, and the Minimum Sugar Selling Price (MSSP). An integrated, commercial-scale lignocellulosic sugar process module based on each pretreatment process serves as the foundation of the cost analysis.

The National Renewable Energy Laboratory (NREL) has developed a methodology for tracking what it refers to as the "State of Technology" (SOT). This SOT methodology for economic and environmental performance is also applied to the CAFI 2 pretreatments. Using well-documented techno-economic models as the basis, experimentally verified data is used within these models to estimate what the commercial-scale costs might be for certain key levels of technical achievement. In this sense, it allows the impact of research progress on economic and environmental improvements to be quantified.

Process data generated by NREL researchers, along with pilot-scale demonstrations provide critical guidance for annual R&D directions. Extensive R&D has been used for the SOT assessment of the biochemical conversion of cellulosic ethanol for years, providing demonstrated guidance to R&D. Therefore, process economics and LCA using the six leading pretreatment technologies are compared with the 2011 NREL SOT, which serves as the reference case.

It is important to note that the sugar cost reported here is not directly comparable to a traditional commodity sugar cost derived from corn or sugarcane. The sugar derived from lignocellulosic biomass contains a mixture of C5 and C6 sugars with 50% moisture, as well as other contaminants from pretreatment and enzymatic hydrolysis, such as salts, soluble lignin, *etc.* This sugar production may require cleanup for sugar upgrading, at an additional cost.

19.2 Methods

19.2.1 Conceptual Process Design

The process described here uses various pretreatment technologies for lignocellulosic biomass (hybrid poplar), followed by enzymatic hydrolysis (saccharification) of the remaining cellulose (and possibly the remaining hemicelluloses) to sugar products. The process design also includes feedstock handling and storage, sugar product partial dehydration, lignin combustion, product storage, and all other required utilities. A dry metric tonne day of hybrid poplar was chosen, with 8406 operating hours per year. Aspen Plus Version v7.2 [ref. 26] was used to calculate the material and energy balances. NREL's in-house databank was used for the properties of the biomass-related components.[29] For biomass feedstocks, although only minimum storage and feed handling are required in this approach, feedstock composition significantly influences the overall analysis. The studied hybrid poplar contains 43% cellulose, 15% xylan and 29% lignin (by weight). Corn stover was used for the 2011 NREL SOT reference case. It contains 35% cellulose, 21% xylan and 21% lignin. Six leading pretreatment technologies were selected for this study using yield data from the CAFI 2 project: AFEX, DA, LHW, Lime, SAA and SO_2. The feedstock and pretreatment chemical costs are listed in Table 19.1 and typical processing conditions for pretreatment are shown in Table 19.2. Detailed process designs were summarized in the publications from previous CAFI projects[8] for each pretreatment technology. Process design was

Table 19.1 Raw material costs in 2007 US dollars.

Raw material	Price
Corn stover	$58.5/dry ton
Hybrid poplar	$61.6/dry ton
Glucose	526.5/dry ton
Sulfuric acid (93%)	$81.4/dry ton
NH_3	$407.0/ton
Lime	$180.9/ton
SO_2	$275.7/ton
Water	$0.2/ton
Power (by-product credit)	$0.06/kWh

Table 19.2 Summary of process conditions.

Pretreatment technology	Temp/°C	Residence time/min	Catalyst	Catalyst loading/g $100 g^{-1}$ DBP
2011 NREL SOT	158	20	H_2SO_4	2
AFEX	180	10	Concentrated NH_3	39
DA	190	1.1	H_2SO_4	2
Lime	160	120	$Ca(OH)_2$	39
LHW	200	10	Water	—
SAA	185	27.5	Aqueous NH_3	15
SO_2	190	5	SO_2	3

incorporated into the process simulation model. Equipment for the pretreatment area was sized from the material and energy balance calculated by simulation. All other areas were derived from the NREL 2011 design report[27] with changes in the conversion yields for each pretreatment/enzymatic hydrolysis combination, utilities, and related storage and handling equipment for the chemicals used for pretreatment.

The 2011 NREL annual SOT assessment is based on pilot plant demonstrations using corn stover as the biomass feedstock for R&D efforts in 2011. It reflects NREL's best estimate of the production costs in a hypothetical n^{th} plant using the current best slate of demonstrated technical capabilities. Dilute acid is used as the pretreatment technology in the SOT assessment and R&D efforts, in a similar process design to the DA pretreatment used for the CAFI 2 project using poplar feedstocks. This pretreatment is compared with all the CAFI pretreatment conditions in Table 19.2.

Pretreatment chemical recovery systems are also included in the analyses when the recycling results in economical benefit for the overall process. The solids level of enzymatic hydrolysis is managed at 20% total solids maximum, to be consistent even as the pretreatment solids level varies with different technologies. The residence time of enzymatic hydrolysis is 3.5 days with respective enzyme loadings for each biomass feedstock.

The diluted sugar production stream from enzymatic hydrolysis is further concentrated by a triple-effect evaporation system, to achieve 50% water in the final product. Additional units to 'upgrade' the sugar product stream are not considered in this study, because this sugar production stream can also be utilized as an intermediate. It is noted that non-sugar compounds are present in the sugar production streams; additional cost for cleanup may be required.

19.2.2   Process Economic Analysis

The process economic analysis includes the following parts.

(1) Variable operating cost: the variable operating cost is based on material and energy balance calculations from process modeling using Aspen Plus simulations. Raw materials include biomass feedstocks, pretreatment

and neutralization chemicals, and others. Major raw material unit costs are listed in Table 19.1. Utilities include steam (both low- and medium-pressure steam), power, water, and nitrogen gas. The biomass delivery cost to the conversion process was assumed to be $61.57 per dry ton for hybrid poplar as woody biomass, and $58.5 per dry ton for corn stover in the 2011 NREL SOT base case, for the cost year 2007.[28] The enzyme cost was calculated with on-site enzyme production, which is sized mainly based on the enzyme loading. Enzyme cost, contributing significantly to the raw material costs, will be discussed in a later section.

(2) Fixed operating cost: salaries were inflated to 2007 dollars using NREL's 2011 design basis.[27] In addition to salary, the general overhead is a factor of 90% applied to the salary total and covers items such as safety, general engineering, general plant maintenance, and payroll overheads. Annual maintenance materials are estimated at 3% of the total project investment, and property insurance and local property tax are estimated at 0.7% of the total project investment based on standard literature assumptions.[27,29]

(3) Equipment cost and installed cost: based on the material and energy balances, the equipment can be sized appropriately. For the NREL SOT case, the process economics model for annual assessment was converted to cellulosic sugar production, with all demonstrated pretreatment and enzymatic hydrolysis data from NREL's 2011 efforts. Equipment costs were developed using a number of sources, including past vendor quotations (for more specialized equipment) from the corn stover ethanol biochemical design report,[27,29] the CAFI 1 process design,[7] the CAFI 3 process design,[8] the United States Department of Agriculture (USDA) corn ethanol model,[30] as well as costing software estimates (for simpler equipment such as distillation columns, pumps, and tanks), and chemical engineering textbooks.[31] Equipment costs for areas outside of pretreatment were derived from the NREL 2011 design case using the power law to adjust for changes in capacity, such as enzymatic hydrolysis, sugar evaporation and lignin combustion. Equipment quotations obtained in earlier or later years were inflated or deflated to year 2007 dollars using chemical indices. All cost numbers stated here are in 2007 US dollars. The scaling exponent for the power law was obtained from the NREL 2002 and 2011 design cases[27,29] for most of the equipment, and from CAFI projects[7,8] for pretreatment reactors and for equipment for recycling pretreatment chemicals. For equipment not listed in the NREL design cases and for which we were not able to get vendor's guidance, the exponent term was assumed 0.60. Standard NREL factors[29] were used to obtain the total project investment from the purchased equipment costs. However, pretreatment reactor installation factors have been re-evaluated to reflect both proper scaling up and the consideration of construction materials, based on common engineering judgments. Operating costs, revenues, and discounted cash flows were obtained by modifying the NREL 2011 n^{th} plant design.

(4) Discounted cash flow analysis: the method for the discounted cash flow calculation in this study assumed 40% equity financing and 3 years construction plus 0.25 years start-up. The plant life was 30 years. The income tax was 35%. Working capital was 5% of fixed cost investment (FCI). The MSSP is the minimum price that the sugar must sell for in order to generate a net present value (NPV) of zero for a 10% internal rate of return (IRR). This makes the MSSP higher than a true cost of production. In order to be consistent with LCA in the following section, the MSSP is reported on a "per kg sugar" basis (with 50% water content).

Several sensitivity cases are analyzed in comparison with the base case scenario for each pretreatment. It should be emphasized that a certain percentage of uncertainty exists around conceptual cost estimates. These values are best used for relative comparisons against technological variations or process improvements. Use of absolute values without a detailed understanding of the basis behind them can be misleading.

19.2.3   Life-Cycle Assessment

This study used the SimaPro v7.3 life-cycle assessment (LCA) modeling software[32] to develop and link unit processes. The Ecoinvent database[33] was used for materials and processes that were not developed by the authors. Life cycle inventory (LCI) data for the conversion processes were based on process modeling outputs from Aspen Plus. A detailed description of the LCA modeling approach can be found in ref. 34. The modeling boundary for this study is from field-to-plant-gate including embodied energy and material flows. The down-stream processes of distribution and end-use are not included. The functional unit is 1 kg of sugars product (five monomer sugars). The reference case evaluated in this study is the 2011 NREL annual SOT assessment (*i.e.*, NREL 2011 SOT), which is based on pilot plant demonstrations using corn stover as a biomass feedstock for year 2011 R&D efforts. In addition to SimaPro quantification of the LCA metrics of global-warming potential (GWP) and fossil energy demand, Aspen Plus was used to estimate direct greenhouse gas (GHG) emissions (CO_2), consumptive water use, and select criteria air pollutant (CAP) emissions (NO_2 and SO_2).

19.3   Results and Discussion

Each pretreatment for TEA was modeled in the framework of NREL's 2011 biochemical conversion design model.[27] The process design in this study is shown in Figure 19.1. For the production of advanced biofuels and bio-products, it is important to define the sugars that can be utilized in the down-stream sugar upgrading. Depending on applications, some of the sugars produced from pretreatment and enzymatic hydrolysis of lignocellulosic biomass can be utilized, but others cannot. For instance, for the cellulosic ethanol

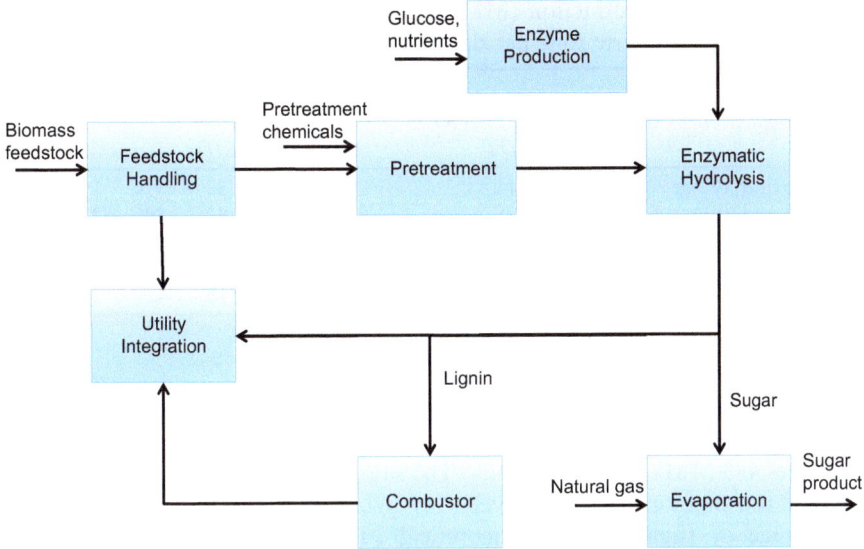

Figure 19.1 Simplified process flow diagram of lignocellulosic biomass to sugar.

conversion process, both xylose and glucose can be fermented to ethanol using *Zymomonas mobilis* (*Z. mobilis*) strains. However, it is still unclear that *Z. mobilis* can convert arabinose, galactose and mannose to the same conversion level of that of xylose. For other sugar-upgrading technologies (such as liquid-phase catalytic reforming), all five sugars (glucose, xylose, arabinose, mannose and galactose) have the potential to be converted to value-added products, potentially including both of the monomer and oligomer forms, as well as other carbonaceous materials. Therefore, four different sugar productions are reported in the cost study:

- xylose and glucose production;
- xylose, glucose monomer and oligomer production;
- five monomer sugar production (the basis for LCA); and
- five monomer and oligomer sugar production.

In the Discussion section of this chapter, the enzyme contribution, capital cost comparison, and LCA-related study are based on five monomer sugar production only. Additionally, no optimization or systematic design was applied to any of the pretreatments for the production of sugars. The sugar yields and process economic data presented here may or may not represent the optimized process results for each of the pretreatments studied.

19.3.1   Sugar Yields

Hybrid poplar has the highest content of cellulose (43%) and lignin on a dry weight basis, but it also has the highest moisture level (50%). The theoretical

Table 19.3 Sugar production (MMkg per year) for a conversion plant using 2000 metric tons per day of biomass feedstock.

Sugar production/MMkg per year	2011 NREL SOT	Pretreatment technology					
		AFEX	DA	LHW	Lime	SAA	SO₂
Glucose and xylose	344	235	368	234	374	197	404
Glucose and xylose & their oligomers	365	248	368	282	396	256	424
Five monomer sugars	370	257	398	250	402	210	430
All mono and oligomer sugars	394	274	398	315	427	288	459

annual sugar production is 495 MMkg (million kg) per year using 2000 dry metric tons per day of biomass feedstock. Comparably, corn stover feedstock produces 483 MMkg sugar annually. The sugar yields are 641 and 625 kg sugar per dry ton of biomass, respectively, for hybrid poplar and corn stover. The percentage of sugar yield to theoretical sugar production is presented in Table 19.3 for glucose, xylose, and the sum of five monomer sugars' yields. The conversions of arabinan, mannan and galactan to their respective monomer and oligomer sugars are assumed to be the same as those for xylan.

Both monomer and oligomer yields of glucose and xylose are shown in Figure 19.2. The sugar yields of glucose and xylose have noticeable variability for all six pretreatments, even if both oligomer and monomer sugars are considered. Compared to the other pretreatments, DA has the lowest yields of oligomer sugars, while LHW has the highest yields of oligomer glucose and xylose. If both monomer and oligomer sugars can be utilized efficiently in the down-stream process, Lime pretreatment will have better sugar yields than DA, potentially resulting in cost-effective sugars as the intermediates. DA, Lime and SO₂ pretreatment technologies have better sugar yields than the other pretreatments, namely AFEX, SAA and LHW. Alkaline pretreatment appears more effective on agricultural residues and herbaceous crops than on woody biomass, as woody biomass containing more lignin. The high lignin content in hybrid poplar probably contributes to the low sugar yield using AFEX or ARP/SAA pretreatments, shown in Figure 19.2.

As stated above, sugar yields have the most significant impact on sugar cost. The sugar yields are shown in Figure 19.3 for both monomers and oligomers. The total sugar yields of both monomers and oligomers are comparable for DA, Lime and SO₂, but sugar yields are considerably lower if using AFEX, SAA and LHW.

19.3.2 Enzyme Loadings

For hybrid poplar in the CAFI efforts, the baseline enzyme formulation was a combination of β-glucosidase and cellulase at a CBU (cellobiose units)-to-FPU (paper filter units) ratio of 2.0 [ref. 35], with a cellulase loading of 15 FPU g^{-1}

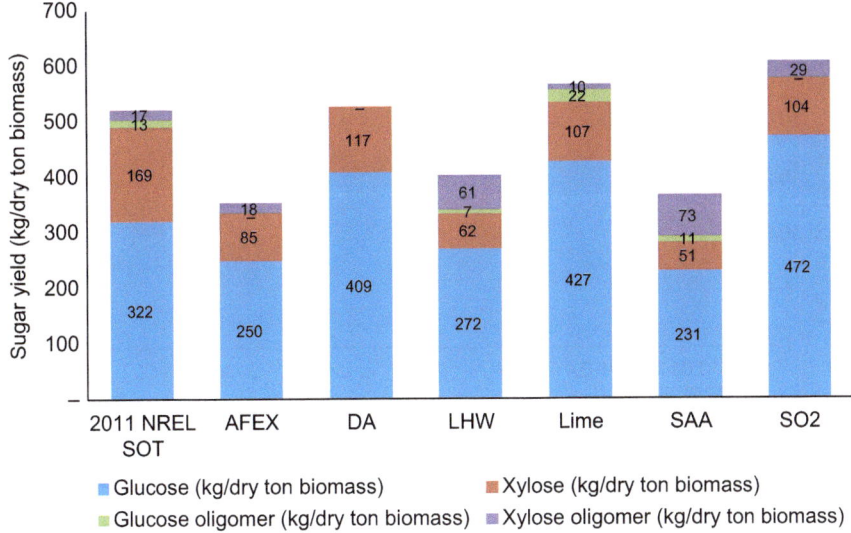

Figure 19.2 Glucose and xylose monomer and oligomer yields (kg per dry ton of biomass).

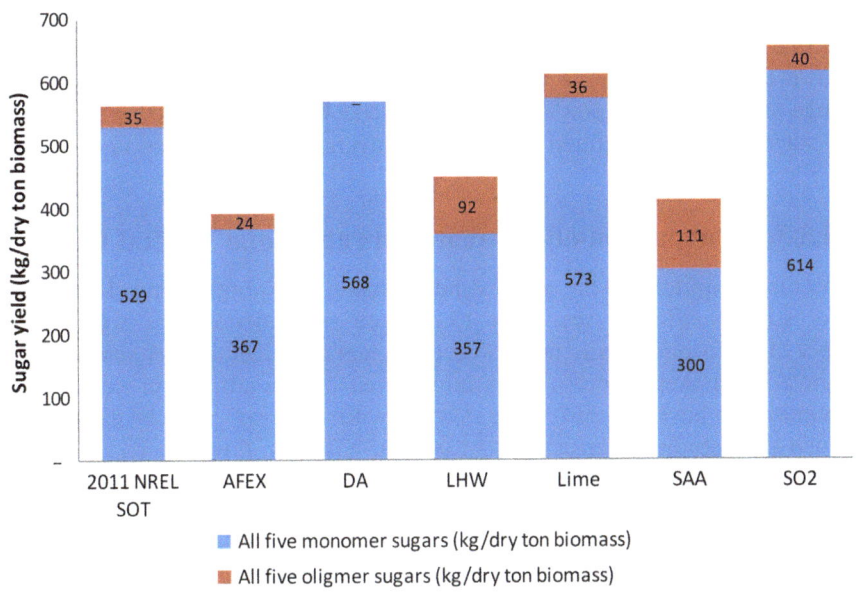

Figure 19.3 Five monomer and oligomer sugar yields (kg per dry ton of biomass).

glucan and a β-glucosidase loading of 30 CBU g^{-1} glucan, equivalent to 29.0 mg protein per g of glucan.[36] The enzyme loading for the 2011 NREL SOT study was 20 mg protein per g of glucan, using Novozyme Cellic CTec2.

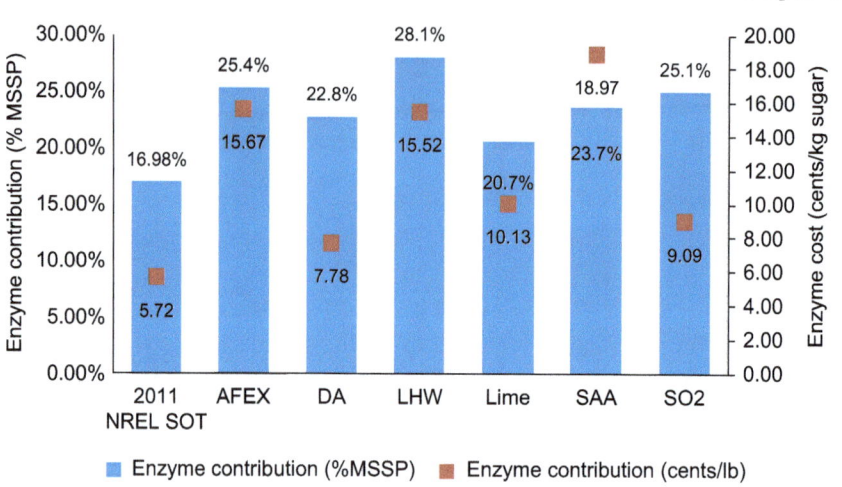

Figure 19.4   Enzyme cost contributions.

The enzyme costs are compared in Figure 19.4, based on the loading used for each feedstock and on on-site enzyme production to fulfill enzyme requirements. The 2011 NREL SOT case as the reference case had an enzyme cost contribution of 6 ¢ per kg of sugar (*i.e.*, 17% of the sugar production cost). It is noted that the enzyme cost varied from 8–16 ¢ per kg of sugar in the CAFI project using hybrid poplar, depending on the pretreatment technology studied. However, it averages 13 ¢ per kg of sugar in this study (Figure 19.4), with a range of contributions to the overall sugar cost of 21–28%. Further reductions in enzyme cost are required to reduce the overall sugar production cost.

19.3.3   Comparison of Pretreatments and Total Capital Costs

The direct capital for the pretreatment section is strongly dependent on the processing conditions (see Table 19.2), such as residence time, solids levels, types of reactors, and any needed pretreatment chemical recovery strategies. It is noted that these process conditions for each pretreatment process are not optimized to have minimized sugar costs, with all sugar yields being based on bench-scale studies. The equipment is sized based on flow rate data from the detailed process simulation tool.[26] Most of the pretreatment reactor costs were referenced from the CAFI 1 project on corn stover[7] and the CAFI 3 project on switchgrass,[8] with proper scaling. The pretreatment installed equipment cost and the total installed equipment cost are illustrated in Figure 19.5, on the basis of five monomer sugars production. The installed equipment cost for pretreatment varies from $10 M to $86 M depending on the technology, compared with a pretreatment capital of $31 M in the 2011 NREL SOT case. LHW has relatively low pretreatment capital cost as well as total capital among all other pretreatments. Lime has the highest capital due to the combined complexity of the Lime recycling scheme, gas processing, and high lime loading. Although a

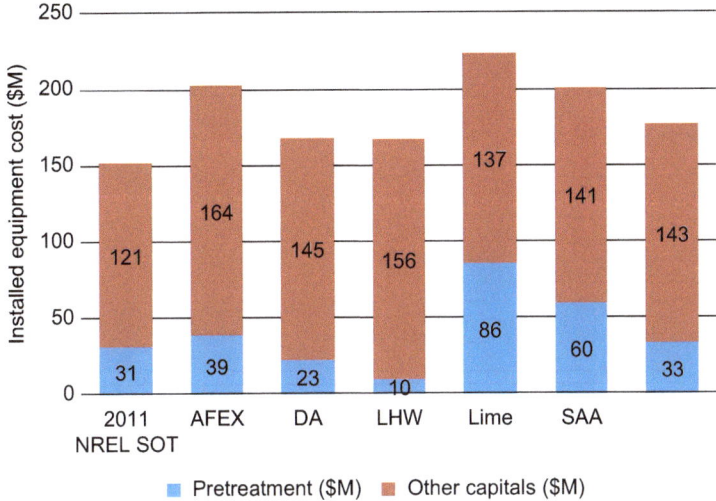

Figure 19.5 Summary of pretreatment capitals and total installed capitals.

single unit of the Lime pretreatment reactor is not expensive compared to the DA pretreatment reactor, the number of reactors required is significant due to the long residence time (4 h). In DA and SO_2, pretreatment reactor cost contributions to total capital are significant due to the exotic materials required for construction for metallurgy. However, the acids are not recycled. In the AFEX, SAA and Lime pretreatment processes, significant portions of equipment cost are associated with the recovery of pretreatment chemicals, as one-pass use of the pretreatment chemical is not realistic due to the high pretreatment chemical usage levels in these processes. It should be noted that the design of these pretreatment chemical recovery and recycle systems is preliminary. Further development of efficient pretreatment chemical recovery configurations may mitigate process challenges, as well as reduce overall capital costs for these pretreatment technologies.

The total installed capital ranges from \$152 M to \$222 M, including the capital costs of enzyme production, enzymatic hydrolysis, sugar evaporation, combustion and utility integration. The 2011 NREL SOT case had the lowest total installed capital of \$152 M. Figure 19.5 illustrates pretreatment and other capital costs.

19.3.4 Comparison of Minimum Sugar Selling Prices

Depending on the potential for sugar utilization, four types of sugar production are studied here for calculating MSSPs.

19.3.4.1 Xylose and Glucose Production Costs

The MSSP is calculated based on the production of glucose and xylose alone, shown in Figure 19.6. The other oligomer sugars of glucose and xylose, minor

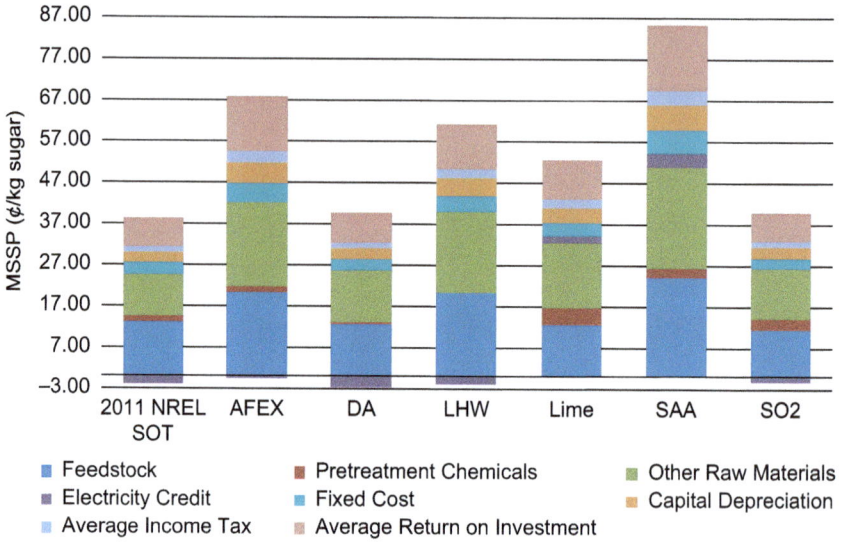

Figure 19.6   Glucose and xylose MSSPs (¢ per kg of sugar).

Table 19.4   Percentage of theoretical glucose, xylose and total sugar yields.

		Pretreatment technology					
	2011 NREL SOT	*AFEX*	*DA*	*LHW*	*Lime*	*SAA*	*SO₂*
Glucose	83%	52%	85%	57%	89%	48%	98%
Xylose	76%	51%	70%	38%	64%	31%	63%
Total sugar	80%	52%	80%	51%	81%	43%	87%

sugars although present in the sugar production stream, are not considered as value-added intermediates for down-stream processing. Significant cost variation is found in this case, simply due to the wide range of monomer glucose and xylose yields for each pretreatment technology (also shown in Figure 19.2 and Table 19.4). The 2011 NREL SOT case has the lowest MSSP of 36 ¢ per kg of sugar using corn stover, comparable to DA in the CAFI study of 37 ¢ per kg of sugar using poplar. Among all studied cases, the highest MSSP wass 85 ¢ per kg of sugar (SAA using poplar), and 54 ¢ per kg of sugar was the average. In Figure 19.6, the contribution from feedstock, pretreatment chemicals, other raw materials, electricity credits, fixed costs, capital depreciation, income tax, and return on capitals to the MSSP is shown. The majority of the contribution for other raw materials is from the natural gas used to evaporate water from the sugar stream to concentrate the sugars, roughly contributing to 30% of the total sugar cost. If a diluted sugar stream can be used in the up-grading process, this cost contribution can be eliminated, so the total sugar cost can be reduced significantly. This is usually true for the liquid fermentation of sugars.

19.3.4.2 Xylose and Glucose Monomer and Oligomer Production Costs

The MSSP is calculated based on the production of total monomer and oligomer glucose and xylose, shown in Figure 19.7. Since oligomer glucose and xylose are considered utilizable, the MSSP variation is reduced significantly for technologies that produce higher oligomer glucose and xylose yields, such as Lime and LHW. For pretreatment producing few oligomer sugars, like DA, the MSSP is only slightly decreased, such as from 36 to 34 ¢ per kg of sugar for the 2011 NREL SOT case. The MSSP was 48 ¢ per kg of sugar on average, ranging from 34 to 66 ¢ per kg of sugar, with the 2011 NREL SOT case having the lowest MSSP of 34 ¢ per kg of sugar.

19.3.4.3 All Five Monomer Sugar Production Costs

If minor sugars (arabinose, mannose and galactose) can be utilized in the downs-tream processing, the MSSP can be calculated with much higher sugar production, shown in Figure 19.8. MSSPs in this case dropped to 50 ¢ per kg of sugar on average, ranging from 34 to 80 ¢ per kg of sugar.

19.3.4.4 All Five Monomer and Oligomer Sugar Production Costs

If all the monomers and oligomers of the five sugars are considered, the MSSP further decreases to 44 ¢ per kg of sugar on average, ranging from 32 to 58 ¢

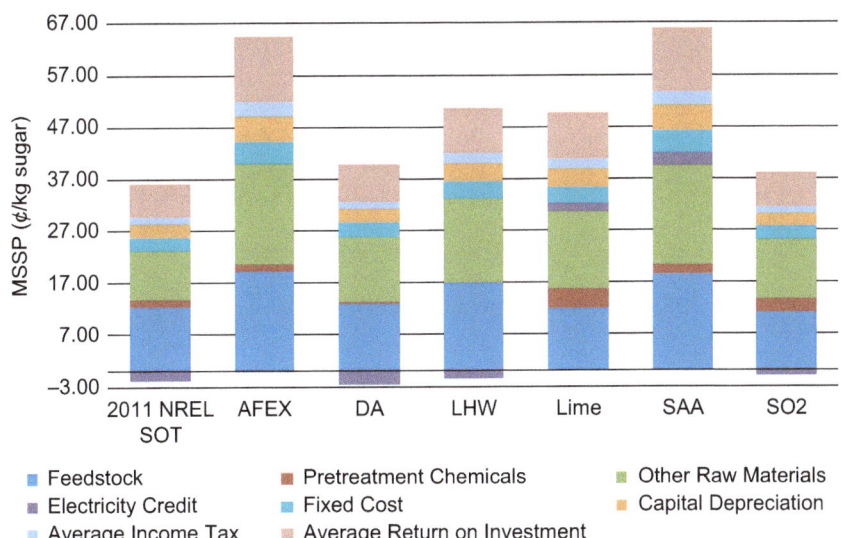

Figure 19.7 MSSPs for glucose and xylose and their oligomers (¢ per kg of sugar).

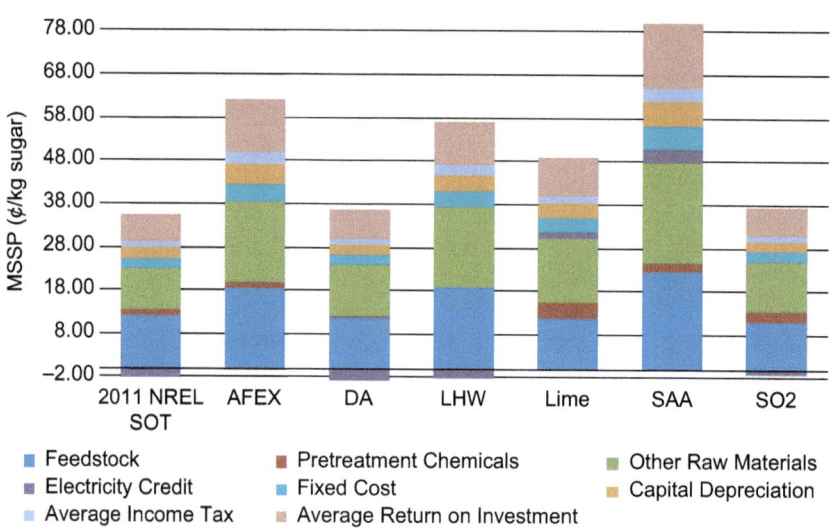

Figure 19.8 MSSPs for all five monomer sugars (¢ per kg of sugar).

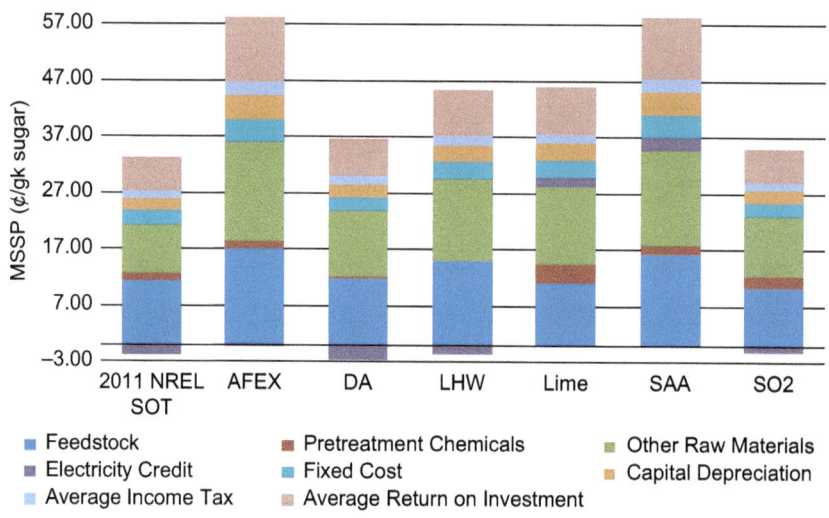

Figure 19.9 MSSPs for all five monomer and oligomer sugars (¢ per kg of sugar).

per kg of sugar. The variation of MSSP by pretreatment technologies and feedstock types diminished significantly, as shown in Figure 19.9. The significantly higher MSSPs for both AFEX and SAA using poplar feedstock are mainly due to generally lower sugar yields and high moisture content in the feedstock stream. The high moisture in the biomass adds complexity in the form of ammonia recycling equipment costs and the energy required for high-purity recycled ammonia.

19.3.5 Direct Plant Emissions and Water Consumption

The direct emissions (CO_2, NO_2, SO_2) and consumptive water use (a.k.a. "makeup water") from the sugar-production processes with various pretreatment technologies are shown in Figure 19.10. The specific CO_2 emission (kg of CO_2 per kg of sugars) is inversely proportional to the product yield. Higher sugar yields provide less unconverted biomass for combusting. The process that releases the most CO_2 is AFEX, at 3.92 kg of CO_2 per kg of sugars. This is followed by LHW and SAA, emitting 3.51 and 3.42 kg of CO_2 per kg of sugars, respectively. The CO_2 emission from the SO_2 pretreatment process is the smallest among the pretreatment technologies, at 1.56 kg of CO_2 per kg of sugars, which is about 40% of the AFEX CO_2 emission. As shown in Table 19.5, CO_2 is generated during cellulase (or enzyme) production and cellulase seed fermentation, but the majority is from combustion. Note that combustion CO_2 includes both biogenic CO_2 (origin from biomass) and fossil CO_2 (origin from natural gas (NG)). NG is the makeup fuel for the burner. It is necessary for the process to generate sufficient heat required for the triple-effect evaporation system to concentrate the product stream to 50% water content. The fossil CO_2 emission rate is 0.056 kg of CO_2 per MJ of NG.

Figure 19.10 also shows the NO_2 emissions. The AFEX and SAA pretreatment technologies emit more NO_2 than other technologies do, 3.22 and 3.89 g of NO_2 per kg of sugars, respectively. This is consistent with the fact that both AFEX and SAA use NH_3 as their pretreatment catalysts. While LHW does not use any NH_3, the process still produces a noticeable amount of NO_2, 2.44 g of NO_2 per kg of sugars. The LHW-emitted NO_2 is essentially thermal NO_2 that is formed through the high-temperature oxidation of the diatomic nitrogen found in combustion air.

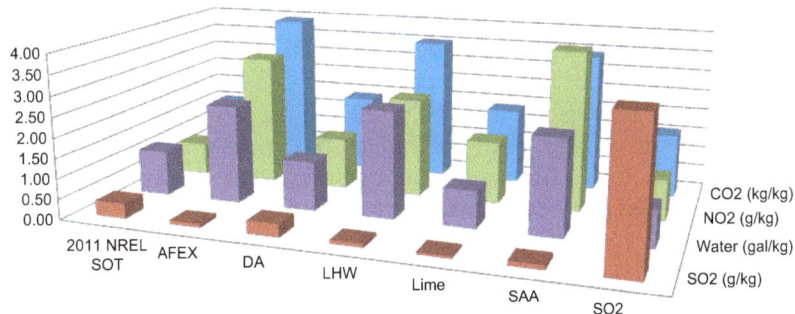

	2011 NREL SOT	AFEX	DA	LHW	Lime	SAA	SO2
■ SO2 (g/kg)	0.37	0.09	0.31	0.09	0.06	0.11	3.54
■ Water (gal/kg)	1.12	2.43	1.21	2.63	0.91	2.34	0.90
■ NO2 (g/kg)	0.83	3.22	1.27	2.44	1.54	3.89	0.98
■ CO2 (kg/kg)	1.52	3.92	1.90	3.51	1.88	3.42	1.56

Figure 19.10 Comparison of direct-process greenhouse gas (CO_2) emission, criteria air pollutant (NO_2 and SO_2) emission, and consumptive water use.

Table 19.5   Direct-process CO_2 emissions from the biochemical plant by process area.

	Pretreatment technology						
	NREL 2011 SOT	AFEX	DA	LHW	Lime	ARP/ SAA	SO_2
	g of CO_2 per kg of sugars						
Combustor stack (biogenic)	449	702	431	673	559	836	33
Combustor stack (fossil)	1018	3063	1395	2687	1220	2396	1087
Cellulase production fermentor vent (biogenic)	52	149	73	148	95	181	86
Cellulase seed fermentor vent (biogenic)	2	6	3	6	4	8	4
Total	1521	3920	1903	3514	1878	3421	1210

SO_2 emissions from the biochemical process depend directly on the amount of H_2SO_4 (in the DA case) or compressed SO_2 (in the SO_2 case) introduced into the pretreatment reactor. The pretreatment catalyst loadings for the DA and SO_2 pretreatment technologies were 2 g of H_2SO_4 per 100 g of dry biomass entering pretreatment (DBP) and 3 g of SO_2 per 100 g of DBP, respectively (see Table 19.2). The corresponding SO_2 emission was 0.31 g of SO_2 per kg of sugars for DA and 3.54 g of SO_2 per kg of sugars for SO_2. None of the other processes use H_2SO_4 or SO_2 as the pretreatment hydrolysis catalyst and thus their SO_2 emissions are relatively small, ranging from 0.06 to 0.11 g of SO_2 per kg of sugars.

The consumptive water use of the pretreatment technologies varies broadly, as shown in Figure 19.10. The LHW process has the highest water usage, 2.63 gal of H_2O per kg of sugars. The water usage for AFEX and SAA is also high at 2.43 and 2.34 gal of H_2O per kg of sugars, respectively. High-water-usage processes require more water removal from the product stream. Currently, a triple-effect evaporation system is adopted to concentrate the product stream to 50% water content. The increase in process heat demand has a direct impact on the CO_2 emissions. Consequently, the profiles of the consumptive water use and CO_2 emissions in Figure 19.10 are rather similar. DA, Lime, and SO_2 pretreatment technologies are more water-efficient, consuming over 50% less water than AFEX, LHW, and SAA pretreatments do.

19.3.6   Life-Cycle Assessment Results

The LCA results for the current evaluated pretreatment technologies are presented in Figures 19.11 and 19.12. The functional units are kg per $CO_{2\text{-eq}}$ per kg of sugars for GWP, and MJ per kg of sugars for fossil energy input. The results are categorized into six areas: (1) direct plant emission; (2) biomass feedstock production; (3) pretreatment; (4) enzyme production; (5) chemicals; and (6) electricity. Areas (2)–(5) are from the associated underlying processes.

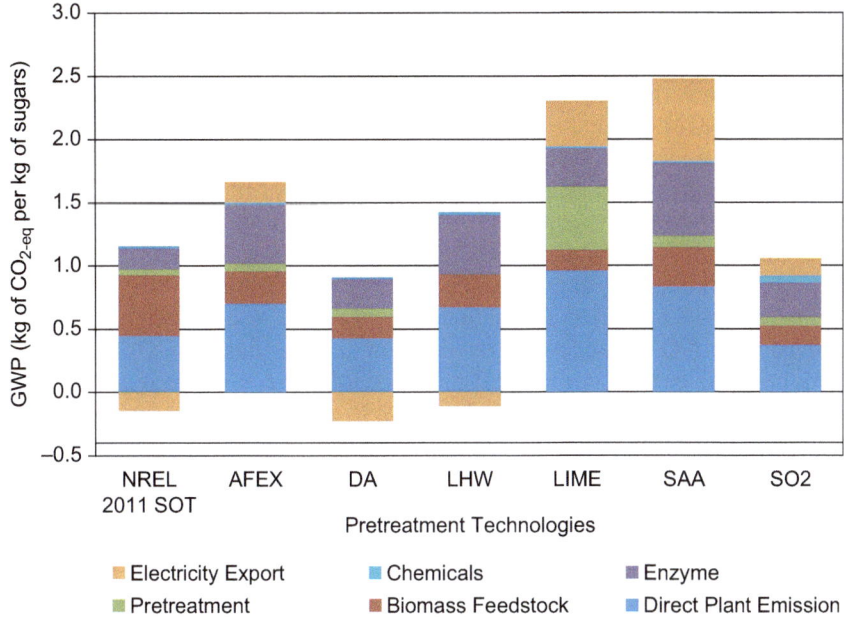

Figure 19.11 Comparison of global-warming potential (kg of CO$_{2\text{-eq}}$ per kg of sugars).

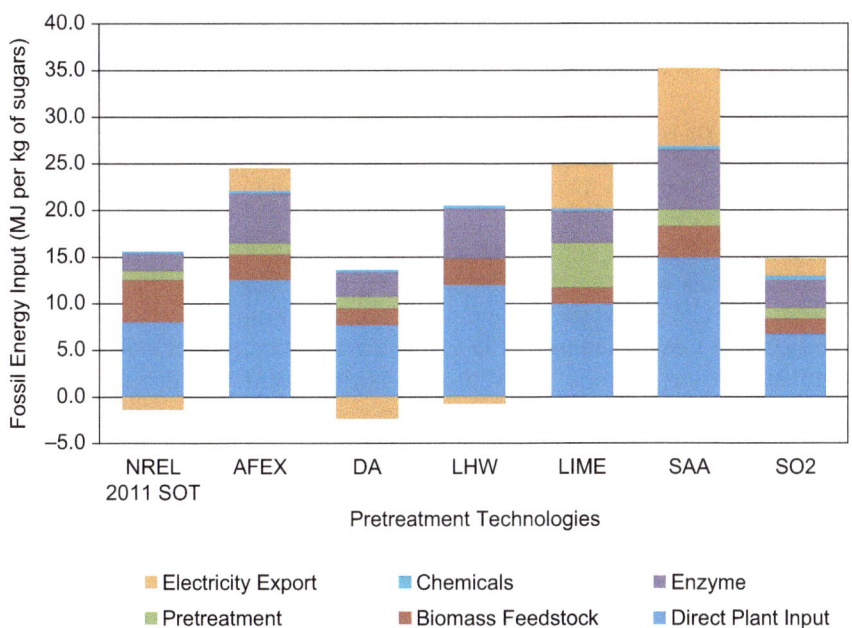

Figure 19.12 Comparison of fossil energy input (MJ per kg of sugars).

19.3.6.1   Direct Plant Emissions

The direct plant emission does not include biogenic CO_2 (*i.e.*, CO_2 absorbed from the atmosphere and incorporated as biomass). With its biomass origin, biogenic CO_2 does not contribute to the increase of greenhouse gases in the atmosphere[37] and is not considered in the IPCC global-warming methodology.[38] Biogenic CO_2 is typically not counted as a contributor to global warming in the IPCC global-warming methodology because it is assumed that the emitted CO_2 was removed from the atmosphere during the same time horizon as the GWP estimate (*e.g.*, 100 annum, 200 annum). Hence, GWP contributed by the direct plant emission for the evaluated pretreatment technologies results from the fossil CO_2 derived from NG (makeup fuel) combustion and the makeup CO_2 (only for the Lime case). Since the Lime pretreatment technology requires makeup CO_2 for its pretreatment step, it has the highest GWP contributed by the direct plant emission, 0.96 kg of $CO_{2\text{-eq}}$ per kg of sugars.

Among the evaluated pretreatment technologies, SAA has the largest total GWP (2.48 kg of $CO_{2\text{-eq}}$ per kg of sugars) and total fossil energy demand (35.13 MJ per kg of sugars). This technology also has the highest fossil energy input contributed by the direct plant input, 14.93 MJ per kg of sugars. Due to high sugar product yield, the DA pretreatment process has the lowest total GWP (0.67 kg of $CO_{2\text{-eq}}$ per kg of sugars) and total fossil energy input (11.19 MJ per kg of sugars). These are better than the NREL 2011 SOT, which had a total GWP of 1.01 kg of $CO_{2\text{-eq}}$ per kg of sugars and total fossil energy input of 14.21 MJ per kg of sugars. The SO_2 process has the highest sugar product yield; it has the second highest total GWP and total fossil energy input, 1.05 kg pf $CO_{2\text{-eq}}$ per kg of sugars and 14.79 MJ per kg of sugars, respectively. The LCA results for the SO_2 process and the NREL 2011 SOT are comparable.

19.3.6.2   Pretreatment

The pretreatment step involves the use of H_2SO_4 and/or NH_3. The Lime pretreatment process also takes in CO_2 and O_2. The GHG emissions and fossil energy demand of the underlying processes associated with the production of these chemicals were quantified. While SAA pretreatment technology has the largest total GWP, it is the Lime pretreatment technology that has the highest pretreatment GWP, 0.50 kg of $CO_{2\text{-eq}}$ per kg of sugars. On the other hand, the pretreatment step for the LHW process is the greenest, with negligible pretreatment GWP. All other processes have relatively low pretreatment GWPs (0.05–0.09 kg of $CO_{2\text{-eq}}$ per kg of sugars).

19.3.6.3   Enzyme Production

The cellulase enzyme is used to hydrolyze cellulose into glucose. The embedded processes (*e.g.*, nutrient and protein production) associated with the enzyme production are quantified. Higher enzyme-production-contributed GWPs and fossil energy inputs are related to processes with lower sugar product yields, namely SAA, LHW, and AFEX (see Table 19.3). The enzyme-contributed GWP for the SAA process is 0.57 kg of $CO_{2\text{-eq}}$ per kg of sugars, more than double than that of the DA process (0.23 kg of $CO_{2\text{-eq}}$ per kg of sugars). Similarly, the enzyme-contributed fossil energy input for the SAA process is 6.49 MJ per kg of sugars as opposed to 2.63 MJ per kg of sugars for the DA process. The NREL 2011 SOT, DA, and SO_2 processes have the lowest enzyme-production-contributed GWPs and fossil energy inputs, attributed to higher sugar product yields.

19.3.6.4 Chemicals

The chemicals used in the current sugar-production processes include boiler water chemicals, cooling water chemicals, and lime for FDU (flue gas de-sulfurization). GWPs and fossil energy inputs contributed by process chemicals are relatively low (*i.e.*, less than 3% of the total). The only exception is the SO_2 pretreatment process. Primarily as a result of its high SO_2 emission, more lime is required for FDU, resulting in a GWP of 0.06 kg of $CO_{2\text{-eq}}$ per kg of sugars (5.3% of the total GWP) and fossil energy input of 0.44 MJ per kg of sugars (still less than 3% of the total).

19.3.6.5 Electricity Export

Lignin solid residuals from the separation of the sugar product stream along with the NG are combusted in a fluidized bed combustor to produce high-pressure steam for electricity production and process heat. The majority of the process steam demand is in the pretreatment reactor and sugar-product-concentration (the triple-effect evaporation system) areas. The excess steam is converted to electricity for use in the plant and for sale to the grid. If the exported electricity co-product is allowed to displace an equivalent amount of grid electricity, a significant amount of GHG emissions can be avoided or credited, assuming an average US electricity grid mixture.

Figure 19.11 shows that processes with electricity GWP credits are NREL 2011 SOT (-0.15 kg of $CO_{2\text{-eq}}$ per kg of sugars), DA (-0.23 kg of $CO_{2\text{-eq}}$ per kg of sugars), and LHW (-0.11 kg of $CO_{2\text{-eq}}$ per kg of sugars). For the NREL 2011 SOT, DA, and LHW processes, which use dilute H_2SO_4 or liquid hot water and heat from steam in the pretreatment step, the biomass is allowed to undergo hydrolysis reactions. The reactions break down lignin in the feedstock, in addition to converting most of the hemicellulose carbohydrates in the feedstock to soluble sugars, primarily xylose, mannose, arabinose, and glucose. Most of the lignin is not solubilized. The solid insolubilized lignin is separated from the sugar product stream and sent to the combustor.

On the other hand, in the alkaline pretreatment processes (AFEX, Lime, SAA, and SO_2), up to 70% of the lignin is solubilized; the liquid-solubilized lignin is present in the liquid sugar product stream. With less lignin available for combustion, these processes require electricity from the grid. Figure 19.11 shows that the electricity-contributed GWPs for AFEX, Lime, SAA, and SO_2

processes are 0.16, 0.35, 0.65, and 0.13 kg of CO_{2-eq} per kg of sugars, respectively. The observation for the GWPs is also true for the fossil energy inputs, as shown in Figure 19.12.

19.4   Conclusion

Sugar costs and life-cycle analysis using the promising pretreatment technologies studied in the second CAFI project are analyzed consistently in this work using the 2011 NREL SOT assessment as the reference case. The pretreatment technologies vary greatly in terms of their process design and projected total capital investment, resulting in significant variation in minimum sugar selling cost and environmental emissions and resources consumption. The sugar cost studies are based on past research, existing databases, and engineering judgments, and include not only pretreatment reactor costs but also pretreatment chemical recovery and recycle equipment costs for the processes that use high pretreatment chemical loadings. In order to have a wider application to sugar upgrading, four different sugar product variations are considered in the cost calculation. These include combinations of glucose monomer, xylose monomer, their oligomers, and minor sugar monomers and oligomers. The minimum sugar selling price has much less variation if all monomer and oligomer sugars are counted as valuable products. It was found that the sugar yield is still the single most important factor in determining projected minimum sugar selling prices for each process, although pretreatment reactor design and pretreatment chemical recycling schemes are significant contributors to overall costs. Process improvements are required to further reduce sugar cost for all cases. Additional sensitivity analysis around sugar yields, capital costs, and energy integration for the overall process, as well as around purification of the sugar production stream, will provide additional insights into the importance of individual factors that may influence costs for each pretreatment approach in a differing manner. This study compiles an inventory of direct emissions (CO_2, NO_2, and SO_2) and consumptive water use from the sugar-production processes with various pretreatment technologies. The corresponding global-warming potential and fossil energy demand of each of the evaluated pretreatment technologies are also quantified. Life-cycle assessment results can help identify the potential environmental impacts associated with identified inputs and releases. When viewed from a life-cycle perspective, the pretreatment technologies that use alkaline chemicals (*i.e.*, Lime and NH_3) for pretreatment hydrolysis exhibit higher global-warming potentials and fossil energy demands than those using acids (*i.e.*, H_2SO_4, SO_2, and H_2O) as pretreatment catalysts. It is not clear if the present life-cycle assessment findings for poplar are feedstock-dependent; the conclusion may be different for other feedstocks such as corn stover and switchgrass.

Acknowledgements

This research was funded under the Office of the Biomass Program of the United States Department of Energy (contract no. DE-FG36-04GO14017).

We would like to acknowledge all the generous help from each CAFI project PI to establish the design basis for each pretreatment method.

References

1. R. D. Perlack, L. L. Wright, A. F. Turhollow, R. L. Graham, B. J. Stokes and D. C. Erbach, *Biomass as Feedstock for a Bioenergy and Bioproducts Industry: the Technical Feasibility of a Billion-Ton Annual Supply*, Technical report DOE/GO-102005-2135, Oak Ridge TN. Available from: http://feedstockreview.ornl.gov/pdf/billion_ton_vision.pdf, 2005.
2. *2011 U.S. Billion-Ton Update: Biomass Supply for a Bioenergy and Bioproducts Industry*, Oak Ridge TN. Available from: http://www1.eere.energy.gov/biomass/pdfs/billion_ton_update.pdf, 2011.
3. M. E. Himmel, S.-Y. Ding, D. K. Johnson, W. S. Adney, M. R. Nimlos, J. W. Brady and T. D. Foust, *Science*, 2007, **315**, 804.
4. M. Galbe and G. Zacchi, *Adv. Biochem. Eng./Biotechnol.*, 2007, **108**, 41.
5. L. R. Lynd, *Annu. Rev. Energy Environ.*, 1996, **21**, 403.
6. R. T. Elander, B. E. Dale, M. Holtzapple, M. R. Ladisch, Y. Y. Lee, C. Mitchinson, J. N. Saddler and C. E. Wyman, *Cellulose*, 2009, **16**, 649.
7. T. Eggeman and R. T. Elander, *Bioresour. Technol.*, 2005, **96**, 2019.
8. L. Tao, A. Aden, R. T. Elander, V. R. Pallapolu, Y. Y. Lee, R. J. Garlock, V. Balan, B. E. Dale, Y. Kim, N. S. Mosier, M. R. Ladisch, M. Falls, M. T. Holtzapple, R. Sierra, J. Shi, M. A. Ebrik, T. Redmond, B. Yang, C. E. Wyman, B. Hames, S. Thomas and R. E. Warner, *Bioresour. Technol.*, 2011, **102**, 11105.
9. United States Government, http://www.epa.gov/otaq/fuels/renewablefuels/index.htm.
10. C. B. Granda, M. T. Holtzapple, G. Luce, K. Searcy and D. L. Mamrosh, *Appl. Biochem. Biotech.*, 2009, **156**, 537.
11. A. Aden and T. Foust, *Cellulose*, 2009, **16**, 535.
12. M. Zhang, C. Eddy, K. Deanda, M. Finkestein and S. Picataggio, *Science*, 1995, **267**, 240.
13. K. Rumbold, H. J. J. van Buijsen, K. M. Overkamp, J. W. van Groenestijn, P. J. Punt and M. J. van der Werf, *Microb. Cell Fact.*, 2009, **8**(64), doi: 10.1186/1475-2859-8-64.
14. B. E. Rittmann, *Biotechnol. Bioeng.*, 2008, **100**, 203.
15. M. A. Rude and A. Schirmer, *Curr. Opin. Microbiol.*, 2009, **12**, 274.
16. A. Carroll and C. Somerville, *Annu. Rev. Plant Biol.*, 2009, **60**, 165.
17. P. Gallezot, N. Nicolaus, G. Fleche, P. Fuertes and A. Perrard, *J. Catal.*, 1998, **180**, 51.
18. G. W. Huber, J. N. Chheda, C. J. Barrett and J. A. Dumesic, *Science*, 2005, **308**, 1446.
19. P. G. Blommel and R. D. Cortrigth, *Production of Conventional Liquid Fuels from Sugars*, http://www.virent.com/resources/white-papers/.
20. J. C. Serrano-Ruiz and J. A. Dumesic, *Energy Environ. Sci.*, 2011, **4**, 83.

21. J. N. Chheda, G. W. Huber and J. A. Dumesic, *Angew. Chem., Int. Edn.*, 2007, **46**, 7164.
22. M. Stocker, *Angew. Chem., Int. Edn.*, 2008, **47**, 9200.
23. J. J. Bozell and G. R. Petersen, *Green Chem.*, 2010, **12**, 539.
24. G. W. Huber, S. Iborra and A. Corma, *Chem. Rev.*, 2006, **106**, 4044.
25. J. B. Binder and R. T. Raines, *J. Am. Chem. Soc.*, 2009, **131**, 1979.
26. AspenPlus™, v7.2, 2007, Aspen Technology Inc., Cambridge.
27. D. Humbird, R. Davis, L. Tao, C. Kinchin, D. Hsu, A. Aden, P. Schoen, J. Lukas, B. Olthof, M. Worley, D. Sexton and D. Dudgeon, *Process Design and Economics for Biochemical Conversion of Lignocellulosic Biomass to Ethanol: Dilute-Acid Pretreatment and Enzymatic Hydrolysis of Corn Stover*, Technical report NREL/TP-510-47764, 2011, Golden CO. Available from: http://www.nrel.gov/docs/fy11osti/47764.pdf.
28. *Biomass Multi-Year Program Plan*, 2011, Office of the Biomass Program, Energy Efficiency and Renewable Energy, US Department of Energy, Available from: http://www1.eere.energy.gov/bioenergy/pdfs/mypp_april_2011.pdf.
29. A. Aden, M. Ruth, K. Ibsen, J. Jechura, K. Neeves, J. Sheehan, B. Wallace, L. Montague, A. Slayton and J. Lukas, *Lignocellulosic Biomass to Ethanol Process Design and Economics Utilizing Co-Current Dilute Acid Pre-hydrolysis and Enzymatic Hydrolysis for Corn Stover*, Technical report NREL/TP-510-32438, 2002, Golden CO. Available from: http://www.nrel.gov/docs/fy02osti/32438.pdf.
30. L. Tao and A. Aden, *In Vitro Cell. Dev. Biol.: Plant*, 2009, **45**, 199.
31. M. Peters and K. Timmerhaus, *Plant Design and Economics for Chemical Engineers*, 4th edn, 1991, McGraw-Hill, New York.
32. SimaPro v7.3, Product Ecology Consultants, Amersfoort, 2011.
33. Ecoinvent v2.2, Swiss Center for Life Cycle Inventories, Duebendorf, 2010.
34. D. D. Hsu, D. Inman, G. A. Heath, E. J. Wolfrum, M. K. Mann and A. Aden, *Environ. Sci. Technol.*, 2010, **44**, 5289.
35. C. E. Wyman, B. E. Dale, R. T. Elander, M. Holtzapple, M. R. Ladisch, Y. Y. Lee, C. Mitchinson and J. N. Saddler, *Biotechnol. Prog.*, 2009, **25**, 333.
36. R. Kumar and C. E. Wyman, *Biotechnol. Prog.*, 2009, **25**, 302.
37. A. B. Biorecro, *Global Status of BECCS Projects*, Global CCS Institute, Canberra, 2010.
38. B. S. Fischer, N. Nakicenovic, K. Alfsen, J. Corfee Morlot, F. de la Chesnaye, J.-C. Hourcade, K. Jiang, M. Kainuma, E. La Rovere, A. Matysek, A. Rana, K. Riahi, R. Richels, S. Rose, D. van Vuuren and R. Warren, *Issues Related to Mitigation in the Long Term Context. Climate Change 2007: Mitigation. Contribution of Working Group III to the Fourth Assessment Report of the Inter-Governmental Panel on Climate Change*, Cambridge University Press, Cambridge, 2007.

Subject Index

Illustrations and figures are in **bold**. Tables are in *italics*.